TROPICAL ECOSYSTEMS
STRUCTURE, DIVERSITY AND HUMAN WELFARE

Proceedings of the
**International Conference on Tropical Ecosystems:
Structure, Diversity and Human Welfare**

15–18 July 2001, Bangalore

www.atb.botany.ufl.edu

Editors

K.N. Ganeshaiah[1,3,5], R. Uma Shaanker[2,3,5], Kamaljit S. Bawa[4,5]

[1]Departments of Genetics and Plant Breeding and [2]Crop Physiology, University of
Agricultural Sciences, GKVK, Bangalore 560 065, India
[3]Jawaharlal Nehru Center for Advanced Scientific Research,
Jakkur, Bangalore 560 064, India
[4]Department of Biology, University of Massachusetts, MA 02125, Boston, USA
[5]Ashoka Trust for Research in Ecology and the Environment,
Hebbal, Bangalore 560 024, India

Originally published by:
Oxford & IBH Publishing Co. Pvt. Ltd., New Delhi, India

Copublished and distributed outside the Indian subcontinent by:

Science Publishers, Inc.

Enfield (NH), USA   Plymouth, UK

Published in association with

Ashoka Trust for Research in Ecology and the
Environment (ATREE), 659, 5th 'A' Main Road,
Hebbal, Bangalore 560 024, India
http://www.atree.org

Association for Tropical Biology, P.O. Box 1897,
Lawrence, KS 66044-8897, USA
http://www.atb.botany.ufl.edu

Sponsored by a grant administered by Winrock International to ATREE with funds
from the Ford Foundation, under the Winrock-Ford Small Grants Program.

Cover designed by N.A. Aravind, ATREE, M. Nageswara Rao, UAS, Bangalore
Cover photo credits: N.A. Aravind, ATREE and K. N. Ganeshaiah, UAS, Bangalore

Copublished and distributed outside the Indian subcontinent by:

SCIENCE PUBLISHERS, INC
Post Office Box 699
Enfield, New Hampshire 03784
United States of America

Internet site: http:/www.scipub.net

Library of Congress Cataloging-in-Publication Data

International Conference on Tropical Ecosystems:Structures, Diversity, and Human
Welfare (2001 : Bangalore, India)
    Tropical ecosystems : structure, diversity, and human welfare / editors, K.N.
Ganeshaiah, R. Uma Shaanker, Kamaljit S. Bawa.
      p. cm.
    Includes bibliographical references.
    ISBN 1-57808-181-5
    1.   Forest ecology--Tropics--Congress. 2. Nature--Effects of human beings
on--Tropics--Congress. 3. Endangered ecosystems--Tropics--Congresses. I.
Ganeshaiah, K N. II. Shaanker, R. Uma. III. Bawa, Kamaljit S. IV. Title.

QH84.5 .I56 2001
577.34--dc21

                                                        2001036509

ISBN 1-57808-181-5

Printed in India

Preface

The tropical and sub-tropical regions of the world, comprising extremely fragile ecosystems, support more than 70 per cent of the world's biota in a complex labyrinth of ecological interactions. These regions are also home to nearly half of the world's human population and thus experience tremendous pressure on their natural resources. In the recent past, there has been a growing concern over the accelerating rates of deforestation in the tropics and the consequent loss of biodiversity.

Reflecting these concerns and in seeking to review the threats and arrive at possible solutions, an International Conference on Tropical Ecosystems: Structure, Diversity and Human Welfare has been organized in Bangalore, India from 15 to 18 July 2001 under the auspices of the Ashoka Trust for Research in Ecology and the Environment (ATREE), Bangalore and the Association for Tropical Biology, USA.

The contributions to the conference are compiled in this volume and reflect an over-arching span of insight into the patterns and processes underlying tropical ecosystems. They represent three major thematic groups, namely, Global change and tropical forest ecosystems; Structure, diversity and function of tropical ecosystems; and Biodiversity hot-spots.

The editors wish to thank all the contributors for complying with the time schedule and for bearing with the persistent requests and queries made by the editorial staff. The papers have been edited for brevity and content, but every attempt has been made to preserve the author's originality of expression. The compilation of this volume owes a lot to the untiring efforts of a number of colleagues and students at the Ashoka Trust for Research in Ecology and the Environment, Bangalore, University of Agricultural Sciences, Bangalore, and the University of Massachusetts, Boston. In particular, we wish to acknowledge the help rendered by C. Seema, N. A. Aravind, N. Ramesh, N. Bhogaiah, N. Laxmikanthaiah, K. Sindhu, Ramesh Kannan, Jagadish Krishnaswamy and Gladwin Joseph (all at ATREE), M. Nageswara Rao and G. Ravikanth (UASB).

We gratefully acknowledge the support from a number of institutions including the Jawaharlal Nehru Center for Advanced Scientific Research, Indian Academy of Sciences, University of Agricultural Sciences, International Union for the Study of Social Insects – Indian Chapter (all at Bangalore), the Organization for Tropical Studies and the National Science Foundation (USA), and the Forest Research Support Programme for Asia and the Pacific, FAO Regional Office (Bangkok).

We thank Mr. M.S. Venugopal and his team for copy-editing and preparing the initial layouts of the volume.

15 July 2001

K. N. Ganeshaiah
R. Uma Shaanker
Kamaljit S. Bawa

CONTENTS

Economic Valuation of Natural Resources: Local and Global Perspectives

Indigenous Knowledge and its Relevance to Conservation and Management of Tropical Forest Ecosystems

Anthropogenic Pressures on Ecosystem Structure and Function

**THEME II:
TROPICAL FORESTS: STRUCTURE, DIVERSITY AND
FUNCTION – PART A**

Plant–Pollinator Interaction

Global Perspectives on Tropical Forest Regeneration

Fire Ecology

Systematics and Evolution of Tropical Plants

Ecological and Bio-Geographical Contrasts among Tropical Rainforests with Conservation Implications

Tropical Forest Canopies

South and South-East Asian Dipterocarp Forests

xxi

THEME I

GLOBAL CHANGE AND TROPICAL FOREST ECOSYSTEMS:
Climate Change and Tropical Forests

Tropical Ecosystems: Structure, Diversity and Human Welfare.
Proceedings of the International Conference on Tropical Ecosystems
K. N. Ganeshaiah, R. Uma Shaanker and K. S. Bawa (eds)
Published by Oxford–IBH, New Delhi. 2001. pp. 3–6.

Effects of forests in the Amazon region on the global carbon budget: Early results from the large scale biosphere atmosphere experiment in Amazonia

Michael Keller

Complex Systems Research Center, University of New Hampshire, Morse Hall,
Durham, NH 03824, USA
e-mail: michael.keller@unh.edu

The forests of the Amazon region of South America cover nearly 6 million square-kilometers, an area nearly twice as large as India. Over the past three decades, the forests of Amazon region have experienced rapid changes owing to human actions. In the Brazilian legal Amazon region alone, official government statistics compiled from exhaustive satellite-based surveys show that approximately 517,000 km^2 have been cleared for agricultural and pastoral use through 1996. Selective timber extraction and fires annually affect similar areas to those suffering clearing for pastures and agriculture.

Using results from my own research and those from colleagues in the LBA project, I will review briefly the trends in land cover and land use that are affecting the carbon budget of the Brazilian portion of the Amazon region. The most important processes involved in land use and land cover changes are conversion of forested land to agricultural and pastoral use

Keywords: Amazon, carbon, deforestation.

(deforestation), secondary succession on abandoned agricultural and pastoral lands, logging, and forest fires. I will also consider the role of the natural forest and ask whether it is plausible that the mature upland forests are acting as a significant carbon sink.

Current deforestation rates in the Brazilian portion of the Amazon (about two-thirds of the entire region) average nearly 20,000 km^2 per year. This activity ultimately could contribute 0.3 to 0.5 Pg-C y^{-1} to the atmosphere, about 20% to 30% of the total land use change contributions (1.6 Pg Pg-C y^{-1}) (Schimel, 1995). The greatest uncertainty is the total carbon stock in the forests that are being cleared (Houghton et al., 2000). The estimation of biomass from local forest surveys incorporates substantial uncertainties (Keller et al., in press) along with the extrapolation of these local surveys to regional scale.

Land clearance is not permanent. Secondary forest areas are frequently used and re-cleared by local populations. Secondary forest regrowth rates depend upon many factors including edaphic conditions and the intensity of former land use (Moran et al., 2000).

Selective logging reduces forest biomass stocks at least temporarily. Both the extent of logging, and the carbon dynamics following logging are poorly quantified. The best available published information is based on informant surveys of total processed log volume from sawmill owners and managers. These data suggest that the area logged annually is rapidly approaching the annual area of land clearance (Nepstad et al., 1999). I will present new data on the carbon losses resulting from logging.

Fire is a constant threat to Amazonian ecosystems. Under natural conditions, forests may burn only on millennial time scales (Sanford et al., 1985; Meggers, 1994). The disturbance of forest by logging and the increased likelihood of ignition of natural forest by fires used for land management have long been recognized as potential threats to Amazon ecosystems, particularly in seasonally dry areas (Uhl and Buschbacher, 1985). Current research indicates that wildfires may be widespread and that they can cause significant tree mortality (Cochrane et al., 2000).

Despite the rapid changes, the Amazon is still mostly forested. How does the world's largest tropical forest contribute to the global budget of carbon? During the 1990s data from several sources indicated that Amazon forests may be a substantial sink for carbon (Grace et al., 1995; Phillips et al., 1998). Eddy covariance-based estimates of carbon uptake from published studies and from ongoing LBA work, suggest that almost all the forest sites are absorbing large amounts of carbon. The diurnal variation of CO_2 exchange, the magnitude of daytime uptake rates (governed by photosynthesis), and the response of fluxes to radiation are remarkably similar for all Amazon

4

forest sites studied so far (Andreae *et al.*, 2001). The preliminary values for total uptake are 5 to 6 Mg C ha^{-1} y^{-1}. Estimates of annual totals vary with analysis methodology. A remaining unknown in the local carbon budget is large scale advection of CO_2, a poorly measurable factor that is usually assumed to be small but could be substantial if strong horizontal gradients of CO_2 concentration exist (Kruijt *et al.*, 2001). Uptake rates vary according to corrections applied to account for potential problems in the estimations of the nocturnal flux. However, there are still serious debates about the validity of these findings. Preliminary data for a forest site outside of Santarem, Para, suggest an annual sink as large as 9 Mg C ha^{-1} y^{-1}. However, when these same data are screened for more turbulent conditions ($u* > 0.3$ m s^{-1}), the apparent sink is reduced to just 1 Mg C ha^{-1} y^{-1}. This correction for calm nights is uncertain though, since these turbulent conditions occurred on only 8% of the night-time periods studied (Michael Goulden, pers. commun.).

Given changes in atmospheric CO_2 and nutrient budgets, it is plausible that tropical forests may be growing more rapidly than they were in the past. Chambers *et al.* (in press) parameterized a model for woody biomass with field data from forest sites outside of Manaus. They increased productivity in their model by 25% over a 50-year period and then maintained productivity at the new increased level. Even with this large step in productivity in their model, the net uptake of carbon in woody biomass only increased by 0.2 to 0.3 Mg-C ha^{-1} y^{-1}. Can carbon be accumulating in other pools or is carbon being exported in other forms? Malhi *et al.* (1999) estimated a net storage of carbon belowground of ~3.6 Mg-C ha^{-1} y^{-1} for a forest site near Manaus based on a combination of eddy covariance results and other ecological studies in their region. However, direct measurements of the rates of carbon cycling through soils show a much more limited capacity for carbon uptake (Trumbore, 2000). For example, a 25% increase in C inputs to oxisols like those found in Manaus results in C storage rates over the first 20 years of only ~0.3 Mg-C ha^{-1} y^{-1}.

Recent inversion models based on global atmospheric CO_2 concentrations generally suggest a small net exchange of carbon in the Amazon region (Rayner *et al.*, 1999; Bousquet *et al.*, 2000). The size of both the carbon source from land use change and the carbon sink from forest growth may cancel out to reach a near zero net exchange. Currently, global models have large bounds of uncertainty in the Amazon region because of the lack of atmospheric sampling stations there. Recent sensitivity studies (Scott Denning, pers. commun.) suggest that weekly aircraft flights to measure CO_2 profiles at two points in the Amazon (one coastal, one interior) could decrease uncertainty by 75% to within a few tenths of a Pg–C. LBA researchers are testing methods for routine implementation of these aircraft measurements.

In the future, LBA will continue to refine the limits of uncertainty on carbon cycling in the Amazon at scales from the stand to the region. This includes continuation of studies described above, and additionally should include new boundary layer atmospheric studies, continental scale atmospheric sampling and improvement of input data for bottom up models.

References

Andreae, M. O. *et al.*, Towards an understanding of the biogeochemical cycling of carbon, water, energy, trace gases and aerosols in Amazonia: The LBA-EUSTACH experiments. *J. Geophys. Res.* (submitted) (2001).

Bousquet, P., Peylin, P., Ciais, P., Le Quéré, C., Friedlingstein, P. & Tans, P. P. Regional changes in carbon dioxide fluxes of land and oceans since 1980. *Science* **290**, 1342–1346 (2000).

Chambers, J. Q., Higuchi, N., Tribuzy, E. S. & Trumbore, S. E. Sink for a century: Carbon sequestration in the Amazon. *Nature* (in press).

Cochrane, M., Alencar, A., Schulze, M., Souza, C., Nepstad, D. C., Lefebvre, P. & Davidson, E. A. Positive feedback in the fire dynamic of closed canopy tropical forests. *Science* **284**, 1832–1835 (1999).

Grace, J., Lloyd, J., Mcintyre, J., Miranda, A. C., Meir, P., Miranda, H. C., Nobre, C., Moncrieff, J., Massheder, J., Malhi, Y., Wright, I. & Gash, J. Carbon dioxide uptake by an undisturbed tropical rain forest in southwest Amazonia, 1992 to 1993. *Science* **270**, 778–780 (1995).

Houghton, R. A., Skole, D. L., Nobre, C. A., Hackler, J. L., Lawrence, K. T. & Chomentowski, W. H. Annual fluxes of carbon from the deforestation and regrowth in the Brazilian Amazon. *Nature* **403**, 301–304 (2000).

Keller, M., Palace, M. & Hurtt, G. E. Biomass in the Tapajos National Forest, Brazil: Examination of sampling and allometric uncertainties. *For. Ecol. Manage.* (in press).

Kruijt, B., Elbers, J., von Randow, C., Araujo, A. C., Culf, A., Bink, N. J., Oliveira, P. J., Manzi, A. O., Nobre, A. D. & Kabat, P. Aspects of the robustness in eddy correlation fluxes for Amazon rainforest conditions. *J. Geophys. Res.* (submitted) (2001).

Malhi, Y., Baldocchi, D. D. & Jarvis, P. G. The carbon balance of tropical, temperate and boreal forests. *Plant, Cell Environ.* **22**, 715–740 (1999).

Manzi, A. O., Randow, C. V., Oliveira, P. J., Waterloo. M. J., Kruijt, B., Culf, A., Elbers, J. A., Zanchi, F. B., Lelis da Silva, R., Gomes, B. M., Kabat, P. & Dias, M. A. F. S. Measurements of energy, water and carbon dioxide fluxes over tropical forest and pasture in South-west Amazonia. *J. Geophys. Res.* (submitted) (2001).

Meggers, B. J. Archeological evidence for the impact of mega-Niño events on Amazonia during the past two millennia. *Climatic Change* **28**, 321–338 (1994).

Moran, E., Brondizio, E. S., Tucker, J. M., Da Silva-Forsberg, M. C., Mccracken, S. & Falesi, I. Effects of soil fertility and land-use on forest succession in Amazônia, *For. Ecol. Manage.* **139**, 93–108 (2000).

Nepstad, D. C., Verissimo, A., Alencar, A., Nobre, C., Lima, E., Lefebvre, P., Schlesinger, P., Potter, C., Moutinho, P. E., Mendoza, Cochrane, M. & Brooks, V. Large-scale impoverishment of Amazonian forests by logging and fire. *Nature* **398**, 505–508 (1999).

Phillips, O. L., Malhi, Y., Higuchi, N., Laurance, W. F., Nunez, P. V., Vasquez, R. M., Laurance, S. G., Ferreira, L. V., Stern, M., Brown, S. & Grace, J. Changes in the carbon balance of tropical forests: Evidence from long-term plots. *Science* **282**, 439–442 (1998).

Rayner, P. J., Enting, I. G., Francey, R. J. & Langenfelds, R. Reconstructing the recent carbon cycle from atmospheric CO_2, $\delta^{13}C$ and O_2/N_2 observations. *Tellus* **51B**, 213–232 (1999).

Sanford, R. L., Saldarriaga, J., Clark, K. E., Uhl, C. & Herrera, R. Amazon rain-forest fires. *Science* **227**, 53–55 (1985).

Schimel, D. Terrestrial ecosystems and the carbon cycle. *Global Change Biology* **1**, 77–91 (1995).

Trumbore, S. E. Constraints on below-ground carbon cycling from radiocarbon: The age of soil organic matter and respired CO_2. *Ecol. Appl.* **10**, 399–411 (2000).

Uhl, C. & Buschbacher, R. A disturbing synergism between cattle ranching, burning practices and selective tree harvesting in the eastern Amazon, *Biotropica* **17**, 265–268 (1985).

Tropical Ecosystems: Structure, Diversity and Human Welfare.
Proceedings of the International Conference on Tropical Ecosystems
K. N. Ganeshaiah, R. Uma Shaanker and K. S. Bawa (eds)
Published by Oxford–IBH, New Delhi. 2001. pp. 7–11.

Regional, spatially explicit assessment of the sensitivity of tropical forests to climate change in the past and future

David W. Hilbert

CSIRO Sustainable Ecosystems, Tropical Forest Research Centre and Cooperative Research Centre for Rainforest Ecology and Management, PO Box 780, Atherton, Qld, 4883, Australia
e-mail: david.Hilbert@tfrc.csiro.au

Climate is not constant in the tropics over long time-scales and .will change rapidly in response to greenhouse gas forcing. Thus, understanding how past climate change affected rainforests is important in order to interpret current ecological and biogeographic patterns. Further the impacts of future climate change need to be assessed and factored into conservation planning. However, evaluating the effects of climate change in tropical regions presents several difficulties compared to analyses of temperate and high latitude areas. These challenges are largely related to the extraordinary species richness in the tropics and the historically small investment of research in tropical forests. Because of these factors, the mechanistic modelling approaches that have been applied in simpler systems are not generally feasible for fine resolution, regional impact studies in the tropics. We have developed empirical methods (Hilbert and Ostendorf, in press) to assess climate change impacts that do not require detailed autecological or biogeographic knowledge of individual species and have successfully applied the method in the humid tropics of North Queensland.

Keywords: Pleistocene, rainforest, neural network, global change, humid tropics.

Our method uses maps of structural/environmental forest classes, spatial estimates of climate variables, maps of soil parent material as an indication of soil fertility, terrain variables from a digital elevation model, and an artificial neural network (ANN) that quantifies the relative suitability of local environments for each of 15 forest classes (Hilbert and van den Muyzenberg, 1999). Inputs to the ANN model include seven climate variables, nine soil parent material classes, and seven terrain variables. The output is a vector of fifteen real numbers [0,1] representing the relative suitability for each forest class (see Table 1 for a list of forest classes). Maps of the environments most suitable to each class are made by supplying all the inputs to the model for each hectare of the region (c. 2×10^6 hectares) and classifying each location as the class with the largest suitability. With today's climate, the model is highly successful at distinguishing the relative suitability of environments for fifteen forest classes with 75% of the region's forest mosaic accurately predicted at a one-hectare resolution. This accuracy is quite high considering the complexity of the vegetation mosaic, the number of forest classes, and the fine scale of spatial resolution. Areas mapped as disturbed, transitional forest types are generally classified by the model as having environments suitable to the forest type they are most likely to become. The approach has high potential for the analysis of climate change impacts as well as inferring vegetation patterns in the past wherever vegetation maps and spatial estimates of climate variables are available.

The ANN model was applied to estimate the spatial distribution and extent of forest environments within the humid tropics in three past climates: (1) the colder and drier climate of the Last Glacial Maximum (LGM) at c. 18000 yr BP; (2) a cooler and wetter climate at c. 7000 yr BP; and (3) a warmer, wetter climate at c. 5000 yr BP. The motivation for this study is that the biogeography and ecology of a region are often as dependent on the area's history as on the current environmental conditions. Thus, the study of past environments and distributions of biota is essential to understand the present-day ecological patterns. Similarly, the analysis of past responses to climatic change provides important insights into the possible impacts of future alterations of climate. By changing the climate variables at all locations it is possible to map the potential distribution of all forest classes in detail over the entire region. A summary of the results is given in Figure 1 where all rainforest classes are aggregated. Tall Open Forest, also known as 'wet sclerophyll', is an ecotonal forest type occurring between rainforest and the dry and open Medium and Low Woodlands class. In the figure, 'other' refers to several minor forest classes (Medium Open Forest and Woodland, Coastal Complexes, and Mountain Rock Pavements). The areas for today's climate were estimated by adding the actual mapped distributions to model estimates for the approximately 25% of the region that is now cleared of forest.

The extent of environments suitable for rainforests and the major sclerophyll forests has changed markedly since the late Pleistocene. At LGM, rainforest

8

environments would have been quite restricted compared to their extent today. Tall Open Forest and Tall Woodland environments had their greatest regional extent at this time. By 7000 yr BP, increased rainfall extended the potential extent of rainforests, especially Simple Notophyll & Simple Microphyll Forests and Thickets. The warm, wet climate at 5000 yr BP lead to the greatest regional extent of Mesophyll Vine Forests but a severe restriction of Simple Notophyll & Simple Microphyll Forests and Thickets to the tops of the highest peaks. The model suggests that Tall Open Forest and Tall Woodlands would have contracted to small, interglacial refugia, at this time (Hilbert *et al.*, 2000). Overall, the model is verified by empirical evidence of rainforest contraction in the late Pleistocene (Hopkins *et al.*, 1993; 1996).

The sensitivity of the region's forests to future climate change has been estimated using a variety of techniques (Hilbert *et al.* in press). One approach is to use the model output to measure the dissimilarity (D) between the environment at each location and the environment that would be most suitable for the forest type that is mapped there. D is found from the angle (γ) between the 'ideal' vector for the mapped forest type and the vector produced by the model for the environment at that location. The angle (radians) between two vectors can be found from

$$\cos\gamma = \frac{e \cdot m}{\sqrt{e \cdot e}\sqrt{m \cdot m}},$$

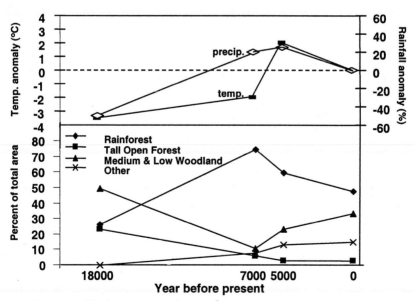

Figure 1. Per cent of the humid tropics (c. 2×10^6 hectares) having environments most suitable to several forest classes in three past climates and today. Compared to today, the climate at 18000 BP was cold and dry, 7000 BP was cool and wet, and 5000 BP was warm and wet.

where (*e*) is the neural net output vector and (*m*) is the 'ideal' vector consisting of 1.0 for the mapped forest type and zeros otherwise. D is then defined as γ divided by $\pi/2$, normalizing the index to the range [0,1]. Values of D greater than 0.5 imply that the local environment is more suitable to some other forest class than for the forest class present there. Since D estimates the mismatch between a location's environment and its current forest class, it can be used as a measure of the likelihood of change in the extant forests that would be produced by rapid climate change. Significantly, Vine Forests with *Acacia* and *Eucalyptus* (the single class in the model representing forests that are converting from one type to another) have by far the largest mean D in today's climate, confirming that D may relate to the propensity for ecological change. Results for three climate change scenarios, all with one degree warming, are presented in Table 1. By this measure, highland rainforests (Simple Notophyll & Simple Microphyll Forests and Thickets) are the most sensitive rainforest class. Mesophyll Vine Forests with Palms and upland rainforests (Complex Notophyll Vine Forests) are also quite sensitive. Thus, one degree warming is likely to cause significant ecological change in the upland and highland rainforests that are the habitat

Table 1. Mean dissimilarity values (D) in today's climate and the mean change in D for three climate change scenarios where rainfall was decreased by 10%, unchanged, or increased by 10%. Temperatures were increased by 1°C in all scenarios. The last column gives the mean change in D for each class, averaged over the three scenarios. Tracey (1982) and Webb (1978) describe the forest classes in detail.

Forest class	D Today's climate	Change in D −10%	Rainfall unchanged	+10%	Mean
Mesophyll Vine Forests	0.282	0.080	0.019	−0.002	0.032
Mesophyll Vine Forests with palms	0.580	0.270	0.346	0.309	0.308
Semideciduous Mesophyll Vine Forest	0.451	0.230	0.139	0.084	0.151
Complex Notophyll Vine Forests	0.381	0.281	0.165	0.215	0.220
Notophyll Vine Forests	0.750	−0.119	0.039	0.250	0.057
Simple Notophyll & Simple Microphyll Forests and Thickets	0.213	0.493	0.390	0.359	0.414
Deciduous Microphyll Vine Thicket	0.572	0.082	0.134	0.428	0.215
Araucarian Vine Forests	0.101	−0.028	0.017	0.188	0.059
Notophyll Semi-evergreen Vine Forests	0.302	0.300	0.062	0.022	0.128
Vine Forest with Acacia and/or Eucalyptus	0.677	0.051	0.005	0.021	0.026
Tall Open Forest and Tall Woodland	0.372	0.334	0.333	0.389	0.352
Medium Open Forest and Woodlands	0.265	0.628	0.572	0.555	0.585
Medium and Low Woodlands	0.221	−0.086	0.001	0.157	0.024
Coastal Complexes	0.426	0.060	0.281	0.538	0.293
Mountain Rock Pavements	0.218	0.258	0.053	0.022	0.111
Overall	0.293	0.135	0.121	0.174	0.143

for the majority of endemic vertebrates in the humid tropics. Lowland rainforests (Mesophyll Vine Forests) are not sensitive to the range of small climate changes studied. Medium Open Forest and Woodlands (often dominated by *Melaleuca* spp.), an uncommon class of sclerophyll forest in the region, are highly sensitive to warming as are Tall Open Forest and Tall Woodlands. The latter has the highest vertebrate richness in the region, including several threatened species. Overall, the region appears to be quite sensitive to climate change. It displays large climate-induced changes in the spatial distribution and extent of rainforests over long time periods and the forests, as they are distributed now, are likely to be strongly affected by rapid climate change in this century.

Acknowledgements

Brett Buckley, Jeroen van den Muyzenberg, Bertram Ostendorf, Trevor Parker and Warwick Sayers provided valuable technical assistance. Discussions with Andrew Graham and Mike Hopkins were instrumental in developing the approach. Vegetation, geology, and digital elevation data were supplied by the Wet Tropics Management Authority.

References

Hilbert, D. W. & Ostendorf, B. The utility of empirical, artificial neural network approaches for modelling the distribution of regional to global vegetation in past, present and future climates. *Ecol. Model* (in press).

Hilbert, D. W., Ostendorf, B. & Hopkins, M. Sensitivity of tropical forests to climate change in the humid tropics of North Queensland. *Aust. Ecol.* (in press).

Hilbert, D. W., Graham, A. W. & Parker, T. A. Forest and Woodland Habitats of the Northern Bettong (*Bettongia tropica*) in the Past, Present and Future: A report prepared for Queensland Parks and Wildlife Service. CSIRO Wildlife and Ecology, Tropical Forest Research Centre, Atherton. Report No. VM1/0300-14 (2000).

Hilbert, D. W. & van Den Muyzenberg, J. Using an artificial neural network to characterise the relative suitability of environments for forest types in a complex tropical vegetation mosaic. *Divers. Distrib.* **5,** 263–274 (1999).

Hopkins, M. S., Ash, J., Graham, A. W., Head, J. & Hewett, R. K. Charcoal evidence of the spatial extent of the Eucalyptus woodland expansion and rain-forest contractions in North Queensland during the late Pleistocene. *J. Biogeog.* **20,** 357–372 (1993).

Hopkins, M. S., Head, J., Ash, J., Hewett, R. K. & Graham, A. W. Evidence of Holocene and continuing recent expansion of lowland rain forest in humid, tropical North Queensland. *J. Biogeog.* **23,** 737–745 (1996).

Tracey, J. G. *The Vegetation of the Humid Tropical Region of North Queensland* (CSIRO, Melbourne, 1982).

Webb, L. J. A general classification of Australian forests. *Aust. Plants* **9,** 349–363 (1978).

Tropical Ecosystems: Structure, Diversity and Human Welfare.
Proceedings of the International Conference on Tropical Ecosystems
K. N. Ganeshaiah, R. Uma Shaanker and K. S. Bawa (eds)
Published by Oxford–IBH, New Delhi. 2001. pp. 12–13.

Tropical forest dynamics and climate: El Niño impact across a rainfall gradient

Richard Condit

Smithsonian Tropical Research Institute, Unit 0948, APO AA 34002-0948, USA
e-mail: condit@ctfs.stri.si.edu

Tropical forest dynamics were examined across a climatic gradient in Central Panama using three inventory plots, one at a site getting 1900 mm rain annually, a second at a site getting 2500 mm, and a third getting 3000 mm. Plots were large enough to estimate demographic patterns for many individual species as well as for the forest as a whole. The strong 1997 El Niño was spanned by censuses at all three sites. The three sites were quite different in species composition. The wettest and driest plots shared few species overall and no common species. The intermediate site shared a few common species with both the wet and the dry sites. Forest-wide growth of large trees was high at the wet and intermediate sites and low at the dry site. But sapling growth was different, highest at the dry site and lowest at the intermediate site. Growth differences were reflected by individual species, for example, saplings of species at the dry site grew faster than saplings of the same species at the intermediate site. Forest-wide mortality was also lowest at the dry site and highest at the wet, but mortality differences were not reflected by individual species. Those species which occurred at both dry and intermediate site tended to have the same mortality rate at both, despite the fact that the dry site had overall lower mortality. Some species performed poorly in terms of demography at one site, but well at another. This

Keywords: Tropical forest, forest dynamics, climate, rainfall gradient, El Niño, Panama.

demonstrates the non-equilibrium nature of the forests, some species declining at one site had stable populations at another. The El Niño elevated growth rates, relative to the period before, at all three sites. This held for most species with samples large enough to evaluate.

We suggest that low turnover in drier forests is mainly an evolutionary response to sporadic recruitment, but that high sapling growth in dry sites is due to greater light penetration to the forest floor. High growth during El Niño was probably caused by greater insolation during the accompanying drought.

Tropical Ecosystems: Structure, Diversity and Human Welfare.
Proceedings of the International Conference on Tropical Ecosystems
K N Ganeshaiah, R. Uma Shaanker and K. S. Bawa (eds)
Published by Oxford–IBH, New Delhi. 2001. p. 14.

Tropical forest response to climate change

F. A. Bazzaz

Department of Organismic and Evolutionary Biology, Harvard University Biological
Laboratory, 16 Divinity Avenue, Cambridge, Massachusetts 02138, USA
e-mail: afs802@bangor.ac.uk

The tropical rain forest is characterized by the highest biological diversity on earth. Although there have been very few studies, it is expected that the species have very narrow niches for them to be packed so closely. Tropical trees have been known to possess leaves that persist for longer times than do temperate trees and experience a near continuous growing season.

The response of tropical forest to climate change is virtually unknown but we can speculate on their possible response. We have conducted a study using *Crecrepia, Halescarpis, Sena* and other early successional species. We have shown that with elevated CO_2 there is a significant change in the architecture of these species. It could be assumed that such changes in the architecture might have a ripple-effect on the competitive interactions among the members of a plant community. Accordingly, based on the nature of such interactions, the overall biological diversity of the community may also be altered. Clearly more studies need to be initiated to examine the response of individual species and plant communities to elevated CO_2 and to understand the responses of tropical forests at large to elevated CO_2 levels.

Keywords: Tropical forest, climate change, *Cecrepia musanga*.

Tropical Ecosystems: Structure, Diversity and Human Welfare.
Proceedings of the International Conference on Tropical Ecosystems
K. N. Ganeshaiah, R. Uma Shaanker and K. S. Bawa (eds)
Published by Oxford–IBH, New Delhi. 2001. pp. 15–19.

Evaluation of land-use effects on carbon stocks and fluxes across the Orinoco llanos

Jose J. San-José[*][†] *and Rubén A. Montes*[**]

[*]Centro de Ecología, Instituto Venezolano de Investigaciones Científicas, Apartado
21827, Caracas 1020-A, Venezuela
[**]Departamento de Estudios Ambientales, Universidad Simón Bolívar. Apartado
89000, Caracas 1080-A, Venezuela
[†]e-mail: jsanjose@oikos. ivic.ve

Mixed tree-grass systems cover twenty to thirty per cent of the Earth's surface (Scholes and Hall, 1996), where trees are the predominant components in savannas and grasses in steppes. The vast cover of these savannas suggests a significant contribution to the global carbon budget and gaseous fluxes. They may account for as much as 500 Tg C, which may contribute more to the missing sink than previously estimated (Scurlock and Hall, 1998). In relation to soil organic matter stocks, the grassland soils represent a carbon pool of 200–300 Pg C (Scurlock and Hall, 1998), which range from 10 to 30 per cent of the world soil carbon (Eswaran *et al.*, 1993).

In addition to the extensive areas occupied by savannas, they include a wide-range of physiognomic types which occur under different climatic and soil features as well as varying levels of human interference. However, the extent to which the heterogeneity of savannas may influence the global carbon budget has scarcely been considered. Currently, human pressure on savanna resources through the effect of fires, cattle raising, agricultural practices and

Keywords: Orinoco llanos, land-use changes, periodic inventories, carbon sink, net flux of carbon.

resource extraction is associated with an overall population density above 24 people per km^{-2} (Weiner, 1979). The impact of careless management practices on savannas and woodlands has resulted in significant degradation of vegetation and soil, leading to a net carbon loss from the ecosystems to the atmosphere, and consequently in many cases to desertification.

There is a controversial discussion on the role of savannas as carbon sink or source in the global carbon budget. Analysis based on land-use change indicated that savannas are sources of carbon (Houghton, 1995), whereas Parton *et al.* (1995) and Scurlock and Hall (1998) have estimated that savannas world-wide are able to sequester carbon up to 500 Tg C per year. These conflicting estimates could be based on different methodological approaches. Thus, the approach based on land-use change addresses only a portion of the net carbon flux, while the evaluation of vegetation inventories represents a complementary approach, which includes inadvertent changes (Houghton, 1995).

The aim of the present work is to inventory the carbon stored in the native Orinoco vegetations and to evaluate the extent to which management practices are affecting the carbon stock and fluxes in this heterogeneous basin. These results may help to increase our understanding of net carbon balance in these grazing lands and to manage the foreseen global change.

The analysis was based on the physiognomic types of vegetation found across the Orinoco llanos (0.39×10^6 km^2), which include: a) the northern region delimited by the foothills of the Coastal Caribbean Range (8°56'N; 67°25'W) and the Orinoco River (7°46'N; 64°25'W) and b) the western region located between the latitudes 3° and 6°N, to the east of the Andes mountains, and between longitudes 71° and 68°W at the western margin of the Orinoco River. The soils encompass mainly Oxisols and Ultisols, in accordance with the US Soil Taxonomy System. In this work a comprehensive approach includes data from vegetation inventories and changes in land-use. The vegetation physiognomic types were grouped into four classes of canopy cover. These classes were: (a) herbaceous savannas with less than 10 per cent tree cover, which are represented by twenty one sites ($n = 21$), (b) tree savannas with 10–50 per cent cover by woody plants and a developed grass layer ($n = 13$), (c) woodlands with 50–100 per cent of tree canopy cover and a sparse graminaceous layer ($n = 6$) and (d) semi-deciduous forests with 100 per cent of tree canopy cover ($n = 5$). At each physiognomic type, site selection was based on a survey covering the landscape heterogeneity. It comprises dissected, conserved and eolian highplains as well as colluvial–alluvial plains, which were formed from detrital materials and accumulated during the lower Pleistocene in a geological formation known as Mesa. Landscape units were selected on the basis of maps, aerial photographs and field trips throughout the Orinoco basin. In addition to the native vegetation, systems modified by human activities were analyzed. For each physiognomic group, rates of land-use change were analyzed from 1982 to 1992 by following the human-induced changes related to clearing the native vegetation for pastures, crops and afforestation. Also, changes in area from crops to cultivated pastures and from pastures to native vegetation (re-growth) were evaluated over the same period.

The changes in area of each physiognomic group due to land-use over 1982–1992 were evaluated by making use of satellite images from LANDSAT (Multi-Spectral Scanner and Thematic Mapper). These images have been used previously to elaborate the vegetation maps and agricultural systems of Venezuela and Colombia, and a land-use map of Venezuela. The resulting cartographic information was corroborated by field trips throughout the Orinoco llanos.

At each site, the carbon stock and fluxes were estimated from the native vegetation inventories, which include the effect of fires and cattle raising. The amount of carbon releases from the native vegetation to the atmosphere through: (a) burning, (b) charcoal and particulate matter from combustion of burned vegetations and (c) decomposition was summed. The net flux of carbon was calculated from this summed value minus carbon uptake from the atmosphere by the native vegetation. Negative values represent withdrawal of carbon from the atmosphere. The turnover rate is considered as the ratio of the annual net carbon exchange to the carbon stock in the system.

The two approaches (changes in land-use and periodical inventories) indi-cated that the Orinoco system is acting as a sink of carbon (-17.53 Tg C yr^{-1}). The gross release due to the total gross emissions was 174.66 Tg C yr^{-1}, whereas gross uptake by the native and cultivated vegetation (-190.40 Tg C yr^{-1}), re-growth of forest vegetation (-0.23 Tg C yr^{-1}), affore-station (-1.43 Tg C yr^{-1}) and accretion in soil carbon (-0.13 Tg C yr^{-1}) was 192.19 Tg C yr^{-1}. The carbon gross uptake was 1.10-fold higher than that of the carbon gross release. These results of carbon exchange are supported by micrometeorological measurements over the Orinoco savannas. Global savannas have been considered as a moderate carbon sinks ranging from 500 to 2,000 Tg C yr^{-1} (Parton et al., 1995; Scholes and Hall, 1996; Scurlock and Hall, 1998). Therefore, the carbon influx to the Orinoco llanos was 0.8–3.5 per cent in relation to those global estimates for savannas and grasslands.

Even though the Orinoco system is a moderate carbon sink and the main carbon source in the tropics is the forest area, the impact of fire and agri-cultural practices on the Orinoco llanos have led to emission of greenhouse gasses, reduction in biodiversity and degradation of land. Heterogeneity of carbon distribution in the Orinoco llanos is increasing under current management activities. In the cultivated areas, the replacement of native vegetation has led to a decline in soil carbon stock over a short period of a few years. This trend could be reversed by protection of the savanna system, which would store carbon in the soil and biomass. It has been calculated that carbon content in a protected savanna can increase from 7,081 to 21,411 g C m^{-2} over 51 years. If the area covered by the Orinoco plains (0.39×10^6 km^2) in northern South America behaved in the same way as this type of protected vegetation, then the sequestered pool of carbon in the restored forests would be 8,300 Tg C. This amount could represent the potential of the Orinoco llanos for storing carbon. However, current carbon stock (1,204.17 Tg C) is 14.5 per cent of the calculated capacity for carbon sequestration. This difference might be due mainly to carbon emissions through the effect of fire on the native vegetation and changes in land use. Fisher *et al.* (1994) have calculated a carbon sink for South American grazing lands (2.5×10^6 km^2) ranging from 100 to 500 Tg C yr^{-1}. Scholes and Hall (1996) have calculated a potential for carbon sequestration in global savannas of 94,300 Tg C.

The carbon turnover time in the Orinoco system was 68 yr (-17.53 Tg C yr^{-1}/ 1,204.17 Tg C = 0.014 yr^{-1}) and consequently, the Basin provides a limited route for CO_2 sequestration. Particularly, in the soil compartment, the value (19.01 Tg C yr^{-1}/1,010.11 Tg C = 0.018 yr^{-1}) is a shorter term (53 years), which relates favorably with the data from Harrison *et al.* (1990) and Lobo *et al.* (1990). They found that turnover time of organic matter increases with soil depth, ranging from several years for litter to 15–40 years in the upper 0.10 m and over 100 years below a depth of about 0.25 m. Batjes and Sombroek (1997) have reported a global mean turnover time for the soil organic matter of 22 years (litter included) and a maximum of up to 5,000 years. The soil turnover time is related to CO_2 sequestration capacity as shown by Batjes and Sombroek (1997). Therefore, the magnitude of the turnover time in the Orinoco soils indicates that the capacity for a relative short-term carbon sequestration does exist.

In the Orinoco llanos, progressively increasing atmospheric carbon emissions due to rapid changes in land-use have resulted owing to overgrazing, frequent and extensive burning and conversion of native vegetations into agricultural lands. On the other hand, the sink capacity of the Orinoco vegetation is limited by climatic conditions, soil type, soil pH, soil minerals and drainage. These findings compare favorably for other systems (Carter *et al.*, 1997), where there are serious limitations for increasing water use efficiency. Therefore, it is necessary to underline the need to monitor carbon storage in soil and identify management practices that may foster carbon sequestration as well as mitigate the emission of carbon. Nilsson and Schopfhauser (1995) and Houghton *et al.* (1993) have proposed options for carbon sequestration. Particularly, in the Orinoco llanos, management practices can be improved to sequester CO_2 using agroforestry and agrosilvopastoral systems (Laarman and Sedjo, 1992; Polley *et al.*, 1997). It is necessary to take into account strategies for acidic and oligothrophic soils based on adaptative combination of grasses, forage/legumes and trees. Therefore carbon sequestration can be achieved by fertilization and management but socio-economic structure capabilities have to be taken into account.

Acknowledgements

This work has been conducted as part of the Savanna Bio-productivity (MAB UNESCO) project of IVIC and was partially sponsored by the National Research Council of Venezuela (CONICIT) and the Man and Biosphere Programme (MAB/UNESCO). We appreciate the technical assistance of M.Sc. R. Bracho and N. Nikonova.

References

Batjes, N. H. & Sombroek, W. G. Possibilities for carbon sequestration in tropical and subtropical soils. *Global Change Biol.* **3**, 161–173 (1997).

Carter, M. R., Angers, D. A., Gregorich, E. G. & Bolindes, M. A. Organic carbon and nitrogen stocks and storage profiles in cool, humidic soils of eastern Canada. *Can. J. Soil Sci.* **77**, 205–210 (1997).

Eswaran, H., van der Berg, E. & Reich, P. Organic carbon in soils of the world. *Soil Sci. Soc. Am. J.* **57**, 192-194 (1993).

Fisher, M. J., Rao, I. M., Ayarza, M. A., Lascano, C. E., Sanz, J. I., Thomas, R. J. & Vera, R. R. Carbon storage by introduced deep-rooted grasses in the South American savannas. *Nature* **371**, 236–238 (1994).

Harrison, A. F., Harkness, D. D. & Bacon, P. J. The use of bomb ^{14}C for studying organic matter and N and P dynamics in a woodland soil. in *Nutrient Cycling in Terrestrial Ecosystems: Field Methods, Applications and Interpretation* (eds Harrison, A. F., Ineson, P. & Heal, O. W.) (Elservier, Barking, 1990).

Houghton, R. A. Effects of land-use change, surface temperature, and CO_2 concentration on terrestrial store of carbon. in *Biotic Feedbacks in the Global Climate System: Will the Warming Speed the Warming?* (eds Woodwell, G. M. & Mackenzie, G. T.) 333–350 (Oxford University Press, New York, 1995).

Houghton, R. A., Unruh, J. & Lefebvre, P. A. Current land cover in the tropics and its potential for sequestering carbon. *Global Biogeochem. Cycles* **7**, 305–320 (1993).

Laarman, J. & Sedjo, R. A. *Global Forests, Issues for Six Billion People* (McGraw-Hill, New York, 1992) 337 pp.

Lobo, P. F. S., Barrera, D. S., Silva, L. F. & Flexor, J. M. Carbon isotopes on the profile of characteristics soils of the south of the state of Bahia, Brazil. *R. Bras. Cienc. Soil.* **4**, 74–82 (1990).

Nilsson, S. & Schopfhauser, W. The carbon-sequestration potential of a global afforestation program. *Climatic Change* **30**, 267–293 (1995).

Parton, W. J., Scurlock, J. M., Ojima, D. S., Schimel, D. S. & Hall, D. O. Group Members Scope Gram. Impact of climate change on grassland production and soil carbon worldwide. *Global Chang. Biol.* **1**, 13–22 (1995).

Polley, H. W., Johson, H. B. & Mayeux, H. S. Leaf physiology, production, water use and nitrogen dynamics of the grassland invader *Acacia smallii* at elevated CO_2 concentration. *Tree Physiol.* **17**, 89–96 (1997).

Scholes, R. J. & Hall, D. O. The carbon budget of tropical savannas, woodlands and grasslands. in *Global Change: Effects on Coniferous Forest and Grasslands Scope* (eds Breymeyer, A. I., Hall, D., Melillo, J. M. & Agren, G. I.) 69–100 (John Wiley and Sons, New York, 1996).

Scurlock, J. M. O. & Hall, D. O. The global carbon sink: a grassland perspective. *Global Change Biol.* **4**, 229–233 (1998).

Weiner, J. S. Human biology. in *Tropical Grazing Land Ecosystem* (UNESCO, Paris, 1979) 655 pp.

Tropical Ecosystems: Structure, Diversity and Human Welfare.
Proceedings of the International Conference on Tropical Ecosystems
K. N. Ganeshaiah, R. Uma Shaanker and K. S. Bawa (eds)
Published by Oxford–IBH, New Delhi. 2001. pp. 20–21.

Estimating carbon dioxide fluxes from soil carbon pools following land-cover change: A test of some critical assumptions for a region in Costa Rica

Jennifer S. Powers

Department of Biology, P.O. Box 90338, Duke University, Durham,
NC 27708-0338, USA
e-mail: jsp8@duke.edu

Each year, tropical deforestation is thought to result in a release of 1.6 Pg carbon to the atmosphere as carbon dioxide. Global assessment of the CO_2 fluxes due to tropical land-cover change only account for uncertainty involved in estimating deforestation rates, not in the distribution of C in different compartments or site-specific effects of forest clearing and cultivation. The objective of this work was to examine how assumptions about pre-disturbance soil C storage and the effects of land-cover change on soil C pools influence calculations of regional soil C storage over time. To accomplish this, I used a geographic information system to apply three models of the effects of land-cover change on soil C storage to two digital pre-disturbance soil C maps that represent a 140,000-ha area in northeastern Costa Rica. I then calculated regional soil C storage for each combination of model and pre-disturbance soil C map for three time periods defined by georeferenced land-cover maps (1976, 1986, and 1996). The region was largely covered in primary forests in 1950, but now exists as a mosaic of forests, pastures, and crops.

Keywords: Soil carbon, land-cover change, Costa Rica.

One pre-disturbance soil C map was generated using values from the literature that are assigned to tropical wet forest, as in the global assessments. A second pre-disturbance soil C map was constructed from linear regression equations developed from field data that related forest soil C storage to terrain attributes from a digital elevation model. The three models of the effects of land-cover change differed in the degree of geographic specificity. The first model assumed a common 20% decrease in soil C stocks following clearing for pasture and a 40% reduction for crop fields, which is similar to global assessments (the global model) (Houghton et al., 1983; Detwiler, 1986). The other two models were derived from extensive field measurements of soil C storage under different land covers in this region, and varied in the extent to which the data were aggregated. In the second model, the observed changes in soil C storage across the region were averaged by land-cover transition and then applied over the entire region (the regional-average model). The last model used linear regression equations that related the changes in soil C storage to terrain attributes such as elevation and topographic relief to derive spatially-explicit estimates of the effects of forest clearing (the geographic model).

Estimated regional soil C (to 0.3 m) under forest vegetation was higher in the map based on field data (10.03 Tg C) than in the map based on life zones (8.90 Tg C). Regional soil C storage declined through time due to the effects of clearing forest for pasture and crops. Estimated CO_2 fluxes depended more on the model of the effects of land-cover change than on pre-disturbance soil C map. Soil C losses under the global model exceeded estimates based on field data by a factor of 10. The small differences in estimated soil C losses between the regional-average model and the geographic model reflect landscape heterogeneity.

References

Detwiler, R. P. Land use change and the global carbon cycle: the role of tropical forests. *Biogeochemistry* **2**, 67–93 (1986).

Houghton, R. A., Hobbie, J. E., Moore, B., Peterson, B. J. & Shaver, G. R. Changes in the carbon content of terrestrial biota and soils between 1860 and 1980: A net release of carbon dioxide to the atmosphere. *Ecol. Mono.* **53**, 235–262 (1983).

Tropical Ecosystems: Structure, Diversity and Human Welfare.
Proceedings of the International Conference on Tropical Ecosystems
K. N. Ganeshaiah, R. Uma Shaanker and K. S. Bawa (eds)
Published by Oxford–IBH, New Delhi. 2001. pp. 22–23.

Salvaging hurricane debris accelerates sequestration of atmospheric CO$_2$ in a tropical wet forest

X. Zou*, M. Warren, Y. Li and J. K. Zimmerman

Institute for Tropical Ecosystem Studies, University of Puerto Rico, PO Box 363682,
San Juan, PR 00936, USA
*e-mail: xzou@sunites.upr.clu.edu

Hurricanes are frequent disturbances affecting a large portion of tropical regions. They cause severe damages to forests by defoliating forest canopy and breaking tree branches and stems (Walker *et al.*, 1992).

We examined the influence of hurricane debris on forest productivity and soil organic carbon pools in a tropical wet forest of Puerto Rico between 1990 and 1998. Plant debris was removed from experimental plots shortly after hurricane Hugo in 1989 (Zimmerman *et al.*, 1995; Walker *et al.*, 1996). We used a randomized block design with treatments randomly assigned to plots of 20 x 20 m. Forest litter production and soil carbon pools were compared between the control and hurricane debris removal treatments.

The removal of hurricane debris increased forest litter production by 20% on average during the nine-year period (Table 1). In 1997, soil organic carbon content in the hurricane debris removal plots was 20% greater than in the control plots for the surface 0–0.25 m layer (Table 1), amounting to an additional sequestration of 1700 kg/ha of atmospheric carbon during the

---

Keywords: Hurricane disturbance, litterfall, soil organic carbon, tropical forest.

Table 1. Carbon in light-fraction (LF, of soil organic matter and total C of soil for the 1–10 cm and 10–25 cm soil depth per square meter in a tabonuco forest in Puerto Rico. Data are means (SE) from 4 replicates. Common letters within a column in each soil depth indicate no significant differences among the treatments by Sheffe's mutiple range test at the significance level of 0.05.

Treatments	C in LF (kg/m^2)	Total C (kg/m^2)
1–10 cm		
Debris removal	0.20 (0.01)[a]	0.52 (0.02)[a]
Control	0.16 (0.01)[b]	0.42 (0.03)[b]
10–25 cm		
Debris removal	0.28 (0.02)[a]	0.53 (0.03)[a]
Control	0.17 (0.04)[b]	0.46 (0.02)[b]

eight-year period. Our results demonstrate that salvaging hurricane debris can significantly accelerate the sequestration of CO_2 from atmosphere in tropical wet forests.

References

Walker, L. R., Brokaw, N. V. L., Lodge, D. J. & Waide, R. B. Ecosystem, plant, and animal responses to hurricanes in the Caribbean. *Biotropica* (Special issue) **23**, 313–521 (1992).

Walker, L. R., Zimmerman, J. K., Lodge, D. J. & Guzman-Grajales, S. An altitudinal comparison of growth and species composition in hurricane-damaged forests in Puerto Rico. *J. Ecol.* **84**, 877–889 (1996).

Zimmerman, J. K., Pulliam, W. M., Lodge, D. J., Quinones-Orfila, V., Fetcher, N., Guzman-Grajales, S., Parrotta, J. A., Asbury, C. E., Walker, L. R. & Waide, R. B. Nitrogen immobilization by decomposing woody debris and the recovery of tropical wet forest from hurricane damage. *Oikos* **72**, 314–322 (1995).

Tropical Ecosystems: Structure, Diversity and Human Welfare.
Proceedings of the International Conference on Tropical Ecosystems
K. N. Ganeshaiah, R. Uma Shaanker and K. S. Bawa (eds)
Published by Oxford–IBH, New Delhi. 2001. pp. 24–26.

Climate change and ecophysiology of *Azadirachta siamensis* and *Pterocarpus macrocarpus* in Forest Plantation, Thailand

S. Kerdkankaew*,[†], J. Luangjame** and P. Khummongkol*

*Environmental Technology Division, School of Energy and Materials,
King Mongkut's University of Technology Thonburi, Thailand
**Royal Forest Department, Thailand
[†]e-mail: sureekaew@hotmail.com

Climite change resulting from rising greenhouse gases, especially carbon dioxide (CO_2) has been a major environmental problem (Taiz and Zeiger, 1998). To abate this problem, it is generally agreed that reforestation at large scale can sequester the excess carbon. However, this efficacy of the sequestration may depend upon the individual species concerned. In this study we examined the CO_2 uptake in plantation of *A. siamensis* and *P. macrocarpus*.

Azadirachta indica A. Juss. var. *siamensis* (Valeton), Thai indigenous neem species, and *Pterocarpus macrocarpus* Kurz are widely planted in Thailand. In 1999, the effect on La Nina phenomenon changed the temperature and rainfall in Thailand.

Keywords: Climate change, CO_2 uptake, *Azadirachta siamensis*, *Pterocarpus macrocarpus*, forest plantation.

24

The field study was located at the Silvicultural Research Center No. 3, in Tha Muang district, Kanchanaburi province –14 01°N latitude, 99 45°E longitude, 60 msl. Two tropical species, *A. siamensis* and *P. macrocarpus* were planted in June 1989 at 4×4 m spacing. The field measurements were taken between 08:00 and 17:00 hr by using Li-6200 portable photosynthesis system (Li-cor Inc., Lincoln, USA). Three trees of *A. siamensis* and four trees of *P. macrocarpus* were randomly selected. Ten mature leaves from east to west in each canopy were chosen and the instantaneous rates of assimilation were estimated by Li-6200 at one-hour interval. The climatic data were recorded at the research center and obtained from the Kanchanaburi meteorological station, about 5-km from the study site.

The CO_2 uptake rates of the two species during 1999 and 2000 are illustrated in Figure 1. The highest CO_2 uptake rates of *A. siamensis* occurred at noon in 1999 and at 9:00 hr in 2000. The rates of *P. macrocarpus* were lower in 1999 than in 2000 and lower than *A. siamensis* as well. The transpiration rates of both plants were significantly lower in 1999 than in 2000. The transpiration rates were significantly higher in *A. siamensis* than *P. macrocarpus*. Corresponding to transpiration patterns, the stomatal conductance of *A. siamensis* was higher. On the contrary, the water use efficiency (WUE) of *P. macrocarpus* was explicitly higher in 1999 than in 2000. In general, the leaves of *A. siamensis* shed in November but those of *P. macrocarpus* shed in February. However, in 1999 leaves of *A. siamensis* and *P. macrocarpus* were shed in November and February, respectively.

Because of the lower temperature in 1999, CO_2 uptake which depended on sunlight or PPFD reached the highest values around noon time for both trees (Williams and Flanagan, 1998; Luangjame, 1999). On the other hand, in 2000

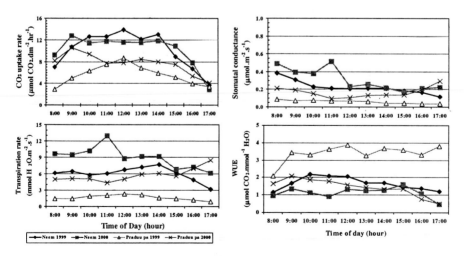

Figure 1. The diurnal courses of ecophysiological parameters in both trees, *A. siamensis* and *P. macrocarpus* during measurements in December 1999 and 2000.

with the normal weather, although sunlight still peaked around noon, the highest CO_2 uptake occurred at 9:00 h due to the midday depressions (Ishizuka and Puangchit, 1996). Generally, in the forest plantation, the maximum CO_2 uptake occurred at sunlight less than 1200 μmol m^{-2} s^{-1} (Baldocchi and Harley, 1995) and at leaf temperature lower than 36°C (Larcher, 1995). Our results indicated that the CO_2 uptake rates of the fast-growing species, *A. siamensis*, were significantly higher than the slow-growing species, *P. macrocarpus*, in both years. Further, *A. siamensis* exhibited less effective change from year to year than *P. macrocarpus* as it could adapt to the lower or higher temperature stress. In fact *A. siamensis* can grow in tropical regions under dry, infertile, or higher temperature conditions. Under these conditions, our study suggests that *A. siamensis* could offer a greater average sequestration of carbon than *P. macrocarpus*.

References

Baldocchi, D. D. & Harley, P. C. Scaling carbon dioxide and water vapour exchange from leaf to canopy in a deciduous forest. II. Model testing and application. *Plant Cell Environ.* **18**, 1157–1173 (1995).

Ishizuka, M. & Puangchit, L. *Seasonal Changes of Leaf Photosynthesis and Stem Respiration of Young Teak Trees in Western Thailand*, Proceedings of the FORTROP'96: Tropical Forestry in the 21st Century. pp. 75–86 (Bangkok, 1996).

Kor-Buakesorn, A. Temperature change land dries-ice melts. *Greenline* **4**, 18–33 (1999).

Larcher, W. *Physiological Plant Ecology* (Springer-Verlag, Berlin, 1995).

Luangjame, J. *Ecophysiological Aspects of Hopea ordorata in the Northeast, Thailand*. Proceedings on *Dipterocarpus alatus* Roxb. and Dipterocarpaceae (eds L. Puangchit & B. Thaiutsa) pp. 37–50 (Bangkok, 1999).

National Research Council. Neem: *A Tree for Solving Global Problems* (National Academy Press, Washington, DC 1992) p. 141.

Taiz, L, & Zeiger. E. *Plant Physiology* (Sinauer Associates, Inc., 1998) pp. 195–244.

Williams, T. G. & Flanagan, L. B. Measuring and modeling environmental influences on photosynthetic gas exchange in *Sphagnum* and *Pleurozium*. *Plant Cell Environ.* **21**, 555–564 (1998).

Tropical Ecosystems: Structure, Diversity and Human Welfare.
Proceedings of the International Conference on Tropical Ecosystems
K. N. Ganeshaiah, R. Uma Shaanker and K. S. Bawa (eds)
Published by Oxford–IBH, New Delhi. 2001. pp. 27–28.

Are Amazonian rainforests absorbing carbon and increasing in biomass? A comparison of evidence and insight from measurements of CO$_2$ fluxes and forest biomass change

Yadvinder Malhi[*][†]*, Oliver Phillips*[**] *and John Grace*[*]

[*]Institute of Ecology and Resource Management, University of Edinburgh,
Scotland, UK
[**]Centre for Biodiversity and Conservation, School of Geography, University of
Leeds, England, UK
[†]e-mail: ymalhi@srv0.bio.ed.ac.uk

Tropical forests are one of the most important biomes on earth, accounting for more than half of global biodiversity, 40% of the carbon held in vegetation biomass, and for 30–50% of total land surface photosynthesis (Malhi and Grace, 2000). Half of the world's tropical forests are in South America, and small changes in the carbon dynamics of this biome can have a significant effect on the global carbon cycle. Here we report some recent results from current field studies of carbon dynamics in a number of forest sites in Brazil, Bolivia and Peru, spanning the entire Amazon basin.

Under equilibrium conditions, mature old-growth forests are expected to be in carbon balance, with the growth of living trees being balanced by the decomposition of dead wood. However, measurements of fluxes of carbon

Keywords: Tropical forest, carbon balance, Amazon, climate change, biomass.

dioxide above the forest at sites in the Brazilian Amazon indicate that the forests are sinks of atmospheric carbon dioxide, probably because of fertilization from enhanced atmospheric CO_2 levels. The magnitude of the sink is greater in the wet season than in the dry season. This seasonal pattern indicates than the growth of eastern Amazonian forests are more limited by drought in the dry season than by lack of light in the wet season. The mean value of the carbon sink measured is approximately 5 t C per hectare per year (Grace *et al.*, 1995; Malhi *et al.*, 1998; Malhi and Grace, 2000).

Where is this absorbed carbon going? We compared the results from flux studies with long-term forest inventories of biomass change. The inventories are indicating that most intact Amazonian forests appear to be increasing in biomass, at a rate of 0.5–1.0 t C ha^{-1} year^{-1}. (Phillips *et al.*, 1998). This biomass change may have substantial impacts on forest ecology and dynamics. However, even including soil carbon dynamics is unlikely to account for the discrepancy between flux studies and biomass studies. This suggests that flux studies are currently overestimating net carbon sinks.

Finally, we report on the Amazon Forest Inventory Network (RAINFOR) that is currently being set up, with the aim of monitoring and understanding the carbon dynamics of Amazonian forests over coming decades (Malhi *et al.*, in review). As part of this network, tree growth bands and automatic weather stations are being set up at a number of forest plots across the Amazon Basin, with the aim of understanding the climatic and nutrient limitations on the growth of tropical forest trees.

References

Grace, J., Lloyd, J., Mcintyre, J., Miranda, A. C., Meir, P., Miranda, H. S., Nobre, C., Moncrieff, J., Massheder, J., Malhi, Y., Wright, I. & Gash, J. Carbon dioxide uptake by an undisturbed tropical rain-forest in Southwest Amazonia, 1992 to 1993. *Science* **270**, 778–780 (1995).

Phillips, O. L., Malhi, Y., Higuchi, N., Laurance, W. F., Núñez, V. P., Vásquez, M. R., Laurance, S. G., Ferriera, L. V., Stern, M., Brown, S. & Grace, J. Changes in the carbon balance of tropical forest: evidence from long-term plots. *Science* **282**, 439–442 (1998a).

Malhi, Y. & Grace, J. Tropical forests and atmospheric carbon dioxide. *Trends Ecol. Evol.* **15**, 332–337 (2000).

Malhi, Y., Nobre, A. D., Grace, J., Kruijt, B., Pereira, M. G. P., Culf, A. & Scott, S. Carbon dioxide transfer over a central Amazonian rain forest. *J. Geophys. Res.* **D24**, 31593–31612 (1998).

Malhi, Y., Phillips, O. L., Baker, T., Almeida, S., Frederiksen, T., Grace, J., Higuchi, N., Killeen, T., Laurance, W. F., Leaño, C., Lloyd, J., Meir, P., Monteagudo, A., Neill, D., Núñez Vargas, P., Panfil, S. N., Pitman, N., Rudas-Ll., A., Salomão, R., Saleska, S., Silva, N., Silveira, M., Sombroek, W. G., Valencia, R., Vásquez Martínez, R., Vieira, I. & Vinceti, B. An international network to understand the biomass and dynamics of Amazonian forests (RAINFOR). *J. Veg. Sci.* (in review, December 2000).

Economic Valuation of Natural Resources: Local and Global Perspectives

Tropical Ecosystems: Structure, Diversity and Human Welfare.
Proceedings of the International Conference on Tropical Ecosystems
K. N. Ganeshaiah, R. Uma Shaanker and K. S. Bawa (eds)
Published by Oxford–IBH, New Delhi. 2001. pp. 31–33.

Returns to investment in conservation: Disaggregated benefit–cost analysis of the creation of a wildlife sanctuary

Sharachchandra Lele*[†], Veena Srinivasan* and Kamaljit S. Bawa**[‡]

*ATREE-ISEC Centre for Interdisciplinary Studies in Environment & Development, Nagarbhavi, Bangalore 560 072, India
**Ashoka Trust for Research in Ecology and the Environment (ATREE), PO Box 2402, HA Farm Post, Bangalore 560 024, India
[‡]Department of Biology, University of Massachusetts, MA 02125, Boston, USA
[†]e-mail: lele@socrates.berkeley.edu

Tropical forest ecosystems provide a wide range of benefits to mankind. Economists have categorised these into use and non-use values, further categorised into direct and indirect use on the one hand and option, bequest and existence value on the other. Estimating the 'total economic value' of forests has become a popular topic of research and discussion in the conservation community. Theory and techniques for estimation of the non-use benefits have dominated recent research. In the process, however, most valuation exercises have overlooked two fundamental issues: a) the 'so what?' problem, i.e. estimating the economic value of something in isolation is not very useful, and b) the 'aggregation problem', i.e. the fact that the supposed 'aggregate social welfare function' has no objective basis. These problems need to be addressed if economic valuation is to provide

Keywords: Conservation, wildlife sanctuary, benefit–cost analysis, Biligiri Rangaswamy Wildlife sanctuary.

meaningful and open-minded input into real-world debates about forest conservation.

The solution to the first problem is to always carry out valuation of two scenarios, viz. the current situation and the most likely alternative land-use or forest-use strategy that could prevail. The solution to the second problem lies in recognising that the marginal utility of one monetary unit of benefit or cost varies from person to person, and that this marginal utility is at least partly related to the economic status of the person. Thus, any aggregation exercise must at least adjust for the disparities in wealth or income across the different beneficiaries of the ecosystem, disparities that can be very high in the case of tropical forest ecosystems, where stakeholders may range from local forest-dwelling communities to global beneficiaries of the forests' carbon sequestration service.

We use the case of the Biligiri Rangaswamy Temple (BRT) Wildlife Sanctuary (WLS) in southern India to illustrate these approaches to addressing the 'so what?' and the 'aggregation' problems. Firstly, we compare the stream of benefits and costs resulting from the management of the BRT forests as a WLS with that which would have prevailed if these forests were managed as Reserve Forests (RF), which is the most likely alternative scenario. Secondly, we categorise the affected population into relatively homogenous income and cultural groups, and the benefits and costs are estimated in a disaggregated manner for each group and then aggregated using various weighting schemes (including income-dependent weights). Furthermore, since the conversion of BRT forests from RF to WLS status occurred in 1977, this partially *ex-post* analysis provides a more realistic appraisal of the incremental benefits and costs of conservation strategies based on the protected area approach than typical *ex-ante* estimates.

We assume that the key users or beneficiaries of the BRT forests are: (i) the forest-dwelling Soliga community that depends upon the forest for fuelwood, grazing, and income through the collection and sale of various non-timber forest products (NTFPs), (ii) the non-Soliga local community that derives fuelwood and grazing benefits, (iii) the farming communities surrounding the BRT forests that directly use the forests for fuelwood collection and grazing as well as indirectly gain from the forest's soil conservation services, (iv) the tourists that visit BRT to enjoy its forest and wildlife, and (v) the global community that benefits from carbon sequestration. Finally, the incremental costs of conservation are borne by the national community of taxpayers.

We make selective use of information from neighbouring forests that remained RFs after 1977 as well as scenario building by experts to construct the trajectory that BRT forests would have followed if they had remained a RF. It appears that, as compared to the WLS scenario, the RF scenario would have led to more rapid degradation of the forest, resulting in lower carbon sequestration, long-term losses to downstream farming communities due to siltation of irrigation tanks, and zero benefits to wildlife tourists, but (at least initially) higher benefits to local communities from heavier NTFP extraction, fuelwood collection, grazing and agricultural expansion, as also greater incomes to government agencies from timber felling. We carry out a detailed sensitivity analysis to allow for uncertainties in our estimation procedure as well as the use of different discount rates over time and across income groups.

Preliminary estimates of benefits and costs suggest that the conversion to WLS seems highly beneficial if aggregate benefit-cost is calculated without adjusting for income disparities, but income-dependent weights can drastically alter the balance, indicating how unfair the distribution of benefits and costs is across different income groups. Furthermore, since returns from NTFP collection and sale form a crucial component of incomes for the poorest group of stake-holders (the Soligas), the costs are less lop-sided than expected only because the WLS authorities permitted NTFP collection, the conversion from RF to WLS was not an unmitigated disaster for the Soligas. This suggests that only a conservation strategy that combines strict protection from external pressures with substantial forest use by forest-dwelling communities can be somewhat socially equitable.

Tropical Ecosystems: Structure, Diversity and Human Welfare.
Proceedings of the International Conference on Tropical Ecosystems
K. N. Ganeshaiah, R. Uma Shaanker and K. S. Bawa (eds)
Published by Oxford–IBH, New Delhi. 2001. pp. 34–35.

Local and global valuation of natural capital: Evidence from indigenous groups of Latin America and implications for conservation and research

Ricardo Godoy, Lilian Apaza, Elizabeth Byron,
Tomás Huanca, William Leonard, Eddy Pérez,
Victoria Reyes-Garcóa, Vincent Vadez, David Wilkie,
Josefien Demmer, Han Overman, Adoni Cubas,
Kendra McSweeney and Nicholas Brokaw*

Sustainable International Development Program, Mailstop 078, Brandeis University,
Waltham, MA 02455-9110, USA
*e-mail: rgodoy@brandeis.edu

Conservation requires convergence in the way different groups in society or the world value natural capital. When outsiders and local users value natural capital in the same way or at the same level, people will likely invest in its preservation. When outsiders and local users attach little objective or subjective value to natural capital, people will likely misuse or deplete the resource. Ambiguous results for conservation arise when outsiders and local users value natural capital in different ways or by different amounts. For

Keywords: Latin America, Amerindian societies, local benefits, global benefits, conservation.

instance, local users might value natural capital for its insurance or for its symbolic value, but outsiders might value the same resource for its objective ecological services. Outsiders might value natural capital for its global ecological services, but local people might value natural capital because it provides food and raw materials for construction. Growing evidence suggests that rural people in developing nations and the rest of the world derive different values from the same natural capital.

If rural people and the rest of the world value natural capital in different ways or receive unequal share of the benefits from the same resource, policy-makers and international development organizations need to answer two questions to ensure that natural capital is conserved or put to its best use for society and the world. First, how big is the gap between the benefits received by local people and the benefits received by the rest of the world? Second, what policies might governments put in place to achieve convergence in values and, in so doing, bring about greater incentives for villagers and outsiders to invest in the preservation of natural capital?

Here we present empirical estimates from two Amerindian societies of the tropical rain forest of Latin America of the objective value to villagers of the forest measured through the consumption and the sale of forest goods. We compare the estimates to each other and, drawing on secondary literature, we compare local benefits to global benefits. In so doing we try to estimate the gap in the value between what villagers and outsiders receive from the same resource. We then discuss the implications of research findings for policy-makers and researchers.

Tropical Ecosystems: Structure, Diversity and Human Welfare.
Proceedings of the International Conference on Tropical Ecosystems
K. N. Ganeshaiah, R. Uma Shaanker and K. S. Bawa (eds)
Published by Oxford–IBH, New Delhi. 2001. pp. 36–37.

Limits of economic valuation and their implications for conservation policy

R. David Simpson

Resources for the Future, 1616 P Street NW, Washington, DC 20036, USA
e-mail: simpson@rff.org

There has been considerable recent interest in the economic valuation of ecosystem services. In addition to merely arousing intellectual curiosity, these issues have increasingly been arising in discussions of biodiversity conservation policy. Diverse natural ecosystems provide a wide array of services to society. Can the value of these services be used to motivate conservation of the habitats that provide them?

Answering this question depends on first answering two others. First, 'What are the values?' As a matter of economic theory, the answer is straightforward. It depends on calculating the incremental contributions of habitat conservation to the production of the array of services human societies value. As a practical matter, however, the problem is far more difficult. Economists frequently exalt prices as conveyors of information. The price of a good that is traded in a market compresses a tremendous amount of information concerning consumer preferences and production possibilities into a compact summary measure. By the same token, the valuation of goods that are *not* traded in markets – and very few ecosystem services currently are – requires far more and better data than we generally have. We can, in some instances, estimate plausible bounds on economic values. Often these

Keywords: Economic valuation, conservation, ecosystem services, markets.

bounds are broad enough to be of little practical use, however. Instances in which statistically defensible value estimates can be derived are extremely rare.

The second question is 'How are values to be realized?' One might argue that, if people really can realise local benefits from wisely managing their ecological assets, they will do so. If this were the case, there would be no need for conservation advocates to be concerned with the valuation of local flood control, pest regulation, pollination, etc. Markets would come into being and solve the problem. There is some evidence that such markets are developing. One can, however, propose a number of reasons for which local people might not fully realize the benefits with which markets in ecosystem services might provide them. Information may not be available to them, or they may have been unable to overcome 'problems of collective action'. These possibilities do not necessarily comprise a reason for *conservation* advocates to turn their attention to ecosystem services, however. There are many reasons. Among these are:

- The link between conservation of natural habitats and ecosystem services is sometimes not clear.
- Partial institutional reforms may have unintended and adverse consequences.
- Conservation objectives may be better achieved by other instruments.
- Designing elaborate approaches, even to admittedly complex problems, may 'oppose the best to the good' by introducing opportunities for political manipulation.

Tropical Ecosystems: Structure, Diversity and Human Welfare.
Proceedings of the International Conference on Tropical Ecosystems
K. N. Ganeshaiah, R. Uma Shaanker and K. S. Bawa (eds)
Published by Oxford–IBH, New Delhi. 2001. p. 38.

The value of functional diversity

Charles Perrings

Department of Environmental Economics and Environmental Management,
University of York, York YO1 5DD, Britain
e-mail: cap8@york.ac.uk

Functional diversity refers to the diversity of species performing a given ecological function. It is a measure of the number of and complementarity between species supporting that function over different environmental conditions. Functional diversity derives its value from the ecological services supported by particular functions, and is increasing in the range of environmental conditions affecting the system. In supporting ecosystem resilience over a range of environmental conditions, the functional diversity of species reduces the risks associated with environmental variation or change. The conservation of functional diversity may therefore be thought of as a means of managing environmental uncertainty. It follows that the value of functional diversity is an insurance value. Because it concerns the ecological services derived from a particular system, it is a local rather than a global value. This has some obvious and important implications for policy.

Keywords: Functional diversity, species diversity, ecological function.

Tropical Ecosystems: Structure, Diversity and Human Welfare.
Proceedings of the International Conference on Tropical Ecosystems
K. N. Ganeshaiah, R. Uma Shaanker and K. S. Bawa (eds)
Published by Oxford–IBH, New Delhi. 2001. pp. 39–40.

Utilization of aquatic weed (*Eichhornia crassipes* (Solm) Nash) as a source of plant nutrients

D. Anusuya* and M. Prasoona

Department of Botany, Bangalore University, Bangalore 560 056, India
*e-mail: anvidhan@vsnl.net

Water hyacinth (*Eichhornia crassipes* (Solm) Nash) is one of the most noxious weeds in India. It hinders agriculture, particularly rice cultivation in low lying areas. Besides pisciculture, water transport, drainage, etc. are also affected adversely. However, the brighter aspect of this weed is its capacity to absorb pollutants and toxicants, and its utility as a mulch and compost (Kamal and Little, 1970) and as an insect growth-inhibitior (Mansoor *et al.*, 1990).

Realizing the potentialities of the diverse uses of water hyacinth and enrichment with beneficial microorganisms, composting of water hyacinth has been recommended by several workers as not only self-remunerative but also possibly financially profitable (Abdalla and Abdel Hafeez, 1969).

Experiments were conducted to assess the role of microbial inoculants, viz. *Phanerocyaetae chrysosporium* and *Aspergillus awamori* in composting of *Eichhornia crassipes* supplemented with 1% rock phosphate. The mineral supplements and the cultures of *Aspergillus awamori* as phosphate solubilizer as well as other fungal cultures were added at the start of the composting. Other fungal cultures were inoculated 17 days after the initial decomposition. The period of composting varied from 60 to 90 days.

Keywords: *Eichhornia*, compost, phosphate, microorganisms.

From the point of view of the quality of phosphocompost, among the four different treatments the best results were obtained in the treatment containing *Trichoderma viride, Aspergillus awamori, Phanerochaetae* and Rock phosphate, where C/N ratio narrowed down with highest amount of N compared to control and other treatments. Inoculation with phophate solubilizing microorganisms (*Trichoderma viride, Aspertillus awamori* and *Phanerochaetae crysosporium*) in the presence of 1 per cent rock phosphate proved to be beneficial in improving the quality of compost at 90 days.

NPK content of the compost at two intervals of composting showed that there was substantial nitrogen enrichment even at shorter intervals of decomposition although corresponding reduction in carbon was relatively low. This showed that the decomposed residue was converted into usable compost material with the help of microbial inoculants.

References

Abdalla, A. A. & Abdel Hafeez. Some aspects of utilization of water hyacinth (*Eichhornia crassipes*), *PANS* **15**, 204–207 (1969).

Kamal, I. A. & Little, E. C. S. The potential utilization of waste hyacinth for horticulture in the Sudan. *PANS* **16**, 488–496 (1970).

Mansoor, A. Siddiqui, M. & Kashkoor Alam. Further studies on the use of water hyacinth in Nematode. *Biol. Wastes* **33**, 71–75 (1990).

Tropical Ecosystems: Structure, Diversity and Human Welfare.
Proceedings of the International Conference on Tropical Ecosystems
K. N. Ganeshaiah, R. Uma Shaanker and K. S. Bawa (eds)
Published by Oxford–IBH, New Delhi. 2001. pp. 41–45.

Silviculture of the economically important woody root hemiparasitic tree *Santalum album* L. (Sandalwood)

K. U. Tennakoon, E. R. L. B. Etampawala and C. V. S. Gunatilleke*

Department of Botany, University of Peradeniya, Peradeniya, Sri Lanka
*e-mail: kushan@botany.pdn.ac.lk

Santalum album (sandalwood) is an economically and culturally important obligate woody root hemiparasitic tree naturalized in the wet and intermediate climatic zones of Sri Lanka. In recent years, there has been a rising interest in planting sandalwood in countries such as Sri Lanka, India, Indonesia, Australia, Hawaii, Fiji, Papua New Guinea and Timor due to the rising world wide demand for santalol oil obtained from the heartwood of sandalwood (Shea *et al.*, 1998). Santalol is used for the preparation of expensive perfumes, cosmetics and medicines. Sandalwood population in Sri Lanka is on the decline due to habitat destruction, over exploitation and its complex silviculture associated with the obligate root hemiparasitism (Tennakoon *et al.*, 2000). Sandalwood plants can live autotrophically (without parasitising a host) up to maximum of one year and for subsequent growth an intimate association with roots of a host plant through haustoria is essential (Tennakoon *et al.*, 1997; Radomiljac, 1998). Like other root hemiparasites, sandalwood abstracts a range of organic solutes and inorganic ions dissolved in the host xylem sap via haustoria attached to host roots.

Keywords: *Santalum album*, sandalwood, hemiparasitic tree, santalol, silviculture.

This study examines a range of biological and functional attributes of *S. album*. It investigates the morphological and genetic variations, seed germination, suitable potting mixtures for initial growth of the pre-parasitic (autotrophic) stage of sandalwood and evaluates different pot-hosts required for nursery establishment of sandalwood.

Twenty seven randomly selected trees in three populations of the intermediate zone of Sri Lanka (Haragama, Hanguranketha and Perawella) were sampled (3 randomly selected plants/plot/population) to select morphologically superior trees. Mean tree height (2.7 ± 0.1 m) and DBH (8.6 ± 0.6 cm) were smaller in the Haragama population compared to the Perawella and Hanguranketha populations (3.6 ± 0.2 m, 3.9 ± 0.6 m and 10.6 ± 0.8 cm, 12.8 ± 1.0 cm respectively). Fruit length, width and seed length varied significantly ($P < 0.01$) among the three populations, but not the leaf parameters (leaf length, width, thickness, petiole length and inter nodal length). Fruit and seed size of the Hanguranketha population was significantly higher ($P < 0.01$) compared to the two populations. These observations suggest that fruit and seed parameters change with the environmental conditions while leaf parameters remain constant.

Genomic DNA extracted from sandalwood plants grown in nine provenances representing different agro-ecological zones of Sri Lanka were used for the polymerase chain reaction (PCR) technique to determine the genetic polymorphism in sandalwood populations. There were no major differences in the RAPD bands of sandalwood leaves, thus suggesting no genetic variation between the different provenances.

Investigations on seed germination included seven treatments and two replicates (36 seeds/replicate). In this experiment, seeds were treated with three gibberellic acid concentrations (250 ppm, 500 ppm and 750 ppm), sandalwood leaf extract, exposure to mild fire and sunlight and soaking in water (control). Seeds treated with 750 ppm gibberellic acid showed over 80% germination (Figure 1). However, the control where seeds were only soaked in water also showed 60% germination success. These germination percentages are much higher than those reported for sandalwood seeds from major sandalwood-growing countries such as India, Indonesia and Australia (Radomiljac, 1998).

The soil substratum for best growth of pre-parasitic (autotrophic) sandalwood plants was examined by raising seedlings in three potting mixtures, comprising sand, top soil and farm yard manure mixed in the ratio of $1:1:1$ (A), $3:1:0$ (B) and $2:1:1$ (C) respectively. The experiment included three treatments, three blocks and two replicates (20 seedlings/replicate). Growth performance (mean leaf number, mean seedling height and mean root collar diameter) of autotrophic seedlings was measured at three-month

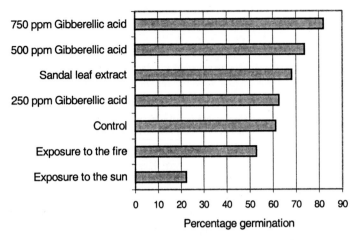

Figure 1. Percentage germination of sandalwood seeds under different treatments.

intervals for a period of nine months. Significantly higher ($P < 0.001$) performance was observed in the substratum with sand, top soil and farm yard manure mixed in the ratio of 1:1:1 (A) compared to plants grown in substrates B and C (Table 1). This clearly showed that pre-parasitic sandalwood seedlings obtain nutrients from the growing medium supplementing the original seed reserves. It also suggests that plants respond better to manure than to sandy soil.

The best light regime for potted sandalwood plants was evaluated by growing seedlings under four light regimes (full sun, 25% shade, 50% shade and 75% shade where the light levels were 1850, 1300, 912, and 450 µmol m^{-2} s^{-1}). These sandalwood plants were grown with a single host *Altenanthera sessilis*. The experiment included four treatments, four blocks and two replicates (20 seedlings/replicate). The best performance of plants in terms of mean seedling height, root collar diameter and number of leaves was observed under 25% and 50% shade levels. Seedlings did not survive under full sun while the performance of plants was significantly poor under 75% shade compared to that under 25% and 50% shade levels. These results indicate that sandalwood plants at the nursery stage should be maintained between 25% (1300 µmol m^{-2} s^{-1}) and 75% (450 µmol m^{-2} s^{-1}) shade levels to ensure optimum growth.

Five potential host species of sandalwood (*Crotalaria juncea*, *Mimosa pudica*, *Altenanthera sessilis*, *Tithonia diversifolia* and *Phaseolus aureus*) were evaluated by growing one-month-old sandalwood seedlings with each of them in polythene bags. As a control, sandalwood seedlings were grown without the host species. The experiment consisted of six treatments, three blocks and two replicates (10 seedlings/replicate). The performance of sandalwood plants (mean leaf number, mean seedling height and mean root collar diameter) was measured at three-month intervals for a period of nine months. *Santalum album* seedling grown with *M. pudica* and *T. diversifolia* was significantly higher ($P < 0.01$) compared to those in the control and also those grown with each of the three remaining hosts (Table 1). Like other root hemiparasites, sandalwood lacks the ability to incorporate soil-borne nitrate and ammonium ions to the organic nitrogen pool of the plant due to the absence of nitrogenase enzyme in its root system (Stewart and Press, 1990; Tennakoon *et al.*, 1997). *Santalum* may be obtaining high levels of nitrogen (and also carbon) from the xylem sap of the nitrogen fixing *Mimosa pudica* and nitrophylous shrub *Tithonia diversifolia*. Surprisingly, seedling performance of *Santalum* grown with the legume *P. aureus* was significantly lower ($P < 0.01$) than that of the control. This can probably be attributed to the short life cycle of the host and its inability to provide required xylem-derived nutrients to *Santalum* via haustoria continuously. Therefore, *Mimosa pudica* and *Tithonia diversifolia* can be recommended as suitable pot-hosts for raising sandalwood plants in pot cultures until they can be transferred to the field. Studies on water and osmotic relations, resource partitioning and the evaluation of long-term hosts for sandalwood field establishment programs are in progress.

Table 1. Performance of nine-month-old sandalwood plants grown in different potting media, under different light regimes and with different hosts (values followed by same letters indicate treatments which are not significantly different).

Experiment	Root collar diameter (mm)	Height (cm)	Mean number of leaves
Potting Mixture			
(Sand : Soil : Farm manure)			
A (1:1:1)	$2.02^b \pm 0.05$	$22.9^b \pm 0.5$	$19^b \pm 1$
B (3:1:0)	$1.67^c \pm 0.06$	$16.6^c \pm 0.4$	$14^c \pm 1$
C (2:1:1)	$2.16^a \pm 0.04$	$25.5^a \pm 0.6$	$21^a \pm 1$
Light regimes			
Full sun (1850 μmol m^{-2} s^{-1})	–	–	–
25% Shade (1300 μmol m^{-2} s^{-1})	$2.01^a \pm 0.09$	$20.3^a \pm 0.8$	$19^a \pm 1$
50% Shade (912 μmol m^{-2} s^{-1})	$1.81^a \pm 0.08$	$22.2^a \pm 1.0$	$19^a \pm 1$
75% Shade (450 μmol m^{-2} s^{-1})	$1.24^b \pm 0.06$	$16.8^b \pm 0.7$	$17^a \pm 1$
Sandalwood : Host associations			
Santalum: Crotalaria juncea	$2.48^b \pm 0.11$	$30.2^b \pm 1.5$	$29^a \pm 2$
Santalum: Mimosa pudica	$3.19^a \pm 0.11$	$37.5^a \pm 1.3$	$33^a \pm 1$
Santalum: Altenanthera sessilis	$3.08^b \pm 0.11$	$33.2^{ab} \pm 1.1$	$29^a \pm 1$
Santalum: Tithonia diversifolia	$2.53^a \pm 0.12$	$34.5^{ab} \pm 1.7$	$29^a \pm 1$
Santalum: Phaseolus aureus	$2.60^b \pm 0.08$	$27.8^{bc} \pm 0.9$	$23^b \pm 1$
Control (*Santalum* grown without a host)	$2.81^a \pm 0.11$	$30.2^b \pm 1.0$	$30^a \pm 1$

Acknowledgements

We are indebted to S. P. Ekanayeke and K. G. S. K. Kapukotuwa for assistance with fieldwork. Financial assistance provided by the Environment Action 1 Project of the Ministry of Forestry and Environment and the Sri Lanka Conservation and Sustainable Use of Medicinal Plants Project is gratefully acknowledged.

References

Radomiljac, A. The influence of pot host species, seedling age and supplementary nursery nutrition on *Santalum album* plantation establishment within the Ord river irrigation area. *For. Ecol. Manage.* **102,** 193–201 (1998).

Shea, S. R., Radomiljac, A. Brand, J. & Jones, P. An overview of sandalwood and the development of sandal in farm forestry in Western Australia. Sandal and its products. *Austr. Centre Int. Agric. Res.* **84,** 9–15 (1998).

Stewart, G. R. & Press, M. C. The physiology and biochemistry of parasitic angiosperms. *Annu. Rev. Plant Physiol. Mol. Biol.* **41,** 127–151 (1990).

Tennakoon, K. U., Ekanayake, S. P. & Etampawala, L. An overview of sandalwood research in Sri Lanka. International Sandalwood Research News Letter Vol. 10, pp. 1–3 (Department of Conservation and Land Management, Australia, 2000).

Tennakoon, K. U., Pate, J. S. & Stewart, G. R. Haustorium related uptake and metabolism of host xylem solutes by the root hemiparasitic shrub *Santalum acuminatum* (R. Br.) DC (*Santalaceae*). *Ann. Bot.* **80,** 257–264 (1997).

Tropical Ecosystems: Structure, Diversity and Human Welfare.
Proceedings of the International Conference on Tropical Ecosystems
K. N. Ganeshaiah, R. Uma Shaanker and K. S. Bawa (eds)
Published by Oxford–IBH, New Delhi. 2001. pp. 46–49.

Relevance of indigenous institutions to management of forest ecosystem: A case study from central India

Alka Chaturvedi*[†], Rucha Ghate**
and Vinayak Deshpande**

*Department of Botany, Nagpur University, Nagpur 440 010, India
**Department of Economics, Nagpur University, Nagpur 440 010, India
[†]e-mail: arunchat@nagpur.dot.net.in

While in case of majority of societies, their relationship with forests has undergone lot of changes (usually from symbiotic to parasitic), few indigenous societies in tropical ecosystems have evolved informal as well as formal norms to sustain forests very successfully. Challenging the modern incentives structured by market forces, technological changes, government forest policies, some indigenous communities are trying to swim against the current strengthening their local institutional structure rooted in traditional knowledge and management systems (Gibson *et al.*, 2000). Their very 'local' nature seems to be effective in preservation of biodiversity due to sustainable use. Several earlier studies have suggested that local people possess more knowledge concerning their local resources, and that communities possess greater organizational skills than is often appreciated by the experts or accepted by (forest) officials (Leach *et al.*, 1999). This paper reflects on Mendha (Lekha) community, with majority of its members belonging to Gond tribe. This community has nurtured its ecologically oriented culture by constantly reforming its traditional panchayat structure with changing time and needs.

Keywords: Forest quality, biodiversity, traditional institutions, homogeneity, India.

Gadchiroli is a declared backward district in a highly developed state of Maharashtra, India (Haeuber and Richard, 1993). It has major share of the forest area and also of tribal population (including primitive tribes). Out of its total geographical area of 14412 km², 93.06 per cent is classified as 'forest' and 38 per cent of its population is 'tribal'. Almost 64 per cent of the state's 'protected forest', where the local communities have comparatively more rights, is in Gadchiroli district. The study becomes relevant because little more than two decades ago this area, which was once known for high quality forest, was in a highly degraded state. Although all the forest in Gadchiroli is under government ownership ever since independence, residents of Mendha village, experiencing the brunt of reduced availability of basic forest products like fuel wood and fodder, decided to revive their age-old management practices and conservation methods. Its impact on the quality of forest is remarkable. Its significance is enhanced by the fact that its very neighbouring village, Lekha, has not gone for any such collective effort. The difference in the quality of the forests belonging to the two villages is stark. The present paper tries to relate the socio-economic aspects of indigenous institutional structure to bio-physical data, reflecting the richness and diversity of the ecosystem.

Survey of the Mendha forest shows low human pressure on the vegetation. It may be due to low population, inaccessible hilly terrain or lesser demand for forest products. The per capita forest area is about 7.05 ha. Woody plants of Mendha forest show a three-storey structure: 1–10 meter category is occupied by *Cleistanthus collinus, Tectona grandis, Madhuca longifolia, Bridellia retusa, Accacia catachue* and *Butea monosperma*; 10–20 meters category shows presence of *Cleistanthus collinus, Tectona grandis, Madhuca longifolia, Boswellia serrata, Buchanania lanzan, Dalbergia* sp., *Diospyros melanoxylon, Erythrina variegata, Holarrhena antidysantrica* and *Terminalia* species repitition. The 20–30 meter high top strata is occupied by *Tectona grandis, Albizia odorattisima* and *Pterocarpus marsupium*. Clumps of *Lagerostromia parviflora, Cleistanthus collinus* and *Bambusa* species are common feature of the forest. *Butea superba* and *Mimosa hamata* are the two main woody climbers.

Saplings are few in number but they belong to a large number of families, Anacardiaceae, Apocyanaceae, Combretaceae, Ebenaceae, Euphorbiaceae, Meliaceae, Mimosoideae, Myrtaceae, Rhamnaceae and Sapindaceae. Ground flora is very sparse. Trees of *Anthocephalus* and *Terminalia* species show the largest DBH, i.e. between 50 and 70 cm. Dominant species of the Mendha forest are *Cleistanthus collinus, Diospyros melanoxylon* and *Pterocarpus marsupium*.

In Lekha forest, large part of the area is along the road side and close to the river, which has resulted in clearing of the forest for cultivation. In Lekha the per capita forest is 0.9 ha. Lekha forest also show three strata in their woody vegetation: 1–10 meter is occupied by *Diospyros melanoxylon, Butea monosperma, Cleistanthus collinus, Anogeissus latifolia, Lannea coromandelica* and *Madhuca longifolia*; 10–20 meter strata shows presence of *Anogeissus latifolia, Boswellia serrata, Diospyros melanoxylon, Pterocarpus marsupium* and *Terminalia* species; 20–30 meter is dominated by *Boswellia serrata, Diospyros melanoxylon, Pterocarpus marsupium, Scheleria oleosa* and *Terminalia* species. Among the sapling flora are *Cleistanthus collinus* and *Mimosa hamata* along with species of family Burseraceae, Sterculiaceae, Ebenaceae, Meliaceae, Apocynaceae, Combretaceae and Fabaceae. Monocots, *Borassus flabellifer* and *Asparagus* are quite common. *Calycoptis floribunda* and *Dioscorea* represent the climbers.

In the Lekha forest species of *Ficus, Tectona grandis* and *Terminalia alata* are tall, reaching up to 25 feet, and over 80 cm in diameter [DBH, *Ficus*, 122.20; *Tectona*, 85 cm; *Terminalia* species, 80 cm]. The forest is dominated by *Cleistanthus collinus, Terminalia alata, Butea monosperma, Diospyros melanoxylon* and *Tectona grandis*.

We examined the hypothesis that, indigenous societies through their traditional institutions can manage their forest sustainably. The major objectives of the study were:

(i) To investigate the impact of socio-economic determinants on forest condition and biodiversity, (ii) To identify the incentives/disincentives leading to activity/inactivity on part of two neighbouring communities as reflected in the quality of forest.

Since the data required for temporal analysis were not available, spatial comparison at a point of time was done. The selected forest condition indicators are: types and number of species, size and density of trees, basal area, and regeneration capacity of the area. Although the two villages selected for analyses were comparable in terms of ecosystem (temperature, rainfall, soil variation, altitude, species, etc.) and location (i.e. distance from the district head quarter, proximity to market and all weather road and the like), the major difference were with respect to population mix and per capita forest area.

The present study adopted the data collection method developed by The International Forestry Resources and Institutions (IFRI) research program consisting of pre-structured research instruments to facilitate collection of information about demographic, economic, and cultural characteristics of communities dependent on forests. It uses rigorous forestry techniques to measure the impact of institutions regulating the use of forest products. Each research instrument includes variables that are used to understand connection between the physical characteristics of the forests and human interaction with the environment.

48

Plots (demarcated areas) were laid in the forest area to study the ecology of forest. For the present study a fixed number of 30 plots was laid in each village. The plots were determined through the grid method and were selected randomly. The points thus determined were actually located in the forest by using the established natural land marks. All the plots were circular in shape, consisting of a circle of 1 meter radius for collecting information on ground flora, 3 meter radius circle for saplings, and a circle with 10 meter radius for collecting data on trees.

The following are the main results of the study:

1. Mendha community has continuously reformed the traditional 'gram-panchayat' structure to incorporate the changing circumstances. It has quite successfully managed its forest despite the fact that they have no authority to form rules, and that they do not own the resource.
2. On the basis of *'Attribute of users'* (Ostrom, 1990; Michael and Ostrom, 1996), identified as factors enhancing the likelihood that forest-resource users will organize themselves, it is clear that the Mendha community enjoys common understanding, regards the salience of the resource, and the users trust each other with reciprocity. These factors were found to be missing in Lekha.
3. Traditional rules regarding forest use, some words in the language, festivals and celebrations, as well as the names and surnames within the community are indicative of the symbiotic relationship between forests and the community.
4. Mendha has sufficient forest area within its village boundary to cater to the forest-based uses of the community.
5. The tradition of collective decision-making and action is not confined to forest-related decisions, but incorporates almost all aspects of social living.
6. Forest quality in case of Mendha was found to be better than Lekha in all aspects: number of trees, variety of species, density, basal area, and regeneration capacity.

Acknowledgements

This paper is the outcome of a pilot study funded by Workshop in Political Theory, Indiana University, Bloomington, US. We are grateful to the funding agency and the other research members of the study group.

References

Gibson Clark, Margaret McKean & Elinor Ostrom (eds). *People and Forests: Communities, Institutions, and Governance* (MIT Press, Cambridge, MA, 2000).

Haeuber and Richard. Development and deforestation: Indian forestry and perspective. *J. Dev. Areas* **27**, 485–514 (1993).

Leach, M., Mearns, R. & Scooners, I. 'Environmental entitlements: Dynamics and institutions in community-based natural resource management. *World Dev.* **27**, 225–247 (1999).

McGinnis, Michael & Elinor Ostrom. Design principles for local and global commons. In *The International Political Economy and International Institutions* (ed. Oran R. Young) Vol. II. Edward Elgar, Cheltenham, UK (1996).

Ostrom, Elinor. *Governing the Commons* (Cambridge University Press, Cambridge, 1990).

Tropical Ecosystems: Structure, Diversity and Human Welfare.
Proceedings of the International Conference on Tropical Ecosystems
K. N. Ganeshaiah, R. Uma Shaanker and K. S. Bawa (eds)
Published by Oxford–IBH, New Delhi. 2001. pp. 50–52.

Worthiness of Budongo Forest Ecotourism project in Uganda: Previous and current local household income and expenditure

Mabe Akhos Wathyso, Obua Joseph and
W. S. Gombya-Ssembajjwe†*

Faculty of Forestry and Nature Conservation, Makerere University, Kampala,
P.O Box 7062, Uganda
*e-mail: fellymabe@usa.net

Ecotourism is a growing industry in many countries (Lindbergh and Huber, 1991). It is a means for achieving modest economic development in rural areas, especially in environmentally or socially fragile areas that cannot support major economic activities (Butler, 1990). Ecotourism is often sought because, relative to other types of development, it is the least harmful to the environment. Romeril (1989) reported that preservation of the existing natural environment must always be considered as a viable option for nature conservation and rural development. Ecotourism can bring many benefits to the host community such as generation of income, employment opportunity, effective conservation and management of the environment, and development of infrastructure. Howard (1995) has shown numerous benefits of ecotourism development to Uganda and, in particular, to conservation and improvement of local people's welfare around Budongo Forest Reserve six years after the Budongo Forest Ecotourism Project (BFEP) was initiated

---

Keywords: Uganda, ecotourism, parish selection, financial values.

(Langoya and Long, 1998). In addition, Langoya and Long (1999) reported that the development of the forest tourism aims at providing a small but regular income for both the local people and the government and to create opportunities for local communities and the Forest Department to work together in the management of the ecotourism resources as part of overall forest resource management strategy. However, Namukwaya (1998), five years after the development of ecotourism project in Budongo forest reported that the majority of the local people were not benefiting sufficiently from the project possibly leading to increased cases of illegal pitsawying.

Budongo Forest Reserve, situated in the north west corner of Uganda, was gazetted as a Central Forest Reserve in 1932. It covers parts of Bujenje, Buliisa and Buruli Counties in Masindi District. It was established to protect its large population of mahoganies and savanna grasslands and woodland which covers 825 km², thus making it Uganda's biggest forest reserve. It is divided into seven management blocks, namely Biiso, Siiba, Nyakafunjo, West Waibira, Busaju, Kitigo and Kaniyo Pabidi. It is adjacent to Murchison Falls National Park, and Bugungu and Karuma Wildlife Reserves in the north of the reserve.

Budongo forest reserve is situated between 1°35'–1°55'N and 31°18'–31°42'E on the edge of the western rift valley in western Uganda. It is a medium altitude moist semi-deciduous forest characterised by minimum temperature between 23 and 29°C and maximum temperature between 29 and 32°C, and by high temperatures from December to February and low temperatures from June to July. There is a moderate relative humidity, 45 to 50%, during the dry season and 75 to 90% during the rainy season. According to Karani *et al.* (1997), the altitude varies between 914 m to 1097 m above sea level. Mean annual precipitation varies from 1397 mm to 1524 mm falling in 100 to 150 days per year. The area is dominated by two rainy seasons separated by two dry ones in December to mid-March and June to July.

Data collection and survey procedure comprised of reconnaissance survey, systematic sampling of households and interviews with Local council Chairpersons/Officials, household heads and market visits to forest products sellers.

It was found that the annual financial values of forest products to a household living in an ecotourism active parish (Biiso, Kihungya, Nyantonzi) ranged from UgShs 36,488 to 111,724 (US$ 24 to 74) whereas a household living where ecotourism is not active (kasenene) had a range of Ugshs 122,197 to 149,929 (US$ 81 to 100). The difference tested at household, village and parish level indicated the statistical significance at parish level. The average minimum and maximum monetary values for all products extrapolated to the entire forest adjacent parishes, ranged from UgShs 517,442,665 to 687,456,323 (US$ 0.5 to 0.6 million per year). The annual variation of forest products consumption to the local people indicated the highest increase in harvesting costs from 1991 to 1994 (UgShs 40,046,188 to 40,709,543). These costs have been decreasing with the ecotourism project and show the highest decrease from 1995 onwards (UgShs 29,367,482 to 20,515,347). The financial and economic analysis with and without the project yielded positive incremental net benefit of UgShs 13,562,539 at 0% discount rate and of UgShs 1,638,877 at 5% discount rate (Conservation discount rate), while the social cost of capital for publicly funded projects in Uganda (12% discount rate) and the commercial bank rate (22% discount rate) produced UgShs 3,351,127 and 8,231,884 respectively.

We conclude that (a) In project area there is a rise in household income by 2%, while the expenditure decreased by 6%. In areas without the project, household income decreased by 12% while the expenditure augmented by 46% and (b) The financial and economic analysis indicated that BFEP was not worthwhile to neither the community nor to the individuals and finally (c) the project has contributed positively towards conservation.

References

Butler, R. W. Alternative Tourism: Pious Hope or Trojan Horse? *J. Trav. Res.* **28**, 40–45 (1990).
Howard, P. C. The Economics of Protected Areas in Uganda: Cost, Benefits and Policy Issues. (Unpublished) M.Sc Dissertation: University of Edinburgh, UK (1995).
Karani, P. K., Kiwanuka, L. S. & Sizomu, K. Forest management Plan for Budongo Forest Reserve. Unpublished Fourth revision (First Draft). Ministry of Nature Resources. Forest Department, Uganda. Plan. Budongo Forest Ecotourism project. Masindi, Uganda (1997).
Langoya, C. D. & Long, C. Local Communities and Ecotourism Development in Budongo Forest Reserve. ODI, London (1998).
Langoya, C. D. & Long, C. Forest Resource Conflict Mitigation Through Ecotourism in Budongo Forest Reserve. ACTS, Nairobi (1999).
Lindberg, K. & Huber, M. J. Economic Issues in Ecotourism Management. in *Ecotourism: A Guide for Planners and Managers*. First Edition (The Ecotourism Society, Washington DC, 1993).
Namukwaya, A. Local Community participation in Ecotourism Development and Conservation of Budongo Forest Reserve, (Unpublished) MSc Thesis: Forestry Department, Makerere University, Kampala (1998).
Romeril, M. Tourism and the Environment: Accord or Disaccord? *Tourism Manage.* **10**, 204–208 (1989).

Indigenous Knowledge and its Relevance to Conservation and Management of Tropical Forest Ecosystems

Tropical Ecosystems: Structure, Diversity and Human Welfare.
Proceedings of the International Conference on Tropical Ecosystems
K. N. Ganeshaiah, R. Uma Shaanker and K. S. Bawa (eds)
Published by Oxford–IBH, New Delhi. 2001. pp. 55–57.

Knowledge and use of biodiversity in Brazilian hot spots

Alpina Begossi*, Natalia Hanazaki and Nivaldo Peroni

Núcleo de Estudos e Pesquisas Ambientais, Universidade Estadual de Campinas
Campinas, S.P. 13081-970, Brazil
*e-mail: alpina@nepam.unicamp.br.

Hot spots in Brazil include a variety of ecosystems, such as areas with mangroves, with savannah or *cerrado*, or with forests. About 68 per cent of the Amazon is geographically located in Brazil in the central, northern and western ranges of the country.

The Atlantic Forest coast is currently represented by 5% of discontinuous forest remnants. Caboclos are rural inhabitants in the Amazon, and, like the caicaras of the Atlantic Forest coast, they descend from a variety of Indian groups and from the Portuguese colonists. Riverine caboclos (ribeirinhos) live along the banks of the rivers, and subsist on economic activities such as artisanal fishing, small-scale agriculture, plant collecting, rubber-tapping, and tourism, among others. Caiçaras and caboclos exhibit a deep knowledge about their environment, and interact with the natural resources, through planting, collecting, and even managing the resources.

In this study we focus on the knowledge that caiçaras and caboclos exhibit which might have implications for management of natural resources. We are interested in sound ecological behaviours and practices, though they may represent incipient forms of local management.

Keywords: Diet, ethnobiology, fishing, human ecology, management.

Data were gathered through interviews with native adults on the communities (Table 1), along with systematic observations. These included samples of fishing trips, using GPS to mark and to map the fishing areas and the spots used in the Atlantic Forest coast. Data on management practices were checked and complemented with systematic observations during fieldwork. Biological material was collected for identification.

Besides a deep knowledge of the local biodiversity, the caiçaras of the Atlantic Forest and the caboclos of the Amazon possess information that may have potential management implications for conservation. For example, the following strategies used by caiçaras or caboclos in their interaction with the natural resources represent local rules that might be used in management planning:

- The use of a high diversity of medicinal plants, which minimize impact per species; medicinal plants may also be part of a cash economy for local or regional markets (Figure 1).
- Management of swiddens plots which increase the availability of palms and the diversity of gardens located in disturbed areas close to houses.
- The use of conservative practices in the extraction of non-timber forest resources, by extracting mature aerial roots or mature leaves, depending on the species, and by retaining reproductive parts of plants.
- The cultivated diversity due to multiple cropping, and the management of crop genetic resources. Examples are given by clonal varieties of species, such as cassava, in which cycles of fallow and swidden interact with seed bank germination.
- The knowledge and use for food of a high diversity of local animals, especially local fish, minimizing impacts per species.
- The occurrence of food taboos, some associated with carnivorous or medicinal animals, which help to conserve animal species.
- The exploitation of the environmental variation, adopting different fishing technologies in specific sites and for different species.
- The informal division of territories or areas used for fishing, in which communities maintain fishing areas close to the home communities, avoiding the overlapping of fishing spots.

Table 1. Areas studied, number of families and interviews

Site	Number of families	Number of interviews
Atlantic Forest		
Islands	165	205
Mainland	224	253
Araguaia river	63	96
Negro River	60	73

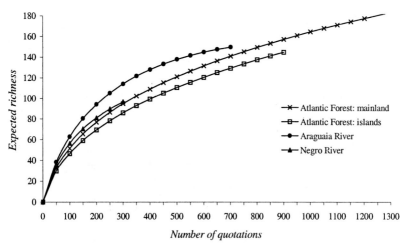

Figure 1. Expected richness by rarefaction curves for medicinal plant quoted at Atlantic Forest (islands and mainland), Araguaia river and Negro river, according to the number of quotations.

Some potential management practices should be stimulated or developed under co-management. Some seem to have temporal stability, such as territorial or informal division of fishing spots in Atlantic Forest sites, and maintenance of the diversity of cassava varieties in both Atlantic Forest and Amazon. In particular, fishing areas can be mapped and reserved for artisanal fishers. Shifting cultivation should be included in management, and not forbidden or excluded from management, as is the current policy of the Brazilian Governmental Environmental Agencies.

Acknowledgments

We are grateful to J. Y. Tamashiro for plant identification; O. Yano for bryophyte identification; J. L. Figueiredo, J. C. Garavello (UFSCar), O. T. Oyakawa (MZUSP), and J. Zuanon (INPA, Manaus) for fish identifications, and E. Camargo for collecting plant specimens at the Negro river. We thank grants from FAPESP (1992–2000), research productivity scholarships (AB) to CNPq (1990–2000), and the grant 97/14514-1 (FAPESP) supporting the travelling expenses of AB to the ATB meeting at Bangalore, India, 15–18 July 2001. Finally, we thank Paulo S. Oliveira and ATB for the invitation to organize the Symposium on Indigenous Knowledge at the ATB 2001 meeting.

Tropical Ecosystems: Structure, Diversity and Human Welfare.
Proceedings of the International Conference on Tropical Ecosystems
K. N. Ganeshaiah, R. Uma Shaanker and K. S. Bawa (eds)
Published by Oxford–IBH, New Delhi. 2001. pp. 58–61.

Changing practice of indigenous knowledge research: Cases and perspectives on sharing across knowledge systems

Fikret Berkes and Iain Davidson-Hunt*

Natural Resources Institute, University of Manitoba, Winnipeg,
Manitoba R3T 2N2, Canada
*e-mail: berkes@cc.umanitoba.ca

The diverse methodologies and practices used in indigenous knowledge (IK) or traditional ecological knowledge (TEK) research reflect the diverse origins of the discipline in ethnoscience and human ecology. They also reflect the diversity of purposes for carrying out IK research (Posey and Dutfield, 1996). Interests in folk taxonomy that used to dominate IK research, especially in terrestrial tropical ecosystems, have been replaced in part with ecological inquiries – indigenous perspectives about human–nature relationships and local understandings of ecosystem processes. These more ecological studies have already had an impact on the discipline and, to some extent, on professional ecologists, by providing insights on ecosystem processes in such diverse geographical regions as the Brazilian Amazon, coastal ecosystems of Oceania, and the Canadian Arctic (Berkes, 1999).

As IK research has come to be taken more seriously in professional circles (witness the TEK special issue in *Ecological Applications*, Ford and Martinez, 2000), many indigenous groups have also realized that

Keywords: Knowledge, research, partnership, intellectual property rights.

'knowledge is power' and have become more guarded about their knowledge. Not only in tropical areas but throughout the world, there is pressure to design IK research projects which are participatory in nature, with the community becoming a partner in a cooperative process of knowledge creation, rather than merely being the object of research (Martin, 1995; Davidson-Hunt, 2000).

Such a partnership also requires agreement on how to make sense of IK findings. IK is contextual knowledge (we define IK/TEK as a knowledge–practice–belief complex, Gadgil et al., 1993; Berkes, 1999). In fact, all systems of knowledge have their own context and assumptions. The researcher comes into a IK project with his/her background knowledge and worldview – usually that of Western science or social science. Thus, the model of participatory research has to address the dilemma of how to combine two different knowledge systems, or ponder over the question whether the two can be combined at all.

The purpose of the paper is to investigate models of participatory research that can combine IK and Western science in ways that do justice to both, while at the same time producing good ecology and social science. We analyze the main features of the partnership research model that we have used in two case studies, and suggest some recommendations regarding the kind of partnerships that may work. Both examples are based on recent research projects carried out with Canadian indigenous groups. One of the cases is in the area of climate change and the other is on non-timber forest products. Neither comes from tropical ecosystems, but the models are applicable anywhere.

The first case is based on the project, 'Inuit observations of climate change', carried out with the Inuvialuit people of Sachs Harbour, in the Canadian Western Arctic. It was a project initiated by the people of Sachs Harbour who wanted their perceptions of climate change recorded and disseminated to the world. Lead by the International Institute for Sustainable Development (IISD), the project started with a planning workshop which asked the people of Sachs Harbour, at the very start of the project, their objectives and what they considered important for the project to focus on. The priority issues, research questions, plans for video documentation, and the overall process for the project were all defined jointly by the project personnel and the community.

Information was collected using a variety of interlinking methodologies: brainstorming workshops, focus groups, video interviews, individual semi-directive interviews, and participant observation. The early versions of the video, along with technical trip reports and newsletters prepared for the community, were used for feedback and for the verification of results. In several of the key areas for climate change, the project invited southern scientific experts to Sachs Harbour to provide opportunity for one-on-one interaction between scientists and local experts on such topics as changes in sea-ice conditions. Repeat visits were made to the community around a yearly cycle, focusing on activities appropriate for that season. Continuity was provided by two members of the team who took part in all the trips and stayed for longer periods (Ford, 1999; Riedlinger and Berkes, 2001).

The second case is based on the project, 'Combining scientific and First Nations knowledge for the management of non-timber forest products', carried out with the Ojibwa people of Shoal Lake, Ontario in southcentral Canada. The project started with joint discussions between the Ojibwa and the researchers, looking for common grounds. It proceeded through a series of meetings with both political leaders and elders of the group, culminating in a negotiated research protocol between the University group and the First Nation. The project itself started with a ceremony establishing proper social relationships among members of the two parties, and ensuring proper ethical attitudes towards the plants to be studied.

Information was collected through interviews with elders and participant observation. Biological data were obtained through sample plots in which non-timber forest products (e.g. berries, medicinal plants and other culturally significant products such as birch bark) were collected and voucher samples taken. Elders took the initiative to take researchers to sites that they deemed important to teach them what they considered important, rather than researchers taking the elders to plot surveys (as originally planned). Results were recorded by audiotape and videotape to be deposited, along with photo records of plants, with the Shoal Lake Ojibwa. Technical reports of the project were presented to the Ojibwa and others through workshops, and verified with individual Ojibwa elders.

The two research projects are different in organizational details, as well as being different in geographic setting and subject matter. But they share a number of features that characterize a particular kind of partnership project. First, the project creates a 'table' of equal partners, a forum in which different objectives can be discussed and the agendas of the parties laid on the table and made transparent. The actual objectives used, research approaches, and rules of conduct are all determined jointly by the community and the researcher. The actual research process has both a science and an IK component, and there is provision for the two systems of knowledge to inform one another and to learn from one another. From the research process, there is feedback to the community in the form of provisional results; there is feedback to the research team in the form of ⋅revised approaches and verification.

The results of the project are shared as previously agreed upon, and the overall results are deposited with the community in culturally appropriate ways. Regarding the published work, the community retains control by indicating when certain kinds of knowledge (such as culturally sensitive information) should be left out. Community people receive credit for their knowledge and retain control over it. Legitimacy and authority of knowledge are not restricted to the researcher and the institution of the researcher but include the local people and their institutions as well.

Such a partnership approach satisfies the requirements for recognizing intellectual property rights of knowledge-holders, and holds potential benefit for the indigenous group in question. Further, the approach has scholarly significance in at least two ways. First, it positively deals with the human ecological dilemma posed by Ingold (2000) 'human beings must simultaneously be constituted both as organisms within systems of ecological relations, and as persons within systems of social relations'. Second, by posing a fundamental challenge to expert-knows-best science, it shows the way to other disciplines, such as sustainability science, on how research can be 'created through processes of co-production in which scholars and stakeholders interact to define important questions, relevant evidence, and convincing forms of argument' (Kates *et al.*, 2001).

References

Berkes, F. *Sacred Ecology. Traditional Ecological Knowledge and Resource Management* (Philadelphia and London, Taylor & Francis, 1999).

Davidson-Hunt, I. Ecological ethnobotany: Stumbling toward new practices and paradigms. *MASA J.* **16**, 1–14 (2000).

Ford, J. & Martinez, D. (ed). Invited feature: Traditional ecological knowledge, ecosystem science and environmental management. *Ecol. Appl.* **10** (2000).

Ford, N. Communicating climate change from the perspective of local people: A case study from Arctic Canada. *J. Dev. Commun.* **1**, 93–108 (1999).

Gadgil, M., Berkes, F. & Folke, C. Indigenous knowledge for biodiversity conservation. *Ambio* **22**, 151–156 (1993).

Ingold, T. *The Perception of the Environment: Essays on Livelihood, Dwelling and Skill* (London and New York, Routledge, 2000).

Kates, R. W. Sustainability science. Statement of the Friibergh Workshop on Sustainability Science. http://sustsci.harvard.edu/keydocs/friibergh.ht_(2001).

Martin, G. J. *Ethnobotany: A Methods Manual* (London, Chapman and Hall, 1995).

Posey, D. A. & Dutfield, G. *Beyond Intellectual Property: Towards Traditional Resource Rights for Indigenous Peoples and Local Communities.* Ottawa: International Development Research Centre (1996).

Riedlinger, D. & Berkes, F. Contributions of traditional knowledge to understanding climate change in the Canadian Arctic. *Polar Record* (in press) (2001).

Tropical Ecosystems: Structure, Diversity and Human Welfare.
Proceedings of the International Conference on Tropical Ecosystems
K. N. Ganeshaiah, R. Uma Shaanker and K. S. Bawa (eds)
Published by Oxford–IBH, New Delhi. 2001. pp. 62–66.

Using local knowledge as a research tool in the study of river fish biology – Experiences from Mekong

John Valbo-Jørgensen* and Anders F. Poulsen

MRC Fisheries Programme, P.O. Box 582, Phnom Penh
*e-mail: bigcatch@bigpond.com

For the rural poor in the Mekong Region, there is no substitute for the income and food generated by the river fisheries. The fish resources must therefore be managed appropriately in order to sustain them for future generations. Riverine fish stocks are extremely resilient and can sustain incredibly high fishing pressures, but the stocks are vulnerable to environmental degradation and especially to water management projects. The Mekong Basin is still relatively unaffected by pollution, and the water quality is good, but increased industrial development and rapidly growing human populations will inevitably lead to a considerable rise in the demand for water, energy and arable land, and the pressure on the natural resources is quickly building up. Destruction of aquatic habitats is already widespread and mainly consists in the conversion of natural floodplains into paddy fields and the cutting of trees and bushes for firewood. Dredging of the riverbed and removal of rocks in order to improve navigation will change water flow, and may also affect spawning grounds of many species. There are plans for a number of dams in the Mekong Basin; some of these are for flood-control, others for irrigation and hydropower. Migratory fish species depend on free movement during their seasonal migrations, and any construction that hinders the migrations will prevent these fish from completing their lifecycles, thus

Keywords: Local knowledge, river fisheries, Mekong, fish migration.

leading to the gradual disappearance of the affected stocks or in some cases even the species.

Appropriate management must take ecological parameters like distribution, spawning grounds and migration patterns of the individual species into account. However, the lack of detailed knowledge on the ecology, and sometimes even the taxonomy of most species makes it difficult to get a manageable picture of the fisheries situation in a vast basin like the Mekong. The majority of Mekong fish species are migratory and are dependent upon both longitudinal and lateral connectivity in order to complete their lifecycles. Most fishery impacts will therefore be of much more than a localised scale, and fisheries studies must consequently cover all ecological units of the basin.

However, incomplete understanding of river fisheries biology, at the administrative level, often hampers proper planning and management. This is partly due to the difficult access to large parts of such river basins, the complex nature of the fisheries, and in many cases a severe lack of research funds. The thousands, or indeed millions, of people who live along the shores of the rivers and rely on the fishery for their daily survival, in contrast, have a very intimate knowledge of the behaviour and biology of the fish. In this situation, gathering the knowledge of the fishers may provide politicians or planners with baseline knowledge in a relatively quick and cheap way.

In this paper, we wish to illustrate the kind of detail that can be obtained by using local knowledge in the study of river ecology. The actual example is a two-year study of fish migration and spawning in the Lower Mekong Basin. The study involved semi-structured interviews with 355 expert fishermen in four countries along 2,400 km of the Mekong mainstream.

It is not possible to give here a detailed account of the migrations by all 49 species covered by the survey for lack of space. However, some general features are summarised.

First, there are two main migration periods. The first is from May to July when the water is rising. During this migration most fish are full of eggs, implying that this is a migration towards the spawning grounds. The second migration period, from October to December, is when the water level is falling rapidly thereby pushing the fish out of the nursery and feeding areas on the floodplain, from where they migrate to their dry season refuges. Figure 1 shows all migration reports per month for all forty-nine species, together with the monthly discharge of the Mekong River.

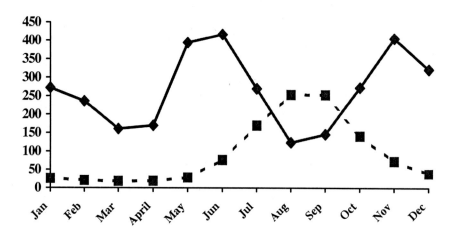

Figure 1. The relationship between fish migration and hydrology. *Dotted line:* Average monthly discharge (m^3/s/100) of the Mekong River at Pakse, Southern Laos (data provided by the MRCS). *Solid line:* Number of migration reports (all stations and all species).

Second, three different migration systems could be identified (Figure 2):
(i) The first system covers the stretch from around the Khone Falls to the Mekong Delta in Vietnam, and includes the Tonlé Sap–Great Lake system. It is the huge annual flood pulse, in southern Cambodia and in the northern Mekong Delta in Vietnam, which drives this system. Species spawning in the mainstream probably spawn in the area between Kratie and Khone Falls, while others migrate into the Se San, Sre Pok and Sekong tributaries. Floodplain spawners spawn from downstream Kratie to the saline intrusion zone in the delta.

During the flood season larvae, juveniles and adults of most species enter the floodplains, where they take advantage of the productive environment. As the water begins to recede at the beginning of the dry season, the fish move back into permanent water bodies, and eventually distribute themselves within dry season refuges, often associated with deep pools in the main river channels.

(ii) The second system extends from just above the Khone Falls to the area where Loei River joins the Mekong; migrations appear to be determined by the presence of large tributaries with extensive floodplains. During the early flood season, fishes move upstream within the Mekong until they reach a tributary, from where they can get access to floodplains. Some migrating fishes may spawn on the floodplain itself, whereas others spawn in the tributary mainstream. Larvae and juveniles originating upstream reach their floodplain nursery areas passively (i.e. drifting with the current), or actively (i.e. as juveniles swimming in from the river). At the end of the flood season, fishes move from flooded areas back to the main river channel. Many species move all the way back to the Mekong mainstream, where they spend the dry season in certain deep pools.

(iii) Lateral movement to floodplains is a less prominent feature of the third system, that constitutes the section from Loei and upstream, because floodplains are scarce in this section. The most conspicuous species involved with this migration system is the Mekong giant catfish,

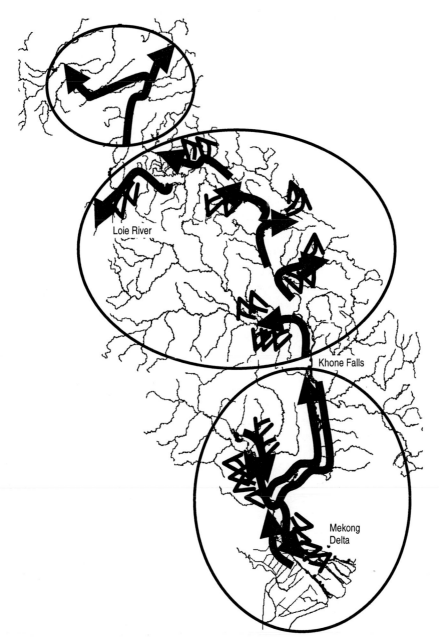

Figure 2. Three overall Mekong migration systems can be delineated: (i) from Khone Falls to the Delta; (ii) from Khone Falls to Loie River; (iii) upstream Loie River. Large arrows: longitudinal migrations, small arrows: lateral migrations.

Pangasianodon gigas. This species is caught every year in northern Laos and Thailand during its upstream spawning migration and its migration is believed to extend into China.

Many fishers were able to provide information about the periods, where the fish have eggs in the abdomen, and in doing so giving us an indication of the spawning period. However, exact spawning grounds turned out to be much more difficult to identify than we expected, probably because the Mekong is very turbid – especially during the flood season where most species are spawning.

This also explains why spawning grounds mainly were reported for species spawning in shallow water on the floodplain. One example is the catfish *Wallago attu*, for which a fisherman in Thailand reported: 'In June–July groups of fishes larger than 2 kg spawn in shallow water on flooded grassland. The eggs stick to the substrate and hatch within 3 days.' Another Thai fisherman reported that *W. attu* is spawning in Huai Kid reservoir near the mouth of the Huai Kid stream, and a Vietnamese fisher stated that it breeds in rice fields.

Other floodplain spawning fish, for which spawning grounds were reported, include *Notopterus notopterus*, cyprinids of the genus *Hypsibarbus*, *Clarias batrachus*, *Wallago leeri*, and *Channa striata*.

Several spawning grounds could however also be identified for the mainstream spawning *Probarbus jullieni* and *P. labeamajor*, probably because these species are spawning in the dry season, when the water is clearer, but also because they make a lot of noise during the spawning performance. A Thai fisherman, for instance, reported observing 'hundred fish gathering near two islands'. Another Thai fisherman had seen three *Probarbus* species migrating together, but spawn separately, during January–February.

Although deep pools were not specifically raised as an issue during the survey interviews, more than 230 records from the survey refer to deep pools as being important habitats for certain fishes. Both relatively sedentary species like *Chitala ornate*, *Wallago attu*, *Mastacembelus armatus* and *Bagarius yarelli*, and highly migratory species, such as the pangasiid catfishes, and the cyprinids *Probarbus* spp., *Catlocarpio siamensis*, *Cirrhinus microlepis*, and *Cyclocheilichthys enoplos* were reported to use deep pools part of the year.

Local knowledge will obviously have a significant position in the study of river ecology in the years to come, because gathering fishermen's knowledge is a quick way to obtain baseline information about the biology of fish. The method is sustainable in developing countries because it is cheap and requires limited training if carried out by experienced fisheries people. However, the method cannot, and should not, stand alone, but should rather be used as the first step in a process enabling researchers to ask more specific questions, which can be answered through more focused studies.

Tropical Ecosystems: Structure, Diversity and Human Welfare.
Proceedings of the International Conference on Tropical Ecosystems
K. N. Ganeshaiah, R. Uma Shaanker and K. S. Bawa (eds)
Published by Oxford–IBH, New Delhi. 2001. pp. 67–69.

Economic development and conservation of biological and cultural diversity in Yunnan Province, China

Rey C. Stendell*†, Richard L. Johnson*,
John P. Mosesso** and Zhang Xia†

*US Geological Survey, Fort Collins, Colorado, USA
**US Geological Survey, Reston, Virginia, USA
†Bio-Resources Innovative Development Office, Kunming, PR China
†e-mail: rey_stendell@usgs.gov

Chinese and American scientists are cooperating to develop concepts, strategies, agreements, and proposals in support of an economic and sustainable ecosystems project in Yunnan Province, People's Republic of China. Yunnan's Provincial Government has initiated a major program to develop and further utilize its biological resources to help improve economic conditions for its citizens. Programs to accelerate economic development by utilizing biological resources, along with population growth, could place unsustainable demands on Yunnan's ecosystems. The Provincial Government is cooperating with the US Geological Survey (USGS) on evaluation and management of biological resources so that development will be compatible with sustainable ecological systems. USGS capabilities in assessing the status and trends of biological resources, conserving and restoring natural ecosystems, estimating economic values and impacts, modeling ecological economic systems, using geographic information systems, and designing education and information systems will assist in this process. This paper provides a status report on this developing cooperative project.

Keywords: China, Yunnan Province, biodiversity, cultural diversity, ecological economics.

China is facing many important problems regarding the balance of conservation and sustainable economic development. This country is one of the most biologically diverse regions of the world. Tropical, subtropical, temperate, and boreal climate zones all occur within its borders. The flora of China is extremely rich and of great value to understand the world's flora, with more than 7,500 species of indigenous trees and shrubs. About 30,000 species of vascular plants (12 per cent of the world's total) occur in China. Moreover, 10 per cent of the world's species of land vertebrates and 14 per cent of all known birds occur there. More specifically, the Yunnan Province is extraordinarily rich in natural resources and biological diversity. This richness is characterized by the abundance of species of plants and animals, many of which are endemic to the Province.

China is a country of great cultural diversity. Han Chinese is the predominant nationality, comprising about 94 per cent of the population. Fifty-five other nationalities, referred to as the 'minorities', reside in more than 60 per cent of the land area, including many in strategic border areas. Yunnan is sometimes referred to as the 'Home of the Minorities'. Of China's 55 minorities, 25 are found in Yunnan, more than any other province.

The biological and cultural diversity found in Yunnan Province are valuable asset to China and the world. Increasing population and economic development are putting increased pressure on these resources. Ecological economics is being used to better understand co-evolving ecological, economic, and cultural systems. There is considerable interest among conservationists and natural resource managers to improve the economic conditions of the rural people, making them less dependent on biological resources for subsistence. The expansion of market economies will have an impact on biological resources and many cultural traditions within local communities.

Yunnan has initiated a program of industrialized development for four types of natural resources: biological, mineral, hydropower, and tourism. Eighteen biological projects, focusing on areas such as medicine, medicinal and economic plants, and mushrooms, have been identified. The Bio-resources Innovative Development Office was established to guide this activity. Projects are selected based on market demand for the products involved. The objective is to create a leading enterprise in the bio-agro industry that will provide power for economic growth and social development. Chinese and USGS economists will work together to identify and evaluate potential development projects for economic viability and ecological sustainability.

A new geographic information system will be used to determine the status and protection afforded to biological diversity on a large spatial scale. The methodology, called Gap Analysis, identifies 'gaps' in biodiversity protection that can be filled by changes in land-use practices or by the creation of new natural reserves. The Gap Analysis Program, managed by the USGS, has two primary goals. The first is to characterize and map vegetation types and naturally occurring vertebrate species that are absent from or under-represented in areas managed for biodiversity. The second goal is to identify areas where additional protection measures would significantly increase the representation of native biodiversity. Beyond identifying and mapping unprotected species and vegetation types, we will use Gap Analysis to spatially track the effects of alternative economic development strategies on both human and natural resources. The multiple objectives of improving economic well-being, conserving native species, and preserving minority cultures will be sequentially optimized, and the results spatially displayed on the Gap Analysis classification structure.

Ecological economic methods, coupled with comprehensive biological information and mapping techniques, can optimally use and conserve Yunnan's cultural and biological diversity to create a sustainable economy.

Tropical Ecosystems: Structure, Diversity and Human Welfare.
Proceedings of the International Conference on Tropical Ecosystems
K. N. Ganeshaiah, R. Uma Shaanker and K. S. Bawa (eds)
Published by Oxford–IBH, New Delhi. 2001. pp. 70–73.

People's Biodiversity Registers: Lessons learnt

Madhav Gadgil

Centre for Ecological Sciences, Indian Institute of Science,
Bangalore 560 012, India
e-mail: madhav@ces.iisc.ernet.in

The People's Biodiversity Register (PBR) programme is an attempt to record people's knowledge and perceptions of the status, uses, history, ongoing changes and forces driving these changes in the biological diversity resources of their own localities. It also aims to document people's perspectives on who the gainers and losers are of ongoing changes in the ecological setting and utilization of biodiversity resources, and how they feel the resources ought to be managed (Gadgil *et al.*, 2000). The endeavour acknowledges that different stakeholders from local communities, as well as outsiders may have divergent understanding and perspectives on these issues and attempts to record these. It is hoped that the information thus recorded would eventually become a part of a broader biodiversity information system. Such an information system could serve to (a) support a decentralised system of management of natural resources, as well as (b) help organise equitable sharing of benefits flowing from commercial utilisation of biodiversity resources and knowledge of their uses (Gadgil and Rao, 1998). PBRs are expected to become an element of an ongoing effort at monitoring of biodiversity resources and to be updated at regular intervals. The process of compilation of PBRs is expected to enhance the public awareness of

Keywords: Biodiversity information system, folk knowledge, participatory appraisal, Convention on Biological Diversity, People's Biodiversity Register.

significance of conservation, sustainable use and equitable sharing of benefits of biodiversity and to catalyse grass-roots action.

A number of groups from the NGO sector as well as the academic community have been involved in preparation of PBRs in different parts of India since 1995. The primary motivation behind the early attempts was to record the rapidly eroding folk knowledge of medicinal uses of plants. Other NGOs have focussed on recording the occurrence and management practices of land races of cultivated crops to support their on-farm conservation, as well as promotion of farmers' rights. Kerala Sastra Sahitya Parishat has prepared PBRs covering all 85 village councils of the district Ernakulam over 1998–1999 as an element of the people's planning movement in the Kerala state.

The most systematic attempt at preparation of PBRs was undertaken by a network co-ordinated through the Indian Institute of Science, Bangalore (India) over 1996-1998 (Srishti Jigyaasa Parivar, 2000). This programme involved collaborators from eight states selected to represent the varied ecological and social regimes of the subcontinent. Of the fifty two Principal Investigators of the programme, fourteen were college teachers, two University teachers, two school teachers. There were four government officials, thirteen NGO workers and six individuals engaged in development activities on their own. The entire programme engaged 350 researchers from all these sectors and 200 assistants from village communities. As many as 1000 villagers had extensive involvement in the programme as local knowledgeable individuals. The methodology of field investigations included the following components: building of rapport with local people, clarifying project rationale and obtaining their approval for the joint studies, identification of different biodiversity user groups, identification of individuals knowledgeable in different aspects of distribution and uses of biodiversity, individual as well as group interviews with members representing different user groups as also knowledgeable individuals, mapping of the landscape of the study site, field visits to representative elements of this landscape along with some user groups members and knowledgeable individuals, discussions with the entire village assembly, and discussions with outsiders affecting the resource use at study site such as nomadic shepherds or artisans, traders and government officials.

The PBR exercise is still in an early experimental phase, yet it is clear that it is of value in several contexts. It can help the local community become aware of specific resources of value being harvested from their locality. In village Mala of Udupi district in Karnataka, the villagers had not realised that the total value of non-timber forest produce being collected annually from their village was around half-a-million rupees. Realisation of this large volume has triggered an interest in sustainable use through mechanisms such as the establishment of a Joint Forest Management Committee. Such a documentation could bring out very useful understanding of ecological processes. For example, villagers from Teligram in Hooglie district of West Bengal contend that fertilisation of fish ponds by poultry waste is responsible for the outbreak of diseases amongst domesticated ducks. The PBR process also helps to record and promote an assessment of possible value of a variety of conservation-oriented traditional resource use practices. For example, the PBR of Mala village of Karnataka has documented the existence of as many as 400 sacred groves in an area of about 50 km^2. The PBR process can play a valuable role in social mobilisation. In many of the study sites the exercise led to public debates on the issues raised by the information and in turn to conservation actions. PBR documentation could potentially serve as a basis for equitable sharing of benefits from commercial application of folk knowledge of uses. At the moment, however, the PBR exercises have not taken up the recording of such novel knowledge of uses, since the PBR documents do not as yet enjoy any legal status. The uses thus far recorded pertain to those that are already common knowledge.

There are thus many positive experiences of the value of the PBR documentation. However, it is clear that the information thus being recorded must be assessed in terms of factual validity and cross-checked against other scientific evidence. The villagers of Mala (Udupi district, Karnataka), for instance, report having seen a snake of the size of a king cobra with a crest shaped like a cock's comb. They also report the use of some herbs as an antidote against poisonous snake-bites. The traditional resource use practices of one of the communities in this village include semi-annual communal hunts during which all the larger mammals that they come across are killed. It is evidently desirable that such claims of knowledge and consequences of practices are carefully assessed.

Ethnobiology has a long tradition of recording folk uses of biological material to serve as a possible starting point for development of new commercial products. Such documentation is not concerned with careful recording of contributions of particular individuals or communities, nor with any sharing of benefits with them. These, on the other hand, are significant concerns of PBR exercise. At the same time, PBR exercises focus on people's perceptions of ecological processes, ecological changes, forces driving ecological changes, gainers and losers of uses of living resources and management options. It is therefore to be expected that the PBR experiment will be opposed by some of the beneficiaries of the current system. In one of Indian states, for instance, a senior bureaucrat blocked the acceptance of a major grant to initiate the preparation of PBRs on a large scale on grounds that this would come in the way of state's implementation of development programmes. There have been other sources of opposition to the PBR experiment as well. These include a fear that recording of folk knowledge of uses may be appropriated by commercial interests without a just sharing of benefits; and that recording of information on occurrence of biological resources would lead to an overexploitation of these resources by commercial interests.

While there exist such barriers to further development of PBRs as a tool of value in many contexts, there are a number of supportive developments, at a global as well as a national scale. At the global scale, the Convention on Biological Diversity, especially article 8(j) recognises the value of folk knowledge and practices and calls for activities which would include many elements of the PBR exercise. At the national level, the Government of India has tabled before the Parliament in the Monsoon 2000 session a Biological Diversity Bill which provides for local level Biodiversity Management Committees. Similarly many villages in India are now involved in managing local forest resources through joint committees involving village communities and forest department. The management plans prepared for such programmes of co-management have many elements in common with PBRs.

There are then a number of supportive developments, globally and within India which suggest that exercises of systematic, periodic documentation of people's knowledge and perceptions of biological diversity and its management such as PBRs may become significant elements of the world-wide attempts at conservation, sustainable use and equitable sharing of benefits of biodiversity.

Acknowledgements

The PBR exercise is a very broad-based co-operative endeavour to which a large number of people have made valuable contributions.

References

Gadgil, M. & Rao, P. R. S. *Nurturing Biodiversity: An Indian Agenda* (Centre for Environmental Education, Ahmedabad, 1998).

Gadgil, M., Rao, P. R. S., Utkarsh, G., Chhatre, A. and members of the People's Biodiversity Initiative. New meanings for old knowledge: The people's biodiversity registers programme. *Ecol. Appl.* **10**, 1307–1317 (2000).

Srishti Jigyaasa Pariwar. Let the people speak, *The Hindu Survey of Environment*. 101–137 (1998).

Tropical Ecosystems: Structure, Diversity and Human Welfare.
Proceedings of the International Conference on Tropical Ecosystems
K. N. Ganeshaiah, R. Uma Shaanker and K. S. Bawa (eds)
Published by Oxford–IBH, New Delhi. 2001. pp. 74–75.

Systems of knowledge: Dialogue, relationships and processes

Kenneth Ruddle

Graduate School of Policy Studies, Kwansei Gakuin University, Japan
e-mail: Ii3k-rddl@asahi-net.or.jp

During the last 20 years, the existence of rich systems of local knowledge, and their vital support to resource use and management regimes, has been demonstrated in a wide range of biological, physical and geographical domains, such as agriculture, animal husbandry, forestry and agroforestry, medicine, and marine science and fisheries.

Local knowledge includes empirical and practical components that are fundamental to sustainable resource management. Among coastal-marine fishers, for example, regular catches and, often, long-term resource availability are ensured through the application of knowledge that encompasses empirical information on fish behaviour, marine physical environments, fish habitats and the interactions among ecosystem components, as well as complex fish taxonomies. Local knowledge is therefore an important cultural resource that guides and sustains the operation of customary management systems. The sets of rules that compose a fisheries management system derive directly from local concepts and knowledge of the resources on which the fishery is based.

Beyond the practical and the empirical, it is essential to recognise the fundamental socio-cultural importance of local knowledge to any society. It is through knowledge transmission and socialisation that worldviews are

Keywords: Local knowledge, design principles, fishing community.

constructed, social institutions perpetuated, customary practices established, and social roles defined. In this manner, local knowledge and its transmission shape society and culture, and culture and society shape knowledge.

Local knowledge is of great potential practical value. It can provide an important information base for local resources management, especially in the tropics, where conventionally-used data are usually scarce to non-existent, as well as providing a shortcut to pinpoint essential scientific research needs. To be useful for resources management, however, it must be systematically collected and scientifically verified, before being blended with complementary information derived from Western-based sciences.

But local knowledge should not be looked on with only a short-term utilitarian eye. Arguments widely accepted for conserving biodiversity, for example, are also applicable to the intellectual cultural diversity encompassed in local knowledge systems: they should be conserved because their utility may only be revealed at some later date or owing to their intrinsic value as part of the world's global heritage.

At least in cultures with a Western liberal tradition, more than lip-service is now being paid to alternative systems of knowledge. The denigration of alternative knowledge systems as backward, inefficient, inferior, and founded on myth and ignorance has recently begun to change. Many such practices are a logical, sophisticated and often still-evolving adaptation to risk, based on generations of empirical experience and arranged according to principles, philosophies and institutions that are radically different from those prevailing in Western scientific circles, and hence all-but incomprehensible to them. But steadfastly held prejudices remain powerful.

In this presentation I describe the 'design principles' of local knowledge systems, with particular reference to coastal-marine fishing communities, and their social and practical usefulness. I then examine the economic, ideological and institutional factors that combine to perpetuate the marginalisation and neglect of local knowledge, and discuss some of the requirements for applying local knowledge in modern management.

Tropical Ecosystems: Structure, Diversity and Human Welfare.
Proceedings of the International Conference on Tropical Ecosystems
K. N. Ganeshaiah, R. Uma Shaanker and K. S. Bawa (eds)
Published by Oxford–IBH, New Delhi. 2001. pp. 76–77.

Land cover changes in the extended Ha Long city area, north-eastern Vietnam during the period 1988–1998

Luc Hens*[†], Eddy Nierynck*, Y. Tran Van**,
Nguyen Hanh Quyen**, Le Thi Thu Hien** and
Le Duc An**

*Human Ecology Department, Vrije Universiteit Brussel, Laarbeeklaan 103,
B-1090 Brussels, Belgium
**Institute of Geography, National Centre for Science and Technology,
Hanoi, Vietnam
[†]e-mail: gronsse@meko.vub.ac.be

The north-eastern province of Quanh Ninh in Vietnam is characterised by rapid economic, social and environmental development. Using LANDSAT TM images, we analysed the land cover changes in this province between the period 1988 and 1998. The changes were classified into three main groups: coastal features, natural land features and human features. These groups were further subdivided into 22 different mapping categories.

The results show that by 1998, 39.9 per cent of the 1988 land cover had changed. The results indicate:

- A rapid expansion of the human features: during these 10 years the area of urban settlements doubled and the area for coal mining activities increased by 75 per cent.

Keywords: Land cover changes, LANDSAT images, Vietnam.

- The coastal area changed in a complex way driven by aquaculture activities, agriculture and mangrove expansion (replanting and natural colonisation of tidal flats without vegetation).
- The original dense forest in the area rapidly declined: of the 2,010 ha cover in 1988, only 335 ha remained in 1998. Dense forests were replaced by degraded and secondary forest.

The results presented here offer a powerful tool for planners to develop a strategic plan for the development of the province in a sustainable manner.

Tropical Ecosystems: Structure, Diversity and Human Welfare.
Proceedings of the International Conference on Tropical Ecosystems
K. N. Ganeshaiah, R. Uma Shaanker and K. S. Bawa (eds)
Published by Oxford–IBH, New Delhi. 2001. p. 78.

Edible invertebrates among Amazonian Indians: A critical review of disappearing knowledge

M. G. Paoletti*[†], E. Buscardo* and D. L. Dufour**

*Department of Biology, Padova University, Padova, Italy
**Department of Anthropology, University of Colorado, Boulder, CO 80309, USA
[†]e-mail: paoletti@civ.bio.unipd.it

For the indigenous populations of Amazonia, invertebrates constitute an important component of the diet. We have information on entomophagy for 39 ethnic groups, about 21.4 per cent of the 182 groups known in the Amazon Basin, but the use of this non-conventional food resource is probably much more widespread. We present here a database of all the information available for each ethnic group regarding the species included in the diet, the scientific and the ethno name if known, the stage of life-cycle consumed, the manner of preparation and, when known, the host plant. This database lists 115 species scientifically identified and 131 ethno names. In addition, we have information about 384 ethno names, with unsecure link to the Linnean taxonomy suggesting that local knowledge is very extensive. The database represents not only an easy-to-consult resource, but also a support for further research. The knowledge of the relations between indigenous populations and ecosystem is indeed the base for the natural and cultural biodiversity preservation.

Keywords: Edible invertebrates, Amazonian Indians.

Tropical Ecosystems: Structure, Diversity and Human Welfare.
Proceedings of the International Conference on Tropical Ecosystems
K. N. Ganeshaiah, R. Uma Shaanker and K. S. Bawa (eds)
Published by Oxford–IBH, New Delhi. 2001. p. 79.

Ethnobotany in Cabo Delgado, Mozambique: Use of medicinal plants

Joaquim Matavele[*][†] and Mohamed Habib*[*]

*INDE (National Institute for Education Development), Maputo-Mozambique
**UNICAMP/Department of Zoology (State University of Campinas), Brazil
[†]e-mail: matavele@inde.uem.mz

Human communities from Cabo Delgado have a long tradition of using medicinal plants. In Mozambique, rural populations, in general, are strongly dependent on the natural resources. One example is the use of surrounding vegetation by people from Cabo Delgado. They use the plants for food, handicraft, construction, as a primary energy source and even for medicine purposes. In this survey, we examined the diversity of plant usage for medicinal purposes by 146 individuals, including adults and young people living in Cabo Delgado Province. This community quoted 16 species of plants, belonging to 13 families. Utilisation by different categories of people based on sex and age was compared and differences were found among some groups. In general older people show a deeper knowledge of medicinal plants than younger ones. Men and women seem to show similar knowledge on medicinal plants.

Keywords: Medicinal plants, Cabo Delgado, ethnobotany, Mozambique.

Tropical Ecosystems: Structure, Diversity and Human Welfare.
Proceedings of the International Conference on Tropical Ecosystems
K. N. Ganeshaiah, R. Uma Shaanker and K. S. Bawa (eds)
Published by Oxford–IBH, New Delhi. 2001. p. 80.

How do groups solve local commons dilemmas? Lessons from experimental economics in the field

Juan-Camilo Cardenas

School of Environmental and Rural Studies, Universidad Javeriana,
K7 # 42-27 P.7, Bogota, Colombia
e-mail: jccarden@javeriana.edu.co

Experimental settings to observe human behaviour in a controlled environ-ment of incentives, rules and institutions, have been widely used by the behavioural sciences for sometime now, particularly in psychology and economics. In most cases the subjects are college students recruited for one to two hour decision-making exercises in which, depending on their choices, they earn cash averaging US$ 20. In such exercises players face a set of feasible actions, rules and incentives (payoffs) involving different forms of social exchange with other people, and that in most cases involve some kind of externalities with incomplete contracts, such as in the case of common-pool resources situations. Depending on the ecological and institutional settings, the resource users face a set of feasible levels of extraction, a set of rules regarding the control or monitoring of individual use, and sometimes ways of imposing material or non-material costs or rewards to those breaking or following the rules. We brought the experimental lab to the field and invited about two hundred users of natural resources in three Colom-bian rural villages to participate in such decision-making exercises and through these and other research instruments we learned about the ways they solve – or fail to solve tragedies of the commons with different social institutions. Further, bringing the lab to the field allowed us to explore some of the limitations of existing models about human behaviour and its cones-quences for designing policies for conserving ecosystems and improving social welfare.

Keywords: Experimental economics, designing policies, conserving ecosystems, social welfare.

Tropical Ecosystems: Structure, Diversity and Human Welfare.
Proceedings of the International Conference on Tropical Ecosystems
K. N. Ganeshaiah, R. Uma Shaanker and K. S. Bawa (eds)
Published by Oxford–IBH, New Delhi. 2001. pp. 81–84.

The Mentawai of Siberut Island, Indonesia: Traditional knowledge and its influence on natural resource extraction and conservation

Stacy Marie Crevello

Louisiana State University, P.O. Box 17587, Baton Rouge,
LA 70802, USA
e-mail: screvel@lsu.edu

Indonesia has been blessed with the second most ecologically diverse rainforests in the world (Siberut ICMP, 1995). Indonesia covering only 1.3% of the earth's landmass holds ten per cent of the world's rainforests and forty per cent of Asia's rainforests. Worldwide concern over the status of rainforests disappearing due to widespread logging has impressed upon the Indonesian government the importance of preserving these natural resources, but the question of how best to effectuate conservation remains. These forests are also home to most of Indonesia's 1.5 million indigenous people who rely on natural resources for survival.

New conservation laws have recognized the importance of local ecological knowledge of indigenous people in management of nature reserves in many nations (Furze *et al.*, 1996; Mackinnon, 1986; Primack and Lovejoy, 1995, Stevens, 1997). This realization, that conservation management should incorporate local people as legitimate partners in conservation and development efforts, is important. Active and meaningful involvement of

Keywords: Traditional knowledge, resource extraction, conservation, national park, Siberut Island, Indonesia.

indigenous people in this process, and in decisions related to it, creates a partnership of equals, as opposed to outside agencies designing and implementing nature reserve regulations and priorities.

This research encompasses a broad but integrated system of human–land relationships. This study analyzes the natural resource use and exploitation of the Mentawai of Siberut Island, Indonesia. The Mentawaians have historically practised conservation by following strict taboos in their traditional religion and 'Adat', the customary law. Recent changes due to modern influences have changed the lifeways of the Mentawaians which may be causing a transformation of former conservation practices. In 1995 the entire western half of Siberut Island was declared a national park with various use zones. This has or will greatly impact the Mentawaians, as a new prototype of conservation will be practised as well as the development incorporated into the conservation plan. It is the goal of the national park on Siberut to catalyze full participation of local communities in planning and implementing park management and local development activities (Barber et al., 1995).

Three study sites on Siberut Island were chosen in order to identify differences in resource use and economic utility of residents living in the national park and those living in the buffer zone (in the southern region of the island). Differences in resource extraction between both groups were identified and used to analyze contrasts in the lifeways of park and buffer zone residents.

Data were collected through qualitative and quantitative methods to determine current land use and forest management practices among the Mentawaians. In-depth individual interviews were conducted to understand traditional belief systems as they pertain to land management. The next phase of research was accomplished by conducting household village-wide surveys. A 1995 survey, made available by Dr. Randal Kramer of Duke University and Directorate-General for Forest Protection and Nature Conservation (PHPA) combined with surveys conducted for this research were used to analyze the current land use situation on Siberut Island.

Rates and yields of resource extraction, ownership of traditional homes (*uma*), expenses and other related variables along with ethnographic research were used to identify similarities and differences among buffer zone and park residents. Mentawai residing in the buffer zone extracted far more forest resources than park residents did (Figure 1).

Based on several factors it can be postulated that the Mentawai in the buffer zone are living in a more traditional manner than those within the boundaries of the national park. Although the traditional customs practised in this region promote conservation, there is a continuance of high resource

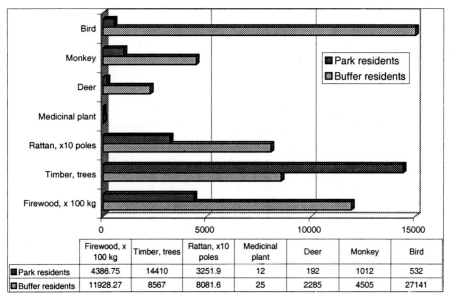

	Firewood, x 100 kg	Timber, trees	Rattan, x10 poles	Medicinal plant	Deer	Monkey	Bird
Park residents	4386.75	14410	3251.9	12	192	1012	532
Buffer residents	11928.27	8567	8081.6	25	2285	4505	27141

Figure 1. Annual Natural Resource Extraction: Siberut Island, Indonesia (1997).

extraction. They rely heavily on the flora and fauna of Siberut for subsistence. The buffer zone residents overwhelmingly reported higher frequencies and yields for extracting forest resources. Buffer zone residents have maintained much of their traditional cultural traits and prefer to stay in their traditional extended family dwellings (*uma*) for extended periods, which are located along the river network of the island. The buffer zone residents, although living on the fringes of the park rely more heavily on the park for subsistence than originally hypothesized.

The Catholic Church has influenced park residents located on the remote western half of the island. The park residents remove large amounts of timber from the forest for construction of churches and other larger homes in government initiated villages. They seem to have partially abandoned their traditional lifestyle in their dress and customs. The churches also act as schools to provide education for the children in the area.

Regardless of the differences in forest exploitation between buffer zone and park residents, the forest is the life force for the Mentawai people of Siberut Island. Extraction of forest resources for survival is an essential part of the daily life of the Mentawai. The forest offers a land base for shifting cultivation and products for subsistence needs and income generation (Silva, 1996).

Because resource exploitation is in direct conflict with conservation laws of *Taman Nasional Siberut*, the park has implemented programs in agroforestry, alternative agriculture, livestock rearing, and community development to provide assistance for alternative income generation, and sustainable extraction of forest resources. If sustainable extraction of natural resources along with conservation are practised and supported by the Mentawai then there is a high probability that *Taman Nasional Siberut* will succeed. This must be a joint effort by policy makers and the local population to achieve the goals of conservation.

Acknowledgements

The US–Indonesia Society and the American Association of Geographers Asian Specialty Group granted funding for this fieldwork. I am grateful to my late advisor Dr. Elvin T. Choong and the Indonesian Institute of Sciences (LIPI); without them this research would not have been possible. I am thankful to my new advisor Dr. Richard Vlosky for providing me the opportunity to attend this conference. I owe a great debt to the Mentawaians of Siberut for their kindness and generosity while conducting my research.

References

Barber, C., Suraya, A. & Purnomo, A. *Tiger by the Tail?* (World Resources Institute, Washington, DC, 1995).

Siberut National Park Integrated Conservation and Development Plan. Directorate General of Forest Protection and Nature Conservation (ICMP). Ministry of Forestry, Republic of Indonesia Jakarta, Indonesia, 1995.

Furze, B. *et. al. Culture, Conservation, Biodiversity. The Social Dimension of Linking Local Level Development and Conservation through Protected Areas* (John Wiley and Sons, Chichester, New York, Brisbane, Toronto, Singapore, 1996).

Primack, R. & Lovejoy, T. *Ecology, Conservation, and Management of Southeast Asian Rainforests* (Yale University Press, New Haven and London, 1995).

Silva, G. Extraction of Non-Timber Forest Products in Pintao, Venezuela. Thesis Louisiana State University. Baton Rouge, 1996.

Stevens, S. *Conservation Through Cultural Survival: Indigenous Peoples and Protected Areas* (Island Press, Washington, DC, Covelo, California, 1997)

Mackinnon, J. *Managing Protected Areas in the Tropics* (IUCN Gland, Switzerland and Cambridge, UK, 1986).

Tropical Ecosystems: Structure, Diversity and Human Welfare.
Proceedings of the International Conference on Tropical Ecosystems
K. N. Ganeshaiah, R. Uma Shaanker and K. S. Bawa (eds)
Published by Oxford–IBH, New Delhi. 2001. pp. 85–88.

Participatory resource monitoring for non-timber forest products in Biligiri Rangaswamy Temple wildlife sanctuary, Karnataka, India

R. Siddappa Setty*,†, K. S. Bawa** and J. Bommaiah*

*Ashoka Trust for Research in Ecology and the Environment,
No. 659, 5th 'A' Main Road, Hebbal, Bangalore 560 024, India
**Department of Biology, University of Massachusetts, Boston, MA 02125, USA.
†e-mail: siddssetts@yahoo.com

Extensive harvesting of plant and animal products from natural ecosystems is one of the principal causes of the loss of biodiversity and degradation of ecosystems. In India, as elsewhere, millions of people derive their livelihoods by harvesting a wide variety of non-timber forest products (NTFPs) from the ecosystem. Although local communities had some sort of monitoring programs in place to estimate production and harvest levels, such systems are gradually breaking down with increasing commercialization and the loss of local control over resources. We initiated a program of participatory resource monitoring (PRM) with the Soligas, the indigenous people of the Biligiri Rangaswamy Temple (BRT) Wildlife Sanctuary, as part of an effort to develop models of sustainable use of forest resources (Bawa, 1999). A detailed description of BRT is available in Ramesh (1989) and Hegde *et al.* (1996). The Soligas derive almost half their income from NTFPs (Lele *et al.*, unpublished data). Of the various products that the Soligas harvest, the most important are *Phyllanthus emblica* and *P. indofischeri* (collectively known as 'amla'), and honey from *Apis dorsata*, or rock bees.

Keywords: Participatory resource monitoring, BRT wildlife sanctuary, NTFP, regeneration.

The participatory resource monitoring program was part of a larger effort to help Soligas derive better economic returns through processing and higher prices of their harvested products, and to attain sustainability by reducing levels of harvest, if such levels were excessive (Bawa, 1999; Bawa et al., 1999).

There were several components of the PRM program. Prior to the harvest, monitoring exercises were conducted to assess traditional knowledge, to explain the goals and to discuss the methods of participatory resource monitoring. Also discussed were methods to estimate the amount of the resource available for harvest. Available resources were estimated in two ways: (a) visually, by the Soliga harvesters, and (b) by us, using a series of transects which were then extrapolated to the entire sanctuary. The more systematic transect method was used to verify visual estimates made by the Soligas.

Pre-harvest discussions were also held with the Soligas to review the harvest techniques to be followed. Prior to the pre-harvest meetings, workshops were held to discuss less-destructive harvest techniques. For example, in the case of amla these included reducing branch breaking, restricting the quantum of harvest, and removing parasites from amla trees.

Post-harvest meetings were held to evaluate the amount of harvest, harvesting techniques and the methods and effectiveness of participatory resource monitoring. The proportion of the available resources, the proportion of branches cut, and the proportion of parasites removed during the harvest were also estimated. The Soliga harvesters also participated in annually evaluating amla regeneration by censussing the number of seedlings, saplings, and adults per unit area. In addition, awareness campaigns in the form of dramas and folk art by the Soligas were held to highlight conservation and natural resource management issues, and workshops were conducted to share the outcome of the PRM program with the remainder of the Soliga community and the state forest department.

In addition to estimating production, extraction and regeneration, the Soligas have prepared resource maps on production, extraction and regeneration. The information collected has been documented in the unit used by the Soligas to process non-timber forest products. Manuals on participatory resource monitoring have been prepared in the local language.

There was a close correspondence between estimates of amla productivity made by us and by the Soligas (Table 1). This indicates that visual assessments made by the Soligas are a reliable means of estimating productivity.

Table 1. Comparison of amla productivity estimated by the transect method, and visually, by the harvesters.

Years	Transect estimates (in tons)	Visual estimates (in tons)
1998–1999	23.2	25.0
1999–2000	148.0	145.0
2000–2001	169.6	135.0

The number of seedlings and saplings (125 individuals/1.75 acre) exceeded the numbers of juvenile (19), and mature (44) trees, indicating that the regeneration has been good.

Our concept of participatory resource monitoring has evolved over the three-year study period in several ways. First, we have now included monitoring of such socioeconomic parameters as income generated from non-timber forest products and the disposition of such income along with biological monitoring. Second, we have suggested to Soligas to undertake monitoring and to document its results in their own way. Third, we have started to devise methods to estimate the effectiveness of participatory resource monitoring. As an example, we have followed the harvesting techniques in areas where participatory monitoring took place versus area where such monitoring did not occur. In the former area branches were cut on 7% of the trees as compared to 52% in the latter area.

A fairly large number of Soligas have participated in the PRM activities. In the three years, 128 pre-harvest and 74 post-harvest group discussions were conducted. Total attendance over this period was 5958, which includes men, women, and children. The Soligas now have a three year record of productivity, extraction, and regeneration, in the form of resource maps. Based on these maps, they can track temporal changes in productivity and can vary the amount harvested accordingly. The Soligas have also started to practice better harvesting techniques. The continuing success of participatory monitoring will be dependent upon the incentives the Soligas receive and the eventual role they play in management of resources. Although the Soligas have started to receive better prices for the raw products they harvest, profits from the enterprise unit set up to process non-timber forest products have declined. The Soligas have also shown disinclination to monitor in the absence of better control over the resources they harvest and in the absence of clear economic benefits from monitoring.

Acknowledgements

This work was funded by a grant from the Ford Foundation. We thank the Karnataka Forest Department, and the Vivekananda Girijana Kalyana Kendra for their collaboration.

References

Bawa, K. S. Prescriptions for Conservation. Biodiversity Conservation Network, Final Stories from the Field. Biodiversity Support Program (Washington DC, 1999), pp. 48–55.

Bawa, K. S., Lele, S., Murali, K. S. & Ganesan, B. Extraction of Non-timber Forest Products in Biligiri Rangan Hills, India: Monitoring a Community-based Project. in *Measuring Conservation Impact: an Interdisciplinary Approach to Project Monitoring and Evaluation* (eds Saterson, K., Margolius, R. & Salafsky, N.) 89–102 (Biodiversity Support Program. World Wildlife Fund, Inc. Washington DC, 1999).

Hegde, R., Suryaprakash, S., Achoth, L. & Bawa, K. S. Extraction of non-timber forest products in the forests of Biligiri Rangan Hills, India. 1. Contribution to rural income. *Econ. Bot.* **50,** 243–251 (1996).

Ramesh, B. R. Evergreen forests of the Biligiri Rangan Hills, South India. Ph.D. thesis (French Institute, Pondicherry, India, 1989).

Uma Shankar, Murali, K. S., Uma Shaanker, R., Ganeshaiah, K. N. & Bawa, K. S. Extraction of non-timber forest products in the forests of Biligiri Rangan Hills, India 3. Productivity, extraction and prospects of sustainable harvest of amla *Phyllanthus emblica* (Euphorbiaceae). *Econ. Bot.* **50,** 270–279 (1996).

Tropical Ecosystems: Structure, Diversity and Human Welfare.
Proceedings of the International Conference on Tropical Ecosystems
K. N. Ganeshaiah, R. Uma Shaanker and K. S. Bawa (eds)
Published by Oxford–IBH, New Delhi. 2001. pp. 89–90.

Diversity of medicinal plants in secondary forest post-upland farming in West Kalimantan

Izefri Caniago

NRM/EPIQ, Jl. Abdurahman Saleh IA No. 1, Pontianak 78124, Indonesia
e-mail: icaniago@ptk.centrin.net.id

This study sought to determine the identity, abundance, and distribution of medicinal plants in Nanga Juoi, West Kalimantan, Indonesia. Nanga Juoi, a village near Bukit Baka/Bukit Raya National Park, provides an ideal site to investigate the ecology and use of medicinal plants based on traditional values and livelihood activities as they relate to medicinal plants. Nanga Juoi is inhabited by the Dayak sub-group known as Ransa. The inhabitants of Nanga Juoi have lived in the general region for generations and continue to observe many traditional practices, including extensive use of medicinal plants.

A variety of sample plot sizes and numbers were established to estimate medicinal plant abundance and distribution in different traditional forest types. In *tempalai* (earliest young secondary forest), where medicinal plants were most abundant, I established 38 plots (2 × 2 m) on a random basis. In *bawas baling* (early young secondary forest), which is smaller in size than *tempalai*, I established 23 plots (2 × 2 m). Eleven plots (10 × 10 m) were established in *bawas beliung* (late young secondary forest), 23 plots (100 × 2 m) in *agung kelengkang* (early old secondary forest), 30 plots (100 × 2 m) in *agung tua* (late old secondary forest), 10 plots (2 × 2 m) plots in *pinggir sungai* (river bench) and 7 transects (1000 × 2 m) were established

Keywords: Medicinal plants, West Kalimantan, traditional values, ecology.

from lowland to highland forest (200–915 m above sea level). When medicinal plants used by villagers were not recorded in any sample plots, information regarding species distribution was collected from the local people.

Over 250 medicinal plant species were recognized and utilized by the people. At least 225 plants were identified and they occurred in 162 genera and 75 families. Medicinal plant species were found to be more diverse in old swidden fallows, while densities were highest in young fallows. The distribution of medicinal species in various forest types is quite interesting. This is demonstrated by the number of medicinal plant species found in *agung kelengkang* which has the highest number of all forest types (79 species) followed by *rimba* (42 species), *tempalai* (37 species), *bawas beliung* (36 species), *bawas baling* (29 species) and *agung tua* (28 species).

Trees are the most important source of medicinal plants (81 species), followed by herbs (65 species), vines (45 species), ferns (14 species), epiphytes (15 species) and aquatic plants (2 species). Medicinal plant species are widely distributed in the various successional forests around Nanga Juoi. Medicinal plants occur in all six vegetation types (i.e., successional stages) following abandonment of shifting cultivation.

Some medicinal plant species are restricted to specific forest types. The number of restricted species is quite different when compared to the general species distribution. For example, while primary forests (*rimba*) do not contain the highest number of medicinal plant species, they have the highest number of individuals with restricted distributions (10 species). When the distribution of medicinal plant species is divided into secondary and primary forest types, 62% of all medicinal plant species occur in secondary forest while only 5% are found in primary forest.

The conservation of medicinal plants in Nanga Juoi will require conserving all the environments in which plants occur. Information from this study could prove useful in forest and protected area management efforts that seek to balance resource use by local inhabitants and forest preservation or exploitation by outside interests.

Anthropogenic Pressures on Ecosystem Structure and Function

Tropical Ecosystems: Structure, Diversity and Human Welfare.
Proceedings of the International Conference on Tropical Ecosystems
K. N. Ganeshaiah, R. Uma Shaanker and K. S. Bawa (eds)
Published by Oxford–IBH, New Delhi. 2001. pp. 93–96.

Plant reproduction in a fragmented landscape

Saul A. Cunningham*,† and David H. Duncan†

*CSIRO Entomology, Box 1700, Canberra ACT, 2601, Australia
†Botany & Zoology Department, Australian National University, Canberra ACT,
0200, Australia
†e-mail: saul.cunningham@cnto.csiro.au

Habitat fragmentation is one of the key threats to global diversity. The impact of fragmentation is expected to be severe in the tropics, where biodiversity is rich, and human populations are rapidly growing. In this paper we discuss our research and briefly review the literature to assess the effect of habitat fragmentation on reproduction by plants in fragments. We discuss results by breaking the process of plant reproduction into different components. We highlight the growing evidence that there are many possible kinds of impact on plant reproduction, and that these have proven easy to detect, suggesting the effects are real and common.

Our research was conducted in woodland vegetation ('mallee') in central New South Wales, Australia. In this region most native vegetation has been cleared for agriculture. What remains are scattered reserves and linear vegetation strips that follow roads and property boundaries. The linear strips, often < 50 m across, have a high edge-to-interior ratio, but broadly similar structure and composition to nearby reserves. We selected pairs of reserves and nearby linear strips, then used replicated pairwise contrasts to determine the effect of fragmentation on plant reproduction.

Keywords: Pollination, flowering, fruit production, seed predation.

We considered six reserve-linear strip pairs in 1997 and seven in 1998 (Cunningham, 2000b).

Flowering is expected to vary with resource availability, which may be altered by fragmentation. Fragmentation increases edge effects, such as elevated light and temperature, and reduced humidity (Matlack, 1993; Young and Mitchell, 1994). In closed forests, edges can be similar to forest gaps, with increased flowering by understorey species. In dry environments edge habitat along roads can receive additional moisture, increasing flowering (Lamont et al., 1994). We detected increased flowering in linear strips in both years (Table 1), which may be due to increased light or water in these fragments.

Changes in flower production are driven by the abiotic environment, whereas other changes are mediated by interactions with animals. Aizen and Feinsinger (1994a) studied 16 plant species in fragmented habitat; nine had lower pollination in small fragments. We found seven additional studies reporting lower pollination rates in smaller populations (7 species), some in an explicit fragmentation context. Only two studies (2 species) provided counter examples, and a further four report no effect (5 species). Of six studies that assess pollinator activity, four report less activity in smaller populations. One study highlighted the complexity of pollinator response, reporting less visits, but of longer duration, to plants in small patches (Klinkhamer and de Jong, 1989). Aizen and Feinsinger (1994b) found a decrease in native pollinators in small fragments but an influx of Africanized bees.

We examined pollination rates for two plant species in small fragments (Table 1) by counting pollen grains adhered to stigmas, and comparing flowers collected in linear strips with those collected in reserves

Table 1. Differences in plant reproduction comparing linear strips to nearby reserves, summarized for four plant species.

Plant species	Flowering	Fruit per flower	Fruit predation	Net fruit production	Flowering	Pollination	Fruit per flower
	1997	1997	1997	1997	1998	1998	1998
Acacia brachybotrya	0	–	–	–	+	–	–
Eremophila glabra	+	–	–	0	+++	–	–
Senna artemisioides	+	+++	0	++	0	n.a.	++
Dianella revoluta	0	–	0	–	n.a.	n.a.	n.a.

"+" indicates the response variable was greater in the linear strip. "–" indicates the response variable was lower in the linear strip. "0" indicates no consistent pattern. Triple symbols indicate a significant effect, $P < 0.05$ (ANOVA, U test for pollination effect). Double symbols indicate a significant effect using ANOVA, $0.05 < P < 0.10$. A single symbol indicates that the difference is in the same direction for the majority of pairwise comparisons. "n.a." indicates data were not collected. For details see Cunningham (2000a, b, and in review).

(Cunningham 2000a). Pollination rates were lower in linear strips for both species (Table 1).

Fragmentation can affect the number of flowers converted into fruits by affecting resource availability, or pollination rates. Aizen and Feinsinger (1994a) found lower fruit per flower in small patches for 5 of 15 species. We found 13 additional studies (16 species) reporting less fruit or seeds per flower in small populations, some of which explicitly implicated habitat fragmentation.

We found only two reports of higher fruit set in small fragments: our own study (*Senna artemisioides,* Table 1) and two species in Aizen and Feinsinger's (1994a) study. Reports for 10 species suggest no fragmentation effect on fruit set per flower (Aizen and Feinsinger 1994a).

We found depressed fruit set per flower in linear strips for *Acacia brachybotrya* and *Eremophila glabra* (Table 1). We were able to link the decline in fruit production per flower to the correlated decline in pollination in small fragments, by demonstrating that fruit production by these species was pollen limited. Although a number of authors have suggested that declining pollination may cause reduced fruiting, few other studies have provided empirical support for the relationship (Cunningham, 2000a).

While much attention has been focused on the possibility of pollination decline, relatively few studies address the possibility that seed predation, another key plant-animal interaction, could be affected by habitat fragmentation. Santos and Telleria (1994) found increased seed predation in small forest fragments because of an increased abundance of mice. We found that pre-dispersal seed predation was lower in small fragments for two plant species (Table 1), but these effects were weak.

We suggest examining the components of plant reproduction separately to help understand mechanisms of change. From a management point of view, however, the focus may be overall reproductive output. Different components of plant reproduction respond to fragmentation differently, making net reproductive outcome difficult to predict. In our study, for example, a decline in the number of flowers converted to fruit in *E. glabra* was counteracted by increased flowering in fragments, so that there was no overall effect on net fruit production (Table 1). *Acacia brachybotrya* sustained less seed predation in fragments, but this effect was less stronger than the decline in fruit production per flower, so that there was nevertheless depressed total seed production in fragments (Table 1).

We found reports for 12 species (8 studies including our own, Table 1) of lower net fruit or seed production in small fragments. For three species (2 studies, inc. Aizen and Feinsinger, 1994a) the net output was higher in small fragments. Our own study found *Senna artemisioides* had higher net seed production in linear strips (Table 1). One might expect that there are many other plant species that are more vigorous in disturbed environments and thus reproduce more in highly fragmented sites, but these species may be less likely to be the focus of fragmentation studies.

The evidence for fragmentation effects on plant reproduction is substantial. Most studies report depressed reproductive output in small and fragmented populations. It will be important to determine if this trend reflects the general pattern, or if researchers have focused on species expected to be vulnerable. Our study, for example, found three species with depressed fruit production per flower and one with greater fruit production per flower in small fragments. Few studies managed to explicitly link effects of fragmentation on reproduction to causal mechanisms. It is important to better understand how seed production responds to altered patterns of pollination, and how this relates to the spatial scale of fragmentation.

Acknowledgements

This research was supported by the Australian Research Council, and the Key Centre for Biodiversity and Bioresources, Macquarie University.

References

Aizen, M. A. & Feinsinger, P. Forest fragmentation, pollination, and plant reproduction in a chaco dry forest, Argentina. *Ecology* 75, 330–351 (1994a).

Aizen, M. A. & Feinsinger, P. Habitat fragmentation, native insect pollinators, and feral honey bees in Argentine chaco serrano. *Ecol. Appl.* 4, 378–392 (1994b).

Cunningham, S. A. Depressed pollination in habitat fragments causes low fruit set. *Proc. R. Soc. Biol. Sci. Ser. B.* 267, 1149–1152 (2000a).

Cunningham, S. A. Effects of habitat fragmentation on the reproductive ecology of four plant species in mallee woodland. *Conserv. Biol.* 14, 758–768 (2000b).

Klinkhamer, P. G. L. & de Jong, T. J. Effect of plant size, plant density and sex differential nectar reward on pollinator visitation in the protandrous *Echium vulgare* (Boraginaceae). *Oikos* 57, 399–405 (1989).

Lamont, B. B., Whitten, V. A., Witkowski, E. T. F., Rees, R. G. & Enright, N. J. Regional and local (road verge) effects on size and fecundity in *Banksia menziesii*. *Austr. J. Ecol.* 19, 197–205 (1994).

Matlack, G. R. Microenvironment variation within and among forest edge sites in the eastern United States. *Biol. Conserv.* 66, 185–194 (1993).

Santos, T. & Tellería, J. L. Influence of forest fragmentation on seed consumption and dispersal of spanish juniper, *Juniperus thurifera*. *Biol. Conserv.* 70, 129–134 (1994).

Young, A. & Mitchell, N. Microclimate and vegetation edge effects in a fragmented podocarp-broadleaf forest in New Zealand. *Biol. Conserv.* 67, 63–72 (1994).

Tropical Ecosystems: Structure, Diversity and Human Welfare.
Proceedings of the International Conference on Tropical Ecosystems
K. N. Ganeshaiah, R. Uma Shaanker and K. S. Bawa (eds)
Published by Oxford–IBH, New Delhi. 2001. pp. 97–100.

Direct and indirect effects of human disturbance on the reproductive ecology of tropical forest trees

Jaboury Ghazoul

Imperial College of Science Technology and Medicine, Silwood Park, Ascot
Berkshire SL5 7PY, UK
e-mail: j.ghazoul@ic.ac.uk

Invertebrates mediate several important ecological processes, including pollination and seed predation, and events that affect invertebrate diversity or behaviour can potentially disrupt forest regeneration processes. This study investigates both direct and indirect impact of logging in Thailand and forest fragmentation in Costa Rica on the pollination and seed production of three forest trees.

Two dipterocarp tree species *Shorea siamensis* and *Dipterocarpus obtusifolius* are common in the dry deciduous forests of Thailand but they differ in that *S. siamensis* is logged by local people while *D. obtusifolius* is not. Logging of the forest therefore resulted in reduced densities of *S. siamensis* and variably isolated individual trees. Although logging had no effect on the number of flower visits to *S. siamensis* by pollinating *Trigona* bees, these pollinators did spend longer periods of time foraging in the canopies of the more isolated trees which were more prevalent in logged areas where tree density had been reduced. Consequently, at logged sites few cross-pollinations were effected and fruit set of *S. siamensis* was considerably lower than at nearby unlogged sites where distances between flowering conspecifics was smaller

Keywords: Dipterocarps, pollination, seed set, regeneration, human disturbance, logging.

(Figure 1). Reduced fruit set has long-term implications for the recovery of *S. siamensis* populations in disturbed areas, and local population genetic structure is likely to be affected as reduced outcrossing rates among trees in disturbed regions result in relatively inbred seed.

Logging can, however, have additional indirect impacts by affecting species or activities that mediate important ecological processes. Thus, extraction of *S. siamensis* also impacted the reproductive ecology of the self-incompatible butterfly- and bird-pollinated tree *D. obtusifolius*. As for *S. siamensis*, pollinator activity at *D. obtusifolius* and the trees' subsequent fruit production were recorded at three sites subject to differing extraction intensities. Disturbance by extraction of *S. siamensis* resulted in a marked increase in the cover of understorey flowering herbaceous plants and a decrease in butterfly–pollinator activity at *D. obtusifolius* trees. The abundance of butterfly pollinators, however, remained similar across all sites. Pollination by birds increased at the disturbed site, but was not sufficient to offset the overall decline in pollinator activity. Decreased pollinator activity in the canopies of *D. obtusifolius* trees at the disturbed site was likely due to a shift in the relative abundance of floral nectar resources from the canopy to the understorey. Thus extraction of *S. siamensis* indirectly affected the pollination of *D. obtusifolius* by causing changes in the foraging behaviour of butterfly pollinators. Despite reduced pollination, fruit set of *D. obtusifolius* was similar at all sites. Pollination by birds or moths at the disturbed site likely compensates for reduced butterfly pollination.

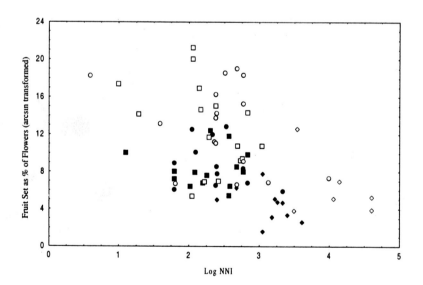

Figure 1. Fruit set as a function of nearest neighbour index. 1996 values are represented by solid symbols, 1997 values by open; circles represent undisturbed site; squares, moderately-disturbed; and diamonds, disturbed site. ANOVA within cells regression: $F = 7.837$, d.f. $=(1, 67)$, $p = 0.0067$. $y = 17.19–2.91x$. Figure from Ghazoul, Liston & Boyle (1998).

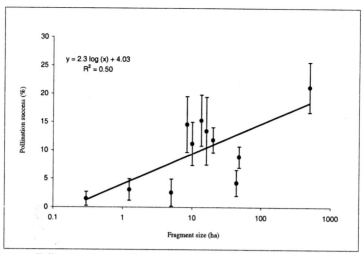

Figure 2a. Pollination success of *A. excelsum* trees in forest fragments of varying sizes. Pollination success quantified as the proportion of flowers in which pollen tubes were detected.

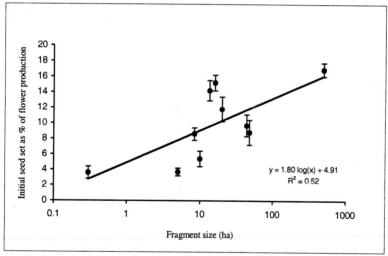

Figure 2b. Initial seed set (two weeks after flowering) of *A. excelsum* trees occurring in variably-sized fragments.

The research in Thailand shows that logging has both direct and indirect effects on the reproductive ecology of forest trees and that these effects can act through changes in the behaviour of pollinators rather than by affecting their numerical abundance. Before generalisations from such a limited research base can be made, it is necessary to seek further information from sites and species far removed from those in Thailand. Research was therefore conducted in Costa Rica where forest fragmentation has restricted the once widespread tree *Anacardium excelsum* to forest patches located in an agriculturally-dominated landscape. As with *S. siamensis*, the abundance of the pollinators, also *Trigona* bees, in the canopies of *A. excelsum* was largely unaffected by fragment size. Nevertheless, pollination success (Figure 2a) and seed production (Figures 2b) of *A. excelsum* were positively correlated with fragment size. We propose that small bees rarely move between forest fragments and gene exchange through pollination occurs predominantly among trees within fragments and, together with likely low genetic variability in small fragments, that this contributes to the observed reduced fertilisation and seed set of *A. excelsum*.

Thus increased tree isolation through selective logging or habitat fragmentation by forest clearance can result in reduced seed set due to changes in the foraging patterns of poorly mobile pollinators. Even if population sizes of the pollinators are maintained following environmental perturbation, this study shows that disturbance may disrupt pollination processes through changes in pollinator foraging behaviour. Forest managers aiming to minimise impacts of timber extraction should be aware of the ecological processes which might be disrupted as a result of their activities. However, these impacts may be very difficult to predict when they are indirectly related to logging. It is likely that among species compensatory mechanisms provide differential resistance to disturbance, and in the forestry and conservation context efforts should be made to determine which tree species are most susceptible to indirect effects of anthropogenic disturbance. More attention needs to be focussed on changes in the behaviour of species involved in key ecological interactions following disturbance events in tropical forests.

Tropical Ecosystems: Structure, Diversity and Human Welfare.
Proceedings of the International Conference on Tropical Ecosystems
K. N. Ganeshaiah, R. Uma Shaanker and K. S. Bawa (eds)
Published by Oxford–IBH, New Delhi. 2001. pp. 101–103.

Impacts of anthropogenic pressures on population dynamics, demography, and sustainable use of forest species in the Western Ghats, India

Aditi Sinha*,‡ and Kamaljit S. Bawa*,†

*Department of Biology, University of Massachusetts, Harbor Campus,
Boston, MA 02125, USA
*,†Ashoka Trust for Research in Ecology and the Environment, No. 659, 5th A
Main, Hebbal, Bangalore 560 024, India
‡e-mail: aditi.sinha@umb.edu

In many parts of the tropics, humans are an integral part of forest ecosystems utilizing the forest for a wide variety of purposes such as extractions of various nontimber forest products (NTFPs). The use and extraction of NTFPs have had historical and current importance in local economies and cultures of indigenous people. Today, with increased market demand, NTFPs have come to play an important role in large-scale commercial income generation and employment in many parts of the world (Nepstad and Schwartzman, 1992; Panayotou and Ashton, 1992; Mendelsohn and Balick, 1995; Runk, 1998). However, commercial extractions of NTFPs can increase anthropogenic pressures on the forest structure and dynamics at various levels. For instance, extractions and other human-induced disturbances such as fire can affect the population dynamics and demography of the harvested species, and ultimately the sustainability of extractions.

Keywords: Nontimber forest products, anthropogenic pressures, harvesting, sustainability, India.

The forest of Biligiri Rangan Hills (BR Hills) in the Western Ghats, south India has had a long history of human influences. An indigenous people, the Soligas, live in this forest and have been extracting NTFPs for centuries. Today, NTFPs play a critical role not just in their subsistence but also in their economy. Apart from extractions of NTFPs, frequent fires and spread of invasive plant species are some of the other predominant anthropogenic pressures on these forests. Over the years, the intensity of use of the forest, and the impacts thereof, has changed. Harvesting of some NTFPs has gone from a subsistence level activity to a commercial one and fires have become more frequent and intense. Yet, there have been very few studies to understand the influences humans have on the forest of BR Hills.

Sustainability of nontimber forest product extractions can be affected not just by harvesting patterns and techniques but also by other human influences such as fire. We assessed the sustainability of fruit extractions by examining the effects of fire regimes and harvesting on the population structure and demography of two NTFP species, *Phyllanthus emblica* and *P. indofischeri*, in the forest of the Biligiri Ranga Swamy Temple Wildlife Sanctuary, BR Hills, south India. Fruits of *P. emblica and P. indofischeri* are harvested by the Soligas and sold to traders through tribal cooperatives. Data on harvesting techniques, recruitment, growth, mortality and fecundity for the two *Phyllanthus* spp. were collected from permanent plots over a period of three years. Matrix population models were used to analyze the demographic data. We compared various demographic indices in space and across plots at different stages of recovery from fire.

The population structures of both *Phyllanthus* spp. were dominated by the sapling classes. We found spatial heterogeneity in the population growth rates with most populations either declining or fluctuating very close to the theoretical stable population growth rate. Certain local populations are threatened as seen from the negative population growth rates, and therefore are clearly in need of protection.

Population growth rates at different stages of recovery from fire ranged from 0.9204 to 1.2190. Frequent fires have negative effects on the population growth rates as well as on the levels of fruit production. Fires occurring at frequencies of 2–3 years significantly lower the population growth rate. Frequent fires also have a negative effect on fruit production and increase the time to maturity by altering growth and survival rates. These results suggest that a longer frequency of fire, i.e. more than 4 years, may be more conducive to the persistence of the two species. Both timing and intensity of fires are also critical in determining fruit production in the following fruiting season.

Removal of fruits from the forest has a smaller effect on population growth rates of *Phyllanthus* spp. than the existing fire regimes. However, the current fruit-harvesting techniques such as lopping of the entire canopy and cutting of trees used by the Soligas have negative impacts on *Phyllanthus* trees. Lopping of major branches of the tree reduces fruit production in the following year (fruit production (log) = 3.20–0.86 (harvesting techniques), $r^2 = 0.27$, $p = 0.03$, $N = 18$). Survival of reproductive adults is an important parameter that determines the rate at which populations grow. Thus, any factor, such as adverse harvesting techniques, that lowers the survival probability of adults will decrease the rates at which the populations grow, thereby compromising the sustainability of fruit extraction of *Phyllanthus* spp.

Our findings are being integrated with other biological and socio-economic research from the wildlife sanctuary to develop guidelines for sustainable use and management of nontimber forest products. Our results also demonstrate how fire regimes can affect sustainability of extraction of nontimber forest products and the need to incorporate various human-generated-physical regimes in assessing sustainability of NTFP extractions.

Acknowledgements

This research was funded by a fellowship awarded to A. Sinha by the Wildlife Conservation Society, New York, and in part by grants from the Ford Foundation, the MacArthur Foundation, International Plant Genetic Resources Institute, and the Biodiversity Conservation Network of the World Wildlife Fund awarded to K. S. Bawa. We are grateful to H. Sudarshan of Vivekananda Girijana Kalayana Kendra (VGKK) for encouraging us to undertake this work and to VGKK, the Karnataka Forest Department and Ashoka Trust for Research in Ecology and the Environment (ATREE), Bangalore, India for logistical support in BR Hills.

References

Mendelsohn, R. & Balick, M. The value of undiscovered pharmaceuticals in tropical forests. *Econ. Bot.* **49**, 223–228 (1995).

Nepstad, D. C. & Schwartzman. S. Non-timber product extraction from tropical forests: evaluation of a conservation and development strategy. in *Non-timber Products from Tropical Forests: Evaluation of a Conservation and Development Strategy* (eds Nepstad, D. C. & Schwartzman, S.) vii–xii (The New York Botanical Garden, Bronx, USA, 1992).

Panayotou, T. & Ashton. P. S. *Not by Timber Alone: Economics and Ecology for Sustaining Tropical Forests*, pp. 70–109 (Island Press, Washington DC, 1992).

Runk, J. V. Productivity and sustainability of a vegetable ivory palm (*Phytelephas aequatorialis*, Arecaceae) under three management regimes in Northwestern Ecuador. *Econ. Bot.* **52**, 168–182 (1998).

Tropical Ecosystems: Structure, Diversity and Human Welfare.
Proceedings of the International Conference on Tropical Ecosystems
K. N. Ganeshaiah, R. Uma Shaanker and K. S. Bawa (eds)
Published by Oxford–IBH, New Delhi. 2001. pp. 104–109.

Assessment of genetic diversity of *Scaphium macropodum* in logged-over forests using isozyme and RAPD markers

R. Wickneswari[†] and Chai-Ting Lee***

*School of Environmental and Natural Resource Sciences, Faculty of Science and
Technology, Universiti Kebangsaan Malaysia, 43600 Bangi, Selangor, Malaysia
**Forest Plantation Division, Forest Research Institute Malaysia, Kepong, 52109
Kuala Lumpur, Malaysia
[†]e-mail: wicki@pkrisc.cc.ukm.my

Genetic diversity is the basis of adaptive flexibility in populations and the ultimate evolutionary potential (Given, 1994; Templeton, 1995; Young *et al.*, 1996). Loss of genetic diversity immediately after harvesting has been reported in *Pinus strobus* (Buchert *et al.*, 1997) and a few tropical rainforest species (Changtragoon, 1997; Wickneswari *et al.*, 2000). Inbreeding as a result of logging has been observed in *Shorea megistophylla* (Murawski *et al.*, 1994). Liengsiri *et al.* (1998) reported a decrease in outcrossing rates of *Pterocarpus macrocarpus* following reduction in population density due to disturbance. Similarly, Lee (2000) showed a decrease in outcrossing rates of *Dryobalanops aromatica* in logged-over forests in Peninsular Malaysia. However, breeding system of *D. aromatica* in Brunei (Kitamura *et al.*, 1994) and *Dipterocarpus obtusifolius* (Chaisurisri *et al.*, 1997) was not affected by logging.

Keywords: Genetic diversity, *Scaphium macropodum,* selective logging, molecular markers, tree density.

Scaphium macropodum (Miq.) Beumee ex Heyne (Sterculiaceae) is locally known as Kembang Semangkok Jantong and is found all over Malaysia (except Perlis), Cambodia, Peninsular Siam, Sumatra and Borneo (Kochummen, 1973); mainly in the lowland forests but occasionally on ridges up to 1200 m and also in swampy areas. It is a light hardwood (Menon, 1993) timber species with high silica content (Kochummen, 1973), suitable for veneer and plywood. This paper reviews the findings of two studies on the impact of logging on genetic diversity of this tropical timber species using isozyme analysis (Wickneswari *et al.*, 1997a,b; Wickneswari *et al.*, 2000) and RAPD analysis (Lee *et al.*, 2001) and a subsequent study on regenerants of *S. macropodum* after logging.

Two study sites were chosen, adopting two different approaches. The first was to look at the immediate effect of logging on the genetic diversity level of *S. macropodum* by studying the same population before and after logging in Compartment 48, Serting Tambahan Forest Reserve. The second approach was to compare the gene pools of three adjacent sub-populations (RS1, RS2 and US) in Pasoh Forest Rerserve with different logging histories, based on the assumption that they were genetically identical before logging. Details of these two approaches including study sites, estimation of disturbance levels, sampling methods, and isozyme and RAPD analysis are given in Wickneswari *et al.* (2000) and Lee *et al.* (2001). Besides adult trees, 79 regenerants from 3 diameter classes, viz. seedlings (33 samples), saplings (31 samples) and poles (15 samples) were sampled from the study site for the first approach for isozyme analysis.

Estimates of genetic diversity measures for *S. macropodum* in Serting Tambahan F.R. and Pasoh F.R. are summarised in Tables 1 and 2, respectively. *t*-tests did not reveal any significant difference in the genetic diversity parameters before and after logging. Mean expected heterozygosity, H_e based on isozyme data for Serting Tambahan F.R. after logging was slightly lower than before logging (6.0%) for the adults. Similarly, mean Shannon's diversity index, H for Serting Tambahan F.R. after logging was slightly lower (9%). However, slight increments in the effective number of alleles, A_e (2.3%) and H_e (7.1%) based on RAPD data were observed after logging, probably due to random retention of more heterogeneous adult individuals.

RAPD analysis revealed a significant reduction ($p \leq 0.05$) for all genetic diversity parameters surveyed for RS1 in Pasoh F.R. Percentage of polymorphic loci decreased by 31.0, while A_e dropped by 7.8%. H_e was also reduced by 27.6%. Meanwhile, no significant change in genetic diversity was detected for RS2 except in A, with 5.3% reduction. There was a significant decrease ($p \leq 0.05$) in H of *S. macropodum* in RS1 (0.998) from Pasoh F.R. compared to US (1.457), which is equivalent to 31.5% reduction. In contrast, RS2 showed a minor reduction of only 3.9%. On the contrary, isozyme analysis revealed an increase in A_e and H_e for both RS1 and RS2 compared to US.

Table 1. Estimates of genetic diversity measures for *S. macropodum* in Serting Tambahan F.R.

Sampling time	Molecular marker	Age cohort*	P	A	A_e	H	**H**	F_{is}
Before logging	Isozyme	Adults (33)	100	3.3	1.71	0.415	–	0.132
	RAPD	Adults (33)	70.2	1.7	1.33	0.198	1.075	–
After logging	Isozyme	Adults (22)	100	3.0	1.64	0.390	–	–0.036
		Poles (15)	100	2.5	1.5	0.318	–	0.382
		Saplings (31)	75	3.0	1.5	0.337	–	0.076
		Seedlings (33)	100	3.0	1.7	0.396	–	0.235
	RAPD	Adults (22)	68.1	1.7	1.36	0.212	0.978	–

P, percentage of polymorphic loci; A, mean number of alleles per locus; A_e, effective number of alleles per locus; H, Nei's (1973) gene diversity; *H*, mean Shannon's diversity index per primer; F_{is}, Fixation index, *Figures in brackets indicate sample size.

Table 2. Estimates of genetic diversity measures for *S. macropodum* in Pasoh F.R.

Compartment	Molecular marker*	P	A	A_e	H	H	F_{is}
US	Isozyme (30)	100	3.5	1.6	0.381	–	–0.037
	RAPD (30)	79.3	1.8	1.4	0.225	1.457	–0.041
RS1	Isozyme (11)	100	3.0	2.2	0.541	–	0.155
	RAPD (11)	54.7	1.5*	1.3*	0.163*	0.998*	0.073
RS2	Isozyme (18)	75	3.0	2.0	0.498	–	0.175
	RAPD (18)	69.8	1.7*	1.4	0.218	1.400	0.143

P, percentage of polymorphic loci; A, mean number of alleles per locus; Ae, effective number of alleles per locus; H, Nei's (1973) gene diversity; *H*, mean Shannon's diversity index per primer; F_{is}, Fixation index; *Figures in brackets indicate sample size. (*) indicates significant difference at $p \leq 0.05$.

The reduction in basal area (trees > 5 cm dbh) of Compartment 48 from Serting Tambahan F.R. after logging was about 57.5%. However, logging did not cause adverse change in genetic diversity of *S. macropodum*. This may probably be attributed to its high abundance in the FMU. *Scaphium macropodum* is one of the most abundant species in Serting Tambahan F.R., with the estimated adult tree (≥ 20 cm dbh) density of 10 trees ha[-1]. Post-harvest inventory revealed that it was reduced to 8 trees ha[-1] after logging, still considerably high. This result concurs with the study using isozyme markers (Wickneswari *et al.*, 1997a). Wickneswari *et al.* (1997a) also demonstrated a higher loss of genetic diversity (23.4% reduction in H_e and 25.0% in A) in *Shorea leprosula*, an important timber species, which is in contrast, of low abundance in the forest management unit (FMU).

Compared to compartment 48 of Serting Tambahan F.R., the mature tree density of *S. macropodum* in US of Pasoh F.R. was only about 2 trees ha[-1]. As the three sub-populations of *S. macropodum* in Pasoh F.R. were from a continuous population, it is assumed that they were genetically identical before logging. Hence, the significant reductions ($p \leq 0.05$) observed across all the genetic diversity parameters for the RS1 (logged in 1951) indicate a substantial genetic diversity loss due to logging. In fact, the mean genetic distance among the three sub-populations of *S. macropodum* in Pasoh F.R. was the highest (0.062) among six species investigated by Wickneswari *et al.* (2000) indicating most probably change in allelic frequencies of sub-populations due to logging. On the contrary, *S. macropodum* from the RS2 which was relatively more disturbed in terms of mean basal area reduction, suffered no genetic erosion. The result however, corroborates with the sample sizes of respective stands. RS1 had the least number of samples available. This suggests that the genetic erosion may have been due to severe population decline after selective logging, resulting in increased genetic drift and reduced outcrossing rates.

Scaphium macropodum has unisexual flowers, i.e. male and female flowers on separate inflorescences on the same tree (Ashton, 1988). Kochummen (1973) reported bees, flies, beetles and butterflies as its pollinators. Isozymes analysis of progeny arrays of three mature *S. macropodum* trees from Serting Tambahan F.R. revealed a substantially high mean t_m value, i.e. 0.92 (unpublished data). Owing to the presumably smaller number of individuals left in RS1 after logging, selfing might have been enhanced and caused a further loss of genetic diversity in that FMU. Further information on the mating system of *S. macropodum*, the behaviour of pollinator, as well as how these factors are affected by logging activities is essential to corroborate this suggestion.

Isozyme analysis using four polymorphic loci (*Pgm-1, Ugp-1, Gpi-1 and Sdh-1*) did not reveal any significant change in the genetic diversity measures of *S. macropodum* in both the regenerated stands of Pasoh F.R. (Wickneswari *et al.*, 1997b). In contrast, significant loss of genetic diversity was detected in RS1 through RAPD analysis. Therefore, this result demonstrates that RAPD markers are more sensitive in detecting change in genetic diversity compared to isozymes markers, in particular when the number of loci surveyed is limited. This is in agreement with Bucci *et al.* (1997) who reported that the sensitivity of RAPD markers in population differentiation was about 8 times greater than that of the allozymes. The reason is RAPDs are primarily from non-coding, repetitive DNA sequences with random distribution (Williams *et al.*, 1990; Plomoin *et al.*, 1995). Besides, RAPDs have a larger coverage of the genome compared to the isozyme markers which are restricted to a limited number of relatively more conserved protein coding loci.

Assessment of genetic erosion should be carried out using more than one molecular marker analysis if one of them shows no significant difference. Results of these studies show that species' vulnerability to the threat of genetic erosion posed by selective logging is highly correlated with its abundance in a particular FMU. Timber species of high abundance and high heterogeneity in a FMU are likely to have higher capacity to buffer genetic erosion compared to those of low incidence. Besides, tree density for the species can be a useful indicator in reflecting the risk of genetic erosion rather than the overall disturbance level based on reduction in basal area of all trees.

More studies involving different species in different forest types are essential to gain better understanding of the effect of a single selective logging event on the genetic resources of tropical forest. If significant genetic erosion is detected, remedial actions such as enrichment planting of affected species could be carried out to mitigate the negative impacts and to avoid risk of irreversible genetic erosion. It is important that protective measures be taken to conserve the tropical genetic resources, despite the fact that in Malaysia, already, 23.5% of the total permanent forest estates (14.20 million ha) are protected and set aside for conservation, water catchment, game and recreation purposes (MPIM, 1997).

Acknowledgements

This work was supported by Center for International Forestry Research (CIFOR) and the International Plant Genetic Resources Institute (IPGRI) in collaboration with the Forest Research Institute Malaysia (FRIM) and Universiti Kebangsaan Malaysia.

References

Ashton, P. S. *Manual of the Non-Dipterocarp Trees of Sarawak.* vol. II (Dewan Bahasa dan Pustaka, Kuala Lumpur, 1988).

Bucci, G., Vendramin, G. G., Leili, L. & Vicario, F. Assessing the genetic divergence of *Pinus leucodermis* Ant. Endangered populations: Use of molecular markers for conservation purposes. *Theor. Appl. Genet.* **95**, 1138–1146 (1997).

Buchert, G. P., Rajora, O. P., Hood, J. V. & Dancik, B. P. Effects of harvesting on genetic diversity in old-growth eastern white pine in Ontario, Canada. *Conserv. Biol.* **11**, 747–758 (1997).

Chaisurisri, K., Wungplong, P., Liewlaksaneeyanawin, C. & Boyle, T. J. B. Impacts of disturbance on genetic diversity of some forest species in Thailand. Paper presented at Wrap-up Workshop of the CIFOR-IPGRI Impact of Disturbance Project, Bangalore, India, 18–22 August 1997.

Changtragoon, S. Impact of disturbance on genetic diversity of *Cycas siamensis* in Thailand. Paper presented at Wrap-up Workshop of the CIFOR-IPGRI Impact of Disturbance Project, Bangalore, India, 18–22 August 1997.

Given, D. R. *Principles and Practice of Plant Conservation* (Timber Press, Portland, Oregon, 1994) pp. 292.

Kitamura, K., Mohamad Yusof, A. R., Ochiai, O. & Yoshimaru, H. Estimation of outcrossing rate on *Dryobalanops aromatica* Gaertn. F. in primary and secondary forests in Brunei, Borneo, Southeast Asia. *Plant Species Biol.* **9**, 37–41 (1994).

Kochummen, K. M. Sterculiaceae (from the genus *Sterculia*). *Tree Flora of Malaya* **2**, 353–372 (1973).

Lee, S. L. Mating system parameters of *Dryobalanops aromatica* Gaertn. F. (Dipterocarpaceae) in three different forest types and a seed orchard. *Heredity* **85**, 318–345 (2000).

Lee, C. T., Wickneswari, R., Mahani, M. C. & Zakri, A. H. Effect of selective logging on the genetic diversity *of Scaphium macropodum. Biol. Conserv.* Accepted (2001).

Liengsiri, C., Boyle, T. J. B. & Yeh, F. C. Mating system *in Pterocarpus macrocarpus* Kurz in Thailand. *J. Heredity* **89**, 216–221 (1998).

Menon, P. K. B. Structure and identification of Malayan woods. Malayan Forest Records No. 25 (1993).

MPIM. Statistics on commodities 1997. Ministry of Primary Industries Malaysia (1997).

Murawski, D. A., Gunatilleke, I. A. U. N. & Bawa, K. S. The effects of selective logging on inbreeding in *Shorea megistophylla* (Dipterocarpaceae) from Sri Lanka. *Conserv. Biol.* **8**, 997–1002 (1994).

Plomoin, C., Bahrman, N., Durel, C. E. & O'Malley, D. M. Genomic mapping in *Pinus pinaster* (maritime pine) using RAPD and protein markers. *Heredity* **74**, 661–668 (1995).

Templeton, A. R. Biodiversity at the molecular genetic level: experiences from disparate macroorganisms. in *Biodiversity: Measurement and Estimation* (ed. Hawksworth, D. L.) 59–64 (Chapman & Hall, London, 1995).

Wickneswari, R., Lee, C. T., Norwati, M. & Boyle, T. J. B. Immediate effects of logging on the genetic diversity of five tropical rainforest species in a ridge forest in Peninsular Malaysia. Paper presented at Wrap-up Workshop of the CIFOR-IPGRI Impact of Disturbance Project, Bangalore, India, 18–22 August 1997a.

Wickneswari, R., Lee, C. T., Norwati, M. & Boyle, T. J. B. Effects of logging on the genetic diversity of six tropical rainforest species in a regenerated mixed dipterocarp lowland forest in Peninsular Malaysia. Paper presented at Wrap-up Workshop of the CIFOR-IPGRI Impact of Disturbance Project, Bangalore, India, 18–22 August 1997b.

Wickneswari, R., Lee, C. T., Norwati, M. & Boyle, T. J. B. Impact of logging on genetic diversity in humid tropical forests. in *Forest Genetics and Sustainability* (ed. Matyas) vol. 63, 171–181 (2000).

Williams, J. G. K., Kubelik, A. R., Livak, K. J. Rafalski, J. A. & Tingey, S. V. DNA polymorphisms amplified by arbitrary primers are useful as genetic markers. *Nucleic Acids Res.* **18**, 6531–6535 (1990).

Young, A., Boyle, T. & Brown, T. The population genetic consequences of habitat fragmentation for plants. *Tree* **11**, 413–418 (1996).

Tropical Ecosystems: Structure, Diversity and Human Welfare.
Proceedings of the International Conference on Tropical Ecosystems
K. N. Ganeshaiah, R. Uma Shaanker and K. S. Bawa (eds)
Published by Oxford–IBH, New Delhi. 2001. pp. 110–113.

Changes in the composition, structure and diversity of small mammals (family Soricidae, Tupaiidae, Sciuridae, Muridae) after selective logging

Nor Azman Hussein and Abdul Rahim Nik*

Forest Research Institute Malaysia (FRIM), Kepong, 52109 Kuala Lumpur,
Malaysia
*e-mail: norazman@nt1.frim.gov.my

Studies on the effects of habitat change especially logging on mammal populations have long been carried out in Malaysia. Some were published as early as the sixties, for example Harrison (1968; 1969) and Stevan (1968); the seventies, Lord Medway and Wells (1971). Most of these studies were conducted over small spatial scales and limited to a short duration. It is feared that the data obtained by these studies could have been affected by the intrinsic cyclic activity patterns of the animals (Smith, 1980; Begon, 1979; Caughley, 1978; Nor Azman *et al.*, 1993).

The present research project aims to study the effect of logging on the diversity of small mammals in the Hill Forest of the Ulu Muda Forest Reserve in the northwest of Peninsular Malaysia, on a large spatial and temporal scale. The main objective of this research is to monitor the population of small mammals in the study area and to relate changes of the small mammal populations to disturbances caused by the conventional selective logging method and the newly proposed modified selective logging system. It is also intended to determine whether small mammals are

Keywords: Small mammals, logging, indicator group, biodiversity, sustainable forest management.

a good indicator group for logging disturbance. This paper addresses the preliminary quantitative assessment of the immediate effects of conventional selective logging on the composition, structure and diversity of small mammal population.

The study site is located in the Ulu Muda Forest Reserve ($5°55'N$, $100°55'E$), which is within the northwest region of the peninsular Malaysia. The total area of the study site is 1987.03 ha which was divided into 3 research compartments. Topography for each of the research compartment ranges from flatland to steep slope and ridges, with altitudes varying from approximately 400 m to about 1,000 m. The vegetation can be generally described as Hill Dipterocarp Forest of northern region of the Peninsular Malaysia. Each of the three research compartments was given different treatments. Plot C25b was a control plot where no logging was carried out, C26/C27 were subjected to a new modified selective logging system designed for the study and plot C28/C29 was logged according to conventional harvesting practices in accordance to the hill forest harvesting guidelines of Peninsular Malaysia. Small mammal trappings were carried out every month for each compartment prior to, during and after logging. Animal caught were identified, measured, marked and released at the same point (Begon, 1979; Caughley, 1978). Vegetation surveys were also carried out for each compartment and the damages to vegetation composition and structure were assessed after logging.

A total of 14 species were recorded throughout the study period. They comprise two species of family Soricidae (*Echinosorex gymnurus, Hylomys suillus*), 1 species of Tupaiidae *(Tupia glis)*, 7 species of Muridae (*Leopoldamys sabanus, Maxomys rajah, Maxomys surifer, Maxomys whiteheadi, Niviventer cremoriventer, Rattus tiomanicus, Sundamys muelleri*) and 4 species of Sciuridae (*Lariscus insignis, Rhinosciurus laticaudatus, Sundasciurus lowii, Sundasciurus tenuis*). Before logging the distribution and composition of these species differed significantly within and between plots, indicating a strong association of ecological features to small mammal diversity under natural habitat. Monthly total catch confirmed the cyclic pattern common to these group of mammals. All plots seemed to follow the same general trend (Figure 1). There was no significant difference in the trend between the compartments after logging. Thus our results indicate that selective logging does not drastically reduce carrying capacity of forest area immediately after logging.

However, diversity and composition structure of small mammals changed drastically immediately after logging. Figure 2 shows the differences in abundance curves for the period before and after logging in compartments C28/C29. The most affected are the dominant species while others seem to

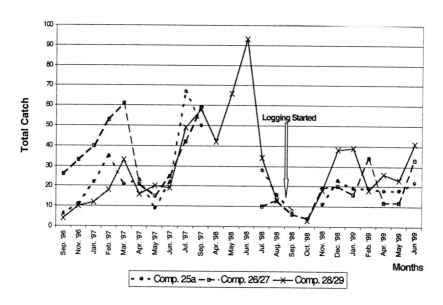

Figure 1. Monthly fluctuation in total catch for the three plots before and after logging.

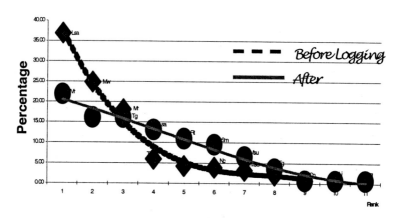

Figure 2. Abundance curves for the period before and after logging in compartments C28/29.

flourish well. Invasion of a shrub species *Rattus tiomanicus* was also observed in the logged plots. It is evident that the logged-over area is going through a transition phase that will determine the outcome of the future population in the area. Factors that determine these small mammals responses to logging disturbances are being investigated. Long-term assessment of these changes and responses is necessary to understand the ecology of forest disturbances by selective logging so that better and comprehensive management guidelines can be proposed for sustainable forest management programs in Malaysia.

Acknowledgements

I wish to record my appreciation to Dato' Dr. Abdul Razak Mohd Ali (Director General of FRIM), Director of Environmental Science, Dr. Lim Boo Liat and Prof. Mohd. Nordin Hj. Hassan for their support to make this study possible. My special thanks go to all the staff of the Zoological unit for their help and cooperation.

References

Begon, M. *Investigating Animal Abundance: Capture-recapture for Biologist* (Edward Arnold, London, 1979).

Briese, L. A. & Smith, M. H. Seasonal abundance and movement of nine species of small mammals. *J. Mammal.* **55**, 615–629 (1975).

Caughley, G. *Analysis of Vertebrate Population* (John Wiley & Sons, Chichester, 1978).

Harrison, J. L. The effect of forest clearance on small mammal. in *Conservation in Tropical South East Asia.* IUCN publication new series. No. 10, 153–156 (1968).

Harrison, J. L. The abundance and population density of mammals in Malayan lowland forest. *Malay. Nat. J.* **22**, 174–178 (1969).

Lord Medway & Wells, D. R. Diversity and density of birds and mammals at Kuala lompat, Pahang. *Malay. Nat. J.* **24**, 238–247 (1971).

Nor Azman Hussein, Ratnam, L. C. & Lim, B. L. Some ecological aspects of small mammals in an old growth forest plantation. *Proceeding of the Conferences on Forestry and Forest Product Research, Forest Research Institute Malaysia.* (Kepong, Kuala Lumpur, 1993).

Smith, R. L. *Ecology and Field Biology* (Harper and Row, New York, 1980).

Steven, W. E. Habitat requirement of Malayan mammals. *Malay. Nat. J.* **22**, 3–9 (1968).

Tropical Ecosystems: Structure, Diversity and Human Welfare.
Proceedings of the International Conference on Tropical Ecosystems
K. N. Ganeshaiah, R. Uma Shaanker and K. S. Bawa (eds)
Published by Oxford–IBH, New Delhi. 2001. pp. 114–115.

Tree diversity and population structure in undisturbed and human-impacted tropical wet evergreen forests of Arunachal Pradesh, Northeast India

Putul Bhuyan*, M. L. Khan*[§] and R. S. Tripathi[†]

*Department of Forestry, North Eastern Regional Institute of Science &
Technology, Nirjuli 791 109, India
[†]Department of Botany, North Eastern Hill University, Shillong 793 022, India
[§]e-mail: mlk@nerist.ernet.in

Species richness, tree density, basal area, population structure and distribution pattern of tree species were investigated in four stands of tropical wet evergreen forest in Arunachal Pradesh, Northeast India. The forest stands were selected based on the disturbance index: (i) undisturbed stand (0% disturbance index), (ii) mildly disturbed (20% disturbance index), (iii) moderately disturbed (40% disturbance index), and (iv) highly disturbed stand (70% disturbance index). These stands are 1 to 5 km apart in the same wet evergreen forest belt.

Species richness varied along the disturbance gradient in different stands. The mildly disturbed stand showed the highest species richness (54) of 51 genera. It was lowest (16 of 16 genera) in highly disturbed stand. In undisturbed stand, 47 species of 42 genera were recorded while in moderately disturbed stand 42 species of 36 genera were found. Tree species diversity ranged from 0.7 to 2.02 in all the stands. The highest species

Keywords: Tree diversity, tropical wet evergreen forest, Arunachal Pradesh, human-impats, population structure.

diversity was recorded in undisturbed stand and the lowest in highly disturbed stand. The values for concentration of dominance were similar in undisturbed stand, mildly disturbed stand and moderately disturbed stand whereas it was lowest in highly disturbed stand. The similarity index value was maximum in undisturbed stand and minimum in highly disturbed stand. The forest stand density and basal area were highest in undisturbed and lowest in highly disturbed stand.

The stands differed with respect to the tree species composition at the family and generic level. Dominance calculated as importance value index (IVI) of different species varied greatly across the stands. *Shorea assamica, Dipterocarpus macrocarpus, Mesua ferrea, Castanopsis indica, Terminalia chebula, Vatica lancefolia* were dominant in all the stands except in the highly disturbed stand. The canopy layer of the strata was occupied by *S. assamica, D. macrocarpusm* and *T. chebula* in all the stands.

Stand densities and species richness consistently decreased with increasing girth class of tree species from 20 to more than 200 cm girth. The highest species density and species richness were represented in the medium girth class (51–110) in all the stands. In undisturbed stand, the highest density was found in 111–140 cm girth class, while in the mildly disturbed stand 51–80 cm girth range recorded the highest density. In the highly disturbed stand no tree was recorded in more than 140 cm girth range. *D. macrocarpus, S. assamica, M. ferrea, C. indica, Canarium resiniferum* and *T. chebula* are the dominant species and are uniformly distributed in all the forest stands. The distributions of these species were greatest in 60–100 cm girth class in all the stands.

Out of the 47 species in undisturbed stand, only 26 were found to be regenerating. Thirteen species showed good regeneration (presence of saplings + seedlings) 8 species had fair regeneration and 5 species showed poor regeneration. No regeneration was recorded for other species. In mildly disturbed stand, out of 54 species, 37 were found regenerating of which 15 species had good regeneration, 8 showed fair regeneration and 14 had poor regeneration. Out of 42 species, 22 were found regenerating in moderately disturbed stand and good regeneration was recorded in 9 species, fair in 7 species and poor in 6 species. No regeneration was recorded in the highly disturbed stand.

The highest shrub density was recorded in undisturbed stand but the shrub species richness was maximum in the mildly disturbed stand. In all the stands *Blastus cochichinensis* and *Litsea salicifilia* dominated over other species. No shrub was recorded in highly disturbed stand. The undisturbed stand recorded the highest herb and vine density while the lowest density was recorded in moderately disturbed stand. Herb and vine species such as *Cyperous rotundus, Forestia glabrata* and *Pteris quadrissmita* were common to all the stands.

Tropical Ecosystems: Structure, Diversity and Human Welfare.
Proceedings of the International Conference on Tropical Ecosystems
K. N. Ganeshaiah, R. Uma Shaanker and K. S. Bawa (eds)
Published by Oxford–IBH, New Delhi. 2001. pp. 116–120.

Temporal vegetation changes in selectively logged and unlogged stands in Sinharaja Forest Reserve, Sri Lanka

M. A. B. N. Gunasekara*, P. M. S. Ashton**, I. A. U. N. Gunatilleke* and C. V. S. Gunatilleke*

*Botany Department, Faculty of Science, University of Peradeniya, Sri Lanka
**School of Forestry and Environmental Studies, Yale University, New Haven, USA
e-mail: nirosha17@hotmail.com

Conservation and management of the world's declining rainforests predominantly depend upon acquiring knowledge on community dynamics, achieved through long-term studies. Understanding temporal changes in species composition and vegetation structure is among the first steps towards developing silvicultural guidelines for sustainable use of these forests.

In the southwest lowlands of Sri Lanka, Sinharaja is the largest remaining rain forest where more than 60% of the tree species are endemic to the island (Gunatilleke and Gunatilleke, 1981). Parts of this forest were selectively logged from 1972 to 1977. During this time, much public protest against logging led to a ban on logging following which the area was declared a conservation forest.

Between 1977 and 1979, the tree vegetation in hundred permanent plots (each 0.25 ha) was enumerated in the undisturbed forest, to investigate the

---

Keywords: Sinharaja, floristic composition, density, basal area, selective logging.

phytosociology at different sites in Sinharaja (Gunatilleke and Gunatilleke, 1985). The sites sampled represented different parts of Sinharaja and its topographical gradient. They included Morapitiya (335–490 m) and Waturawa (520–610 m) on its western side, Sinhagala (a small hill ranging between 550 and 730 m) and Warukandeniya (520 and 610 m) in its central part and Deniyaya (670 and 915 m) on its east. After these plots were established and sampled, the Morapitiya site alone was selectively logged.

In 1999, the Morapitiya, Waturawa and Sinhagala plots were re-enumerated to understand changes in vegetation structure and floristic composition over a period of 20 years in two unlogged sites (Sinhagala – 5.5 ha, and Waturawa – 5 ha) and the selectively logged Morapitiya (4.5 ha) site. The study also compared the growth rates of different species, their recruitment and mortality, in the logged and unlogged stands.

At both censuses the highest number of species was recorded at Sinhagala (121–124) followed by Waturawa (117–119) and Morapitiya (116–114). Over the twenty years, species composition changed most at Waturawa followed by Sinhagala and Morapitiya. Among the species recorded in 1979, 11, 8 and 6 species were not found after twenty years while 13, 11 and 4 additional species were recorded in 1999 in Waturawa, Sinhagala and Morapitiya respectively. This change was predominantly restricted to the rare understorey species. The dominant families and species based on density and basal area common to the three sites and recorded at both census were as follows: In the canopy, Clusiaceae, Dipterocarpaceae, *Mesua nagassarium* and *Shorea worthingtoni*; in the sub-canopy, Myristicaceae and *Myristica dactyloides* and in the understorey, Clusiaceae, Annonaceae, *Garcinia hermonii*, and *Xylopia championii* were the dominants.

Overall density declined during this period by 73, 65 and 49 individuals per ha at Waturawa, Sinhagala and Morapitiya respectively. The number of individuals in the 30–89 cm girth classes of canopy species increased at Morapitiya but declined in Waturawa and Sinhagala. Girth class distribution of individuals in each site showed a typical reverse 'J' curve at both census (Figure 1). Basal area of canopy species as expected decreased much more at Morapitiya (-7.063 m^2 per ha) compared to Sinhagala (-0.249 m^2 per ha) and Waturawa ($+ 0.986$ m^2 per ha) (Figure 1).

The number of stems recruited during these two decades was highest at Morapitiya (29%) compared to that at Waturawa (13%) and Sinhagala (6%). In all the sites, more than 90 per cent of recruits were in the lowest size class, 30–59 cm gbh. Among recruits, the understorey and canopy species were high (Figure 2). Correspondingly, the proportion of dead and missing individuals followed the same order (31%, 22% and 19% respectively). Mortality was high in the lowest size classes of the understorey species at Sinhagala compared to that in the other two sites (Figure 2).

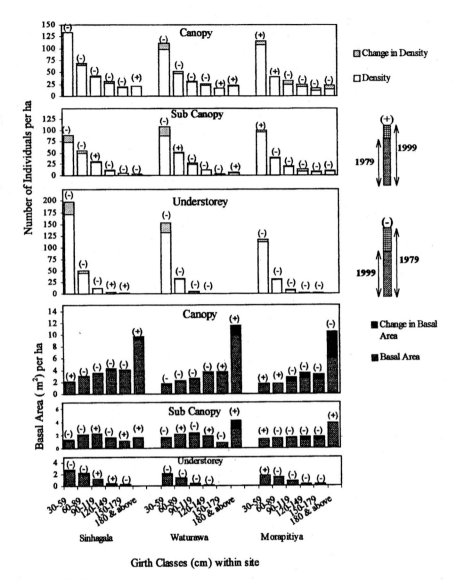

Figure 1. Girth class distribution of individuals >30 cm GBH and strata wise changes in density (per ha) and basal area (m² per ha), in each girth class at each site over twenty years. Positive and negative changes observed at the second census (1999) are indicated by plus (+) and minus (–) signs at the top of each bar respectively.

The structural changes observed during this period at Sinhagala and Waturawa may be attributed to natural gap dynamics rather than to anthropogenic factors. The higher mortality in the smallest size class 30–59 cm girth at Sinhagala may be related to the steeper topography at this site. Moreover, the higher elevational gradient may increase physical damage to

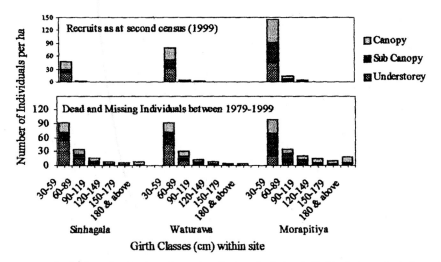

Figure 2. Girth class distribution of recruits, dead and missing individuals in each stratum at each site. Distribution of dead and missing individuals refers to their initial size classes as at first census.

individuals in the understorey due to the uprooting of large trees that fall on them. The changes in floristic composition at Waturawa are mostly due to the death of secondary and forest fringe species, while those at Morapitiya are largely due to logging. Generally, the number of individuals of rare species decreased at all sites.

The results indicate that the level of logging at Morapitiya has neither changed the species composition very much nor the vegetation structure of the woody species > 30 cm gbh. Any drastic changes that might have been occurred at the time of logging appear to have been repaired and the vegetation restored to its original state after 20 years. If the gaps formed in Morapitiya due to logging were considerably large, pioneer species would have invaded them initially. Subsequently, they may have completed their life cycle, giving way to more shade-tolerant species. Alternatively, if the gaps were not too large, instead of pioneers, through the release of advanced regeneration, rapid growth of rain forest species would have taken place. The second census data do not indicate which of the processes actually took place.

From the results, it is evident that most of the species in the forest are regenerating well at Morapitiya and the intensity of selective logging has not been detrimental to the plant diversity.

Acknowledgements

Assistance given by Messrs T. M. N. Jayatissa, T. M. Ratnayaka, S. Harischandran, W. A. Tennakoon, A. H. M. A. K. Tennakoon, M. A. Gunadasa during the recensus is gratefully acknowledged. We thank MacArthur Foundation, USA, and the National Science Foundation, Sri Lanka for financial assistance.

References

Gunatilleke, I. A. U. N. & Gunatilleke, C. V. S. The floristic composition of Sinharaja – A rain forest in Sri Lanka with special reference to endemics and dipterocarps. *Malay. For.* **44,** 386–96 (1981).

Gunatilleke, I. A. U. N. & Gunatilleke, C. V. S. Phytosociology of Sinharaja – a contribution to rain forest in Sri Lanka. *Biol. Conserv.* **31,** 21–40 (1985).

Tropical Ecosystems: Structure, Diversity and Human Welfare.
Proceedings of the International Conference on Tropical Ecosystems
K. N. Ganeshaiah, R. Uma Shaanker and K. S. Bawa (eds)
Published by Oxford–IBH, New Delhi. 2001. pp. 121–124.

Protected areas as refugias for genetic resource: Are sandal genetic resources safe in our sanctuaries?

M. Nageswara Rao[†]*, M. Anuradha[†],
K. N. Ganeshaiah[†,‡] and R. Uma Shaanker[§,‡,#]

[†]Department of Genetics and Plant Breeding and [§]Crop Physiology, University of
Agricultural Sciences, GKVK Campus, Bangalore 560 065, India
[‡]Jawaharlal Nehru Centre for Advanced Scientific Research, Jakkur,
Bangalore 560 065, India
*Ashoka Trust for Research in Ecology and the Environment, No 659, 5th A Main,
Hebbal, Bangalore-24, India
[#]e-mail: rus@vsnl.com

Sandal (*Santalum album* L.) is one of the most economically important tree species indigenous to peninsular India. The species is extracted for its heartwood and oil. The sandal genetic resources in the country are threatened by a variety of factors including logging of the trees, poaching and due to large scale changes in land use. Analyzing long-term records of extraction, in Karnataka, India, Meera *et al.* (2000) and Nageswara Rao *et al.* (1999) showed a substantial decline in the total quantity of sandalwood extracted over the years. The decrease in availability of sandal is also mirrored in the reduction of sandalwood supplied to the factories (Meera *et al.*, 2000). It is feared that unless these losses are checked, it could lead to an irreversible loss of the sandal genetic resources from the country (Nageswara Rao *et al.*, 1999; Uma Shaanker *et al.*, 2000).

Keywords: Sandal, threats, sanctuary, genetic diversity, protected areas.

Among the various approaches to mitigate such loss and conserve the genetic resources, it is proposed that protected areas in the form of sanctuaries and national parks could offer a refugia for the conservation of genetic resources. However, the protected areas might not necessarily address the conservation concerns of all the species as they often are established based on the presence of some charismatic large mammal species (Rodgers and Panwar, 1988). We examined the impact of extraction pressure of sandal on the regeneration and genetic diversity of populations in the core, buffer and periphery of the Biligiri Rangan Temple (BRT) Wildlife Sanctuary, Karnataka, in the Western Ghats. The three zones were assumed to offer decreasing levels of immunity to poaching (from core to periphery) due to the decreasing levels of protection.

The results on size class distribution of sandal populations in BRT Hills along the three study zones suggest that, there was a higher proportion of large stemmed trees, in the core compared to the buffer and periphery (Figure 1).

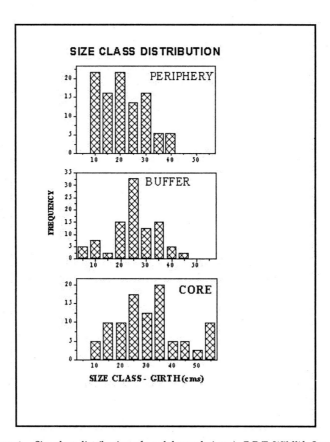

Figure 1.   Size class distribution of sandal populations in B.R.T. Wildlife Sanctuary.

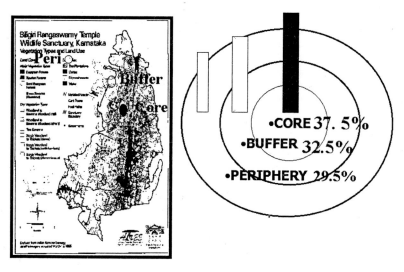

Figure 2. Genetic diversity study in BRT Wildlife Sanctuary.

The study of genetic diversity based on four enzyme systems namely, phosphogluco mutase (PGM), phosphogluco isomerase (PGI), malate dehydrogenase (MDH) and 6-phosphogluconate dehydrogenase (PGDH), indicated that the percentage observed heterozygosity decreased from the core (37.5) to buffer (32.5) to the periphery (29.5) (Figure 2). Compared to the expected levels of heterozygosity, the populations in the periphery were deficient in heterozygotes while that in the core had an excess of heterozygotes.

Thus the present study showed that the populations of sandal within the core limits of the sanctuary seem to be well protected as was evident from the maintenance of a relatively large proportion of large sized individuals and a higher extent of genetic variability. In contrast, populations of sandal from the periphery of the sanctuary that do not receive protection were not only impoverished for want of large sized trees but had a lower level of genetic variation. In other words, it appears that the sanctuary has been effective in protecting the sandal genetic resources and can in fact serve as *in situ* sites of conservation of the genetic resources. We propose that monitoring of key plant resources in the protected areas compared to that in adjoining buffer areas can provide a good indicator of the status of the protected areas in conserving the genetic resources.

Acknowledgement

This work was supported from grants from the IPGRI-Rome Forest Genetic Resources Program.

References

Meera, C., Nageswara Rao, M., Ganeshaiah, K. N., Uma Shaanker, R. & Swaminath, M. H. Conservation of sandal genetic resources in India: I Extraction patterns and threats to sandal resources in Karnataka, *My Forest*, **36,** 125–132 (2000).

Nageswara Rao, M., Padmini, S., Ganeshaiah, K. N. & Uma Shaanker, R. Sandal genetic resources of Souh-India: Threats and conservation approaches. in *National symposium on role of plant tissue culture in biodiversity conservation and economic development.* (Kosi-Katarmal, Almora, UP, (1999) p. 63.

Rodgers, W. A. & Panwar, H. S. *Planning a wildlife area in India,* Vol 1, (The report, Wildlife Institute of India, Dehra Dun, 1988).

Uma Shaanker, R., Ganeshaiah, K. N. & Nageswara Rao, M. Conservation of sandal genetic resources in India: problems and prospects. in: *International Conference on Science and Technology for Managing Plant Genetic Diversity in the 21ˢ Century,* (Kuala Lumpur, Malaysia, 2000) pp 73.

Tropical Ecosystems: Structure, Diversity and Human Welfare.
Proceedings of the International Conference on Tropical Ecosystems
K. N. Ganeshaiah, R. Uma Shaanker and K. S. Bawa (eds)
Published by Oxford–IBH, New Delhi. 2001. pp. 125–128.

Anthropogenic pressures in a tropical forest ecosystem in Western Ghats, India: Are they sustainable?

N. A. Aravind*, Dinesh Rao*, G. Vanaraj*, J. Poulsen[@],
R. Uma Shaanker** and K. N. Ganeshaiah[#,†]

*Ashoka Trust for Research in Ecology and the Environment (ATREE), #659 5th
'A' Main, Hebbal, Bangalore 560 024, India
[@]Center for International Forestry Research (CIFOR), P. O. Box 6596 JKPWB,
Jakarta 10065, Indonesia
**Department of Crop Physiology and [#]Department of Genetics and Plant Breeding,
University of Agricultural Sciences, GKVK, Bangalore 560 065, India
[†]e-mail: kng@vsnl.com

Several million of humans still thrive in the forest of tropics and use resources therein for their livelihood. Owing to the fact that they have been with, and depending upon the forest for several centuries, it is believed that their harvesting patterns could represent a good model for sustainable use of forest resources. As of yet however, there are no identified models of such sustainable use, though rigorously sought for (Uma Shanker, *et al.*, 1996). In this paper we report the impact of one such indigenous community on the structure and composition of the forest ecosystem using indicator species birds, butterflies (Kremen, 1992, 1994) and vegetation of a wildlife sanctuary in South India.

Biligiri Rangaswamy Temple Wildlife Sanctuary (77°–77°16′E and 11°47′–12° 9′N) is along the eastern outgrowth of Western Ghats. The sanctuary houses

Keywords: Anthropogenic pressures, BRT Sanctuary, birds, butterflies, Western Ghats, biological indicators.

about 4500-odd indigenous tribes, Soligas in about 25 settlements (Ganeshaiah and Uma Shaanker, 1998) spread along the sanctuary. Soligas derive most (~70%) of their needs from the forests alone (Hegde *et al.*, 1996).

Disturbance gradients (d1 to d5; d1 representing highest disturbance and d5 the least disturbance) were identified based on interviews with local residents and also personal observations. Two transects of 600 m length were laid at a distance of 150–200 m with in each disturbance regime. Per cent cut and broken stems, which has shown to be a reliable indicator of disturbance (Ganeshaiah, *et al.*, 1998) were recorded in them.

Birds were sampled during early hours of the day between 6:30 hr and 10:00 hr. All birds seen and heard within 50 m on either side of the transect were recorded. Butterflies seen within 10 m on either side of the transect were recorded between 8:30 hr and 11:30 hr. Vegetation was sampled in 600 x 20 m strips under three habit categories namely, trees, shrubs, and herbs. All the trees (>10 cm DBH) were enumerated in the 50 m × 10 m segments at 50 m intervals along the transect. All trees in the entire 600 m × 20 m strip were enumerated. Shrubs (> 1 cm and < 10 cm DBH) were sampled in 10 m × 10 m quadrats within 50 m × 20 m segment and herbs (< 1 cm DBH) were enumerated in two sub-quadrats (1 m × 1 m) within 50 m × 20 m segment.

In general the five levels of disturbance segregated clearly on the basis of their proximity to the settlements. The most disturbed sites were located immediately surrounding their settlements and least disturbed were farther away from them.

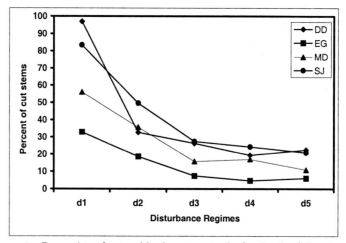

Figure 1. Proportion of cut and broken stems in the five levels of disturbance of the four vegetation types.

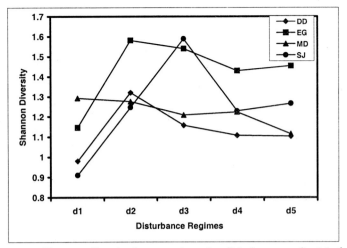

Figure 2. Shannon diversity for trees along the disturbance gradient in four vegetation types.

There was a steep decrease in per cent cut and broken stems from d1 to d2 in all the vegetation types. However, from the second disturbance level (d2) onwards there was little change in the damage caused to the tree stems. In general the five categories chosen did represent the five levels of disturbance in all the four vegetation types (Figure 1). The impact of human disturbance was also visible on the density of trees in the transect. In all the four vegetation types highly disturbed zones had 200 trees per ha, which increased abruptly to around 300 (moist deciduous forest) and more than 1200 trees (evergreen forest) at second level of disturbance (d2). There was in general a decrease in density of trees with the disturbance. The similar pattern was also seen for the Shannon diversity for trees (Figure 2).

The impact of disturbance did not show any discernible pattern for birds and butterfly diversity. However, within scrub jungle and dry deciduous forests, less disturbed sites showed high diversity for butterflies.

Thus our study suggests that the pattern of resource use by *Soligas* affects the forest in and around their settlements and the impact decreases with the distance from the settlements. Whether or not the reduced level of forest use at farther distances from the settlements is sustainable is not clear and requires long-term monitoring . The response to disturbance also appears to be taxa dependent. The mobile taxa such as birds and butterflies exhibit relatively more resilience to disturbance than the sessile plants.

Acknowledgements

This work is funded by Center for International Forestry research (CIFOR), Indonesia. We also thank Karnataka Forest Department for cooperation rendered during the course of this study.

References

Ganeshaiah, K. N. & Uma Shaanker, R. (eds). *Biligiri Rangaswamy Temple Wildlife Sanctuary – Natural History, Biodiversity and Conservation* (ATREE and GKVK, Bangalore, 1998).

Ganeshaiah, K. N., Uma Shaanker, R., Murali, K. S., Uma Shanker & Bawa, K. S. Extraction of non-timber forest products in the forest of Biligiri Rangan Hills, India. 5. Influence of dispersal mode on species response to anthropogenic pressures. *Econ. Bot.* **52**, 316–319 (1998).

Hegde, R., Suryaprakesh, S., Achoth, L. & Bawa, K. S. Extraction of non-timber forest products in the forests of Biligiri Rangan Hills, India. 1. Contribution to rural income. *Econ. Bot.* **50**, 243–251 (1996).

Kremen, C. Assessing the indicator properties of species assemblages for natural areas monitoring. *Ecol. Appl.* **2**, 203–217 (1992).

Kremen, C. Biological inventory using target taxa: A case study of the butterflies of Madagascar. *Ecol. Appl.* **4**, 407–422 (1994).

Uma Shanker, Murali, K. S., Uma Shaanker, R., Ganeshaiah, K. N. & Bawa, K. S. Extraction of non-timber forest products in the forest of Biligiri Rangan Hills, India. 3. Productivity, extraction and prospects of sustainable harvest of Amla *Phyllanthus emblica* (Euphorbiaceae). *Econ. Bot.* **50**, 270–279 (1996).

Tropical Ecosystems: Structure, Diversity and Human Welfare.
Proceedings of the International Conference on Tropical Ecosystems
K. N. Ganeshaiah, R. Uma Shaanker and K. S. Bawa (eds)
Published by Oxford–IBH, New Delhi. 2001. pp. 129–132.

Impact of human-induced disturbance on the diversity of dung beetles (Coleoptera: Scarabaeidae) and ants (Hymenoptera: Formicidae)

Priyadarsanan Dharma Rajan*[†], K. N. Ganeshaiah**,
R. Uma Shaanker**, T. M. Musthak Ali**,
A. R. V. Kumar** and K. Chandrashekara**

*Ashoka Trust for Research in Ecology and the Environment, # 659, 5th A Main,
Hebbal, Bangalore 560 024, India
**University of Agricultural Sciences, GKVK, Bangalore 560 065, India
[†]e-mail: priyan@atree.org

Disturbance, whether natural or anthropogenic, is an important factor that affects the ecosystem functions. Although insects contribute to the major share of biodiversity, impact of disturbance on its diversity is poorly understood (Hamer and Hill, 1999). Insects are known to respond differently to disturbance dynamics (Schowalter, 1985). Some studies have shown that disturbance causes a reduction in diversity of insects (Daily and Ehrlich, 1995; Hill *et al.*, 1995) while others suggest that disturbance results in an increase in diversity (Kremen, 1992; Hamer *et al.*, 1997). In this study we have attempted to assess the response of dung beetles and ants to disturbance in a wildlife sanctuary of South India.

The study was conducted in the moist deciduous forests of Biligiri Rangan Temple (BRT) Wildlife Sanctuary, located in the Chamaraja Nagar District of

Keywords: Dung beetles, ants, South India, BRT wildlife sanctuary.

Karnataka State, India (77°–77°16'E and 11°47–12°9'N). The sanctuary is inhabited by indigenous people called Soligas who depend on this forest for collecting NTFPs, fuel wood and for grazing their cattle.

The major criterion for grading disturbance was based on the assumption that the most proximal sites to the human activity are likely to be more disturbed (Murali *et al.*, 1996). Accordingly 10 sites of varying disturbances from the settlements were identified on their proximity to Kannery Colony, one of the major settlements in this area. Another set of five highly disturbed sites identified in our earlier studies (Murali *et al.*, 1996) were also earmarked for this study. Number of cut stems and frequency of cattle dung in the study plot were also recorded. On each of these sites, grids of 50 m × 50 m were marked for sampling dung beetles and ants.

Eight baits of one litre dung were placed randomly in each grid. Two pats each were retrieved after 6, 24, 48 and 72 h. Dung beetles were recovered from the dung pats by floating and hand picking. Sugar solution (30%), dried coconut scrapings and egg white were used as baits for sampling ants. Three baits of each type were kept equidistant in the grids and the ants were collected after 1, 6, 24 and 48 h using renewed baits. Collected insects were preserved in 70% alcohol and later identified and counted. Shannon diversity index, avalanche index and Evenness index were calculated for ants and dung beetles with the data collected.

Diversity and species richness of ants and dung beetles were found to increase with the disturbance (see Figure 1 and 2). The most disturbed sites

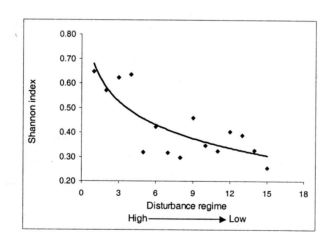

Figure 1. Shannon diversity of dung beetles at different disturbance regimes ($R^2 = 0.6499$).

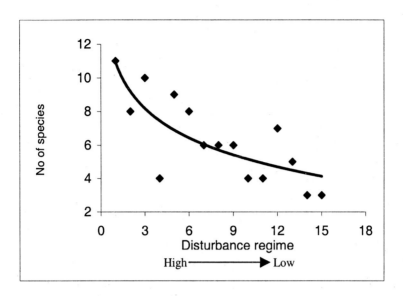

Figure 2. Species richness of ants at different disturbance regimes ($R^2 = 0.6095$).

showed high diversity for both ants and dung beetles compared to the relatively less disturbed sites. 616 dung beetles representing 18 species and 7 genera were collected from the highly disturbed sites where as 373 individuals representing 17 species and 4 genera were collected from the sites less disturbed. Similarly, 2340 individuals of ants representing 21 species and 13 genera were collected from the highly disturbed sites while 3861 ants (12 species and 7 genera) were collected from the areas of low disturbance. 887 dung beetles belonging to 14 species and 7 genera and 3530 individuals of ants representing 16 species and 11 genera were recovered from the sites of medium disturbance.

Thus it appears that human settlements and surrounding areas harbour more species of dung beetles and ants. Owing to high concentration of cattle in and around the settlements the dung pats will be highly concentrated, offering substantial resource base for the dung beetles. Similarly the high concentration of human excreta in and around the settlements is also likely to facilitate the higher dung beetle diversity and richness. The agricultural activities near the settlements perhaps offer a good supply of resources to ants facilitating their abundance and diversity.

Other studies also have shown that disturbance can enhance species diversity and the relative abundances of the species present, by increasing spatial and temporal heterogeneity of ecosystems (Grubb, 1977; Platt and Weis, 1977; Tilman 1982).

References

Daily, G. & Ehrlich, P. R. Preservation of biodiversity in small rain forest patches: rapid evaluations using butterfly trapping. *Biodiver. Conserv.* **4**, 35–55 (1995).

Grubb, P. J. The maintenance of species richness in plant communities: the importance of the regeneration niche. *Biol. Rev. Cambridge Phil Soc.* **52**, 107–145 (1977).

Hamer, K. C., Hill, J. K., Lace, L. A. & Langan, A. M. Ecological and biogeographical effects of forest disturbance on tropical butterflies of Sumatra, Indonesia. *J. Biogeogr.* **24**, 67–75 (1997).

Hamer, H. & Hill, J. K. Scale dependent effect of habitat disturbance on species richness in tropical forests. *Conserv. Biol.* **14**, 1435–1440 (1999).

Hill, J. K., Hamer, K. C., Lace, L. A. & Banham, W. M. T. Effects of selective logging on tropical forest butterflies on Buru, Indonesia. *J. Appl. Ecol.* **32**, 754–760 (1995).

Kremen, C. Assessing the indicator properties of species assemblages fornatural areas monitoring. *Ecol. Appl.* **2**, 203–217 (1992).

Murali, K. S., Uma Shankar, Uma Shaanker, R., Ganeshaiah, K. N. & Bawa, K. S. Extraction of non-timber forest products in the forests of Biligiri Rangan Hills, India. 2. Impact of NTFP extraction on regeneration, population structure and species composition. *Econ. Bot.* **50**, 252–269 (1996).

Platt, W. J. & Weis, I. M. Resource partitioning and competition within a guild of fugitive prairie plants. *Am. Nat. III* 479–513 (1977).

Schowalter, T. D. Adaptations of insects to disturbance, in *The ecology of Natural Disturbance and Patch Dynamics* (eds Pickett, S. T. A. & White, P. S.) 235–252 (Academic Press, San Diego, 1985).

Tilman, D. *Resource Competition and Community Structure* (Princeton Univ. Press, Princeton, New Jersey, 1982).

Land use and Forest Cover Change and the Consequences on Tropical Biodiversity

Tropical Ecosystems: Structure, Diversity and Human Welfare.
Proceedings of the International Conference on Tropical Ecosystems
K. N. Ganeshaiah, R. Uma Shaanker and K. S. Bawa (eds)
Published by Oxford–IBH, New Delhi. 2001. pp. 135–139.

Influence of land use decisions on spatial biodiversity

Alka Chaturvedi*, Rucha Ghate** and
Phani Kumar Garalpati*

*Department of Botany, University Campus, Nagpur 440 010, India
**Department of Economics, University Campus, Nagpur 440 010, India
e-mail: arunchat@nagpur.dot.net.in

The unsustainable rate of forest exploitation in tropical regions is a matter of great concern for the governments and environmentalists alike. Though deforestation can be attributed to several factors like population growth, technology, government policy, commercial logging, poverty, etc., the apparent impact is – extension of cultivable land at the cost of forest land. Although the figures of encroachments on the land under forests are soaring, there also are studies indicating existence of communities that are consciously maintaining and managing the forests within their village boundaries, at their own initiatives (Gadgil and Berkes, 1991; Clark *et al.*, 2000). The present paper discusses cases of two neighbouring villages in Brahmapuri taluka, Chandrapur district of Maharashtra. One of these, Saigata, has preferred sustainable ecological use of forest to immediate benefits. The other village, Maral Mendha, has heavily encroached upon the forest land, discounting the future heavily. Such contradictory approaches towards forests by communities living in similar circumstances have been intriguing researchers all over the world. Difference in the two approaches towards land use, reflects on the quality of the resource in terms of diversity of species, regeneration capacity of

Keywords: Land use, biodiversity, discount rate, central India.

land, etc. The two villages selected for this study represent the two diverse strategies and its impact on the forest thereby.

Ecological and taxonomical observations have been made to study the influence of anthropogenic pressure on the forest of the two villages in Central India. Forest map of Saigata (1977–1978) shows a total forest area of 280.13 hectare. The entire area has been divided into a number of segments by the villagers for better management. 'Gawatache beed' is kept for the fodder and thatching material. This area shows pressure of grazing animals. Woody plants are not in existence. Only few saplings of *Terminalia* sp. and of *Tectona* are visible. Ground vegetation is of herbs. Dominant plants belong to family Poaceae, Fabaceae and Acanthaceae. The soil is shallow, moderately drained, and reddish brown in colour.

'Andhari Van' is completely prohibited for any fresh wood cutting activity. Only dead, dried wood is allowed to be collected for the purpose of fuel wood. Village map of 1977–1978 shows an area of 280.13 hectares under forest with a per capita forest area of 0.65 hectares. The soil of this forest is well drained, differentiated into A and B horizons. Thick litter of leaves, moderately high temperature and humidity in and around soil, under the forest cover leads to growth of micro fauna and rapid development of soil. Phanerophytic vegetation shows two-storey structure. The top strata is dominated by species of family Verbenaceae, Combretaceae, Meliaceae and Myrtaceae. Average height of the plants in this strata is between 6 and 10 m and DBH between 15 and 19 centimeters. The second strata is made-up of Euphorbiaceae, Rubiaceae and Anacardiaceae. Natural regeneration of woody species is quite clear by the high density of saplings, belonging to family Combretaceae, Ebenaceae, Euphorbiaceae, Fabaceae, Tiliaceae and Verbenaceae.

The rest of the Saigata forest consists of 'Kotwal Raan, Tiwas Van, Mothya devacha Ban, Koseraab and Mukhiache Raan'. These segments show minor anthropogenic activities, such as collection of fuel wood, fodder, material for thatching, minor forest products, Tendu leaves, fruits, minor timber and gitty. The soils range from sandy loam to silty clay and are shallow to very deep and reddish to dark brown in colour. Ground vegetation is sparse. Saplings are large in number. The most abundant species is *Cleistanthus collinus* followed by *Anogeissus latifolia, Terminalia alata, Lagerostroemia parviflora* and *Buchanania lanzan.* Thorny bushes also form a substantial percentage in the sapling flora, *Maytenus emerginata* with 46.66% density followed by *Xeromphis spinosa* (53.33%), *Mimosa hamata* (60%), and species of *Ziziphus* (86%). Tree flora is dominated by family Anacardiaceae, Combretaceae, Euphorbiaceae, Meliaceae, Myrtaceae, Fabaceae, Rubiaceae and Verbenaceae. The tallest trees belong to *Anogeissus latifolia* and *Tectona grandis* ranging between 15 and 20 m in height. The second strata of 1–10 m is formed by *Lagerostroemia parviflora,*

Cleistanthus collinus, Butea monosperma, Buchanania lanzan, Zizyphus glaberrima, Xeromphis spinosa and *Grewia hirsuta.* The highest DBH is observed in *Terminalia alata, Lagerostroemia parviflora* and *Albizia odoretissima* followed by *Tectona grandis, Anogeissus latifolia* and *Buchanania lanzan.* Besides these, a large number of woody climbers are also present.

Village Maral Mendha is situated on the same landform and in identical climatic conditions that occur in Saigata. Forest map of 1977–78 shows a total forest area of 71.46 hectares which is divided into Protected Forest (70.88 ha) and Zudpi Forest (00.63 ha). Per capita forest is 0.12 ha. However, the present survey of the area shows a significant change in the status of the forest. Hardly any woody species is remaining. In span of 25 years the entire forest is converted into crop lands. Soil is in poor quality, with extensive soil erosion. The impact of heavy grazing and extreme damage by insects is visible. Land is highly degraded. Paddy cultivation in major part of the area has led to leaching and soil erosion. High temperature and exposed surface along with other conditions has resulted in poor development of soil. There is no tree cover and no growth of saplings. Vegetation is mostly in form of ground flora. Dominant species belong to family Asteraceae, Fabaceae and Poaceae, while thorny shrubs and saplings are present but are low in density.

Based on these insights, a study was initiated to address the impact of land use patterns on the biodiversity and forest quality. It was hypothesized that *de facto* land use decision by local people sharply affects the biodiversity and forest quality at micro level.

The objectives of the study were: (i) To examine and compare the impact of local land use decisions on the quality of surrounding forest in case of two villages with opposite development strategies, (ii) To study the effectiveness of locally developed institutions on the use of forest products as reflected in the forest condition and biodiversity and (iii) To assess the impact of forest use within a village on biodiversity.

Since the data required for temporal analysis are not available, spatial comparison at a point of time is done. The selected forest condition indicators are: types and number of species, size and density of trees, basal area, and regeneration capacity of the area. Although the two villages selected for analyses are comparable in terms of ecosystem (temperature, rainfall, soil variation, altitude, species, etc.) and location (i.e. distance from the district head quarter, proximity to market and all weather road and the like), the major difference is regarding population mix (caste combination) and per capita forest area. Both the populations are heterogenous but some members of Saigata community belong to one tribe. Also, Saigata (0.65 ha) has much more per capita forest area than Maral Mendha (0.12 ha).

The present study has adopted the data collection method developed by The International Forestry Resources and Institutions (IFRI) research program consisting of pre-structured research instruments to facilitate collection of information about demographic, economic, and cultural characteristics of communities dependent on forests. It uses rigorous forestry techniques to measure the impact of institutions regulating the use of forest products.

For the present study 30 plots were determined for Saigata and 5 plots for Maral Mendha villages. The plots were determined using grid method and were selected using random numbers. All the plots are circular in shape, consisting of a circle of 1 meter radius for collecting information on ground flora, 3 meter radius circle for saplings, and a circle with 10 meter radius for collecting data on trees.

The following are broad results of the study:

1. Restricted activity of human beings and animals; unhampered natural regeneration of plants (biotic factors); mostly flat geomorphology, well-developed soil with humus (edaphic factors), moderately high temperature, adequate water regime and light (climatic factors) are responsible for a good forest with greater bio-diversity in forest of village Saigata.

2. However, in village Maral Mendha, increased pressure of biota, poor soil conditions and comparatively harsh climatic conditions are responsible for low bio-diversity.

3. Difference in the caste-combination in the two villages has played a major role in evolving the shared understanding amongst the users of forest. While Saigata, with major component of tribal population, ascribes low discount rate and has adopted sustainable resource use strategy, Maral Mendha has discounted future heavily and has gone for converting forest land into agriculture land.

4. Evolving the local leadership in Saigata with sufficient skills of organization and use of participatory method has contributed to the success of collective action. Its absence in Maral Mendha due to lack of common understanding has led to massive degradation of forest land.

Acknowledgements

The present paper is the outcome of a pilot study funded by Workshop in Political Theory, Indiana University, Bloomington, US. We are grateful to the funding agency and the other research members of the study group.

References

Gadgil, M. & Fikret Berkes. Traditional resource management systems. *Res. Manage. Optimization*, **8**, 127–141 (1991).

Gibson, Clark. Dependence, scarcity, and the governance of forest resources at the local level in Guatemala. in *The Commons Revisited: An Americas Perspective* (eds Joanna Burger, Richard Norgaard, Elinor Ostrom, David Policansky & Bernard Goldstein) (Island Press, Washington, DC).

Gibson Clark, Margaret McKean & Elinor Ostrom (eds). *People and Forests: Communities, Institutions, and Governance* (MIT Press, Cambridge, MA, 2000).

Tropical Ecosystems: Structure, Diversity and Human Welfare.
Proceedings of the International Conference on Tropical Ecosystems
K. N. Ganeshaiah, R. Uma Shaanker and K. S. Bawa (eds)
Published by Oxford–IBH, New Delhi. 2001. pp. 140–142.

Forest types in the Southern Yucatan Peninsula: Effects of long and short term land use on regional biodiversity

Diego R. Pérez-Salicrup

Departamento de Ecología de los Recursos Naturales, Instituto de Ecología,
Universidad Nacional Autónoma de México. A.P. 27-3 (Xangari), CP. 58089,
Morelia, Michoacán, México
e-mail: diego@oikos.unam.mx

Forests in the Southern Yucatan Peninsular Region (henceforth SYPR) have experienced two major waves of human activity (Kepleis and Turner, 2001; Turner *et al.*, in press). The first one was that of the Ancient Maya, who developed a very elaborate culture largely dependent on extensive corn (*Zea mays*) agriculture. Mayans inhabited the SYPR for about 1000 years, and suddenly abandoned the region, migrating north, at ca. 800 A.D. The second wave of human activity came in the 20th century, and has included the exploitation of the timber tree species *Swietenia macrophylla* and *Cedrella odorata* (Flachsenberg and Galletti, 1998; Argüeyes, 1991; Snook, 1998); the extraction of sap from *Manilkara zapota* (Lundell, 1933; Dugelby, 1998), for the production of chewing gum; and the establishment of immigrant farmers from other parts of Mexico who have engaged in shifting agriculture (Hernández-Xolocotzi, 1959; Barrera *et al.*, 1977). In this study, I investigated tree composition and tree community structure of forests in the SYPR, to elucidate the legacies of long (Maya) and short-term (20th century) land use on these forests.

Keywords: Biodiversity, land-use, Mexico, tropical forests, Yucatán.

I established a network of over 150 circular 500 m^2 permanent plots located in six regions within the SYPR. These regions encompass gradients of rainfall, soil depth, elevation, hurricane frequency, and 20th century land use. In each plot, we marked, identified and measured all trees $\geq$ 10 cm dbh (diameter at 1.3–1.5 m, or above buttresses). I used tree diversity as surrogate for biodiversity in SYPR, although I recognize that not all taxonomic groups correlate in their response to human land uses. Using NMS ordinations and MRPP test, the differences among the major natural forest types, and modern successional stands in SYPR were evaluated.

Consistent with the findings of other botanists working in the region (Miranda, 1958; Flores and Carbajal, 1994), I identified two major types of forest in the SYPR, Low Statured Forests (LSF) and Mid Statured Forests (MSF). These forest types can be readily identified based on the abundance of tree species growing in one or the other forests. These forests are determined mostly by the depth and degree of flooding of soils and structurally are quite distinguishable. The two natural forest types, MSF and LSF, share more than 80% of their tree species.

Recent successional forests are quite distinguishable from MSF and LSF in terms of species composition during the first 25 years of succession. However, at 25–30 years after agricultural abandonment, they acquire a tree composition which is not different from that of mature MSF.

Because LSF occur on areas unsuitable for shifting agriculture, and because LSF seldom have Maya ruins, it appears that LSF were not used in Maya agriculture. Hence this forest type might have served as repository for tree species diversity during the demographic peak of Maya land use. Consequently, MSF, which occurs on agriculturally suitable soils and with a high presence of Maya ruins, might have regenerated from the tree species pool of LSF. This scenario could explain the high similarity between the two major natural forest types. In this regard, we suggest that the LSF should receive a high conservation priority in SYPR, because they might represent the repositories of biodiversity in that region.

After 25–30 years of succession, fallow fields have a tree species composition not distinguishable from MSF. Such rapid gain of species can be the result of two important factors. First, this is the first agricultural cycle in over 1000 years, so conditions for tree species reestablishment are ideal, in part because there is presumably little soil nutrient depletion and also because agricultural fields still occupy small extensions imbedded in a large matrix of forest. Second, these forests experienced conditions created by shifting agriculture in times of the Maya. Therefore, the tree species present in the SYPR are very resilient to such conditions. Tree species that are negatively affected by shifting agriculture might have become locally extinct since the time of the Maya.

Human land use, particularly non-timber extraction, has had little effect on the region's biodiversity for most of the twentieth century. However, this scenario has changed dramatically with the arrival of mechanized agriculture in the 1970s. Today, conservation efforts need to promote human activities that promote a dynamic mosaic of forest patches of different successional ages. This scenario would promote the conservation of the region's biodiversity.

References

Argüeyes, L. A. Plan de manejo forestal para el bosque tropical de la empresa ejidal Noh-Bec. Tesis. Departamento de Bosques. Universidad Autónoma de Chapingo. Texcoco, Edo. De México. México, pp 125 (1991).

Barrera, A., Gómez-Pompa, A. & Vázquez-Yanes, C. El manejo de las selvas por los Mayas: sus Implicaciones Silvícolas y Agrícolas. *Biotica* **2,** 47–61 (1977).

Dugelby, B. L. Governmental and customary arrangements guiding chicle latex extraction in Petén, Belize. in *Timber Tourists and Temples: Conservation and Development in the Maya Forest of Belize, Guatemala, and Mexico* (eds Primack, R. B., Bray, D., Galletti, H. A. & Ponciano, I.) 155–177 (Island Press, Washington DC USA, 1998).

Flachsenberg, H. & Galletti, H. Forest management in Quintana Roo, Mexico. in *Timber Tourists and Temples: Conservation and Development in the Maya Forest of Belize, Guatemala, and Mexico* (eds Primack, R. B., Bray, D., Galletti, H. A. & Ponciano, I.) 47–60 (Island Press, Washington DC USA, 1998).

Flores, J. S. & Carvajal, I. E. Tipos de Vegetación de la Península de Yucatán. Etnofloroa Yucatanense Fascículo 3. Universidad Autónoma de Yucatán. México, p 135 (1994).

Hernández-Xolocotzi, E. La agricultura en la Península de Yucatán. in *Los Recursos Naturales del Sureste y su Aprovechamiento* (ed. Beltrán, E.) 3–57 (Tomo II. IMRNAR. México D. F. México, 1959).

Kepleis, P. & Turner B. L. II. Integrated land history and global change science: the example of the Southern Yucatán Peninsular Region project. *Land Use Policy* **18,** 27–39 (2001).

Lundell, C. L. Chicle exploitation in the sapodilla forest of the Yucatán Peninsula. *Field and Laboratory* **2,** 15–21 (1933).

Miranda, F. Estudios acerca de la vegetación. in *Los Recursos Naturales del Sureste y su Aprovechamiento* (ed. Beltrán, E.) 215–271 (Tomo II. IMRNAR. México D. F. México, 1958).

Snook, L. K. Sustaining harvests of Mahogany (*Swietenia macrophylla* King) from Mexico's Yucatán Forests: past, present, and future. in *Timber Tourists and Temples: Conservation and Development in the Maya Forest of Belize, Guatemala, and Mexico* (eds Primack, R. B., Bray, D., Galletti, H. A. & Ponciano, I.) 61–80 (Island Press, Washington DC USA, 1998).

Turner, B. L. II, Cortina Villar, S., Foster, D., Geoghegan, J., Keys, E., Klepeis, P., Lawrence, D., Macario Mendoza, P., Manson, S., Ogneva-Himmelberger, Y., Plotkin, A. B., Pérez-Salicrup, D. R., Roy Chowdhury, R., Savitsky, B., Schneider, L, Schmook, B. & Vance, C. In Press. Deforestation in the Southern Yucatán Peninsular Region: An Integrative Approach. Forest Ecology and Management.

Tropical Ecosystems: Structure, Diversity and Human Welfare.
Proceedings of the International Conference on Tropical Ecosystems
K. N. Ganeshaiah, R. Uma Shaanker and K. S. Bawa (eds)
Published by Oxford–IBH, New Delhi. 2001. pp. 143–144.

Advancing frontiers in tropical biology: The future of remote sensing research in tropical environments

G. Arturo Sanchez-Azofeifa and Benoit Rivard*

Earth Observation Systems Laboratory, Earth and Atmospheric Sciences
Department, University of Alberta, Edmonton, Alberta, Canada T6G 2E3
*e-mail: arturo.sanchez@ualberta.ca

The science of Land Use and Land Cover Change (LUCC) is concerned with identifying and quantifying changes in the landscape, requiring an understanding of what has existed in the landscape, what currently exists, how the landscape will look in the future (prognostic component), and the socio-economic forces that drive the change between these states (people and pixels). As land cover change monitoring platforms become more complex with the integration of new earth-observation systems, satellites ranging from fine resolution (e.g. IKONOS) to middle resolution (e.g. Landsat 7 and MODIS) and to large resolution (e.g. SPOT/Vegetation), new opportunities to advance the frontier of tropical biology and research in conservation biology are more evident. These opportunities are not only enhanced by an improvement in spatial resolution but also by an enhancement in spectral resolution (multi-spectral to hyper-spectral). Most of the research in the use of these platforms have been restricted to the temperate and boreal regions, with little or no work in tropical ecosystems. Research efforts based on phenology and changes in leaf area index (LAI), normalized derived vegetation index (NDVI) and single ratio analysis have been almost entirely implemented in non-tropical ecosystems (Chen and

Keywords: Land cover, remote sensing, tropical dry forests, Costa Rica, India.

143

Cihlar, 1996; Frazer et al., 2000). The very few examples of the applicability of these approaches to tropical ecosystems have yielded limited results (Steininger, 1996).

In this paper, we discuss the challenges involved in tropical remote sensing research in the context of current advances in earth observation systems platforms and methodologies for analysing the information embedded in them. We focus our paper on specific case studies at the Santa Rosa National Park in Costa Rica, a tropical dry forest ecosystem in Costa Rica, Central America, and the Western Ghats, India. In these studies we contrast the needs and priorities in tropical remote sensing research with those suggested by Chilar (2000) for land cover mapping. We recognize that priorities suggested by Chilar (2000) may be relevant in the context of remote sensing research in boreal ecosystems, but may be inappropriate in tropical environments. We argue that the advancement of research in remote sensing in tropical environment may need innovative approaches that take into consideration the complexity, structure and diversity of land cover types in the tropical regions. Furthermore, we suggest that new remote sensing approaches in tropical environments must take into consideration factors associated with the biophysical dimensions of land cover change (e.g. forest age and phenology), the diversity of species distribution in tropical ecosystems, the ability of current remote sensing platforms to discriminate between forest classes and specific biodiversity-relevant crops (e.g. shaded coffee), and the strong need to integrate the human dimension of land cover change into remote sensing analysis (socializing the pixel). We suggest that advances in remote sensing research in tropical environments must not only provide sound and scientific approaches to resolve pure remote sensing questions, but contribute to the goals of developing countries to develop approaches aimed at valuing services provided by nature, appropriating funds to protect nature and the conservation of biological resources (Daily, 1996).

References

Chilar, J. Land cover mapping of large areas from satellites: Status and research priorities. *Intl. J. Remote Sensing* **21** (2000).

Chen, J. M. & Chilar, J. Retrieving leaf area index of boreal forest using Landsat TM images. *Can. J. For. Res.* **55**, 153–162 (1996).

Daily, G. C. (ed.) *Nature's Services: Societal Dependence on Natural Ecosystems* (Island Press, Washington, DC, 1996) pp 392.

Frazer, G. W., Trofymow, J. A. & Lertzman, K. P. Canopy openness and leaf area in chronosequences of costal temperate rainforests. *Can. J. For. Res.* **30**, 239–256 (2000).

Steininger, M. K. Tropical secondary forest regrowth in the Amazon: age, area and change estimation with thematic mapper data. *Intl. J. Remote Sensing* **17**, 9–28 (1996).

Tropical Ecosystems: Structure, Diversity and Human Welfare.
Proceedings of the International Conference on Tropical Ecosystems
K. N. Ganeshaiah, R. Uma Shaanker and K. S. Bawa (eds)
Published by Oxford–IBH, New Delhi. 2001. pp. 145–149.

Vegetation classification and land-cover change in the Kalakad– Mundanthurai Tiger Reserve, south India, as inferred from satellite imagery: Implications for conservation of a bio-diversity hotspot

Mahesh Sankaran

Department of Biology, Syracuse University, Syracuse, NY 13244, USA
e-mail: msankara@mailbox.syr.edu

Changes in vegetation cover arising from human land-use patterns, deforestation and succession impact ecological processes at various spatial and temporal scales. The ability to detect/monitor such change is important for designing effective management strategies and for assessing efficacy of extant conservation measures. Here, remotely sensed data were used to classify vegetation and determine land-cover change over 8 years (1990–1998) at the Kalakad–Mundanthurai Tiger Reserve (KMTR), India. KMTR is situated at the southern tip of India's Western Ghats Mountains; one of the global biodiversity 'hotspots' (Myers, 2000). KMTR is a center of high plant diversity (Ganesh *et al.*, 1996) and also harbors some of the largest contiguous tracts of non-equatorial tropical evergreen forests in India (Ramesh *et al.*, 1997).

Keywords: Kalakad–Mundanthurai, remote sensing, NDVI, land-cover change, savanna.

It is estimated that 47 km^2 of forest area in the Agastyamalai Biosphere region (~1657 km^2, including KMTR which covers ~900 km^2) was lost between 1920 and 1960, a further 156 km^2 being lost to plantations, reservoirs and human encroachments between 1960 and 1990 (Ramesh *et al.*, 1997). The latter occurred despite the area being protected in some form between 1960 and 1990 (Menon & Bawa, 1997). Although protection measures have been enhanced at KMTR following its declaration as a tiger reserve in 1988, human demands on KMTR have also increased given the high human population densities and growth rates in the region (Cincotta *et al.*, 2000). Consequently, regular monitoring of land-cover change is critical for assessing conservation efficacy. This study (i) classified vegetation types within the reserve, (ii) quantified land-cover change between 1990 and 1998, (iii) identified areas/habitat types vulnerable to change, and (iv) estimated fire-regime characteristics at KMTR.

The analysis employed a Landsat-5 Thematic Mapper image from 1990 (March 06, path 143/row 54, bands 2–5) and two Indian Remote Sensing Satellite images from 1998 (Mar 08: IRS 1D-LISS-III and April 20: IRS 1C LISS-III; path 101/row 68, 4 bands corresponding to TM bands 2–5). A supervised maximum-likelihood classification was conducted on a multi-temporal composite of both 1998 IRS images (8 bands) to generate a vegetation map of the area. Classification categories (Table 1) differed from previous studies (Ramesh *et al.*, 1997, Menon & Bawa, 1997) in that subclasses of evergreen forests (dry evergreen, low elevation evergreen, etc.) were not distinguished, but 'closed' deciduous forests were differentiated from open-moderately closed savannas. Savannas are ecologically distinct from other habitats at KMTR because of the frequent fires they support. Distinguishing savannas from closed deciduous forests was therefore a priority given the high incidence of fires in KMTR (Kothari *et al.*, 1989).

Vegetation cover change detection was based on ratios of normalized difference vegetation index (NDVI) images from March 1990 and 1998. Negative values in images were first assigned the value 0.01 (to avoid errors in detecting direction of change) and images ratioed. Following this, a log-transformation was carried out to produce an image with normally distributed pixel values in which zeroes indicated no change, while positive and negative values indicated increases and decreases in NDVI, respectively. In the absence of ground-truth data from 1990, two arbitrarily chosen thresholds of 1 and 2 standard deviations above and below the mean were used to operationally define pixels showing significant changes in vegetation status between years.

Results from the classification procedure are listed in Table 1. Wet forests (including tea plantations and *Ochlandra* reed patches) and savannas were the most widespread habitat types, covering 40% and 36% of the reserve area, respectively (Table 1). Wet forests predominate in high elevations along the western boundary while savannas dominate low-elevations along

Table 1. Classification accuracy and areas of different land-cover types at KMTR. (Water bodies and clouds excluded for calculation of per cent coverage of different vegetation classes). Overall classification accuracy based on 4150 training and reference pixels, respectively, was estimated at 94.4% with a Kappa coefficient of 0.936.

Land-cover type	Producer's accuracy	User's accuracy	Area (in km^2)	% of reserve area
Water	95.65	100	3.61	–
Cloud	100	100	2.24	–
Cloud shadow	99.53	100	2.93	–
Degraded/bare ground	99.57	99.15	12.16	1.45
Short grassland	92.98	99.07	2.48	0.3
Dry deciduous/plantations	93.68	92.23	65.67	7.84
Tall-grass savanna	94.49	93.12	305.35	36.47
Moist forests	87.74	92.69	120.87	14.44
Wet forests	98.59	97.21	330.79	39.51

the eastern boundary. Moist and 'closed' dry-deciduous forests occupy the central portion of the reserve, and degraded/sparsely vegetated areas mostly occur along the eastern boundary in areas of high human impact. For wet and moist forests, 90% of the spatial coverage was accounted for by a single large patch in each habitat. The two largest patches of savanna habitats similarly accounted for 87% of their total spatial coverage.

A change detection threshold of 2 standard deviations revealed significant increases and decreases in NDVI between 1990–1998 in 4.4 km^2 and 2.62 km^2 of the reserve, respectively. Of areas showing increases, 89% comprised regions sparsely vegetated/bare in 1990 but supporting vegetation in 1998. Most such patches were those that had burnt in 1990. Besides these, the remaining increase occurred in savanna habitats (10.2%). Decreases in cover primarily occurred in areas classified as savannas (55.4%) and forests (40.47%) in 1990. Fires were again major contributors to decreased NDVI in savannas. Of the change detected in forests, 24.6% resulted from conversion of 'closed' forests to 'open' savannas, while another 10.7% of the change resulted from tree loss along rivers and streams following a flood in the early 1990s. Other habitat types showed minimal change. Besides reductions resulting from fires and floods, most of the remaining loss occurred in areas of high human impact within and near reserve boundaries.

A less stringent criterion of 1 standard deviation around the mean NDVI ratio revealed increases and decreases in cover for 36.13 km^2 and 17.54 km^2 of the reserve, respectively. Besides re-vegetation following 1990-fires, this criterion additionally identified increases in NDVI in both forests and savannas (67% of total area), reflecting succession to later seral stages in these areas. Similarly, besides loss of cover due to fires, the less stringent criterion also identified additional areas where forests had converted to savannas. Most additional areas identified as showing increases and decreases in NDVI were again concentrated along the eastern boundary and around human settlements in the reserve.

Most burns were small (< 10 ha), with average sizes of about 28.4 ha (0.7–203.8 ha). Fires in KMTR are mostly ground-fires restricted to savannas (unpubl. data from 44 'natural' fires), and the spatial extent of savannas can be treated as a measure of the reserve area with the potential to support fires, i.e. 36.5% or 305 km^2. About 60–72 fires were estimated to have burned annually at KMTR between 1997 and 1998, suggesting that about 1704 to 2045 ha, representing 2–2.3% of the reserve area and 6–6.6% of the area covered by savannas, burnt each year.

Results indicate no drastic loss of forest cover in KMTR between 1990 and 1998. This is not surprising since stringent protection measures have been in place since 1988. The use of 1 and 2 standard deviation thresholds indicated significant reductions in cover in 0.3% and 2.1% of the total reserve area, respectively. Although savanna fires accounted for a major part of this loss, cover-loss was also evident along the eastern boundary and around human settlements within the reserve. Although not drastic in the sense of widespread deforestation, these areas are clearly under threat of deterioration from human activity.

The single large patch of wet-evergreen forest (~ 295 km^2) is of particular conservation relevance since it represents one of the least fragmented tracts of non-equatorial tropical evergreen forest in India (Ramesh et al., 1997) and constitutes the watershed for several streams and rivers in the area. This region, traditionally assigned a high conservation value (Ramesh et al., 1997; Menon and Bawa, 1997), has been relatively stable over the past decade, at least in terms of spatial extent. It may be concluded that protection measures have been effective at maintaining important watershed and ecosystem functions associated with this habitat to the degree that it has not witnessed major changes in spatial extent.

Savannas, on the other hand, which showed the most change over the 8-year period, have been thus far largely ignored from a scientific/management viewpoint at KMTR. Based on floral and faunal richness and endemism, previous studies have deemed these areas of low conservation value in KMTR (Ramesh et al., 1997; Menon & Bawa, 1997). Although depauperate relative to evergreen forests in terms of tree species richness or endemism, savannas are nevertheless important from a conservation viewpoint. The large spatial extent and high site-to-site heterogeneity in species composition of savannas (Sankaran, unpublished) suggest that they are important repositories of species in KMTR (235 plant species recorded from approx. 1 ha across the reserve). Savannas also typically support higher herbivore densities compared to wet/moist forests (Eisenberg & McKay 1974), and are, therefore, important from a herbivore management viewpoint, which is in turn, a prerequisite for predator conservation at KMTR.

Management strategies would benefit by focussing attention on savannas in peripheral areas of the reserve. While this does not imply that other habitats such as wet evergreen forests are free from threat, savannas are under more of a threat currently because of their proximity to human settlements (Ramesh *et al.*, 1997). Although widespread dominance of unpalatable tall-grasses currently renders large tracts of savannas in KMTR a poor herbivore habitat, these areas have the greatest potential to be effectively managed for herbivore and predator populations. Long-term experimental studies detailing the effects of changes in fire regimes or fire-suppression on plant species composition and diversity, nutrient cycling and herbivore-use patterns are critical in this regard.

References

Cincotta, R. P., Wisnewski, J. & Engelman, R. Human population in the biodiversity hotspots. *Nature* **404**, 990–992 (2000).

Eisenberg, J. F. & McKay, G. M. Comparison of ungulate adaptations in the New World and Old World tropical forests with special reference to Ceylon and the rainforests of Central America. in *The Behaviour of Ungulates and its Relation to Management* (eds Walther, V. & Geist, F.) Vol. 2 (Unwin Bros. Ltd., Gresham Press, Surrey, England, 1974).

Ganesh, T., Ganesan, R., Soubadra Devy M., Davidar, P. & Bawa, K. S. Assessment of plant biodiversity at a mid-elevational evergreen forest of Kalakad–Mundanthurai Tiger Reserve, Western Ghats, India. *Curr. Sci.* **71**, 379–392 (1996).

Kothari, A., Pande, P., Singh, S. & Variava, D. *Management of National Parks and Sanctuaries in India: A status report.* Indian Institute of Public Administration, New Delhi, India (1989).

Menon, S. & Bawa, K. S. Application of geographic information systems, remote-sensing, and a landscape ecology approach to biodiversity conservation in the Western Ghats. *Curr. Sci.* **73**, 134–145 (1997).

Myers, N., Mittermeier, R. A., Mittermeier, C. G., Da Fonseca, G. A. B. & Kent, J. Biodiversity hotspots for conservation priorities. *Nature* **403**, 853–858 (2000).

Ramesh, B. R., Menon, S. & Bawa, K. S. A vegetation based approach to biodiversity gap analysis in the Agastyamalai Region, Western Ghats, India. *Ambio* **26**, 529–536 (1997).

Tropical Ecosystems: Structure, Diversity and Human Welfare.
Proceedings of the International Conference on Tropical Ecosystems
K. N. Ganeshaiah, R. Uma Shaanker and K. S. Bawa (eds)
Published by Oxford–IBH, New Delhi. 2001. pp. 150–152.

Biodiversity conservation in productive forest landscapes

Robert Nasi

Biodiversity and Managed Forests, CIFOR, PO Box 6596 JKPWB,
Jakarta 10065, Indonesia
e-mail: r.nasi@cgiar.org

Attempts to pursue biodiversity objectives in both protected areas and in production forests have often failed because the attribution of costs and benefits was unfair and regulations proved unenforceable. Most people would agree that biodiversity is most likely to be maintained if local benefits are maximised and local costs are minimised. This paper will argue that various sorts of multiple-use forests are likely to be the best option for biodiversity conservation in many situations where poor people live in proximity to forests rich in biodiversity. It is inevitable that timber extraction will be a major element of this multiple-use in many forest areas. We will further argue that there are no fundamental technical obstacles to meeting many biodiversity objectives in forests managed for timber. The diversity of forests and the people who depend upon them is so great that it is neither desirable nor possible to develop broadly generalised prescriptions for management. The extent of the trade-offs in reconciling global and local values is such that even with optimal management arrangements some form of compensation or subsidy to forest-dependent stakeholders will often be unavoidable.

There is a strong emerging consensus that it is imperative to conservation that efforts extend beyond protected areas (Cabarle, 1998). Even the most

Keywords: Biodiversity, productive landscapes, logging, tropical forests.

ambitious exponents of biodiversity protection only advocate the allocation of around 10% of forests to parks and reserves and obviously the fate of much biodiversity will depend upon what happens to the remaining 90% of the forest estate. A tropical landscape containing a matrix of protected old-growth forests, logged and secondary forests, and agricultural land can probably conserve most of the existing biodiversity (Chazdon, 1998). All types of land use have validity and finding the proper balance among these uses is the key to improve resource and biodiversity conservation.

The process of negotiating a diversified forest estate will also define where direct financial revenues may be sought, and where a subsidy may have to be directed, how much land to allocate to different uses and how to regulate the use of that land. If society decided to shift the balance of land-use allocations, it could do so through zoning and other forms of regulations, tax incentives and disincentives, and other forms of direct and indirect subsidies.

A very large proportion of the world's forests are used for the production of timber, and this situation is likely to persist. Production forests provide habitats for many, often the majority, of the plant and animal species found in more pristine forests. However, the management of a forest for production generally has some sort of modification or degradation of the forest. Though this may be challenged (Bowles et al., 1998; Rice et al., 1997; 1998), we believe that significant gains in conserving biodiversity can be made by promoting sustainable forest management (Grieser-Johns, 1997; Cabarle, 1998).

However, measures designed to enhance biodiversity values may increase costs or reduce yields. Fairly allocating among forest stakeholders of the costs and benefits associated with tradeoffs related to biodiversity conservation may represent the biggest challenge to improving management. Stakeholders range from local inhabitants of forests, loggers and to downstream fishermen, more distant national governments, and citizens living in far away countries. Who should bear the costs associated with the loss of biodiversity, or the loss of production? 'Markets' that would provide a mechanism for paying for many of the benefits of biodiversity conservation are lacking.

How can tradeoffs be negotiated between those who are near the forest and those who are distant? Between those who are powerful and those who are not? Between those who are linked together within a national political system and those who are linked across borders through a global market system? The ecosystem management and adaptive co-management are interesting approaches to reach consensus and negotiate among all the stakeholders. Additional development of these mechanisms is necessary to provide for the representation of the stakeholder community to local, regional, national, and global perspectives (Szaro *et al.*, 1999).

The main reason for losing biodiversity in the tropics is not a lack of technical knowledge about forest management but rather the absence of the institutional and political framework under which social choices can be negotiated and biodiversity-friendly management practices can be implemented. In order to sustainably manage forests, a basic enabling framework of institutions, policies, and laws has to be in place. This includes (ITTO, 1998) among others institutional arrangements, sound policy and legal frameworks and appropriate economic environment. Proper institutions are a basic requirement for action towards sustainable forest resources development. Over the longer term, countries will need to develop the ability to learn and institutionalize new roles and new performance standards with respect to sustainable forest management. Of critical importance will be two-factors: first, the enhancement of the role of groups outside government such as non-governmental organizations and the private sector; and second, the shift to a more cross-sectoral approach to the design and implementation of sustainable forestry practices.

References

Bowles, I. A., Rice, R. E., Mittermeier, R. A. & Da Fonseca, G. A. B. Logging and tropical forest conservation. *Science* **280,** 1899–1900 (1998).
Carbarle, B. J. Logging on in the rain forests: Response to Bowles *et al.* 1998. *Science* **281,** 1453–1454 (1998).
Chazdon, R. L. Topical forests-log 'Em or Leave 'Em. *Science* **281,** 1295–1296 (1998).
Grieser Johns, A. Timber production and biodiversity conservation in tropical rain forests. *Cambridge Studies in Applied Ecology and Resource Management* (Cambridge University Press, UK, 1997).
ITTO. Criteria and Indicators for Sustainable Management of Natural Tropical Forests. ITTO Policy Development Series 7, Yokohama, Japan (1998).
Rice, R. E., Gullison, R. E. & Reid, J. W. Can sustainable management save tropical forests. *Sci. Am.* **276,** 34–39 (1997).
Rice, R., Sugal, C. & Bowles, I. *Sustainable Forest Management: A Review of the Current Conventional Wisdom* (Conservation International, Washington, DC, 1998).
Szaro, R. C., Sayer, J. A., Sheil, D., Snook, L. & Gillison, A. *Biodiversity Conservation in Production Forests* (Center for International Forestry Research, Bogor, 1999).

Tropical Ecosystems: Structure, Diversity and Human Welfare.
Proceedings of the International Conference on Tropical Ecosystems
K. N. Ganeshaiah, R. Uma Shaanker and K. S. Bawa (eds)
Published by Oxford–IBH, New Delhi. 2001. pp. 153–154.

Modelling tropical biodiversity and species distribution using environmental factors derived from global remote sensing data

Sassan Saatchi* and Donat Agosti**

Jet Propulsion Laboratory, California Institute of Technology, 4800 Oak
Grove Drive, Pasadena, CA 91109, USA
**American Museum of Natural History, New York, USA
*e-mail: saatchi@congo.jpl.nasa.gov

Conservation of biodiversity on a global scale and particularly in tropical regions requires datasets that can provide information about the geographical distribution of species, environmental factors that define the resilience of ecosystems and species habitats, and processes that create or change the habitats. In this paper, we present the results of a model developed to map the distribution of species in tropical regions based on environmental parameters derived from remote sensing and GIS data. These parameters range from climate conditions such as temperature and rainfall, and landscape and vegetation properties such as elevation, slope, vegetation type, structure, moisture condition, seasonality, and productivity. The model is based on a stochastic decision rule approach that associates probabilities to species presence on a geographical pixel based on the existing specimen data from museums and environmental parameters. The application of the model to tropical regions of south America and central Africa is presented in this paper. The model maps the presence of species on

Keywords: Biodiversity, remote sensing, decision rule model, tropics.

gradients of climate and landscape variabilities. These variabilities often coincide with natural factors that provide the environmental stability for existing biological communities and species abundance. Although the model concentrates on the deterministic relationships between habitats and the environment, the renewable parameters from remote sensing data allow the model to incorporate the spatio-temporal heterogeneities and dynamics as a result of changes in landscape and climate conditions. We discuss the application of the model for conservation planning such as the development of protected areas and biological corridors.

Acknowledgement

This work was carried out at the Jet Propulsion Laboratory, California Institute of Technology, under contract from National Aeronautic and Space Administration.

Tropical Ecosystems: Structure, Diversity and Human Welfare.
Proceedings of the International Conference on Tropical Ecosystems
K. N. Ganeshaiah, R. Uma Shaanker and K. S. Bawa (eds)
Published by Oxford–IBH, New Delhi. 2001. pp. 155–158.

Forest cover changes and its implication on wildlife diversity in Langat Basin, Malaysia

Saiful Arif Abdullah* and Mohd Nordin Hj. Hasan

Institute for Environment and Development (LESTARI), Universiti Kebangsaan
Malaysia, Bangi, 43600 Selangor Darul Ehsan, Malaysia
*e-mail: saiful@pkrisc.cc.ukm.my

The Langat Basin is a large-scale ecosystem that is undergoing rapid development. The Basin is situated adjacent to the Klang Valley (the most urbanised region in Malaysia) and has an area of about 2900 km^2 (Nordin and Azrina, 1998). Many new Federal Government development projects such as the new administrative capital of Malaysia (Putrajaya), the Kuala Lumpur International Airport (KLIA), a new intelligent city (Cyberjaya), the Multimedia Super Corridor (MSC) were initiated in 1996 and are all located within this Basin.

The Basin had experienced considerable depletion of forest cover over the last three decades with the large-scale conversion of forest areas into agricultural plantations that began in the early 1960s (Wong, 1974). The recent development of the projects mentioned above marked a new phase of development that further threatens mainly the remaining peat swamp and mangrove forests of the coastal plain, vestiges of lowland dipterocarp forest in the middle section of the Basin and to a lesser extent the hill dipterocarp forest areas to its north and northeast. With loss of forest cover there is at present less habitat for wildlife.

Keywords: Forest, wildlife, ecosystems, Langat Basin, Malaysia.

In 1974 forests in the Langat Basin constituted approximately 43% of the total land cover (Table 1) (Anon, unpublished). By 1981, forest cover had declined a further 5% and by 1995 only 31.2% of the Basin was under forest cover. A year later only 25.1% of land in the Basin was covered by forest. The rate of forest cover loss during the period from 1981 to 1995 (24%) was almost five times higher than the rate of loss over the period 1974 to 1981 (5%).

Table 1 shows that reduction of forest cover was concomitant with expansion of agricultural and urban areas in the Basin. During the period between 1974 and 1995, urban areas in the Basin increased more than five-fold (571%). At the same time agricultural areas increased by about 17%. The rate of spread of urban cover in the Basin increased significantly during the period between 1981 and 1995 (250%) being almost threefold the rate during the period between 1974 and 1981 (92%).

Depletion of forested areas in the Basin left fragmented forest patches of varying sizes (Schelhas and Greenberg, 1996). In the Langat Basin the largest tract of forest is the dipterocarp forest (lowland and hill), which is mostly located to the north and northeastern sections of the Basin. This forest is contiguous with forests on the hinterland of Peninsular Malaysia along the Main Range. In the past three decades most of the lowland dipterocarp forests in the Basin have been converted into other land uses such as for rural human settlements, urban as well as industrialized areas

Mammalian wildlife diversity (mammals) in the Basin was assessed from published and unpublished data (Lim, 1997) and the findings were correlated to the reductions in forest areas described above for the years 1950, 1975 and 1987/1988. The study revealed that, in 1950 a total of 85 species from 22 families of mammals have been recorded (Table 2). The number of species recorded represented 38% of the total number of mammals species that have been recorded in Peninsular Malaysia (Medway, 1983). Twenty-five years later, the number of species recorded was approximately 36% lower. By 1987/1988, the number of species recorded represented only 15% of the total number of mammals known to exist in Peninsular Malaysia.

Table 1.   Land use changes in the Langat Basin, 1974 to 1995.

Land-use category	1974 (km²)	1981 (km²)	1995 (km²)
Forest	1274.7	1210.6	915.1
Urban	20.2	38.8	135.7
Agriculture	1477.9	1539.2	1735.5
Others	164.9	149.1	151.4
Total	2937.7	2937.7	2937.7

Source: Anon (unpublished).

Table 2. Number of species recorded in the Hulu Langat Forest Reserve in 1950, 1975 and 1987/1988.

Order	Family	Number of species		
		1950	1975	1987/1988
Insectivora	Erinaceidae	2	1	–
	Soricidae	4	1	1
Dermoptera	Cynocephalidae	1	–	–
Chiroptera	Pteropodidae	7	5	4
	Emballonuridae	2	1	1
	Nycteridae	1	–	–
	Megadermatidae	1	–	–
	Rhinolophidae	12	9	6
	Vespertionidae	5	5	2
Scandentia	Tupaiidae	3	2	1
Primates	Lorisidae	1	1	–
	Cercopithecidae	3	3	2
	Hylobatidae	2	–	–
Pholodita	Manidae	1	–	–
Rodentia	Sciuridae	15	8	4
	Rhizomyidae	1	–	–
	Muridae	10	10	7
	Hystricidae	2	1	1
Carnivora	Mustelidae	2	1	–
	Viverridae	6	3	2
	Felidae	2	1	–
Artiodactyla	Tragulidae	2	1	–
	Total	85	53	32

Source: Lim (1997).

Forest clearance and fragmentation in the Basin is bound to affect not only the availability of areas that can act as wildlife habitats but also the quality of wildlife that can survive. Further, forest clearance is anticipated to lead to a disruption of the resilience of the Basin ecosystem and result in the emergence of many of the ecosystem distress syndromes normally associated with degraded ecosystems. Such effects will ultimately affect the health of the Langat Basin ecosystem.

Acknowledgement

This paper synthesises the results of studies by researchers from the Institute for Environment and Development (LESTARI), Universiti Kebangsaan Malaysia and elsewhere.

References

Anon. Unpublished. Land resource in Langat Basin In: MATREM 2 Training Course Module: Incorporating ecosystem health concepts in environmental Management: Langat Basin Pilot Study.

Lim, B. L. Small mammals study (1950–1988) in relation to environmental changes in the Langat Basin. Report submitted to the UNDP/ISIS Programme Research Grants for small-scale projects on the environment and development (1997).

Medway, L. *The Wild Mammals of Malaya (Peninsular Malaysia) and Singapore*. 2nd edition (Oxford University Press, 1983).

Nordin, M. & Azrina, L. A. Training and research for measuring and monitoring ecosystem health of a large-scale ecosystem: The Langat Basin, Selangor, Malaysia. *Ecosys. Health* **4**, 188–190 (1998).

Schelhas, J. & Greenberg, R. Introduction: The value of forest patches. in *Forest Patches in Tropical Landscapes* (eds Shelhas, J. & Greenberg, R.) (Island Press, 1996).

Wong, K. H. *Land use in Malaysia*. (Ministry of Agriculture, Kuala Lumpur, 1974).

Tropical Ecosystems: Structure, Diversity and Human Welfare.
Proceedings of the International Conference on Tropical Ecosystems
K. N. Ganeshaiah, R. Uma Shaanker and K. S. Bawa (eds)
Published by Oxford–IBH, New Delhi. 2001. pp. 159–161.

Flora, forest classification and diversity on Betung Kerihun National Park

Albertus Tjiu*[†]*, Herwasono Soedjito**,*
Tukirin Partomihardjo**, Syahirsyah* and Lisa Curran

*WWF Betung Kerihun National Park Project, Pontianak, Indonesia
**Herbarium Bogoriense, LBN-LIPI, Bogor, Indonesia
[†]e-mail: wwf@pontianak.wasantara.net.id

Indonesia is an archipelago country with five large islands. Indonesian Borneo (Kalimantan) is the world's third largest island with a land area of 451,865 km^2. It is located at the eastern edge of the Sunda shelf as part of the Sundaic portion of the Old World Tropics. The Betung Kerihun National Park (BKNP), one of the six national parks in Kalimantan, is the largest conservation area in West Kalimantan. It was established on 5 September 1995 by the Ministry of Forestry of Indonesia, within the administrative district of the Upper Kapuas Hulu. The Park is located between 112°15′E to 114°10′E and 0°40′N to 1°35′N, covering an area of almost 800,000 ha. Based on recorded climatology data (1976–1995), the average rainfall in this area has remained relatively stable ranging from 4400 to 4.620 mm per year; such rainfall areas could be classified as type A.

Until the 20th century, approximately 90% of Borneo was under natural forest cover. At present, forests still cover more than 60% of the land area, though most have been disturbed by human activities. Borneo is the centre of the distribution for the paleotropical Dipterocarpaceae, a family of trees with 267 known species (34% of which are endemic) and 59 genera unique to

---

Keywords: Forest classification, Indonesia, National Park, Kalimantan.

the island. Dipterocarps dominate Borneo's lowland and hill forests and form the bulk of the variable timber species. On lowland podzolic soils mixed forests are dominated by the family Dipterocarpaceae; these extend into hilly areas forming dipterocarp forest of a different species composition on slopes from 300 to 900 m a.s.l. Above 1000 m, the forest canopy is lower and a submontane forest type is increasingly common above 1300 m.

An ecological study was initiated in Betung Kerihun National Park, Hulu Bungan Kapuas, Embaloh, Sibau rivers and its tributaries on May–June, November 1996–January 1997, July–August 1997, and September–October 1997. Embaloh, where a large altitudinal gradient provides ecosystems from lowland dipterocarp to montane forest, contains a history of traditional agriculture and provides a diversity of forest successional stages. The Sibau site possesses a long history of undisturbed lowland forest. The Bungan site has a distinctly different geological history than the above areas; in addition to lowland dipterocarp forest, this region contains both the highest elevational zone (1,960 m) and a limestone forest on Mt. Kerihun.

The study involved the establishment of a number of ecological plots to record and identify the flora and their characteristics. Forty nine plots, each of 10×50 m, were established across an altitudinal gradient from 150 to 1150 m a.s.l. The plots were distributed either along the slopes or across ridges depending on the nature of the terrain. Each plot was further divided into ten subplots of 5x10 m each. Poles tied with yellow or orange flags were planted at the corners of each main plot and subplot to mark out the boundaries. In each main plot, all trees attaining the diameter of 10 cm dbh and over were measured and identified. Whenever necessary, leaf samples were collected for identification in the herbarium. These were pressed in old newspapers back at the camp and preserved in 55% methylated spirit or 75% alcohol in plastic bags.

Over 2, 749 trees (> 9.5 cm dbh) and 695 species representing 156 genera and 63 families were recorded; 118 species remain unidentified. Members of the Dipterocarpaceae are common accounting for 844 individuals (30.7%) with total basal area 84.2 m^2 (48%). The most common families are Myrtaceae (296 individuals, 4 genera, 28 species), Euphorbiaceae (185 individuals, 20 genera, 73 species), Guttiferae (121 individuals, 4 genera, 33 species), Myristicaceae (93 individuals, 4 genera, 28 species) and Burseraceae (87 individuals, 3 genera, 30 species).

Overall, we delineated six major forest formations in Betung Kerihun National Park: lowland dipterocarp forest, hill dipterocarp forest, summit ridge forest, submontane or lower montane forest, montane forest and alluvial forest. Within some areas, old secondary forest occurred, but *Shorea* (Dipterocarpaceae) is the single most dominant genera. The importance of this site for plant biodiversity and conservation will be discussed in light of land use patterns across West Kalimantan.

References

Kuswanda, M., Paul Chai, P. K. & Nengah Surati Jaya, I. ITTO Borneo Biodiversity Expedition 1997. Scientific Report. International Tropical Timber Organization. Yokohama, Japan (1999).

Partomihardjo, T., Albertus, Syahirsyah, Herwasono Soedjito. Flora Pohon Dan Tipe Hutan Taman Nasional Bentuang Karimun Kalimantan Barat. in Proceeding: RPTN Bentuang Karimun 2000–2024. Project PD 26/93 Rencana Pengelolaan Taman Nasional Bentuang Karimun. WWF Indonesia–PHPA–ITTO–ISBN 979-95102-3-6, Jakarta (1999).

Tropical Ecosystems: Structure, Diversity and Human Welfare.
Proceedings of the International Conference on Tropical Ecosystems
K. N. Ganeshaiah, R. Uma Shaanker and K. S. Bawa (eds)
Published by Oxford–IBH, New Delhi. 2001. pp. 162–163.

Consequences of forest conversion to coffee plantations on litter beetle and ant communities

Smitha Badrinarayanan[*,†]*, Jagdish Krishnaswamy*[*]*,
Sharachchandra Lele*[‡] *and K. Chandrashekara*[§]

*Wildlife Institute of India, P.O. Box 18, Chandrabani, Dehradun 248 001, India
[†]Institute for Social and Economic Change, Nagarbhavi, Bangalore 560 072, India
[§] Department of Entomology, University of Agricultural Sciences, GKVK Campus,
Bangalore 560 065, India
[†]e-mail: bsmitha@yahoo.com

There have been dramatic changes in land-use over the past few centuries. The effects of these alterations on larger organisms such as mammals and birds have been studied to a certain extent. In contrast, almost no information exists about changes that have occurred in highly diverse and functionally important groups such as the invertebrates (Watt *et al.*, 1997).

The changes in litter ant and beetle communities resulting from conversion of forests to two types of coffee plantations were studied in the humid tropical forest belt of Chikmagalur district in the Western Ghats of Karnataka, India. Four replicate blocks with homogeneous topography, each containing a control forest and two adjacent coffee plantations, one with polyculture shade trees and the other with monoculture silver-oak (*Grevillea robusta*) shade trees, were chosen for the study. Pitfall traps ($n = 7$) were placed along 2 transects in each of the land-use treatments. They were used to sample for arthropods occurring in the litter. There were two rounds of

Keywords: Tropical forest conversion, coffee plantations, diversity, ants, beetles.

sampling at the beginning and end of the dry season. Beetles and ants captured in the traps were removed and sorted to the level of morphospecies. Microclimatic parameters and litter characteristics were measured at each site. Composition and structure of the vegetation was assessed at each of the sites. Measures of diversity and several non-parametric estimators of diversity were used to compare forests and the two types of coffee estates for their ant and beetle communities.

Preliminary analyses indicate that there are large differences in the composition of the litter beetle and ant communities in the forests and coffee estates. Beetle morphospecies richness and Shannon's diversity are highest for the forests, and lowest for the monoculture-shade coffee plantations, with intermediate richess in the polyculture-shade coffee plantations. The number of ant morphospecies observed in the forests and coffee estates is almost the same. The ant communities occurring in the monoculture-shade coffee plantations are dominated by fewer species, leading to a lower Shannon's diversity than in the other two land-use types. The total number of individuals of beetles is highest in the forests, followed by polyculture-shade coffee plantations, and much lower in the monoculture coffee. The total number of ant individuals, however, is much higher in the silver-oak shade coffee plantations than in the polyculture-shade coffee plantations and lowest in the forests. Estimators of diversity that give importance to rare species, such as Chao1 and Chao2 (Anderson and Ashe, 2000), give similar trends for beetles as the Shannon's index, although not as marked. These estimators of diversity for the ants have the highest values in coffee polyculture-shade plantations, followed by the monoculture-shade plantations.

Conversion of forests, especially to monoculture shade coffee, is causing the local extinction of many beetle species and the proliferation of a few ant species. The ecological implications of such changes are currently unknown. Incentives to coffee planters and other measures to maintain polyculture shade and protect remnant forests within a mosaic of land-uses are urgently required.

Acknowledgements

We thank the Ashoka Trust for Research in Ecology and the Environment (ATREE) for facilitating this study and providing infrastructure and computer support. We also thank Dr G. K. Bhat of J. C. B. M. College, Sringeri for providing laboratory space, facilities and support.

References

Anderson, R. S. & Ashe, J. S. Leaf litter inhabiting beetles as surrogates for establishing priorities for conservation of selected tropical montane cloud forests in Honduras, Central America (Coleoptera; Staphylinidae, Curculionidae). Biodiver. Conserv. 9, 617–653 (2000).

Watt, A. D., Stork, N. E., Eggleton, P., Srivastava, D., Bolton, B., Larsen, T. B., Brendell, M. J. D. & Bignell, D. E. Impact of forest loss and regeneration on insect abundance and diversity. in Forests and Insects (eds Watt, A., Stork, N. E. & Hunter, M.) 273–286 (Chapman and Hall, London, UK, 1997).

Harnessing Market Forces for Biodiversity Conservation

Tropical Ecosystems: Structure, Diversity and Human Welfare.
Proceedings of the International Conference on Tropical Ecosystems
K. N. Ganeshaiah, R. Uma Shaanker and K. S. Bawa (eds)
Published by Oxford–IBH, New Delhi. 2001. pp. 167–171.

Challenges and possibilities in the development of non-timber forest products for conservation and sustainable development: Experience from West Kalimantan, Indonesia

R. Rudijanta Utama

Yayasan Dian Tama, Jl. Cendrawasih 53 B, Pontianak, 78111, Indonesia
e-mail: diantama@pontianak.wasantara.net.id

The condition of the forests of West Kalimantan is a cause of great concern. A particularly long dry season in 1997 resulting from the climate change brought by El Niño caused severe wildfires, destroying wide swaths of forested land in the province. The on-going economic crisis in Indonesia and the weak condition of the government have resulted in a dramatic increase in illegal logging. Over-extraction of several commodities, such as gold, has also led to the destruction of forests.

However, the awareness of the necessity of maintaining the forest's continued viability has given rise to a variety of innovative efforts to prevent the destruction of forest resources and encourage sustainable development practices. The development of non-timber forest products (NTFPs) represents one promising strategy for meeting the twin goals of conservation and development. In West Kalimantan, Yayasan Dian Tama (YDT) has gained

Keywords: Sustainable development, non-timber forest products, West Kalimantan.

significant practical experience and learned several lessons from projects in the well-known Social Forestry Development Project (SFDP) and in the Danau Sentarum National Park, that focus on NTFPs as a significant source of alternative income for local communities, enabling them both to improve their well-being and protect the forest upon which they depend for their livelihood.

Collaboration with local communities must be rooted in their basic needs and must start with what they can accomplish by themselves. Beginning from the fulfillment of basic needs, local communities can improve their overall quality of life. Community empowerment should take the form of gradual, participatory education and must emphasize a harmonious synergy between humanity and environmental sustainability.

Marketing strategies based on readily available natural resources need to be conceptualized and developed with the goal of forest conservation that is appropriate to local communities. An inventory of available natural resources and assessment of the skills and knowledge of local communities will become the basis for enterprise-building. This kind of inventory-taking often requires significant effort and comes with a high price tag. Technical problems concerning land management claims frequently occur as the official borders of different villages and hamlets often overlap with those delineated by customary law and authoritative ownership is often difficult to establish. This complicates the collection of data where plants sometimes have different names in different areas. An accurate inventory of plant species such as bamboo, rattan, and resin-producing trees relevant to NTFP development necessitates a careful examination and comparison across geographic areas. This, in turn, often requires special skills and knowledge, which may be expensive. Furthermore, the need to differentiate between the *total* availability of a given resource in a particular area and its *practical* availability represents another difficulty.

In addition to these issues, a crucial factor in the process of strategic commodity selection is the current and future availability of natural resources selected for NFTP development. For example, in the SFDP area, which comprises some 102,000 hectares in the Sanggau region of West Kalimantan, there are 10 different forest types due to variation of relief and soil, 800 different tree species with a widely varying distribution. Without including other plant species such as epiphytes, grasses, etc., there are over 1,400 known local uses of trees in the SFDP area. From a single given species useful products may be developed from the roots, stems, flowers, leaves, bark, or seed, and may be in fresh, dried, or powdered form (medicinal products), be reshaped (baskets and other handicrafts), or be refined (resins, essential oils). This picture points to the wide potential of just one species for NTFP development. However, while there may be much potential in one species, in the development of NTFPs its short as well as long-term availability must be factored into business plans.

Frequent changes in fashion mean that maintaining the profitable production of consumer goods such as women's handbags made of rattan or bamboo requires creativity. Given the market fluctuations, demand can change dramatically. For instance, it may rise suddenly 10 to 100 times or may disappear altogether due to a change in the style preferences of consumers. By contrast, it is much easier to make a business plan that includes quality, price, and availability for commodities that are not dependent on following the latest fashion trends, such as honey, resins, or charcoal made from forest waste products.

Quality control measures represent another factor that NTFP enterprises must confront. Whereas previously quality control took place at the factory level, given changes in the market, it must also take into account quality standards at an international level. For example, based on thinking from developed countries, the International Standards Organization (ISO) has delineated clear, uniform standard procedures for the production of NTFPs as well as other products. Buyers look for an assurance of quality as well as an assurance that the product is in line with conservation goals. This is not a bad idea, but from a practical business level perspective the process of securing assurance that a product was produced at a sustainable harvest level can be difficult and expensive. The concept of Forest Steward Council (FSC) is only feasible for large-scale enterprises, but the majority of NTFP enterprises are small to medium scale. Indigenous people in local communities have traditionally harvested products from the forest in a sustainable manner. This practice, even though not officially certified, is in line with FSC concepts. The problem is that if a commodity becomes quite valuable, people from outside communities that have traditionally managed it might introduce unsustainable extraction practices.

Among the many technologies that are used in the world of NTFPs, charcoal represents an especially interesting case as it can be made from almost any forest waste product, such as coconut shells, sawdust, or unused bamboo. Among other applications, it can be used as biomass energy or as a material for soil improvement for agriculture or forestry. Initial investment is relatively cheap and it can be made nearly anywhere. Its production opens space for productively using material that would otherwise be burned or go unused. This is one way many NTFPs (as well as timber) may follow a SCIMPO strategy (single commodity input multi-product output), which makes efficient and environmentally-friendly use of forest products.

Table 1. Initial matrix for selection of NTFPs for development

Potential NTFP	Market scope: Potential buyers available		
	Locally	Within country	Internationally
There is potential product There is technical skill There is selling	(e.g. tapped rubber)	(e.g. forest-derived sweeteners)	(e.g. crumb rubber)
There is potential product There is technical skill No selling	(e.g. damar resin)	(e.g. medicinal plant material)	(e.g. medicinal plant material)
There is potential product No technical skill No selling	(e.g. essential oil)	(e.g. activated carbon)	(e.g. bamboo activated carbon)

Many forest products must undergo a lengthy processing chain before they can be marketed as a final product. Other NTFPs, however, require relatively little processing from the harvest to final product stage. One person may manage the whole process, as is the case for many handicrafts made from bamboo or rattan. Those products which require many processing steps may involve many segments of communities and are better suited to sub-sector development. Such products may also involve the use of alternative technologies, result in specialization, be more dynamic and of higher quality than NTFPs that do not undergo such processing, and be sold at more competitive prices. These commodities may have a stronger impact on local communities because, in addition to involving many people and strengthening skills at different stages in production, they also are likely to attract the attention of government, business, and research actors.

Table 1 represents a useful tool for assessing the possible range of NTFPs that may be selected for development. It can help us understand more about what products are potentially available and how to shape appropriate production and marketing strategies. In addition, the table can serve as a basis for beginning sub-sector analysis of a particular commodity.

In addition to the issues discussed above, one of the major conclusions drawn from the YDT experience is that NTFP development and conservation planning must be an interactive process leading to better long-term policy (de Jong and Utama, 1998). In terms of the movement of NTFP marketing strategies, government regulation – or lack thereof – play a crucial role. Clear communication is the key to productive cooperation so that all parties can understand and help each other.

Acknowledgements

I thank Ganesan Balachander and Wil de Jong for continuing support, the staff at Yayasan Dian Tama and P. D. Dian Niaga, the BioDiversity Conservation Network. I also thank Maura Kondo and Daniel Miller for translation and editorial assistance.

References

De Jong, W. & Utama, R. Turning ideas into action: Planning for non-timber forest product development and conservation. *Incomes from the Forest: Methods for the Development and Conservation of Forest Products for Local Communities* (eds Eva Wollenberg & Andrew Ingles) 43–55 (1998).

171

Tropical Ecosystems: Structure, Diversity and Human Welfare.
Proceedings of the International Conference on Tropical Ecosystems
K. N. Ganeshaiah, R. Uma Shaanker and K. S. Bawa (eds)
Published by Oxford–IBH, New Delhi. 2001. pp. 172–175.

Linking mountains, markets and biodiversity: The Mountain Institute and Ecotourism

Nandita Jain

The Mountain Institute, P.O. Box 2785, Kathmandu, Nepal
e-mail: njain@mountain.org

The mission of The Mountain Institute is to conserve mountain environments and cultural heritage while improving the livelihoods of mountain people. The Mountain Institute (TMI) brings over 25 years of field-based, participatory and scientific experience to address the challenges and opportunities facing mountain cultures, communities and conservation. The Mountain Institute programs operate in the Himalaya, Andes and Appalachain mountain ranges. TMI provides active advisory support to mountain programs around the world covering a range of topics and activities.

For developing countries, their natural and cultural heritage is a primary attraction for increasing numbers of international and domestic visitors. Tourism associated with natural and protected areas, and cultural experiences continues to be a growing sector in the global tourism industry, with an estimated $110 billion in receipts for adventure travel alone in 1998. Growth forecasts ranging from 10% to 30% also make nature-related tourism among the fastest growing sectors in the global economy. Together with related conservation-linked enterprise development, ecotourism thus

Keywords: Mountains, ecotourism, markets, biodiversity, incentives.

represents a potentially significant source of revenue and opportunities that would address both economic and conservation problems of mountain and other communities. However, unchecked and unmanaged tourism has been shown to increase threats to fragile ecosystems, such as those in mountain regions.

For TMI, Ecotourism is responsible tourism that seeks to minimise negative impacts, generates economic benefits for participants, provides an ideal visitor experience and is supported by enabling policy frameworks. Our approach focuses on rigorous economic and environmental analyses with active and meaningful participation by key stakeholders from local villagers to policy makers and the commercial sectors. Key elements of the TMI ecotourism model are capacity growth, generating benefits, monitoring and mitigating ecological impacts, and policy reform.

TMI's ecotourism programs are characterised by collaboration and partnerships across sectors and scales of operation. While this creates a broad-based sense of ownership, it is also a challenging structure to participate in and to manage. Without clear definitions of responsibilities at the start, such arrangements have led to confusion, mistrust and miscommunication both within a participating group and among groups. With a focus on generating economic incentives to conserve, it has also been a challenge to make sure that enterprise activities are financially viable, both within the project and when replicated without project support. The critical element here has been to address issues of capacity in business planning, almost irrespective of the scale of operation. Even in the urban areas of mountain regions, larger tour operators and travel companies lack the skills and access to information that would improve their viability. In the absence of business planning skills, the immediate tendency in such cases is to reduce costs, and that invariably increases environmental costs even though careful business planning with a responsible attitude towards environmental concerns could be profitable. A related issue is vulnerability in the global market of tourism where customer tastes and decisions change quickly due to a variety of reasons (conflicts, devaluation of currencies, policies of neighbouring countries and so on). For suppliers of ecotourism products in remote and marginal areas, the impacts can be acute and unpredictable. The challenge is to develop and implement strategies that reduce or avert risks associated with product and market development. In some cases, the efforts of private entrepreneurs have been undermined by government-supported tourism operations where there is often little or no incentive to be profitable nor environmentally responsible. The challenge here is to integrate both financial and environmentally accountability for funds spent, or to reform policies so that the private sector has more opportunity to develop ecotourism products.

TMI Ecotourism Programs

India
- Sikkim Biodiversity and Ecotourism Project – working across community, private and public sectors in capacity building, economic development, monitoring and applied research, policy and planning.
- Eco-Development – Design of Ecotourism components at national and protected area levels (1994-1997)
- Kalimpong and Darjeeling, West Bengal – Ecotourism Planning Support for local organisations and entrepreneurs
- Ladakh – Planning for Community-Based Tourism (2001)

Nepal
- Makalu Barun Conservation Program – tourism development and management
- Langtang Ecotourism Project – ecotourism development and capacity building (1995–1999)
- Sustainable Tourism Network – co-founded a network of professional and practitioners in Nepal

China
- Qomolongma National Nature Preserve, Tibet Autonomous Region – Tourism Plan
- Peak Enterprise Program – Ecotourism training for tourism entrepreneurs
- Qomolongma Conservation Project – Community-Based Tourism activities

Peru
- Huascaran National Park – Multi-Stakeholder Ecotourism Plan (1997)
- Huayhuash – Ecotourism and Community-Based Tourism planning and implementation
- Huascaran National Park – Community-Based Tourism Projects with local stakeholders

International
- International Courses in Community-Based Tourism for Conservation and Development – 1999, 2000 in Nepal, 2001 in Thailand, in partnership with RECOFTC. Participants have subsequently initiated activities in Vietnam, China, Kyrgyzstan, India, Nepal, Bhutan and Indonesia.
- Community-Based Mountain Tourism – E-Conference on Mountain Forum, 1998
- Himal-Andes Ecotourism Exchange supported by Swiss Development Cooperation, 1999
- Transboundary Ecotourism Exchange – Nepal/China, 1998 (with ICIMOD)

One key success is that the hypothesis of economic benefits serving as incentives that lead to conservation action does have value, and can provide a framework for influencing local action and policy reform. For a porter in Sikkim, economic benefit from trekking tourism was a reason not to cut trees along the trail for a bonfire for visitors. Safeguarding the resources on which tourism depends has provided the reason for establishing local community-based organisations that manage and monitor local natural and cultural resources. In some cases the power that comes from increased understanding and having access to information, has enabled local groups to engage governments in debate over the direction of policy development for tourism and conservation. A focus on skill development and capacity growth has left a lasting legacy of local action in some sites. Individual entrepreneurs, especially those who made independent investments in 'greener practices' have profited from the marketability of responsible tourism approaches. Local organisations have continued to provide training to service providers, such as porters, with independent funding sources. Exchanges between tourism trade associations have continued, providing mutual support for responsible tourism development and marketing between countries. TMI's experience in ecotourism over the past decade indicates that there is a growing demand for methods that help communities and organisations plan for responsible tourism that results in conservation and economic benefits. The response and follow-up to the international courses in Community-Based Tourism, with over 50 professionals implementing tourism programs in eight countries, is encouraging and exciting. For people in marginal areas, yet rich in natural and cultural resources, tourism is one of several options that can contribute to conservation and development.

Tropical Ecosystems: Structure, Diversity and Human Welfare.
Proceedings of the International Conference on Tropical Ecosystems
K. N. Ganeshaiah, R. Uma Shaanker and K. S. Bawa (eds)
Published by Oxford–IBH, New Delhi. 2001. pp. 176–178.

Bioprospecting as a mechanism to conserve biodiversity and indigenous knowledge systems

Henk van Wilgenburg

Department of Pharmacology, Academic Medical Centre, University of Amsterdam,
1105 AZ Amsterdam, The Netherlands
e-mail: h.vanwilgenburg@amc.uva.nl

Bioprospecting has been practised for many years in different forms. In more recent times, it has attained significance with regard to the issue of sharing the benefits arising from bioprospecting. Article 8(j) of the Convention on Biological Diversity (CBD, 2000) calls on the Contracting Parties to respect, preserve and maintain the knowledge, innovations and practices of indigenous and local communities embodying traditional lifestyles. It also calls for the equitable sharing of benefits arising from the utilisation of such knowledge, innovations and practices. The issue of integrating equity principles in benefit sharing arrangements has been under the consideration of the Contracting parties since the 3rd meeting of COP held at Buenos Aires in 1966.

Certain critical issues remain unresolved, particularly in relation to how to go about legalising and formalising the bioprospecting process in a way which ensures that there is full and prior informed consent of fair and equitable benefit sharing with the originator of the knowledge and source that enable the bioprospecting. The absence of an internationally agreed methodology

Keywords: Bioprospecting, biodiversity, biopiracy, traditional medicines.

for sharing economic benefits from the commercial exploitation of biodiversity with the primary conservers and holders of traditional knowledge and information is leading to a growing number of accusations of biopiracy committed by business and industry in developed countries. Biodiversity in both developing and developed countries has been accessed for long time, for various purposes, by outside researchers, private companies as well as local communities, with little or no returns to conservation activities.

In the past, bioprospecting in developing countries has been the preserve of field researchers in universities and botanical gardens and until recently, most bioprospectors in developing countries have been individual professors and collectors. Bioprospecting is of particular importance in the field of traditional medicine. A large portion of the population in a number of developing countries still relies on traditional healthcare and local medicinal plants. There has been a recent growth of interest in traditional medicine from the international pharmaceutical industry in Europe and America. Traditional medicine has come to be viewed by the pharmaceutical industry as a source of qualified leads in the identification of bioactive agents for use in the production of synthetic modern drugs. In the industrialised world only a small portion of medicinal and aromatic plants are cultivated, most being imported from Asia and Africa. Therefore, bilateral collaboration on this issue of bioprospecting is of mutual importance and has to be elaborated upon within the context of biodiversity conservation. Bioprospectors express optimism that they can help by encouraging biodiversity conservation and capacity building in developing countries. Many indigenous people and local communities, however, are sceptical of existing bioprospecting agreements.

The most important underlying factors leading to loss of biodiversity, both directly and indirectly, include habitat loss due to human population pressure, increased urbanisation and over-exploitation. Increased demand often also leads to increased incidences of non-sustainable harvesting techniques. In general, the more lucrative it is to harvest a species, the more collectors will start harvesting and more effort that is put in harvesting. This can eventually lead to economic extinction of the desired plant. An example of cutting trees for bark is the yohimbe (*Pausinystalia johimbe*). It was estimated that in Cameroon 98% of these exploited trees are probably felled (Sunderland *et al.*, 2000).

Loss of biodiversity results in the loss of medicinal and aromatic plant species. The private sector involved in the business of herbal drugs should take responsibility. Governments can provide incentives for the involvement of the private sector in biodiversity conservation. There is a need to continue debate with true stakeholders–practitioners of traditional medicine, representatives of medical community, the pharmaceutical industry, intergovernmental organisations, etc. Lasting solutions can only be found if all stakeholders work together in good faith towards a common understanding and solution of problems. Many steps in the chain are necessary to make sustainable development a reality, from pure conservation to pure commercialisation. Conservation should also include the involvement of traditional medicinal practitioners and medicinal plant sellers in assessing rarity, as they are very aware of the availability of these plants. Other steps are in education and training, research and agroforestry and agriculture. In the value chain of products are many different interests including suppliers of raw materials, importers, producers and final retailers. Local production of pharmaceuticals with strict quality control would reduce the cost of medication whilst providing alternatives to plant-based medicines. More processing and local added value should be done in-country to obtain greater benefits for the local communities. The development of bioprospecting partnerships with developing countries should ensure maximum value addition to the region's indigenous plants.

An example of a successful initiative is the International Cooperative Biodiversity Groups (ICBG) program linking the discovery of drugs and genes for agriculture with biodiversity conservation. A combination of approaches includes technology transfer to countries in Latin America, the development of scientific infrastructure for drug and gene discovery, the training of scientists, the involvement of laboratories in research, and the equitable sharing of revenues among a diversity of institutions. For example, in Panama any royalties received from commercialisation will be shared among (i) a fund of the Panamanian government that supports the national park system, (ii) a Panamanian foundation funding conservation and sustainable development projects, and (iii) the three collaborating scientific institutions. In collaboration with Conservation International indigenous communities are helped to record their ethnobotanical traditions and provide educational and internship programs for indigenous participants. Using ecological insight for collection, promoting infrastructure development, and returning benefits to a broad range of institutions will have maximal impact on conservation both in the short- and long-term.

References

Convention on Biological Diversity (CBD), Nairobi, Kenya, 15–26 May 2000.
Sunderland, T., Tchoundjeu, Z. & Ngo-Mpeck, M. *The exploitation of Pausinystalia johimbe.* Medical Plant Conservation. Newsletter of the Medicinal Plant Specialist Group of the IUCN Species Survival Commission. Vol. 6, 21 (2000).

Tropical Ecosystems: Structure, Diversity and Human Welfare.
Proceedings of the International Conference on Tropical Ecosystems
K. N. Ganeshaiah, R. Uma Shaanker and K. S. Bawa (eds)
Published by Oxford–IBH, New Delhi. 2001. pp. 179–183.

The Forest Stewardship Council – Certification as a market-based mechanism for promoting responsible forestry

H. Cauley, H. Melchior, E. Lobecker, J. O'Connor,
B. Addlestone, W. Wilkinson and F. Arens

Forest Stewardship Council, Washington, DC, USA
e-mail: fscoax@fscoax.org

Founded in 1993, The Forest Stewardship Council (FSC) is the only international, independent, third party, forest certification scheme in the world. It is the most concrete result to follow the United Nations Conference on Environment and Development (UNCED) 1991 Rio Earth Summit.

As a point of departure for understanding the FSC and its context, one must first consider the movement from which it was born: the movement to define and implement 'sustainable development' principles and apply them to landscape conservation, 'forest certification' (Vallejo and Hauselmann, 2000).

Though the ideas behind sustainability had been gestating since the early 1970s, it was not until the late 1980s and early 1990s that workable definitions of the concept began to emerge. A seminal document for the sustainability movement, The Brundtland Report (1987), defined sustainable development

Keywords: Forest Stewardship Council, certification, responsible forestry.

as 'development that meets the needs of the present without compromising the ability of future generations to meet their own needs.'

Building upon this work five years later, the Forest Principles that emerged from Rio's UNCED process applied these concepts to natural resources, asserting that 'forest resources and forest lands should be sustainably managed to meet the social, economic, ecological, cultural, and spiritual needs of present and future generations'. This was to form the core of the FSC concept – a voluntary, non-governmental, market-based approach.

This theoretical balancing act needed a real-world structure to develop broad-based credibility, and it has come to us through the movement toward forest certification. Interestingly, forest certification predates the Rio Summit, with the 1989 founding of the Rainforest Alliance's SmartWood Program.

That program, later to become one example of the work under the FSC's governance, is the first instance where forest certification developed 'as an economic policy instrument to achieve environmental and economic objectives'. A program of forest certification, such as that which the FSC directs, includes 'a process by which an independent party (third party) assesses whether the forest management practices, in a specific management unit, fulfill a set of requirements', in the FSC's case, it's *Principles and Criteria for Forest Management*.

This independent forest assessment and monitoring process can be considered one of the means by which forest certification achieves its ends: 'to improve the environmental, social, and economic quality of forest management' around the world, and 'to ensure benefits for the managers and/or owners of certified forests'. Another important means it uses to reach these and other desirable ends is through the promotion of a credible brand that represents the quality of forest management in the marketplace.

The FSC is the only organization that uses a 'chain of custody', or an assurance tracking system, that monitors the flow of wood from a certified forest to finished products. The purpose of this monitoring is to ensure that products bearing the mark of the certification system (in the FSC's case, its 'checkmark and tree' brand) did, indeed come, in sufficient part or entirely, from a forest managed to the FSC's *Principles and Criteria for Forest Management*.

Within this framework, the FSC stands out as the first worldwide example of a program that brings all of the various ideas behind responsible forestry and forest certification together. The FSC is an independent, non-profit organization that defines and promotes responsible forestry. As a standard-setting body for a forest certification system, the FSC, through its *Principles and Criteria for Forest Management*, has established the highest standards for environmentally, economically, and socially responsible forestry. The FSC forest management standards consider key environmental issues such as minimizing clear-cuts, striving to eliminate pesticide use, and the protection of forests with high conservation value (ones that contain certain rare and endangered species), while remaining economically attractive to business.

The FSC mission statement remains virtually unchanged since 1993:

- The Forest Stewardship Council shall promote environmentally appropriate, socially beneficial, and economically viable management of the world's forests.
- Environmentally appropriate forest management ensures that the harvest of timber and non-timber products maintains the forest's biodiversity, productivity, and ecological processes.
- Socially beneficial forest management helps both local people and society at large to enjoy long-term benefits and also provides strong incentives to local people to sustain the forest resources and adhere to long-term management plans.
- Economically viable forest management means that forest operations are structured and managed so as to be sufficiently profitable, without generating financial profit at the expense of the forest resource, the ecosystem, or affected communities. The tension between the need to generate adequate financial returns and the principles of responsible forest operations can be reduced through efforts to market forest products for their best value.

The FSC is a membership-based organization. The FSC's unique membership system includes a balance of interests – economic, environmental, and social – that have impact on and influence over forestry, and these three 'chambers' jointly govern the organization. Its membership has grown over the past two years from 250 to over 500 members from over 50 countries, largely due to the increased interest in National Initiatives and standards development.

Also, there has been a significant growth in the number of FSC National Initiatives. There are presently 40 FSC national operations around the world, in countries with many diverse regions including Africa, Asia-Pacific, Europe, North, Central and South America. The products of their work, regional or national standards, have been approved by the FSC, including standards from diverse regions and countries, such as Belgium, Bolivia, the Maritimes region of Canada, Germany, the UK, and Sweden.

Finally, there has been progress in regard to the number of certification bodies performing forest management and chain-of-custody audits. In 1998, the FSC had accredited five certification bodies, two in the US, two in the UK,

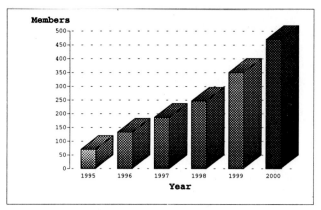

FSC membership growth.

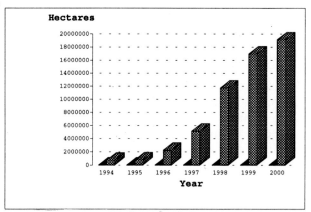

Hectares certified.

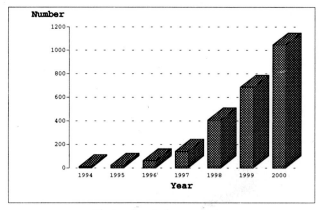

Chain-of-custody companies.

and one in The Netherlands. In less than two years, the FSC has 10 accredited certification bodies with newcomers considerably expanding the FSC's geographic range and including bodies in Canada, Germany, South Africa, Switzerland, and a third in the UK. There are also currently six applicant certification bodies, which are based in Canada, France, Italy, and the UK. Work is underway in several countries in Africa, Asia, and Latin America, namely Cameroon, Ghana and Nigeria, Indonesia and Malaysia, Brazil, Colombia, Costa Rica, and Ecuador. Increasing the number of accredited certification bodies in all parts of the world will ensure the ability of the FSC to grow with credibility.

The FSC membership is positioning the organization for continuing success in promoting responsible forest management. Key elements of the FSC vision for the future are:

- Continue to be inclusive of all groups interested in promoting responsible forest management;
- Maintain the balance of interests in the social, economic, and environmental considerations of responsible forest management;
- Safeguard the integrity of the brand by ensuring that standards are rigorous, accreditation of certifiers is consistent and thorough, and addressing the concerns of stakeholders in a fair, timely, and transparent manner; and,
- Expand FSC's efforts to realize forest values in a socially responsible manner by developing standards for the harvesting of non-timber forest products and carbon sequestration.

Reference

Uncited quotations in the body of this essay come from this publication, unless otherwise noted. Vallejo & Hauselmann, *Institutional Requirements for Forest Certification: A Manual for Stakeholders*, Working Paper 2, June 2000, Deutsche Gesellschaft fuer Technische Zusammenarbeit (GTZ) GmBH Forest Certification Project.

Tropical Ecosystems: Structure, Diversity and Human Welfare.
Proceedings of the International Conference on Tropical Ecosystems
K. N. Ganeshaiah, R. Uma Shaanker and K. S. Bawa (eds)
Published by Oxford–IBH, New Delhi. 2001. pp. 184–186.

Markets for environmental goods and services

Ganesan Balachander

The Nature Conservancy, Center For Compatible Economic Development, Leesburg,
Washington DC, USA
e-mail: gbalachander@tnc.org

Traditional approaches to conserving natural ecosystems have failed to successfully conserve the natural resource base. Such approaches relied heavily either on government regulations or on a forest or coastal community's voluntary code of natural resource protection. Government regulation has often proven difficult to enforce, has bred corruption, and been difficult to finance sustainably. Communities on the other hand have had difficulty managing the commons in the absence of good leadership, organization and homogeneous stakeholder groups. They often lack the necessary knowledge, capital, or legal rights to take full advantage of market opportunities where they exist. Against a backdrop of increasing concern about the global loss of forests and degradation of marine ecosystems, is a growing recognition of the value of ecosystem goods and services. *Can markets, under an appropriate institutional and regulatory framework, play a role in enhancing incentives for conservation?*

Mainstream markets still do not adequately recognize and reward the local communities for a host of environmental services generated by forests and marine resources that benefit society at large. But in the past decade, there has been a proliferation of new initiatives aimed at achieving conservation

Keywords: Ecosystem services, green enterprises, sustainable development.

while enhancing benefits for local communities through market-based mechanisms. Examples include biodiversity conservation, which protects medicinal plants for pharmaceutical or traditional use; watershed protection and attendant benefits for agriculture; and carbon sequestration to ameliorate global warming. These and other initiatives represent the following categories of instruments as summarized by the Katoomba Group organized by Forest Trends:

- Markets for environmental goods and services, with a number of buyers and sellers, approximating an efficient market
- Direct financial resource transfers to communities around forests and coastal resources which provide environmental services, and
- Indirect payments, through a price premium or cost subsidy, for products produced in ways that conserve or enhance environmental services.

Given the past experience, one may be skeptical as to why the private sector, driven primarily by the bottom line, would pay attention to sustainable development and biodiversity. The answer lies in the drivers for private sector investment in sustainable development, which are (1) markets, (2) cost savings, and (3) government policy.

The first driver is market demand for green products, which include certified organic agriculture, ecotourism, and certified forest products. While the volume is still small relative to the overall size of the market, their production is growing rapidly.

The second driver is internal to the firm and can produce cost savings through more efficient processes, sometimes referred to 'ecoefficiency'. Some far-sighted companies are also adopting responsible practices that enhance their image in the marketplace. These may include promotion of small community-owned enterprises that will also benefit the company's business.

The third driver is government policy in the form of regulations, tax, and subsidy signals. High interest rates, land tenure and allocation, road building and agricultural subsidies are examples of bad government policies that affect private sector investment and adversely affect the quality of the environment.

There have been a number of initiatives responding to this new opportunity. Carbon banking is one such initiative. The still unratified Kyoto protocol sets limits on carbon emission into the atmosphere, which has been implicated in global warming. To be in compliance, a power producing utility could choose to install more efficient equipment or invest in a forest conservation project in a tropical country, which will act as a sink for the carbon. The paper also provides examples of resource transfers for watershed protection (Quito water fee), venture capital for ecoenterprises, and green bonds.

Given the relative newness of 'green enterprises' that seek to achieve the triple bottom line of economic viability, environmental sustainability, and social equity, there are however, still a number of questions that need to be addressed in developing markets for environmental services:

- In what markets do forest services deliver benefits?
- What are the rights and responsibilities of suppliers and consumers?
- What support services are required to enable the market?
- Can the service be defined in a way that it can be measured and monitored?

There are still problems in estimating the financial value of environmental services, and in ensuring that the functioning of markets does not harm the poor. The assignment of property rights over natural resources is also key to the development of markets. Finally, all well-functioning markets depend on a strong institutional foundation, and this is even more important in establishing and maintaining markets for environmental services.

Tropical Ecosystems: Structure, Diversity and Human Welfare.
Proceedings of the International Conference on Tropical Ecosystems
K. N. Ganeshaiah, R. Uma Shaanker and K. S. Bawa (eds)
Published by Oxford–IBH, New Delhi. 2001. pp. 187–190.

A computer-aided participatory approach to monitor the local agrobiodiversity through market analysis of a fair price market in Madurai

B. Anitha, J. Arvindh*, V. Saravanan and
D. Winfred Thomas

Department of Botany, The American College, Madurai 625 002, India
*Department of Computer Science, Madurai Kamaraj University College,
Madurai 625 002, India

Agriculture and farming form the backbone of Indian economy. Though farmers in the country contribute to a major portion of the national economy, the socio-economic status of the farmers remains the poorest. In the Indian system of marketing, middlemen play a prominent role and they are the real beneficiaries. Uzhavar Sandai (US) is a giant fair price market, which has been recently established in Tamil Nadu to eliminate the middlemen. It aims at maximizing benefit for farmers and making the farm product available to consumers at a fair price. Both farmers and customers meet at this platform and share their benefits. Uzhavar Sandai in Madurai links 1066 farmers from thirteen taluks belonging to three districts with 8500 customers of Madurai

Keywords: Market analysis, fair price market, vegetables, monitoring, agrobiodiversity.

city. It has a record of selling vegetables worth Rs 2,51,675 in a day. The potential of US could be used for many scientific and management studies.

This study was undertaken to explore the possibilities of designing a participatory monitoring method to understand the status of local agrobiodiversity. Market analysis was coupled with participatory monitoring of biodiversity methods. Random selection method was used to collect data from the farmers and customers on demand, supply and selling potential of vegetables. Along with this area of cultivation, pathogen infestation and other related ecological and spatial data have been collected and analyzed.

Based on these information, we have developed a computer-aided database. The database could be used as a tool to understand the problems and the prospects. This package is designed to analyze the available data, link the agencies that collect relevant data and update the seasonal changes. Many users could share this database. It may be used to monitor the local biodiversity as well as help both the buyers and sellers of US to overcome the difficulties.

Fairs trade organizations (FTO) in general directly link low-income producers with giant market forces. Consumers of FTOs are needed to be educated about fairly traded products and its significance in local market. The salient features of a FTO are, fair wages and benefit sharing with community, cooperative work places, consumer education, environmental sustainability, financial and technical support, and public accountability. All over the world small enterprises are vanishing due to increase in globalization. In this juncture FTOs are gaining momentum and have attempted to compete with the multinational corporations.

Table 1. Successful benefit sharing agency – Madurai Uzhavar Sandai.

Rank	Vegetable	Mean quantity per day (kg)	Mean market rate (Nov. 99 to Nov. 00) (Rs)	Mean unit price for three months (Dec. 00–Feb. 01) (Rs)		
				Central market	Uzhavar Sandai	Local market
1	Tomato	2219	9.25	5.8	6.9	8.15
2	Aggregated onion	1256	9	7.35	8.45	9.55
3	Brinjal	1154	7.75	7.6	9.25	11.5
4	Bellary onion	1047	6.15	8	8.5	9.5
5	Potato	1020	8.75	6.35	7.5	8.7
6	Cauliflower	830	4.75	3.9	4.8	5.5
7	Ladies finger	662	7.25	6.7	7.8	9.65
8	Carrot	566	9	6.6	8.1	10.75
9	Cabbage	531	6	4.8	6.45	7.9
10	Chilly	336	13.5	8	9.55	12

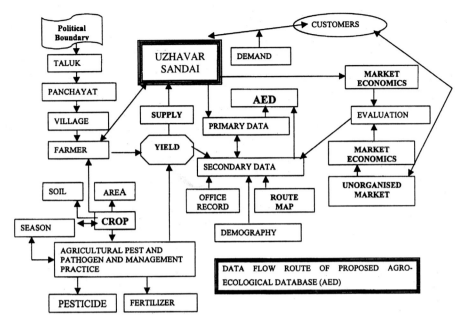

Figure 1.

In Tamil Nadu the concept of FTOs is being transformed into Uzhavar Sandai-related activities. Madurai US is one of the successful organizations in promoting the salient features of FTOs. Currently, US is projected as a benefit-sharing agency for the local farmers (Table 1). If the design of US is modified it could also be made to serve those institutions that are interested in monitoring the local biodiversity. We have identified a set of informants who are supplying primary data to the US. We have also created proper linkages between informants and the agencies that pool the data. Through this mechanism we have promoted continuous flow of data (Figure 1). If this data is processed and suitable inferences are synthesized, the US can become a decision support system for local people, conservation biologists, bureaucrats and any other agencies that are in search for agrobiodiversity-related data.

Modern agriculture scientists are towards precision farming. Unpredictable change in climatic conditions is a great challenge for them. Data from any source could be used to understand the relationship between climatic changes and crop response. Agriculture and land resource information system gives leads to many novel endeavours in crop genetics (Broten, 2000). If the data generated by the US is systematically documented, it could be also used in modelling plant growth patterns with reference to the change in the environmental conditions.

Acknowledgements

We are thankful to our principal Dr. D. Samuel Sudanandha and Prof. John Jebaraj Doraisamy, Head, PG Botany Department for allowing us to carry out this project. We also acknowledge Uzhavar Sandai authorities for helping us in data collection.

Reference

Broten, D. M. Agricultural and land resources information system. *GIS Dev.* IV: 22–25 (2000).

THEME II

TROPICAL FORESTS: STRUCTURE, DIVERSITY AND FUNCTION – PART A
Plant–Pollinator Interaction

Tropical Ecosystems: Structure, Diversity and Human Welfare.
Proceedings of the International Conference on Tropical Ecosystems
K. N. Ganeshaiah, R. Uma Shaanker and K. S. Bawa (eds)
Published by Oxford–IBH, New Delhi. 2001. pp. 193–195.

Forest disturbance and native bees in Brazil's Atlantic coastal forest

Anthony Raw*,† and Maria Elisa da Silva Santos**

*Departameno de Ciências Biologicas, Universidade Estadual de Santa Cruz,
Ilhéus, Bahia, Brazil 45650-000
**Departameno de Educação, Universidade do Estado da Bahia, Senhor de Bonfim,
Bahia, Brazil 45650-000
†e-mail: tonyraw@uol.com.br

Originally Brazil's Atlantic forest covered an area of 1,200,000 km² (Brown and Brown, 1992). In Bahia State it is now a mosaic of remnants of primary forest (3.5% of the original cover), cabruca (primary forest providing shade for cocoa), secondary forest, saplings, cocoa and other plantations, pasture, mangrove and dunes. We examined the effect of disturbance in the forest on the behaviour and ecology of one of the most important groups of pollinators, namely the Euglossini. Large bees are important pollinators of neotropical trees and vines (Frankie *et al.*, 1983, pers. obs.). These bees include members of the tribe Euglossini. The Euglossini possess several characteristic features that make them highly suitable to study:

1. Adults of both sexes are keystone species (and visit different flowers).
2. Almost all the species inhabit the forest canopy (which often suffers severe damage).
3. There are adequate number of species (> 40 species in the cocoa region).
4. The use of scents facilitates easy collecting.

Keywords: Forest fragments, land use, Euglossini, Brazil, Atlantic coastal forest.

5. The collecting method is standardized.
6. Only the males are attracted to the baits (so collecting has limited impact on the bees' populations).

Five traps were used at each of 39 sites, each trap with one scent: methyl salicilate, eugenol, 1,8-cineole, benzyl acetate and vanillin. A total of 1,912 individuals and 32 species were recorded. The bees were most attracted to benzyl acetate (50% of the visits) and methyl salicilate (25%). Most localities had substantial faunas of euglossine bees and not all species are present at all localities.

Our results demonstrate the effects of vegetation type on the number of euglossine bees (Figure 1). The greatest abundance and species diversity were in large areas of primary forest. The diversity diminished progressively with greater disturbance to the original vegetation; thus it was less in primary forest fragments followed by secondary forest and cabrucas with fewest in highly disturbed and open areas. In fact, Powell and Powell (1987) reported that the number of euglossine bees were sharply reduced in small forest remnants.

Several factors explain the differences in diversity and abundance of these bees in the vegetation types. Species vary in their preferences for vegetation types; some are totally or principally restricted to primary forest, some tolerate a little disturbance, some are generalists occurring in all the vegetation types, while a few are more common in disturbed areas. The

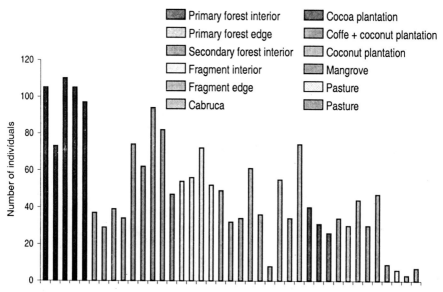

Figure 1. Effect of 13 vegetation types on the number of euglossine bees at 39 localities in southern Bahia State, Brazil.

number of bees in forest fragments and cabrucas varied considerably which could be due to the 'quality' of the vegetation or of the types of neighbouring vegetation. There were fewer individuals and species at the edges than in the forest interiors, with slightly fewer 30 m from the edge and far fewer 5 m from the edge.

The effects of fragmentation also depend on the distance between fragments and on the types of intervening land cover. Some euglossine species move readily across open areas (Raw, 1989) so the mosaic of forest fragments is likely to affect them less than it does those species which do not move across fragments (Becker et al., 1991).

The results demonstrate that in a well-defined taxonomic and ecological group the species differ substantially in the ways forest disturbance affects them and in their ability to deal with the changes in land use.

Acknowledgements

The research was conducted while the author held a visiting research fellowship of the Brazilian Science Research Council (CNPq) and a visiting professorship at the Universidade Estadual de Santa Cruz. The research is part of the Projeto RestaUna financed by the Biodiversity Programme (PRONABIO/PROBIO) of the Ministry of the Environment (MMA) with the support of CNPq, the Global Environmental Facility (GEF) and World Bank (BIRD). I thank Mr Helfred Hess and the Mr Mario Bunchaft, both of Ilhéus for permission to conduct research on their properties.

References

Becker, P., Moure, J. S. & Peralta, F. J. A. More about euglossine bees in Amazonian forest fragments. *Biotropica* **23**, 586–591 (1991).

Brown, K. S. Jr. & Brown, G. G. Habitat alteration and species loss in Brazilian forests. in *Tropical Deforestation and Species Extinction* (eds Whitmore, T. C. & Sayer, J. A.) 119–142 (Chapman & Hall, London, 1992).

Frankie, G. W., Haber, W. A., Opler, P. A. & Bawa, K. S. Characteristics and organization of the large bee pollinator system in the Costa Rican dry forest. in *Handbook of Experimental Pollination Biology* (eds Jones, C. E. and Little, R. J.) 411–447 (Van Norstrand Reinhold Co. Inc. New York, 1983).

Powell, A. H. & Powell. G. V. N. Population dynamics of male euglossine bees in Amazonian forest fragments. *Biotropica* **19**, 176–179 (1987).

Raw, A. The dispersal of euglossine bees between isolated patches of eastern Brazilian wet forest (Hymenoptera, Apidae). *Rev. Bras. Entomol.* **33**, 103–107 (1989).

Tropical Ecosystems: Structure, Diversity and Human Welfare.
Proceedings of the International Conference on Tropical Ecosystems
K. N. Ganeshaiah, R. Uma Shaanker and K. S. Bawa (eds)
Published by Oxford–IBH, New Delhi. 2001. pp. 196–197.

Spatial and temporal foraging dynamics of giant honey bees in an Indian forest

Puja Batra

Department of Zoology, Michigan State University, East Lansing MI 48824, USA
e-mail: batrapuj@msu.edu

Honey bees (genus *Apis*) are ubiquitous and important pollinators in tropical Asia. In India, up to three sympatric species of honey bees may exist in any given location; thus investigation of their foraging and pollination dynamics is vital for understanding plant–pollinator relationships in the Paleotropics. I report here the results of a study on the spatial and temporal dynamics of foraging in the giant honey bee, *Apis dorsata*, in the Biligiri Rangan Hills, a wildlife sanctuary in southern India that is comprised of a heterogeneous mix of habitat types.

The foraging ranges for pollinating insects have scarcely been studied in a manner that captures variation between individuals or colonies, variation that may occur due to changes over time in flowering conditions, and variation that may occur due to spatial heterogeneity in the landscape. This type of detailed information is of relevance to studies of plant–pollinator coevolution, effects of habitat fragmentation, and studies of pollinator-mediated gene flow and population genetic structure. Honey bees are extremely amenable to investigation of spatial foraging patterns, due to their communication system known as the 'dance language', which foragers use to convey information about the distance and direction of a food source to their

---

Keywords: *Apis dorsata*, pollinators, India, spatial, foraging, Biligiri Rangan Hills.

nestmates. By observing dances performed by returning *A. dorsata* foragers, I constructed 'forage maps' of their foraging locations and inferred their typical flight range and foraging area throughout the flowering season. Results indicate that 95% of all foraging flights occurred within a 3 km radius from the nest. Neither temporal scale nor spatial heterogeneity influenced flight range; that is, flight range did not vary across years, weeks or in different areas of the forest. Instead, colonies appear to shift their foraging dynamically throughout the season relative to each other such that they do not forage in the same patches as other colonies in the aggregation.

Using a GIS database on the landscape and forest structure and composition of BR Hills, I examined the forage maps in a more spatially explicit manner. Analysis of habitats used by *A. dorsata* colonies indicates that they do not consistently exhibit a preference for foraging in any specific habitat type. The GIS approach further allows me to relate foraging movements to spatio-temporal variation in resources when combined with flowering phenology data for known food plants.

In India and other parts of its range, giant honey bees often aggregate their nests such that a single tree or cliff side may harbour more than one hundred colonies. Furthermore, these locations are recolonized every year. These nesting aggregations are likely to be important 'epicenters' of pollinators, with hundreds of thousand of potential pollination events radiating out from them every day during the flowering season. The possible implications of this phenomenon for forest and population genetic structure are discussed.

Tropical Ecosystems: Structure, Diversity and Human Welfare.
Proceedings of the International Conference on Tropical Ecosystems
K. N. Ganeshaiah, R. Uma Shaanker and K. S. Bawa (eds)
Published by Oxford–IBH, New Delhi. 2001. pp. 198–199.

Pollinators as sensitive bio-indicators of environmental conditions: Ecological and practical concerns

Peter G. Kevan

Department of Environmental Biology and Botany Department, University of
Guelph, Guelph, Ontario N1G 2W1, Canada
e-mail: pkevan@uoquelph.ca

Recently pollinators have been presented as an important and possibly jeopardized component of agricultural and other ecosystems. International recognition is growing for the importance of pollinators in sustainable productivity for food, fibre, building materials, and in nature. Nevertheless, little is known about pollinators as indicators of environmental conditions, and even less about the importance of pollinator assemblages (diversity and abundance) in ecosystem function.

It is well known that pesticides, parasites, pathogens, and predators can have adverse effects on pollinators. Most information comes from concern for health of honeybees, especially European races of *Apis mellifera*. There is much less information on other commercially useful pollinators, such as leafcutting bees, orchard bees, and bumblebees. The adverse effects of pesticides on assemblages of pollinators have been documented in a few places, notably in eastern Canada for the insecticide, Fenitrothion, on blueberry pollinators and pollination. The effects of industrial pollution on pollination have been investigated mostly from the viewpoint of pollutant accumulation (heavy metals, organic chemicals, radioisotopes, etc.) by

Keywords: Pollinators, bio-indicators, Canada, pesticide.

honeybee foraging. In the latter situation, bees are useful as bio-samplers of the environment.

Habitat destruction and fragmentation are major concerns for the conservation of pollinator diversity and abundance, and of the plants that depend on them for sexual reproduction. Evidence is accumulating to indicate fragmented and degraded landscapes, compared to more pristine areas, support fewer individuals of fewer species of pollinators, and that there are concomitant declines in plant diversity (from specific to genetic) and abundance.

A number of species of pollinators have been introduced, accidentally or purposely, into areas beyond their natural ranges. The consequences of some such introductions seem to have been beneficial to agricultural production, but only one example (the oil palm pollinating weevil *Elaeidobius kamerunicus* in Southeast Asia) is available of an introduction made with appropriate diligent regard for the potential for negative effects. Other introductions seem to have had negative consequences for the natural environment into which the exotics have escaped.

The importance of assemblages of pollinator, i.e. their diversity and abundance, is now being investigated with respect to ecosystem function. The data available suggest that diverse assemblages are better pollinators, even of single species of crop, than are just a few species even with equivalent total abundance. At the level of sexual reproduction by plants in communities, diversity and abundance of pollinators surely contribute to ecosystem stability. A variety of studies from the Americas (Brazil, Mexico, USA, and Canada) suggest that guilds of pollinating insects (e.g. diurnal bees, crepuscular sphinx moths) in more or less unstressed situations, form natural taxocenes with the expected log–normal relationships between diversity and abundance. When environmental stress occurs, the log–normal relationship should be upset, as has been found from analyses of the data from blueberry fields unaffected and affected (during and after) by insecticidal sprays. Thus, pollinators may represent highly sensitive bellwethers for assessing the state of environments ranging from agricultural to natural.

Pollinators are crucial to terrestrial ecosystem productivity from pristine to highly managed. Within most terrestrial ecosystems, it seems that pollinator interactions with plants are robust and resilient. Pollinators seem to be useful in assessing the state of the environment in various ways from bio-samplers to bio-diversity. There are increasing number of instances in which pollinators and pollination seem jeopardized and the red-flag has been raised internationally, particularly with respect to food and fibre security for human beings. A recent review indicates that pollinator shortages have existed for thousands of years. The economic consequences for world trade and food supply are complex, but no matter how those consequences play out, the consumer suffers through higher prices or lack of commodity.

Tropical Ecosystems: Structure, Diversity and Human Welfare.
Proceedings of the International Conference on Tropical Ecosystems
K. N. Ganeshaiah, R. Uma Shaanker and K. S. Bawa (eds)
Published by Oxford–IBH, New Delhi. 2001. pp. 200–201.

Pollination in the Australian tropics – Diversity and threats

Caroline L. Gross

Ecosystem Management, University of New England, Armidale, NSW, 2350,
Australia
e-mail: cgross@pobox.une.edu.au

The wet-tropical rainforests of north Queensland are notable for their high floristic diversity with 516 plant genera and over 1160 species of plants, of which more than half are endemic to Australia. Some of the world's most important ancestral links in the history of plant evolution are found here. Nowhere else in the world is there such a concentration of primitive flowering-plant families (13 of the 19 known families of primitive flowering plants). Floral diversity is paralleled by richness of floral visitors, knowledge of which is still far from complete.

Interest in the pollination ecology of Australian tropical plants has its origins in the 1800s and, although there are many unique and interesting features of Australian tropical rainforests, there are less than twenty-five published studies to date. However, from only a handful of studies it is apparent that a variety of plant–pollinator systems occur with entomophily being the predominant strategy. In particular, beetle pollination has been well documented for several plant species. Long-term community level studies are required for a greater understanding of the dynamics of plant–pollinator systems in the tropical rainforests of Australia.

Keywords: Queensland, Australia, pollination ecology, tropical rainforests, fragmentation.

The impact of fragmentation on the plant–pollinator relationship is poorly understood for most plant and animal species in the Australian tropical rainforests. There are data emerging from studies undertaken in the sub-tropical rainforests of New South Wales which show that *Macadamia* trees in small fragmented patches produce more fruit than trees of the same species found in larger, continuous tracts of rainforest. This information places a new value on these fragment patches.

Managed honeybees also pose a threat to tropical rainforests in Australia. *Melastoma affine* (Melastomataceae) is a pioneer shrub of the rainforest margin in the Australasian region. It is a species that obligatorily relies on native bees to effect pollination – and about eight species are involved with four species being the most frequent pollinators. All of these bee species are buzz pollinators. Hives of the introduced honeybee, *Apis mellifera*, are often placed adjacent to rainforests in northern Australia. A study has shown that honeybees rarely pollinate *Melastoma affine* and in fact these bees only collect previously deposited pollen from stigmas and thereby bring about a reduction in plant fitness through reduced seed-set. The same study revealed that honeybees were quite aggressive when they encountered native bees at flowers. In 91% of interactions between honeybees and native bees, native bees were disturbed from foraging at flowers by honeybees to the extent that honeybees would sometimes aggressively wrestle native bees away from the flower. Tighter controls are needed on the placement of honeybee hives near natural rainforest systems.

Tropical Ecosystems: Structure, Diversity and Human Welfare.
Proceedings of the International Conference on Tropical Ecosystems
K. N. Ganeshaiah, R. Uma Shaanker and K. S. Bawa (eds)
Published by Oxford–IBH, New Delhi. 2001. pp. 202–203.

Diversity, structure, and variations in pollination systems of the seasonal dry forest of Costa Rica

Gordon Frankie* and S. Bradleigh Vinson

University of California, Division of Insect Biology, 201 Wellman Hall, Berkeley,
California 94720, USA
*e-mail: frankie@nature.berkeley.edu

In 1996 our research group initiated two survey studies, which are ongoing, to develop a community picture of a lowland seasonal dry forest in NW Costa Rica. Published and unpublished work on flowering phenology and pollinators formed the foundation for this new effort. One of the goals of these surveys was to assess the conservation status of selected pollinators, especially bees. The study site is a 10 × 10 km area in the vicinity of Bagaces, a small town in the Tempisque region of Guanacaste Province. The area is mostly wooded savanna, with some riparian forest, oak forest, and dry deciduous forest. The site has received light to moderate disturbance over the past 20 years.

Past phenological work in this area has focused on trees and to some extent, shrubs. In the new survey we were interested in recording the blooming periods of trees versus non-trees of most flowering plant species in the 100 km^2 site. The total number of plants recorded as cf January 2001 is slightly over 450 species (the area is thought to have 500–525 flowering plant species). Although there is always some flowering of all plant life forms

Keywords: Costa Rica, seasonal dry forest, diversity, pollinator, phenology, disturbance.

throughout the year, there was a tendency for many tree species to flower in the dry season and non-trees to peak their flowering during the rainy season.

We assigned each plant species to a particular system. This provided a good assessment of the relationship each plant has with its most likely pollinator. The diversity was high with 11 recognized systems. The four most common among trees and non-trees were the large bee (~ 1.2 cm in body length), small bee/generalist, moth, and bat systems. These four accounted for almost 90% of the known systems.

The four common systems were organized temporally in the following manner. Large bee flowers bloom mostly in the long dry season, and their flowers are characteristically large, colourful, and zygomorphic in shape. Moth flowers bloom mostly in the wet season. Small bee/generalist and bat flowers can be found blooming throughout the year. Flowers of the small bee/generalist system were small, mostly cream coloured, and often radially symmetrical.

The survey also revealed that there is considerable cross over of pollinators between different pollination systems. The most obvious was that of numerous bee species using residual floral resources from nocturnal systems (e.g. bat and moth flowers). Cross overs have been known for years, but not to the extent currently being recorded. Also, much of this cross over activity occurs just prior to and during sunrise.

A second type of variation is that of intraspecific differences in attraction of pollinators to flowering populations of plants. Stated differently, not all individual plants of a species population attract pollinators equally. This type of variation appears to be widespread in the dry forest. From the standpoint of conservation, this may indicate that we have been overestimating the amount of floral resources available to pollinators in an area, and this could be very important in disturbed environments.

Our survey includes a large monitoring component, and much of this research has focused on bees. We began this type of work in 1972, and recent new sampling strongly indicates that we are losing pollinating bees in the dry forest. Various types of human development projects and disturbances (e.g. fire) seem the likely factors causing the decline of bee pollinators in this area.

We conclude that pollination biologists must become more proactive in outreach efforts to alert decision makers and the public about these declines. Several suggested recommendations for outreach are offered.

Tropical Ecosystems: Structure, Diversity and Human Welfare.
Proceedings of the International Conference on Tropical Ecosystems
K. N. Ganeshaiah, R. Uma Shaanker and K. S. Bawa (eds)
Published by Oxford–IBH, New Delhi. 2001. pp. 204–206.

Consequences of tropical dry forest fragmentation on pollinator activity, reproductive success and mating patterns of trees in Costa Rica and México

Mauricio Quesada*, Kathryn E. Stoner[†], Jorge A. Lobo**,
Eric J. Fuchs**, Yvonne Herrerias* and Victor Rosas*

*Estación de Biología Chamela, Instituto de Biología, Universidad Nacional
Autónoma de Mexico, A. P. 21, San Patricio, Jalisco, Mexico 48980
**Escuela de Biología, Universidad de Costa Rica, San Jose, Costa Rica
[†]e-mail: kstoner@cariari.ucr.ac.cr

Over the last decade several studies have proposed that tropical trees are particularly vulnerable to forest fragmentation because they are mainly pollinated by animals, occur at low densities, present self-incompatibility systems and express high rates of outbreeding. Forest fragmentation is likely to decrease pollinator activity and this in turn may decrease pollen flow, increase endogamy and eventually produce a high differentiation among remnant populations.

In this study we determine the effects of forest fragmentation in conjunction with the effects of flowering phenology on the abundance of pollinators, natural pollination, reproductive success and the genetic structure of trees of

Keywords: Fragmentation, tropical dry forest, pollination, mating patterns, Costa Rica, Mexico, reproductive success.

the family Bombacaceae in the dry forests of Costa Rica and Mexico. First, we determined the relationship between the abundance of bat pollinators with respect to the phenology of trees in continuous populations in Guanacaste, Costa Rica and Chamela, Mexico. We also determined the relationship between the species of pollen carried by bats and the species of bats during a year of study in Chamela. In addition, we videotaped flowers of trees using night vision cameras to quantify bat visitation rate to flowers and the species of pollinators. Second, we evaluated the relationship between the deposition of pollen with respect to the production of fruits and seeds in natural populations of *Pachira quinata* and *Ceiba grandiflora*. Finally, we compared fruit set and genetic relatedness within progenies in two conditions to evaluate the effect of forest fragmentation on the reproduction of *Pachira quinata* and *Ceiba grandiflora*: (a) isolated trees were separated by more than 500 m from other adult conspecifics and located in disturbed sites and (b) trees from continuous populations consisted of groups of 20 or more reproductive individuals per hectare surrounded by undisturbed mature forest and were located within the Guanacaste Conservation Area, Costa Rica and the Chamela–Cuixmala Biosphere, Mexico. In addition, to evaluate flowering phenology on the same tree species, individuals were classified as: (a) trees with synchronous flowering and (b) trees with asynchronous flowering, based on the date of their flowering peak.

The results of our study indicate that there is a relationship between the abundance of nectivorous bats and the flowering pattern of Bombacaceous trees in Guanacaste and Chamela. In Chamela, the pollen from trees of the family Bombacaceae is found on nectivorous bats during six months of the year. *Glossophaga soricina* and *Leptonychteris curasao* are the most important pollinators of Bombacaceous trees at Chamela. *Glossophaga soricina* is significantly affected by forest fragmentation. In continuous populations of *Ceiba grandiflora*, 10% of the flowers develop into fruit. On average, flowers with more than 450 pollen grains on the stigma always develop into fruit whereas flowers with less than 100 pollen grain abort. The population of *Ceiba grandiflora* studied is predominantly outcrossing and seeds within fruits are sired by several donors, indicating that the pollinators visit several trees within one night. Similarly, in continuous populations of *Pachira quinata*, only 6% of the flowers develop into mature fruits. Flowers with more than 400 pollen grains on the stigma always developed into mature fruits whereas flowers that received less than 200 grains never matured fruits. Half of the pollen grains transferred to a flower stigma germinated and developed pollen tubes to the base of the style. The number of pollen grains on a stigma explained 34% of the variation in seed number per fruit and the number of seeds produced per fruit is positively correlated with the size of the seeds. In *Pachira quinata*, 6% of the flowers set fruit in trees from continuous populations whereas only 3% of the flowers developed fruit in isolated trees. However, total fruit production per tree is apparently not affected by habitat fragmentation because isolated trees tend to produce more flowers. Seed production per fruit was not affected by forest fragmentation. Fruit set is not affected by the flowering phenology of trees but it is mainly influenced by factors associated with forest fragmentation. A genetic analysis revealed that the progeny of trees from continuous populations is produced under higher levels of outcrossing and is sired by more pollen donors than trees in isolation. The same effect is observed in trees with synchronous flowering. The degree of relatedness of the progeny of isolated trees is greater than the progeny of trees in continuous populations. The mating patterns of trees in the family Bombacaceae are also affected by the species and behaviour of pollinators.

Tropical Ecosystems: Structure, Diversity and Human Welfare.
Proceedings of the International Conference on Tropical Ecosystems
K. N. Ganeshaiah, R. Uma Shaanker and K. S. Bawa (eds)
Published by Oxford–IBH, New Delhi. 2001. pp. 207–210.

Do 'pollinator flags' aid in floral visitation in *Mussaenda frondosa*?

Vinita Gowda*, Merry Zacharias, Namita Vishveshwara and Renee M. Borges

Centre for Ecological Sciences, Indian Institute of Science, Bangalore 560 012, India
*e-mail: vinita@ces.iisc.ernet.in

Floral colour signals have been known to be one of the essential visual cues that pollinators use during foraging. Flower colours result from selective absorption of the ambient light by the tissues of the floral petals, and its reflection at cellular levels. Apart from floral petals, other potentially important visual signals are given by pollinator flags (Stiles, 1982). These are defined as extrafloral elements of plants which attract potential pollinators to flowers, are usually closely associated with the flower and are temporally coincident with flowering. They provide visual stimuli but are quantitatively unrelated to the food resources for the pollinator (Stiles, 1982). We investigated the effect of pollinator flags on floral visitation in the climbing shrub *Mussaenda frondosa* (Rubiaceae). This is a distylous plant in which flowers are bisexual, tubular, small, bright orange in colour and in which usually one sepal in a group of flowers become enlarged, white, petaloid in shape, and is persistent. We conducted preliminary experiments to test the effect of bracts on floral visitation, to find out whether the modified sepal or putative pollinator flag may aid in visual attraction for the potential pollinator species. We observed floral visitation to plants from which bracts were removed and compared this with visitation to these plants prior to bract removal.

Keywords: Distyly, bract, southern birdwing, nectar robber.

We conducted this study at Karinja (approx. 12°52′N, 75°E) situated in Dakshina Kannada district of Karnataka State in the Western Ghats of India, where a good population of *M. frondosa* was found. Karinja is an open mixed forest surrounded by plantations. We selected 11 flowering plants (5 pins and 6 thrums) in an area of 1 km² such that they were located at accessible and clearly visible locations. We documented visitation to plants from 7:00 h to 17:30/18:00 h in 30 min durations from 7:00 h to 13:00 h and then at 20 min durations from 13:40 h to 17:20 h such that every plant was observed in the time slot of 7:00 to 18:00 h at least twice each for a duration of 20 or 30 min in a day. Observations on such intact plants continued for 5 days. On the 6th day the bracts were removed from each plant, collected for measurement of bract area, and 5 more days of observation was carried out on these bractless plants in the above mentioned manner. The duration, frequency and total number of flowers visited as well as visitors in the vicinity were also recorded during the period of observation for plants with and without bracts. The visitors to the plant were the following – birds: Tickell's flower pecker *Dicaeum erythrorhynchos*, purple-rumped sunbird *Nectarinia zeylonica*; butterflies: blue mormon *Papilio polymnestor*, blue tiger *Tirumala limniace*, clipper *Parthenos sylvia*, common rose *Pachliopta aristolochiae*, common mormon *Papilio polytes*, crimson rose *Pachliopta hector*, southern birdwing *Troides minor*; carpenter bee *Xylocopa* sp.

Based on visitation frequency, we found that butterflies constituted the major visitors (42.6%), followed by carpenter bees (36.1%), southern birdwing (14.8%), and sunbirds (6.56%). We grouped visitors into the following categories: carpenter bee (*Xylocopa* sp.), birds, butterflies (which included all the above butterflies except southern birdwing), and the southern birdwing. We placed the southern birdwing in a separate category because, of all the butterfly visitors to *Mussaenda*, it was the only true tree canopy butterfly which descended to feed on *M. frondosa* flowers only during our observation period. Carpenter bees acted as nectar robbers, since they punched a slit (4–5 mm long) in the base of the floral tube and fed on the nectar. These slits were later also used by birds and ants to reach the nectar.

We calculated visitation rate (visits/minute/plant) for each type of visitor. We then looked at whether the visitation rates per plant (pooled over the observation days) differed between pre- and post-bract removal experiments. We found that visitation rates (all visitors included) were significantly higher in the pre-bract removal than post-bract removal stage (Table 1, Figure 1). We then examined differences in the pre- and post-bract removal visitation rates for the four groups individually, and found that visitation rates of the butterflies and birdwings were significantly higher in the pre-bract removal stage (Table 1). However, there was no significant difference in the visitation rate of bees and birds between the two experimental treatments (Table 1). We then examined whether the total bract display per plant (total area of all bracts per plant summed in cm²) affected the visitation rate. We observed that

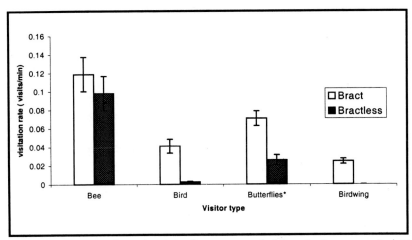

Figure 1. Visitation rate (visits/min) for four groups of visitors in the pre- and post-bract removal experiments (means ± SD; n = 10 plants). *Excludes birdwing.

Table 1. Comparison of visitation rates (visits/min) in plants with and without bracts for the different groups of visitors (results of Wilcoxon matched pairs signed-ranks tests).

Visitor groups	n	Z	P level
All groups combined	10	2.497271	0.012520
Butterflies excluding birdwing	10	2.701130	0.006914
Southern birdwing	10	2.201398	0.027715
Carpenter bee	10	0.420084	n.s.
Birds	10	1.603567	n.s.

there was no relationship between total visitation rate (all visitors pooled) and total area of bract display per plant (Kendall's $\tau = 0.1111$, $n = 10$ plants, $P > 0.05$). A similar non-significant result was observed for the different groups individually. Since the bees did not show significant difference in their visitation rates between the pre- and post-bract removal experiments we examined whether the total number of flowers affected bee visitation rates in the pre- and post-bract removal periods. We found that there was no significant effect of number of flowers on the visitation rate in either of the two experiments (pre-bract removal: Kendall's $\tau = 0.359573$, $P > 0.05$; post-bract removal: Kendall's $\tau = 0.368035$, $n = 10$ plants, $P > 0.05$).

We conclude that the birdwings and other butterflies may be using bracts as a visual cue more than the bees and birds to locate and possibly distinguish *M. frondosa* from the other neighbouring vegetation. Their visitation was hence significantly affected when the bracts were removed. The southern birdwing did not descend on to any of the bractless plants while the other butterfly species showed decreased visitation to the bractless plants. Since visitation by bees and birds did not show any effects of bract removal, they may be using cues other than bracts to locate the flowers of *M. frondosa* (Lehrer, 1997; Menzel *et al.*, 1997). Papilionid butterflies have been demonstrated to have colour vision (Ilse and Vaidya, 1995; Kinoshita *et al.*, 1998; Kelber and Pfaff, 1999) and bees to have trichromatic vision (Chittka, 1997). The visual cues provided by *M. frondosa* via bracts are quantitatively much more than those provided by flowers, though to fully understand the qualitative role of floral and other colours, we need to understand the colour vision systems of insects as well as the spectral reflectance of the total floral display including bracts. This will enable us to conclusively state *how* the pollinator flags aid in visitation and in distinguishing *M. frondosa* from the surrounding vegetation.

References

Chittka, L. Bee color vision is optimal for coding flower color, but flower colors are not optimal for being coded – why? *Israel J. Plant Sci.* **45**, 115–127 (1997).

Ilse, D. & Vaidya, V. G. Spontaneous feeding response to colours in *Papilio demoleus* L. *Proc. Indian Acad. Sci.* **43**, 23–31 (1955).

Lehrer, M. Honeybee's use of spatial parameters for flower discrimination. *Israel J. Plant Sci.* **45**, 157–167 (1997).

Kelber, A. & Pfaff, M. True colour vision in the orchard butterfly, *Papilio aegeus*. *Naturwissenschaften* **86**, 221–224 (1999).

Kinoshita, M., Shimada, N. & Arikawa, K. Colour vision of the foraging swallowtail butterfly *Papilio xuthus*. *J. Exp. Biol.* **202**, 95–102 (1998).

Menzel, R., Gumbert, A., Jan, K., Avi, S. & Misha, V. Pollinators' strategies in finding flowers. *Israel J. Plant Sci.* **45**, 141–156 (1997).

Stiles, E. W. Fruit flags: Two hypotheses. *Am. Nat.* **120**, 500–509 (1982).

Tropical Ecosystems: Structure, Diversity and Human Welfare.
Proceedings of the International Conference on Tropical Ecosystems
K. N. Ganeshaiah, R. Uma Shaanker and K. S. Bawa (eds)
Published by Oxford–IBH, New Delhi. 2001. pp. 211–213.

Insect-induced anther dehiscence, pollination and maximisation of maternal rewards

Susheel Verma, Rani Mangotra and A. K. Koul

Department of Botany, University of Jammu, Jammu 180 001, India
e-mail: eremurus@rediffmail.com

Association of insects with sexual reproduction of flowering plants is a textbook knowledge. However, this interaction is limited to pollen, seed and fruit dispersal. In angiosperms pollen is released from anthers either through longitudinal slits, apical pores or valvular openings (Endress and Hufford, 1989; Endress and Stumpf, 1999). Edwards and Jordan (1992) and Armstrong (1992) have brought to light some peculiar mechanisms of anther dehiscence. In *Incarvillea emodi*, a highly endangered endemic Bignonian lithophyte, the insects are found involved in inducing the release of pollen. The male reproductive apparatus of the companulate hermaphrodite flower consists of four didynamous stamens which are held in pairs. They keep tightly appressed to the style along dorsal side of the corolla tube. Each anther is bithecous with the two lobes widely divergent at maturity. In all four stamens the connective projects beyond pollen sacs and bifurcates into finger-like structures. Anther lobes are sacciform. A prominent suture from where pollen grains escape runs from the tip towards the base of the anther covering almost 3/4th of the anther lobe. In line with the suture but near its base is present a stiff spine which keeps initially bent towards the anther tip.

Keywords: Anther dehiscence, pollination, maternal rewards, *Incarvillea emodi.*

Flowers are herkogamous with a spathulate bilobed stigma positioned much above both pairs of stamens. The stigma exhibits thigmotropy; its lobes close when they are touched. A ring-shaped nectary, which secretes copious amount of nectar is present at the base of the ovary in all flowers. Flowers of *I. emodi* take almost 24 h to open. At the time of anthesis, anthers of each pair of stamens are held together in lock and key manner by the finger-like extension of the connective. The style is appressed to the corolla tube when the flowers are in the process of opening. The anthers are still intact with the anther spine bent towards the tip of the anthers. No amount of pressure exerted manually to the spine in this posture induces pollen release.

The anther spine changes posture as flowers age. After 24 h of anthesis, it stands at right angle to the anther surface. In this posture, even a mild touching of the spine opens the suture and causes explosive discharge of pollen. Since anther dehiscence requires mechanical disturbance of the spine, pollen discharge cannot take place if insects do not visit the flowers. In the event of failure of insect visitation, the anthers do not dehisce and the flowers behave as functionally male sterile, despite the presence of well-differentiated pollen grains inside the pollen sacs. Later on, the spine bends towards the anther base, whether pollen discharge has taken place or not. Once it has bent towards the base, no amount of pressure applied to it induces pollen discharge. In such flowers pollen sacs dry without releasing pollen. The pollen grains lose viability and become dry. The impact of spine on anther dehiscence and pollen release was also checked through manual experiments.

The visiting insects on their way to the nectar gland also induce folding of stigma lobes whereby autogamy is thwarted. In order to understand the nature of stimulus for stigma closure, the stigma lobes were pollinated manually with alien pollen, heat killed self and viable self-pollen grains. Results of these experiments indicate two different stimuli, touch and pollination induce folding of stigma lobes in *I. emodi*. Touching induces abrupt folding. The lobes reopen after some time. Stigmas pollinated with critical load of fertile self pollen, which is always more than the number of ovules present in the pistil, close permanently, otherwise they keep open till this load gets deposited. This is a device which ensures maximum maternal benefit.

Flower opening, anther dehiscence and stigma receptivity are overlapping events in this species. Flowers take 24 h to open fully, the spine becomes erect at that particular time and the stigma lobes are wide and fully receptive. In case insects do not visit the flowers, reproduction fails because the pollen does not get released from anthers. Insect visitation to flowers is thus an obligatory requirement. As the plant grows in isolated patches of ones or twos in the Siwalik ranges of Himalaya in India, in crevices of sand stone rocks, this type of relationship between the flower and insect makes reproduction difficult and accounts for very low number of plants in nature.

Acknowledgements

The authors thank Drs. Namrata Sharma and Veenu Koul for fruitful comments and healthy discussion and Head, Department of Botany for providing necessary facilities for carrying out this piece of work.

References

Armstrong, J. E. Lever action anthers and forcible shedding of pollen in *Torenia* (Scrophulariaceae). *Am. J. Bot.* **79**, 34–40 (1992).

Edwards, J. & Jordan, J. R. Reversible anther opening in *Lilium philadelphium* (Liliaceae): A possible mean of enhancing female fitness. *Am. J. Bot.* **79**, 144–148 (1992).

Endress, P. K. & Hufford, L. D. The diversity of stamen structures and dehiscence patterns among Magnolidae. *Bot. J. Linn. Soc.* **100**, 45–85 (1989).

Endress, P. K. & Stumpf, S. Non-tetra sporangiate stamens in angiosperms: structure, systematic distribution and evolutionary aspects. *Bot. Johrb. Syst.* **112**, 193–240 (1990).

Tropical Ecosystems: Structure, Diversity and Human Welfare.
Proceedings of the International Conference on Tropical Ecosystems
K. N. Ganeshaiah, R. Uma Shaanker and K. S. Bawa (eds)
Published by Oxford–IBH, New Delhi. 2001. pp. 214–218.

Relative impact of floral visitors: A continuum from very harmful parasites to very beneficial mutualists

Steven J. Stein* and Donna L. Hensley

Biology Department, Eastern Washington University, Cheney, WA, 99004, USA
*e-mail: sstein@mail.ewu.edu

Ecologists have always been known to have vigorous debates over two contrasting or dichotomous theories or points of view. These intense battles often rage for decades and may hinder more than help us understand the complexities of the natural world. We might develop a much better understanding of natural principles and processes by thinking more about continuums and less about dichotomies in ecology.

Pollination ecologists are beginning to recognize and investigate the continuum that exists from very harmful parasites to very beneficial mutualists. Some nectar robbers destroy flowers and therefore eliminate any chance of seed set while others may only steal nectar and may have anywhere from a very little to a very large impact on seed set (Inouye, 1983). Pollinators may deliver anywhere from 1 to 100 per cent of the right species of pollen to a plant and may therefore have anywhere from a very minor to a very major positive impact on seed set.

Clearly, a great range or continuum exists in terms of the relative effectiveness of floral visitors on the reproductive success of plants. To

Keywords: Pollination ecology, pollinators, relative effectiveness, continuums, floral visitors.

determine the relative effectiveness of floral visitors we must quantify not only the abundance of each visiting species (population level) but also the efficiency of individuals of each species in both removing and depositing pollen (individual level). Most pollination studies simply quantify visitation and do not take the efficiency of individuals into account which could lead to inaccurate conclusions concerning the relative effectiveness of floral visitors. Jennersten and Morse (1991) showed that the efficiency of individuals may be quite important as diurnal insects visited five times as many flowers as nocturnal insects, but removed ten times as much pollen. However, Olsen (1997) showed that the effectiveness of pollinator species was dictated by their relative abundance.

The purpose of this study was to examine the relative effectiveness of floral visitors as pollinators of the tropical milkweed, *Asclepias curassavica*. The following questions were addressed: (1) What is the frequency distribution of floral visitors? (2) What is the relative efficiency of pollinaria removal? (3) How many pollinaria are on the bodies of floral visitors? (4) What is the relative efficiency of pollinaria delivery? (5) What is the relative effectiveness of floral visitors?

The study was conducted in northern Costa Rica, on the wet Atlantic slope, just north of Rincon de la Vieja Volcano, near the village of Gavilan. Study sites were in abandoned agricultural fields and along dirt roads.

The tropical milkweed, *A. curassavica*, is widely distributed in disturbed areas of the neotropics from near sea level to 2000 m. Its umbel-like inflorescences usually contain from 7 to 12 elaborate, five-parted, orange and yellow flowers. Each flower has five cups, which contain abundant nectar that is readily available to a wide variety of visitors, and five pollinaria which are generally transported on the legs of insects.

The frequency distribution of floral visitors was quantified from January to March 1998. We observed groups of inflorescences throughout the day and night and recorded the species of every visitor and the time of every visit.

The relative efficiency of pollinaria removal and delivery was quantified for each of the 29 species of floral visitors. Using a hand lens we checked every flower on a group of inflorescences to detemine if it contained its original set of five pollinaria and if any pollinaria had been placed in the receptive parts of the flower. We removed any flower without a complete set of pollinaria or with any pollinaria in its receptive parts. We then waited for any visitor to land on any inflorescence to drink nectar. After the visitor departed we used a hand lens to determine how many pollinaria had been removed and deposited by that individual visitor on each inflorescence. This procedure was conducted for 10–29 times for each common species to determine the mean number of pollinaria removed and deposited for each species.

We quantified the mean number of pollinaria on the body of every species of floral visitor throughout the duration of the study. We caught at least 23 individuals of every common insect species with a fine net and used a hand lens to count the number of pollinaria on each individual.

We estimated the relative effectiveness of floral visitors on the reproductive success of *A. curassavica* by multiplying the mean number of visits/day/inflorescence by the mean number of pollinaria delivered for each species. This gave us the total number of pollinaria delivered/day/inflorescence for each visiting species and we multiplied this number by the mean inflorescence life to get the total number of pollinaria delivered/inflorescence life for each species.

Twenty-nine species of insects and birds visited the inflorescences of *A. curassavica* from January to March 1998, with 16 species being fairly common (Table 1). Wasps were the most common group of visitors and they visited two times more often than bees and five times more often than both butterflies and flies. There was great variation in visitation frequency among species (Table 2, visits/day/infl.) with the most common visitors in descending order being black and white wasp, little black bee, brown ab wasp, and black fly. Big black wasp visited only 1/6th as much as black and white wasp.

There was wide variation in the efficiency of pollinaria removal among species (Table 2, pollinaria removed/ind.). Brown ab wasps removed an average of 2.0 pollinaria per visit, big black wasp removed 1.6 per visit, black and white wasp removed 0.53 per visit, little black bee removed 0.50 per visit and black fly did not remove any pollinaria at all.

Table 1. Common and scientific names of the common floral visitors on *Asclepias curassavica*.

Wasps		*Butterflies*	
Big black wasp	*Polistes eythrocephalus*	Anartia	*Anartia fatima*
Black and white wasp	*Polybia* sp.	Hesperid	Hesperidae
Brown ab wasp	*Polybia rejecta*	Monarch	*Danaus plexippus*
Black and yellow wasp	*Agelia myrmrmecophila*		
		Cockroaches	
Bees		Small grey cockroach	Blattoidea
Little black bee	*Trigona* sp.		
Green bee	Halictidae	*Ants*	
Honey bee	*Apis meliferae*	Small ant	Formicidae
Giant black bee	Xylocopidae	Medium ant	Formicidae
Flies		*Hummingbirds*	
Black fly	Diptera	Green thorntail	*Discosura conversii*

Table 2. Relative effectiveness of floral visitor species as pollinators of *Asclepias curassavica*.

Floral visitor	Visits/ day/infl.	Pollinaria removed/ ind.	Pollinaria on body/ ind.	Pollinaria delivered/ ind.	Total pollinaria delivered/ day/infl.	Total pollinaria delivered/ infl. life
Big black wasp	0.20	1.61	7.92	1.44	0.288	1.44
Brown ab wasp	0.88	2.00	3.87	0.30	0.264	1.32
Black and white wasp	1.17	0.53	1.55	0.11	0.129	0.65
Hesperid	0.09	0.58	1.67	0.16	0.14	0.07
Anartia	0.18	0.57	0.32	0.05	0.009	0.05
Monarch	0.01	1?	0.62	0	0	0
Little black bee	0.98	0.50	1.83	0	0	0
Black and yellow wasp	0.25	0.67	1.13	0	0	0
Green bee	0.26	0.13	0	0	0	0
Black fly	0.42	0	0	0	0	0
Giant black bee	0.00	2	1	0.50	0	0
Ants	0.60	0	0	0	0	0
Cockroaches	0.08	0	0	0	0	0
Hummingbirds	0.08	0	0	0	0	0

The mean number of pollinaria on the body of visiting species ranged from 7.9 on big black wasp to none on many species (Table 2, pollinaria on body/ind). There was also a great variation in the efficiency of pollinaria delivery with individual big black wasps delivering by far the most with 1.44 pollinaria per visit (Table 2, total pollinaria delivered/ind.).

The relative effectiveness of floral visitors varied dramatically (Table 2, total pollinaria delivered/infl. life). The most effective species were big black wasp which delivered 1.44 pollinaria per inflorescence life and brown ab wasp which delivered 1.32 pollinaria per inflorescence life. Black and white wasp delivered 0.65 pollinaria per inflorescence life while little black bee, which had more pollinaria on its body than black and white wasp, did not deliver any pollinaria.

Our results suggest that to more fully understand pollination ecology we need to understand the movement of pollen and not just the visitation of potential pollinators. For example, black and white wasp visited *A. curassavica* inflorescences six times more often than big black wasp, and therefore one might conclude that black and white wasp would be a much more important pollinator to the plant. However, each individual big black wasp removed three times as many pollinaria, had five times more pollinaria on its body, and delivered 13 times more pollinaria than each black and white wasp. So even though big black wasp was not a very common visitor, it was the most effective pollinator in terms of plant reproductive success and over twice as effective as black and white wasp.

We have shown that a continuum exists from very harmful parasites to very beneficial mutualists in this system. We feel that more ecological studies should be designed to investigate the entire continuums that exist in nature and not just the two extremes. This would move us away from the dichotomous or contrasting theory approach and could help us to develop a much better understanding of the complexities and subtleties of ecological principles and processes.

The population sizes of many pollinators throughout the world have been declining due to habitat alteration, pesticide poisoning, and the introduction of alien pollinators. These declines can have a major negative affect on plant reproduction which will likely lead to cascading effects through ecosystems. Our results suggest that conservation efforts should be geared towards improving conditions for the most valuable pollinators of keystone plant species. These efforts could help to keep the populations of these important plant species viable, which in turn would better be able to support the great number of animal species that depend on these plants. These kinds of conservation efforts could have an especially large impact in the tropics where a great number of species are rapidly going extinct.

References

Inouye, D. W. The ecology of nectar robbing. *The Biology of Nectaries* (eds Bentley, B. & Elias, T.) 153–173 (Columbia University Press, New York, 1983).

Jennersten, O. & Morse, D. H. The quality of pollination by diurnal and nocturnal insects visiting common milkweeds, *Asclepis syriaca. Am. Midl. Nat.* **125,** 18–28 (1991).

Olsen, M. O. Pollination effectiveness and pollinator importance in a population of *Heterotheca subaxillaris* (Asteraceae). *Oecologia* **109,** 114–121 (1997).

Global Perspectives on Tropical Forest Regeneration

Tropical Ecosystems: Structure, Diversity and Human Welfare.
Proceedings of the International Conference on Tropical Ecosystems
K. N. Ganeshaiah, R. Uma Shaanker and K. S. Bawa (eds)
Published by Oxford–IBH, New Delhi. 2001. pp. 221–223.

Forest edges and regeneration processes: Beyond edge effects

Carla Restrepo

167 Castetter Hall, Department of Biology, University of New Mexico,
Albuquerque, NM 87131
e-mail: carlae@sevilleta.unm.edu

Forest edges strongly influence the structure of landscapes as they develop over time: they may advance, recede, or remain unchanged, and in doing so they affect regeneration processes both of forest and disturbed land. Although recent work suggests mechanisms that can influence the development of edges (Restrepo *et al.*, 1999; Restrepo and Vargas, 1999; Sizer and Tanner, 1999), we still lack a theory that links three key components of forest edges, namely edge structure, process, and function. The objectives of this paper are three-fold. First, I discuss recent work that evaluates the effects of forest edges on seed dispersal and plant establishment. Second, I draw upon findings from this work to develop a model that links forest edge structure, process, and function. Lastly, I put forward the idea that isolated trees in disturbed areas – 'regeneration nuclei', forest edges resulting from anthropogenic activities, and biome ecotones belong to a class of ecological phenomena that I refer to as ecological surfaces. In doing so, I outline general principles that can guide future efforts aimed at restoring degraded land, managing forest fragments, and monitoring ecotone shifts due to climate change.

Seed dispersal and plant establishment are two processes that influence the recruitment and distribution of plants (Harper, 1994). In tropical

Keywords: Forest edges, seed dispersal, plant establishment, edge development.

regions, these two processes are mediated in most instances through plant–animal interactions. At the Reserva Natural La Planada, a Neotropical montane site, I designed a large-scale, well-replicated study ($n = 6$ forest edges) to evaluate changes in fruit abundance and the distribution of frugivorous birds, seed germination and predation rates, and seedling growth from forest edge to forest interior (Restrepo et al., 1999; Restrepo and Vargas, 1999). This work showed that these processes were highly dynamic and complex and that the response variables behaved in different ways. In particular, I found that (1) edge effects change within and among years, (2) factors other than edges, such as the behaviour of animals, can influence responses observed from forest edge to forest interior, (3) small-scale natural disturbances, such as treefall gaps, can modulate edge effects, and (4) the variance for almost all response variables was highest at the forest edge. The lack of consistent results in this and other studies led me recently to explore the possibility that forest edges may exhibit universal properties such as physical interfaces do.

Forest edges, like a variety of physical interfaces, are composed of many interacting particles; specifically, forest edges represent collections of individuals within populations and species within assemblages. Here I entertain the idea that forest edges can be studied in the same fashion as these interfaces are. In doing so, I want to take advantage of theoretical and experimental work aimed at understanding the origin, maintenance, dynamics, and structure of physical interfaces (Barabási and Stanley, 1995). First, I establish whether forest edges roughen over time. Second, I develop a model that may account for the observed patterns. For the six forest edges mentioned above, I used data on the position (x, y coordinates) of *Palicourea gibbosa* and *Faramea affinis* to calculate the roughening exponent [curve $\log l$ vs $\log w$ L (l, t)]. l is the size of a window of length l and w the width of the interface for each l. The advancing front of *Palicoura gibbosa* and *Faramea affinis* at the six forest edges was characterized by a roughening exponent of 0.09 ± 0.04 (average ± 1 SD, $n = 6$ edges). Similar values for this exponent have been reported for the Eden model, a model developed to understand cell growth. Roughening of these edges may result from a combination of birth (Restrepo and Vargas, 1999), growth (Sizer and Tanner, 1999), and death (Gascon et al., 2000) processes and the occurrence of inhomogeneities; the latter contributing to the aggregation of resources and thus, of individuals. Although the data that we used for our analyses have limitations, they are useful to illustrate the possibility that forest edges roughen, or develop into self-affine structures, in the same fashion that other interfaces do. The exponents describing these processes may be used as diagnostic features of the 'health', and thus function, of forest edges.

Other ecological surfaces that play a fundamental role in regeneration processes include isolated trees in disturbed areas – 'regeneration nuclei' and biome ecotones. Whereas the former structure landscapes at small scales, the latter do it at large scales. I will argue that shifts in biome ecotone due to climate change will constrain regeneration processes operating at smaller scales.

Acknowledgements

I am greatly indebted to Natalia Gomez, Sylvia Heredia, Arlex Vargas, Fernando Lozano, Pilar Amezquita, and Madhur Anand for their help at various stages of development of this work.

References

Barabási, A.-L. & Stanley, H. E. *Fractal Concepts in Surface Growth*. (Cambridge University Press, New York, USA, 1995).

Gascon, C., Williamson, G. B. & daFonseca, G. A. B. Ecology: Receding forest edges and vanishing reserves. *Science* **288**, 1356–1358 (2000).

Harper, J. L. *Population Biology of Plants* (Academic Press, London, UK, 1994).

Restrepo, C., Gómez, N. & Heredia, S. Anthropogenic edges, treefall gaps, and fruit-frugivore interactions in a Neotropical montane forest. *Ecology* **80**, 668–685 (1999).

Restrepo, C. & Vargas, A. Seeds and seedlings of two neotropical montane understorey shrubs respond differently to anthropogenic edges and treefall gaps. *Oecologia* **119**, 419–426 (1999).

Sizer, N. & Tanner, E. V. J. Responses of woody plant seedlings to edge formation in a lowland tropical rainforest, Amazonia. *Biol. Conserv.* **91**, 135–142 (1999).

Tropical Ecosystems: Structure, Diversity and Human Welfare.
Proceedings of the International Conference on Tropical Ecosystems
K. N. Ganeshaiah, R. Uma Shaanker and K. S. Bawa (eds)
Published by Oxford–IBH, New Delhi. 2001. pp. 224–227.

Rainforest fragmentation reduces understorey plant species richness in Amazonia

Julieta Benítez-Malvido[*][†] *and Miguel Martínez-Ramos*[**]

[*]Departamento de Ecología de los Recursos Naturales, Instituto de Ecología,
Universidad Nacional Autónoma de México (UNAM), Antigua Carretera a
Pátzcuaro No. 8701, Ex-Hacienda de San José de la Huerta, Morelia, Michoacán,
México, C.P. 59180
[**]Biological Dynamics of Forest Fragments Project, Instituto Nacional de
Pesquisas da Amazônia (INPA), CP. 478, Manaus AM 69011-970, Brazil
[†]e-mail: jbenitez@ate.oikos.unam.mx

Global biodiversity loss is closely related to the destruction and fragmentation of the tropical rainforests (TRF). While plant species diversity is rapidly lost in forest areas cleared for agriculture and cattle raising activities, the rate at which plant species diversity changes in the remaining fragmented forest is poorly known. Forest fragmentation in the tropics severely affects large trees (Laurance *et al.*, 1997; 1998a; 1998b; 2000; Curran *et al.*, 1999; Gascon *et al.*, 2000), but its effect on other life stages and plant life forms is poorly understood (Benítez-Malvido, 1998; Scariot, 1999). In the long term, tree species persistence in forest fragments will depend on the availability of seeds, seedlings, and saplings. In Central Amazonia, 9 to 19 years after fragmentation, we recorded species richness and net seedling recruitment rate in forest fragments of 1, 10, and 100-ha in area, and continuous forest. The study was conducted at the experimentally fragmented landscape 80 km north of Manaus, Brazil,

---

Keywords: Amazonia, forest fragmentation, species richness, understorey plants.

which is composed of several forest fragments of various sizes that were isolated between 1980 and 1990 (Bierreggard *et al.*, 1992). To assess plant recruitment after isolation, in May 1993 we removed manually all understorey plants that were smaller than 1 m tall from the 1 m² plots. Six years and five months later (October, 1999), all new seedlings recruited into the plots were counted, grouped into different life forms and classified into distinct morpho-species. To assess differences in species richness of recruited plants between continuous and fragmented forest, we constructed mean species–area accumulation curves after 500 randomisations of sample quadrat order using the program Estimates (Collwell and Coddington, 1994; Colwell, 1997). For all life forms, the resulting curves showed that species richness was far from being completely recorded in the continuous forest with our sampling effort (Figure 1). In contrast, the curves referring to the fragments showed an asymptotic trend, particularly in the 1-ha fragment. It is known that the observed species-area accumulation curves underestimate species richness (Collwell and Coddington, 1994). Therefore, in an attempt to discover true species richness we used non-parametric methods, selecting those that

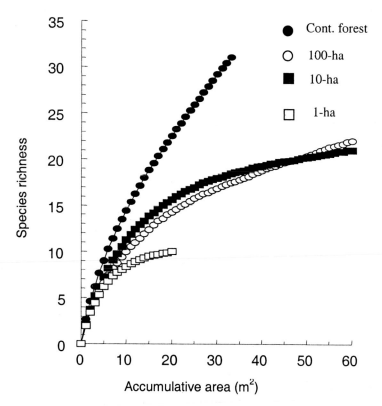

Figure 1. Species–area accumulation curves for seedlings recruited into forest fragments and continuous forest, near Manaus, Brazil. Two standard errors have the size of figure symbols.

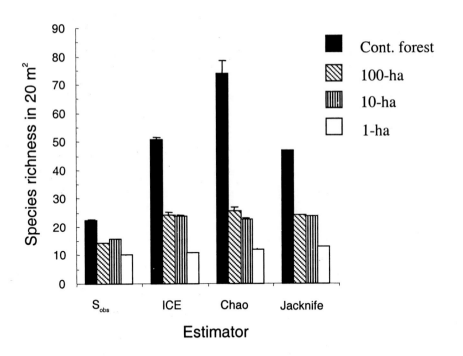

Figure 2. Species richness (mean ± 1 SE) of seedling recruited into continuous forest and forest fragments. The observed species richness is indicated by S_{obs}, and to the right three different non-parametric estimators are provided.

have shown to be the best for TRF tree seedling communities (Chazdon *et al.*, 1998). The non-parametric estimators magnified the differences in species richness between continuous and fragmented forests. In most cases, species richness declined with fragment size and was independent of seedling abundance. The reduction in species richness from continuous forest to fragments was in the order of 1.2 to 11.5 times, depending on the life form and the species richness estimator (Figure 2). The total number of recruited individuals was 40% less than that previously present for all life forms but lianas. Overall, the life form composition and structure of the regenerative plant pool in fragments shifted towards a species-poor seedling community, dominated by lianas, characteristic of disturbed forests (Laurance *et al.*, 2000). Losses of understorey diversity, but especially that of tree seedlings, threaten the maintenance of rainforest biodiversity and compromises future forest regeneration.

References

Benitez-Malvido, J. Impact of forest fragmentation on seedling abundance in a tropical rain forest. *Conserv. Biol.* **12**, 380–389 (1998).

Bierreggard, R. O., Lovejoy, T., Kapos, V., dos Santos, A. & Hutchings, R. The biological dynamics of tropical rain forest fragments. *BioScience* **42**, 859–866 (1992).

Chazdon, R., Colwell, R. K., Denslow, J. S. & Guariguata, M. Statistical methods for estimating species richness of woody regeneration in primary and secondary rain forests of NE Costa Rica. in *Forest Biodiversity Research, Monitoring and Modeling: Conceptual Background and Old World Case Studies* (eds Dallmeier, F. & Comiskey, J. A.) 285–309 (Parthenon Publishing Group, Paris, 1998).

Colwell, R. K. & Coddington, J. A. Estimating terrestrial biodiversity through extrapolation. *Philos. Trans. R. Soc. London* **345**, 101–118 (1994).

Colwell, R. K. Estimates: Statistical Estimation of Species Richness and Shared Species from Samples (Version 5, User's Guide and application published at: http://viceroy.eeb.uconn.edu/estimates (1997).

Curran, L. M. I., Caniago, G. D., Paoli, D., Astianti, M., Kusneti, M., Leighton, C. E., Nirarita, H. & Haeruman. Impact of El Niño and logging on canopy tree recruitment in Borneo. *Science* **286**, 2184–2188 (1999).

Gascon, C., Williamson, B. & da Fonseca, G. A. B. Receding forest edges and vanishing reserves. *Science* **288**, 1356–1358 (2000).

Laurance, W. F., Laurance, S. G., Ferreira, L. V., Rankin-de Merona, J. M., Gascon, C. & Lovejoy, T. E. Biomass collapse in Amazonian forest fragments. *Science* **278**, 1117–1118 (1997).

Laurance, W. F., Ferreira, L. V. Rankin-de Merona, J. M., Laurance, S. G., Hutchings, R. W. & Lovejoy, T. E. Effects of forest fragmentation on recruitment patterns in Amazonian tree communities. *Conserv. Biol.* **12**, 460–464 (1998a).

Laurance, W. F., Ferreira, L. V., Rankin-de Merona, J. M. & Laurance, S. G. Rain forest fragmentation and the dynamics of Amazonian tree communities. *Ecology* **79**, 2032–2040 (1998b).

Laurance, W. F., Delamônica, P., Laurance, S. G., Vasconcelos, H. L. & Lovejoy, T. E., Rainforest fragmentation kills big trees. *Nature* **404**, 836 (2000).

Scariot, A. Forest fragmentation effects on palm diversity in central Amazonia. *J. Ecol.* **87**, 66–76 (1999).

Tropical Ecosystems: Structure, Diversity and Human Welfare.
Proceedings of the International Conference on Tropical Ecosystems
K. N. Ganeshaiah, R. Uma Shaanker and K. S. Bawa (eds)
Published by Oxford–IBH, New Delhi. 2001. pp. 228–230.

Forest rehabilitation in the Asia-Pacific region

Peter D. Erskine and David Lamb*

Rainforest Co-Operative Research Centre, Department of Botany, University of
Queensland, St Lucia, 4072, Australia
*e-mail: p.erskine@botany.uq.edu.au

Large areas of tropical forest land in the Asia-Pacific region have been degraded. The last 50 years have seen rapid changes in the region with the area of cleared agricultural lands and degraded forest land now larger than the area of intact tropical forest. This degradation has reduced the biodiversity, structure and productivity of these areas. In many cases it has also caused a change in the protective function of vegetation and led to increased erosion and river sedimentation. For example, Thailand lost more than 60% of its forest cover in the last half of the 20th century and has experienced devastating floods and landslides as a result of deforestation (Leungaramsri and Rajesh, 1992).

Throughout the Asia-Pacific region monoculture plantations of exotic *Pinus*, *Eucalyptus*, *Acacia*, rubber trees and oil palms have been increasingly established to provide a regular source of paper pulp, firewood, building products and/or oils. These industrial plantations provide some protection against soil erosion but they simplify landscapes and generally have low biodiversity. The rapid turnover of plantation trees, weed control, limited establishment opportunities for native plants and the inability of native wildlife to use these areas contribute to a further biological impoverishment of these landscapes. There are however, several ways by which these areas

---

Keywords: Degradation, rehabilitation, plantations, mixed species, wildlife.

might be reforested that could result in higher levels of biodiversity and a higher level of protection for watersheds. In this paper we describe some of the different approaches to rehabilitation that are currently being undertaken in Australia, Thailand and Vietnam.

One approach is primarily aimed at accelerating the rate of biodiversity restoration without necessarily obtaining a commercial return. Degraded sites in Thailand and Australia have been rehabilitated by planting groups of relatively fast-growing species which are considered ecosystems building blocks in the 'Framework Species Method' (Goosem and Tucker, 1995; Forest Restoration Research Unit, 1998). The advantages of this method are that it only involves one planting and it should be a self-sustaining as it relies on the local gene pool to increase species and life form diversity. A disadvantage of this method is that its dependence on nearby vegetation being close enough to provide a seed source means that it can only be used effectively to connect and enhance forest boundaries. Another approach is to use a large number of mature phase canopy species with the 'Maximum Diversity Method' (Goosem and Tucker, 1995). The advantages of this method are that it can be utilised in areas that are isolated from intact forest areas and it should rapidly recreate as much as possible of the original diversity. Disadvantages of this method are primarily the costs associated with seed collection, raising seedlings, and the intensive maintenance required because of the slow growth of mature phase species.

Developing methods of farm forestry that provide for timber production and biodiversity are an alternative approach currently being undertaken in Australia and Vietnam. This involves establishing plantations of high-value rainforest mixed-species plantations. Such methods should potentially allow relatively large areas to be rehabilitated but will also require some trade-offs between production and conservation. The choice of tree species to use in the mixtures will affect the commercial yield of the plantation because of competitive interactions and the value of the timber produced. Increased productivity of mixed species stands depends on finding species to include in the mix that are complementary to each other. The factors guiding a choice include matching differences in rooting and crown architecture, differences in phenology or any other differences that reduce between-species competition. Ideally the species should have similar growth rates so that one does not outstrip the growth of the other(s). Trials have been established to generate some generic guidelines for assembling species mixtures. A variety of species have been used and these differ in several of the attributes mentioned above (growth rates, crown architecture, etc.). The design of the trials has involved creating mixtures by planting alternate rows of various combinations. Some of these trials are now three years old and are showing improved growth of both species in some combinations, improved growth of one of the species in other combinations or poorer growth in yet other combinations.

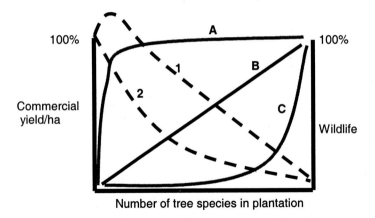

Figure 1. Mixtures with different numbers of tree species and some hypothetical commercial yields or wildlife benefits. The dashed lines indicate some hypothetical alternatives (1 or 2) of commercial yield returns and the solid lines demonstrate some hypothetical consequences for wildlife establishment (A, B or C).

Further information can also be gained from older mixture trials. Trials in both Queensland and Vietnam are particularly useful since many inter-specific interactions only take place once canopy closure occurs and trees become well established. Although there has been a long tradition of assessing (and minimising) inter-tree competition in plantation monocultures, there has been less work done on devising ways of doing this in mixtures. One approach that has been applied to these older trials is to devise Competition Indices (CI) and assess the growth responses of different species to increasing levels of competition. A complementary pair of species are those where the response to increasing competition was similar.

Some of the potential trade-offs between commercial yield from mixtures and the resulting biodiversity are illustrated in Figure 1. Similarly, the different food and habitat resources that are provided by tree species selected for mixed plantations will also result in different rates of wildlife colonisation (Figure 1). Thus, the key component of this work is to designing plantation species mixtures that are complementary, biologically enriched and stable over time.

References

Forest for the Future: Growing and Planting Native Trees for Restoring Forest Ecosystems (Biology Department, Science Faculty, Chiang Mai University, Thailand) (1998).

Goosem, S. & Tucker, N. I. J. *Repairing the Rainforest: Theory and Practice of Rainforest Re-establishment* (Wet Tropics Management Authority, Cairns, Queensland, Australia) pp 72 (1995).

Leungaramsri, P. & Rajesh, N. *The Future of People and Forests in Thailand after the Logging Ban* (Project for Ecological Recovery, Bangkok, Thailand, 1992) pp. 202.

Tropical Ecosystems: Structure, Diversity and Human Welfare.
Proceedings of the International Conference on Tropical Ecosystems
K. N. Ganeshaiah, R. Uma Shaanker and K. S. Bawa (eds)
Published by Oxford–IBH, New Delhi. 2001. pp. 231–234.

Regeneration dynamics of a wet evergreen forest, southern Western Ghats, India

R. Ganesan*, T. Ganesh, M. Soubadra Devy and P. Davidar[†]

Ashoka Trust for Research in Ecology and the Environment, No. 659, 5th A Main Road, Hebbal, Bangalore 560 024, India
[†]Salim Ali School of Ecology, Pondicherry University, Pondicherry 605 014, India
*e-mail: rganesan@atree.org

The evergreen forests of Agasthyamalai in south Western Ghats, India have the highest percentage of endemic species in the Western Ghats. Despite its richness and importance, there are very few studies on the long-term demography and population changes of the tree species in the evergreen forests of Agasthyamalai (Parthasarthy, 2001; Ganesh *et al.*, 1996).

Between 1993 and 1994 three 1-ha plots were established in an undisturbed wet evergreen forests at Kakachi, Kalakad–Mundanthurai Tiger Reserve (KMTR) of the Agasthyamalai range. The site is located at 1300 m asl and receives an annual rainfall of over 3000 mm. The vegetation is characterized by three dominant tree species, *Cullenia exarillata, Palaquium ellipticum* and *Aglaia bourdillonii*. The principal objective of this study was to measure the changes in plant diversity, recruitment and mortality of the species. All trees above 10 cm dbh at 1.3 m above ground were measured and tagged.

Keywords: Western Ghats, diversity, endemic, regeneration, mortality.

Recensus was done in 1999 for all trees and new and dead trees were recorded separately. Annual mortality, recruitment and turnover were calculated.

A total of 67 tree species >10 cm dbh from 52 genera and 31 families were recorded from the 3 ha of forests. The most common family was Lauraceae with 12 species followed by Euphorbiaceae (7) and Myrtaceae (6). Seventeen families had only one species. Genus with single species represented over 90% of the total genera.

Of the 67 species, 48% were endemic to the Western Ghats. A smaller proportion of species (9%) was endemic to southern Western Ghats and the Agasthyamalai region. Majority (71%) of the endemic species were canopy trees and the rest, 29% were understorey species.

A total of 2108 stems were sampled in the 3 ha at an average of 731 stems/ha. Only three species, *Agrostistachys borneensis* (19%), *Cullenia exarillata* (16%) and *Palaquium ellipticum* (13%) represented 45% of the stems while the other 64 species accounted for the remaining 52%. Over 3.5% of the stems were dead during the first enumeration. Majority (70%) of the species had less than 10 individual in the 3-ha plot and only 10% (7 spp.) had over 100 individuals.

Endemic species of the Western Ghats accounted for 51% (1077) of the stems. Of these, 14% were endemic to the Agasthyamalai region. Endemic species and non-endemics species significantly differed among canopy and understorey trees (χ^2: 813.51, $P < 0.0001$, df = 1). Endemic canopy species on an average were more common than non-endemics whereas in the understorey, non-endemic species were more abundant (Table 1).

Total basal area of all species was 58.74 m^2/ha. Two dominant species *Cullenia exarillata* and *Palaquium ellipticum* accounted for over 58% of the basal area while all other species individually accounted for less than 5%. Endemic species accounted for 78% of the basal area. Local endemic species constituted less than 5% of the basal area.

Table 1. Species richness and density of endemic and widely distributed species in 3 one-ha plots.

Category	Endemic		Non-endemic		Total	
	Species	Density	Species	Density	Species	Density
Canopy	23	977	23	309	44	1376
Understorey	9	100	13	721	22	821
Total	32	1077	36	1030	66	2197

The overall species diversity was $H' = 2.85$ and evenness was 0.671. Morisita Horn similarity index showed over 75% similarity between plots.

Table 2. Recruitment and mortality rate of tree species in the three 1-ha plots.

	Total	Plot 1	Plot 2	Plot 3
Recruitment	0.73	0.30	1.21	0.78
Mortality	0.54	0.57	0.60	0.48
Turnover	0.66	0.44	0.91	0.63

The overall recruitment of trees was 0.73% per year and mortality 0.55% per year resulting in an annual turnover of 0.63%. There was considerable variation in both recruitment and mortality across plots. Recruitment was lowest in the plot (0.305) that was homogeneous with respect to slope, while it was maximum in plot that had a high degree of habitat heterogeneity and high slope factor (1.22). However these variations were less pronounced for mortality (Table 2).

Recruitment, mortality and turnover in endemic species were comparable to the general patterns observed. However, when endemic species are compared with the non-endemic species, the latter showed less turnover than the former. The recruitment was on the whole comparable with endemic species but mortality was significantly less for widely distributed species than endemic species (Mann–Whitney $P < 0.05$). The trend in the three parameters across plots was similar in both groups of species (Table 2). This may be due to the high mortality rate of climax endemic species such as *Cullenia exarillata, Palaquium ellipticum* and gap species *Litsea wightiana*.

Forest change and stability may have important implications for the maintenance of biodiversity. Small scale disturbance could lead to enrichment of diversity while no or minimal disturbance may actually maintain low levels of diversity. This is evident from the three small plots studied in Kakachi. One of the plots that had high levels of habitat heterogeneity had more number of species than others which had more uniform conditions. The change in species richness could be attributed to natural disturbances such as tree falls and possibly occasional landslides following heavy rains. The maintenance of diversity could therefore be largely due to small or infrequent levels of natural disturbances.

Acknowledgements

Financial support was available from the MacArthur Foundation grant to Prof. Kamal Bawa, Univ. Massauchussetts, USA. We thank Tamil Nadu Forest Department for the permission to conduct the study and the Tamil Nadu Electricity Board at Upper Kodayar for logistic support.

References

Ganesh, T., Ganesan, R., Soubadra Devy, M., Davidar, P. and Bawa, K. S. Assessment of plant biodiversity at a mid elevation evergreen forest of Kalakad–Mundanthurai Tiger Reserve, Western Ghats, India. *Curr. Sci.* **71,** 379–391 (1996).

Parthasarathy, N. Changes in forest composition and structure in three sites of tropical evergreen forest around Sengaltheri, Western Ghats. *Curr. Sci.* **80,** 389–393 (2001).

Tropical Ecosystems: Structure, Diversity and Human Welfare.
Proceedings of the International Conference on Tropical Ecosystems
K. N. Ganeshaiah, R. Uma Shaanker and K. S. Bawa (eds)
Published by Oxford–IBH, New Delhi. 2001. pp. 235–238.

Forests in transition: Regeneration processes in second-growth and mature lowland rain forests of northeastern Costa Rica

Robin L. Chazdon

Department of Ecology & Evolutionary Biology, U-43, 75 North Eagleville Road,
The University of Connecticut, Storrs, CT 06269-3043, USA
e-mail: chazdon@uconn.edu

Second-growth forests are becoming increasingly important throughout the world's tropics, as reservoirs for biodiversity and as sources of forest products. Successful restoration efforts also depend on our knowledge of secondary forest regeneration. Regeneration studies in tropical forests have strongly focused on factors that maintain equilibrium species diversity and promote species coexistence in old-growth forests. Two paradigms dominate our current understanding of old-growth forest dynamics: 1) canopy gaps are critical foci for tree recruitment for most species (Hartshorn, 1978); and 2) seedling and sapling regeneration are mediated by density-dependent processes (Connell *et al.*, 1984; Harms *et al.*, 2000; Wills *et al.*, 1997). Yet few studies have examined regeneration processes during the transition from young, even-aged, less diverse, second-growth stands to multi-aged, highly diverse, old-growth stands. Our studies during the last 8 years suggest that these paradigms do not generally apply to seedling, sapling, or tree recruitment in second-growth stands ranging in age from 12 to 30 years old.

Keywords: Gap dynamics, tree recruitment, shade-tolerance, pioneer species, vertical structure.

In 1993 we initiated comparative studies of forest structure, species composition and light environments in wet, lowland, second-growth forests of northeastern Costa Rica. These stands have regenerated rapidly on clear-cut land that was burned and planted in pasture grasses and used briefly for cattle grazing prior to abandonment. These second-growth stands illustrate a 'best-case' scenario for forest regeneration because of even and rapid regeneration and adequate seed dispersal from nearby mature forest fragments and reserves. Forest structure, canopy tree species composition, and processes of seedling, sapling, and tree recruitment vary considerably between 'transitional' second-growth stands vs 'mature' old-growth stands. Species richness of trees above 10 cm dbh was substantially lower in 15–20-year-old second-growth stands compared to old-growth stands, despite similar tree basal area (Guariguata et al., 1997). Although mean levels of light availability in the understorey did not differ, second-growth stands had a more homogeneous light environment and showed a smaller spatial scale of light variation compared to the old-growth stands (Nicotra et al., 1999). Seedling density showed significant spatial autocorrelation in both types of stands, but showed a two-scale spatial pattern in old-growth stands, possibly reflecting the influence of local seed shadows (Nicotra et al., 1999). Compared to old-growth stands, second-growth stands had few canopy palms (Guariguata et al., 1997), generally lacked complete canopy gaps, and had more numerous, small openings compared to old-growth stands, permitting slightly increased levels of diffuse light in the understorey (Nicotra et al., 1999).

Vertical patterns of forest structure and species composition also vary between transitional and mature forests. Mature forests show a greater representation of canopy species in regeneration classes within the understorey. In transitional forests, relatively few tree species occur as both canopy trees and seedlings (Guariguata et al., 1997). Second-growth stands showed no linkage between canopy tree density or basal area and understorey light conditions measured below 4 m height (Kabakoff and Chazdon, 1996; Montgomery and Chazdon, in press), whereas old-growth stands showed greater linkage between understorey light conditions and local canopy trees. Overall, these results suggest that regeneration processes within the understorey are more coupled with characteristics of the forest canopy in old-growth stands compared to developing second-growth stands.

Detailed studies of regeneration dynamics initiated in 1997 in four 1-ha monitoring plots in secondary forests aged 12 to 25 years confirm these results. Canopy cover is dense and homogeneous with few or no canopy gaps; light availability in the understorey is low (usually less than 2% full sun) with weak or no spatial structure. The canopy is dominated by long-lived pioneer species that, in general, are not recruiting as seedlings or saplings. Some of these canopy trees exhibit little or no diameter growth, indicating senescence. Rates of seedling, sapling and small-tree recruitment are high despite the lack of high-light availability; new recruits are dominated by a group of shade-tolerant species, including three species of canopy palms. These shade-tolerant species, also common in mature forests, are capable of rapid height growth in the forest understorey and subcanopy. Above 10–15 m, tree recruits may be exploiting increased light availability in subcanopy gaps, zones beneath the canopy where crowns are absent. Subcanopy gaps may also be created by the death of small, short-lived pioneer trees that become overtopped and senesce.

Many woody species recruiting as seedlings and saplings in these young stands are not currently represented by fruiting canopy trees and must therefore be dispersed into the plots from outside, primarily by animal-dispersal agents. Hence, seed shadows are more highly dispersed than in old-growth stands and few species reach high seedling abundance. The potential for density-dependent effects on seedling survival and growth is therefore reduced during the transition from second-growth to old-growth.

When large (> 320 m^2) gaps are created experimentally in 20-year-old stands, seedling recruitment favours species currently dominating the canopy as well as species that are not currently represented in the canopy, increasing the overall abundance and species richness of regeneration compared to non-gap controls (Dupuy and Chazdon, unpublished data). In these gaps, the potential for density-dependent effects on seedling growth and survival increases. The presence of large canopy gaps increases recruitment of light-demanding pioneer species, whereas recruitment of shade-tolerant tree species and canopy palms is favoured in stands lacking large gaps.

Regeneration processes in transitional second-growth forests differ from those described for old-growth forests in this region in several fundamental ways: 1) tree recruitment occurs primarily in the absence of gaps; and 2) density-dependent processes do not appear to limit seedling or sapling regeneration. Moreover, the success of newly recruiting tree species is enhanced by rapid vertical growth rather than by tolerance of suppression and slow growth. In mature-phase patches of old-growth forests, these growth and recruitment processes are also likely to be important (Connell *et al.*, 1997; Lieberman *et al.*, 1995).

As with mature forests, management of transitional forests through logging or thinning will have major implications for future tree species composition and for diversity of forests in the future. If young forests with a structure similar to those described here are left unmanaged and protected from intervention, shade-tolerant tree and palm species will recruit slowly into the canopy and will gradually displace senescent pioneer trees. Tree species composition will more closely resemble that of mature forests, but within-stand heterogeneity will remain low. If logging or other disturbances create large gaps and release seeds stored in the seed bank (Dupuy and Chazdon,

1998), the canopy will continue to be dominated by fast-growing, light-demanding species that are already present in the canopy. Many of these fast-growing species have high commercial timber value as well as medicinal value (Chazdon and Coe, 1999). These results have important consequences for future uses of secondary forests as reservoirs for biodiversity and as sources of forest products.

Acknowledgements

I thank the US National Science Foundation and the Andrew W. Mellon Foundation for financial support. The Organization for Tropical Studies provided invaluable logistical assistance and Orlando Vargas helped with species identification. The work discussed here was done in collaboration with Julie Denslow, Manuel Guariguata, Juan Dupuy, Adrienne Nicotra, Rebecca Montgomery, Braulio Vilchez, Alvaro Redondo, and Deborah Lawrence with the assistance of Marcos Molina, Jeanette Paniagua, and Juan Romero.

References

Chazdon, R. L. & Coe, F. G. Ethnobotany of woody species in second-growth, old-growth, and selectively logged forests of northeastern Costa Rica. *Conserv. Biol.* **13**, 1312–1322 (1999).

Connell, J. H., Lowman, M. D. & Noble, I. R. Subcanopy gaps in temperate and tropical forests. *Austr. J. Ecol.* **22**, 163–168 (1997).

Connell, J. H., Tracey, J. G. & Webb, L. F. Compensatory recruitment, growth, and mortality as factors maintaining rain forest tree diversity. *Ecol. Mongr.* **54**, 141–164 (1984).

Harms K. E., Wright, S. J., Calderón, O., Hernández, A. & Herre. E. A., Pervasive density-dependent recruitment enhances seedling diversity in a tropical forest. *Nature* **404**, 493–495 (2000).

Dupuy, J. M. & Chazdon, R. L. Long-term effects of forest regrowth and selective logging on the seed bank of tropical forests in NE Costa Rica. *Biotropica* **30**, 223–237 (1998).

Guariguata, M. R., Chazdon, R. L., Denslow, J. S., Dupuy, J. M. & Anderson, L. Structure and floristics of secondary and old-growth forest stands in lowland Costa Rica. *Plant Ecol.* **132**, 107–120 (1997).

Harms, K. E., Wright, S. J., Calderon, O., Hernandez, A. & Herre, E. A., Pervasive density-dependent recruitment enhances seedling diversity in a tropical forest. *Nature* **404**, 493–495 (2000).

Hartshorn, G. S. Treefalls and tropical forest dynamics. *Tropical Trees as Living Systems* (eds Tomlinson, P. B. & Zimmerman, M. H.) 616–638 (Cambridge, Cambridge University Press, 1978).

Kabakoff, R. P. & Chazdon, R. L. Effects of canopy species dominance on understorey light availability in low-elevation secondary forest stands in Costa Rica. *J Trop. Ecol.* **12**, 779–788 (1996).

Lieberman, M., Lieberman, D., Peralta, R. & Harshorn, G. S. Canopy closure and the distribution of tropical forest tree species at La Selva, Costa Rica. *J. Trop. Ecol.* **11**, 161–178 (1995).

Montgomery, R. A. & Chazdon, R. L. Forest structure, canopy architecture, and light transmittance in old-growth and second-growth tropical rain forests. *Ecology* (in press).

Nicotra, A. B., Chazdon, R. L. & Iriarte, S. Spatial heterogeneity of light and woody seedling regeneration in tropical wet forests. *Ecology* **80**, 1908–1926 (1999).

Wills, C., Condit, R., Foster, R. B. & Hubbell, S. P. Strong density- and diversty-related effects help to maintain tree species diversity in a neotropical forest. *Proc. Nat. Acad. Sci. (USA)* **94**, 1252–1257 (1997).

Tropical Ecosystems: Structure, Diversity and Human Welfare.
Proceedings of the International Conference on Tropical Ecosystems
K. N. Ganeshaiah, R. Uma Shaanker and K. S. Bawa (eds)
Published by Oxford–IBH, New Delhi. 2001. pp. 239–242.

Density dependence among tropical trees: Case studies from Barro Colorado Island, Panama

S. Joseph Wright

Smithsonian Tropical Research Institute, Apartado 2072, Balboa, Ancon, Panama
e-mail: wrightj@tivoli.si.edu

Negative density dependence occurs when nearby conspecifics impair performance either through allelopathic interaction, intraspecific competition, or pest facilitation. Negative density dependence constrains locally abundant species. However, this may facilitate species coexistence by opening space for otherwise less successful species.

Three methods have been used to detect density dependence among tropical forest plants. The most frequent contrasts juvenile performance for a range of juvenile densities or for a range of distances from reproductive conspecifics. The Janzen–Connell hypothesis has motivated this approach. The second method enumerates all conspecifics for a number of plots and then contrasts their performance among plots with different conspecific density. The third method enumerates all plants in a single plot and then contrasts performance among species with different density. This between-species analysis has been called a 'community compensatory trend' (or CCT). All three methods have been applied to trees on Barro Colorado Island (BCI), Panama. I will critique and synthesize the disparate results of these analyses.

Keywords: Janzen-Connell hypothesis, seed ecology, seedling ecology, tree diversity.

Negative density dependence and life history variation contribute to a dramatic increase in diversity during seedling recruitment on BCI (Harms *et al.*, 2000). The seed rain is heavily dominated by a few species. In contrast, the diversity of first-year seedling recruits is much greater and closely approximates tree diversity. The low diversity seed rain reflects limited seed dispersal and heavy contributions by nearby trees at small spatial scales and the massive production of minute seeds by a few gap-dependent species at larger spatial scales. We identified seeds (S) and first-year seedling recruits (R) to species for 200 census stations and four years ($N = 386,000$ seeds and 13,000 recruits). To quantify life history variation, we calculated the ratio of recruits to seeds for each species ($\Sigma R : \Sigma S$, where sums are over the 200 stations). We multiplied the observed seed rain by $\Sigma R : \Sigma S$ for each species to predict recruit diversity after removing life history differences. Predicted recruit diversity fell half way between observed seed diversity and observed recruit diversity. To evaluate density dependence, we fit the following function $R = c \times S^b$ for each species, where c and b are fitted constants. Negative density dependence ($b < 1$) was evident for all 53 species examined, and the median b-value was just 0.23. Subsequent analyses using a variety of functional relationships and statistical procedures confirm that density dependence was strong and pervasive during seedling recruitment (J. Hillerislambers, unpublished analyses). We substituted observed seed rain into the best-fit function for each species to predict recruit diversity after removing both density-dependence and life history differences. Predicted recruit diversity now closely matched observed recruit diversity. Density-dependent seed mortality enhanced seedling diversity.

Proximity to conspecific adults continues to impair performance for saplings up to 4 cm in diameter. Hubbell *et al.* (1990) contrasted saplings located under conspecific trees with saplings located under heterospecific trees. A nearby conspecific tree reduced the growth of 1 to 2 cm and 2 to 4 cm diameter saplings for nine of the 11 species examined (significant for five and three species, respectively). Reduced sapling growth will reinforce reduced seedling recruitment near conspecific adults. These two studies collectively demonstrate the severe deleterious effects of a nearby conspecific adult for plant recruitment.

The second method used to detect density dependence has provided mixed results for BCI trees. For each of the 48 most abundant species, Hubbell and Foster (1986) fit the densities of juveniles (J) and adults (A) in 50 one-ha plots to the following model: $J = a_0 + a_1 A + a_2 A^2$, where a_0, a_1 and a_2 are fitted constants. Strong negative density dependence was evident for the most abundant tree species ($a_1 < 0$ for *Trichilia tuberculata*), weaker negative density dependence was evident for 20 species ($a_1 > 0$; $a_2 < 0$), and there was no evidence for density dependence for 27 species. Density dependence characterized 43% of the most abundant tree species. Plot-based analyses document negative density dependence but only for a handful of the most abundant tree species on BCI.

This conclusion contrasts strongly with the results of a Monte Carlo simulation for BCI trees. Wills *et al.* (1997) detected significant negative density-dependent recruitment for at least 80% of the 84 most abundant tree species. Wills and Condit (1999) performed a second Monte Carlo simulation for 100 BCI species with similar results. I will show that their Monte Carlo simulations are unable to distinguish density-dependent recruitment from spatially aggregated recruitment. Spatially aggregated recruitment has many potential causes, including limited seed dispersal and narrow regeneration requirements (e.g., tree fall gaps). Spatially aggregated recruitment is the more parsimonious explanation for the significant Monte Carlo results of Wills and associates. Plot-based analyses (method 2) provide no unequivocal evidence for negative density dependence among the many rare tree species found on BCI.

The third method used to detect density dependence was designed to capitalize on this very limitation. Connell *et al.* (1984) reasoned that density dependence should favour rare species and handicap abundant species. Therefore performance and density should be inversely related in *interspecific* comparisons for single large plots. Such inverse relationships or 'community compensatory trends' (CCTs) must be interpreted carefully. A CCT may occur if rare species are associated with a rare microhabitat that favours growth and survival. Tree falls provide an example. Tree falls are rare relative to shaded understorey, light-demanding species only recruit in tree falls and tend to be rare, and high resource availability favours survivorship and growth in tree falls. In this example, light-demanding species are rare as adults and perform well as juveniles contributing to a CCT for reasons unrelated to density dependence. Microtopography may create edaphic microhabitats that share the three critical attributes of tree falls. Plant habitat associations with tree falls, moist microhabitats, and/or nutrient-rich microhabitats may all contribute to an apparent CCT for reasons unrelated to density dependence. This problem could be avoided by restricting analyses to species with similar habitat associations. I will present such analyses for BCI trees. The results are generally negative; an apparent CCT disappears when plant habitat associations are controlled.

So far I have considered just the detection of density dependence. The implications of density dependence for plant diversity are largely unexplored for tropical forests (but see Harms *et al.*, 2000). Negative density dependence always slows population growth. If tropical forests are non-equilibrial communities, then slower population trajectories will delay competitive exclusion and sustain diversity at least temporarily. If tropical forests are at equilibrium, however, a tougher standard applies. Negative density dependence must regulate populations near observed levels to contribute to diversity at equilibrium. For long-lived organisms like trees, population regulation can only be evaluated by extrapolating current performance into the future. These extrapolations have been made using population projection matrices with density-dependent growth, survivorship, and recruitment functions determined empirically in the field. The results have generally been negative; density dependence is insufficient to regulate populations at observed levels (Hubbell *et al.*, 1986).

To summarize, density dependence may contribute to tree diversity in two ways on BCI. First, juvenile performance is greatly reduced in the dense aggregations of seeds and seedlings near seed-bearing conspecifics. This contributes to a dramatic increase in diversity during seedling recruitment on BCI (Harms *et al.*, 2000). Second, negative density dependence also affects the most abundant species at slightly larger spatial scales (up to at least 1 ha). This has the potential to prevent or delay further increases in abundance among the most successful species, thereby preserving space for less successful species. Perhaps not surprisingly, unequivocal evidence of negative density dependence is lacking for the many rare species on BCI. Negative density dependence enhances diversity during recruitment and possibly by checking the most successful species. Negative density dependence is unlikely to explain the coexistence of the many rare species that constitute the bulk of tree diversity in tropical forests. Space and time preclude reviews of many relevant studies from other tropical forests; however, these studies generally support the conclusions drawn for BCI (Wright, 2001).

References

Connell, J. H., Tracey, J. G. & Webb, L. F. Compensatory recruitment, growth, and mortality as factors maintaining rain forest tree diversity. *Ecol. Mongr.* **54**, 141–164 (1984).

Harms, K. E., Wright, S. J., Calderón, O., Hernández, A. & Herre, E. A. Pervasive density-dependent recruitment enhances seedling diversity in a tropical forest. *Nature* **404**, 493–495 (2000).

Hubbell, S. P., Condit, R. & Foster, R. B. Presence and absence of density dependence in a neotropical tree community. *Philos. Trans. R. Soc. London B* **330**, 269–281 (1990).

Hubbell, S. P. & Foster, R. B. Biology, chance, and history and the structure of tropical rain forest tree communities. in *Community ecology* (eds Diamond, J. & Case, T. J.) 314–329 (Harper and Row, New York, 1986).

Wills, C. & Condit, R. Similar non-random processes maintain diversity in two tropical rainforests. *Proc. R. Soc. London B* **266**, 1445–1452 (1999).

Wills, C., Condit, R., Foster, R. B. & Hubbell, S. P. Strong density- and diversity-related effects help to maintain tree species diversity in a neotropical forest. *Proc. Nat. Acad. Sci. USA* **94**, 1252–1257 (1997).

Wright, S. J. Chance, competition, density dependence and the coexistence of tropical tree species. *Oecologia*, in press (2001).

Tropical Ecosystems: Structure, Diversity and Human Welfare.
Proceedings of the International Conference on Tropical Ecosystems
K. N. Ganeshaiah, R. Uma Shaanker and K. S. Bawa (eds)
Published by Oxford–IBH, New Delhi. 2001. pp. 243–246.

Density dependence and seedling regeneration – The case for Borneo

David R. Peart*·†, Campbell O. Webb** and Arthur G. Blundell‡

*Department of Biological Sciences, Dartmouth College, Hanover NH 03755, USA
**Current address: Department of Ecology and Evolutionary Biology,
Yale University, New Haven, CT 06520-8106, USA
‡Current address: National Center for Environmental Assessment,
American Association for the Advancement of Science, US Environmental
Protection Agency, 808 17th St NW, 4th Floor, Washington DC 20006, USA
†e-mail: david.r.peart@dartmouth.edu

The great species richness of tropical forests occurs within the habitat matrix defined by trees, and may be largely dependent on tree species diversity. Thus, our ecological understanding of much of earth's biodiversity, and potentially our strategies for its conservation, may depend on solving the puzzle of how so many tree species can be maintained. Depending on its parameters, stabilizing (negative) density dependence could damp population increases, limit dominance, and 'rescue' small populations. Richness may then be limited only by the increased likelihood of stochastic extinction events for small populations. Whatever other mechanisms promoting coexistence exist, their contributions can be enhanced by negative density-dependent forces in population dynamics.

In assessing density dependence relevant to coexistence, it is important to identify the dependent and independent axes of the relationship to be tested.

Keywords: Coexistence, species diversity, tropical forest dynamics, population regulation.

The independent axis of primary interest is adult density. Because adults occupy the space-limited canopy, and because reproduction depends on attainment of a place in the canopy, canopy recruitment as a function of canopy tree abundance is the critical relationship to evaluate. Density-dependent dynamics affecting canopy recruitment can occur at any prior stage of the life cycle. Thus, there are several relevant dependent variables, including per capita seed production, seed survival, and both growth and survival in the vegetative stages. Density dependence at any stage will produce density dependence in canopy recruitment, unless counterbalanced by opposite density dependence at other stages. Because there are a variety of ecological mechanisms that may respond positively or negatively to density, different trends may occur at different points in the life cycle.

We summarize results from a site in remote moist rain forest in Gunung Palung National Park, Indonesian Borneo (West Kalimantan, Indonesia). We describe two studies, one at the community level (Webb and Peart, 1999) and one on a single population (Blundell and Peart, in prep.). In both studies, we assessed evidence for density dependence in patterns of population and community structure (static data), as well as the critical performance (dynamic) measures. Dynamic studies are necessarily limited in time. But patterns in community and population structure that are consistent with the hypothesized dynamics suggest that those dynamics have been operating over a longer period, in the past. So a combination of dynamic and static analyses showing consistent results provides the most compelling evidence.

In the community study, we established a stratified random sample of canopy tree plots over a 150-ha study area, and conducted demographic studies in seedling plots nested within them. We tested the relationship between 19-month seedling survival and the overall density and basal area of conspecific adults. Seedlings of the more common species suffered significantly higher mortality, on average, than those of comparatively rare species. The estimated trend implies a 10-year survival rate of 12% and 72% for seedlings of the most and least common species, respectively. Such a density-dependent trend necessarily increases diversity of a cohort of seedlings over time, for any diversity measure that incorporates both richness and evenness (e.g. the Shannon–Wiener index).

In the community structure (static) data, both number of species and evenness increased monotonically with plant size, from seedlings to adult trees, in comparison of samples of equal numbers of individuals. This is the trend predicted under the hypothesis of density dependence acting at the community level, i.e. mortality falling most heavily on the most common species, and it is consistent with the measured trends in seedling mortality. For detailed analysis at the population level, we chose one of the most abundant species at the study site, *Shorea quadrinervis*, a member of the Dipterocarpaceae, the dominant family in rain forests over much of Southeast

Asia. The density of *S. quadrinervis,* like that of most rain forest tree species, is spatially variable. Yet, density variations were not associated with any physical variation in the habitat that we were able to identify. Areas of high and low conspecific adult density did not differ in forest structure (total tree density, total basal area, or total seedling density), nor in slope, aspect or light availability in the understorey. We mapped all *S. quadrinervis* adults in 75 ha, and conducted demographic analyses of conspecific juveniles in both high and low adult density areas over 2 years.

Density-dependent trends differed dramatically over the stages of the juvenile vegetative part of the life cycle. First, seedling recruitment was positively density-dependent, i.e. the number of small seedlings (< 25 cm tall) per adult was greater in areas of high than low adult density. This is a destabilizing trend, which would tend, over time, to increase adult density where it is already high. But strong counteracting dynamics occur; seedlings performed much better in areas of low than high adult density, both in terms of growth and survival. The positive density dependence in seedling recruitment was therefore due to factors acting at the seed stage, i.e. in seed production or seed predation.

The poorer performance of individuals in high adult density areas was associated with poorer foliar condition (fewer leaves per unit height). Leaf and meristem damage increased with adult *S. quadrinervis* density and this damage reduced juvenile performance, so herbivory contributed to (and possibly caused) the density-related trends in juvenile dynamics. The strong negative density-dependent dynamics in the seedling stage reduced the densities of surviving juveniles in high (relative to low) adult density areas. Consequently, the relation between juvenile density and adult density changed markedly with juvenile size. While small seedlings were much denser in areas of high than low adult density, abundances became similar over areas of high and low adult density by the time seedlings were 75 cm tall. Thus, the number of potential canopy recruits per capita adult was much lower where adults occurred at high density. We evaluated the impact of these juvenile dynamics at the population level, using stage-based Lefkovitch matrix models. Finite population growth rate (lambda) declined significantly with adult population density over the study plots. In summary, four lines of evidence (juvenile dynamics, population size structure, individual condition and population modelling) all point to the effectiveness of negative, stabilizing density dependence in the population dynamics of *S. quadrinervis,* in space and time.

Overall, evidence from this Bornean study site indicates that stabilizing density dependence occurs widely over species, can have a strong influence on population dynamics, increases diversity over the juvenile stages, and promotes the maintenance of tree species diversity in the canopy. We cannot review here the interesting results on density-dependent dynamics emerging from other rain forest sites. But there is a clear need for a comparative and

245

critical evaluation of the evidence, of the mechanisms responsible, and the potential contribution of density-dependent dynamics to the stabilization of populations and the coexistence of tree species.

Acknowledgements

We thank the Indonesian Institute of Science (LIPI), the Conservation Agency of the Department of Forestry (PHPA), the Center for Research and Development in Biology (PPPB) and the Pontianak Office of Nature Conservation (KSDA), for their assistance and access to the Gunung Palung National Park, and Mark Leighton for pioneering and developing the research site. Funding sources included the National Science Foundation (Graduate Fellowship GER 9253849 to C.O. Webb and NSF DEB 9520889 to D.R. Peart); the US State Department (Fulbright Award to A.G. Blundell); the National Geographic Society; Conservation, Food, and Health Foundation, Inc.; Sigma Xi; and the Explorer's Club. The views expressed are the authors' own and do not represent official EPA policy.

References

Blundell, A. G. & Peart. D. R. In prep. Density-dependent population dynamics of a dominant rain forest canopy tree.

Webb, C. O. & Peart, D. R. Seedling density dependence promotes coexistence of Bornean rain forest trees. *Ecology* **80,** 2006–2017 (1999).

Tropical Ecosystems: Structure, Diversity and Human Welfare.
Proceedings of the International Conference on Tropical Ecosystems
K. N. Ganeshaiah, R. Uma Shaanker and K. S. Bawa (eds)
Published by Oxford–IBH, New Delhi. 2001. pp. 247–250.

Biomass dynamics of regenerating Amazonian forests

Rita de C. G. Mesquita

Ecologia – INPA, C. P. 478, Manaus, Amazonas, Brasil 69011-970
e-mail: rita@buriti.com.br

The most obvious impact of forest fragmentation and deforestation in the Amazon is the rapid decline in aboveground biomass and the release of large amounts of carbon to the atmosphere through burning (Skole and Tucker, 1993; Laurance *et al.*, 1997). There is a lot of controversy about the contribution of secondary forests to global carbon budgets (Phillips *et al.* 1998), but it is widely accepted that the regrowth vegetation function as an important carbon sink (Fearnside and Guimaraes, 1996; Johnson *et al.*, in press). Secondary forests of central Amazonia rapidly accumulate aboveground biomass in the first few years of forest regeneration (Mesquita, 1995), but other sites within Amazonia under more severe climate or land-use histories recover significantly slower (Vieira, 1996; Uhl *et al.*, 1989). Species composition and function also may be severely arrested for many years (Uhl *et al.*, 1988). Land-use history can have a dramatic effect on rates of forest recovery, but the extent of history effect is relatively poorly understood (Moran *et al.*, 1994).

Our studies of the dynamics of regenerating forests focus on (1) the effects of land-use history and age on biomass accumulation and species richness, and (2) rates of change in biomass through time.

Keywords: Amazonia, secondary forests, land-use history, biomass, forest recovery.

We studied a chronosequence of secondary forests varying from 1 to 20 years, under three different land-use types. Some sites were clear-cut and never burned (cut), others were clear-cut, burned once or twice and planted with perennials (plantation), and others were planted with pastures and regularly burned for several years before abandonment (pasture). We estimated biomass and species composition in 100-m transects using non-destructive methods. We re-measured transects once a year to quantify rates of change in biomass. We identified and measured diameter and frequency of all plant species found.

In their first 15 years secondary forests can accrue about 40% of the biomass found in primary forests of the same region. Biomass accumulation was affected by both age ($p = 0.001$) and land-use history ($p = 0.06$). Biomass accumulation rates can be as high as 20 ton dry wt/ha/yr, but the overall average for areas ranging from 2 to 20 years was 8.3 ± 6 ton dry wt/ha/yr (Figure 1). This value is at least 3 times higher than most estimates for primary forests.

Species composition and richness were significantly affected by age ($p = 0.001$), but the history effect was not significant. However, there is a trend that more heavily used sites have fewer species (Figure 2). At any rate, only about 55 species were found in 20-year-old regrowth forests, a small fraction of that found in mature forests.

Our results suggest that secondary forests are important carbon sinks in the tropics. They can recover up to 40% of the losses of biomass due to deforestation in their first 15 years.

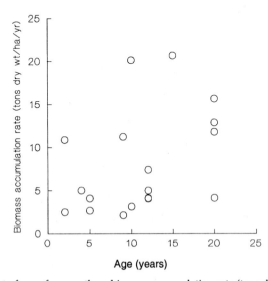

Figure 1. Effect of age of regrowth on biomass accumulation rate (tons dry weight/ha/yr).

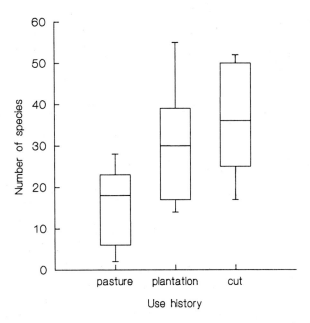

Figure 2. Effect of land-use history on number of species found on secondary forests of central Amazonia ranging from 1 to 20 years old.

However, only a fraction of the original species pool returns in the same time period. Land-use history can be important, but we failed to detect a clear effect of land-use on species richness. Land-use history has a significant effect on biomass accumulation, and possibly on other aspects of forest dynamics as well. A broader range of use-histories and ages should clarify the trends found here.

Acknowledgements

I thank Bruce Williamson and Marcelo Moreira, collaborators in this project. Field research was financed by the NASA-Long-term Biosphere-Atmosphere Experiment in the Amazon, the Biological Dynamics of Forest Fragments Project (BDFFP) administered jointly by the Instituto Nacional de Pesquisas da Amazonia (INPA) and the Smithsonian Institution, and by Brazil's Conselho Nacional de Desenvolvimento Cientifico e Tecnologico (CNPq).

References

Fearnside, P. M. & Guimaraes, W. Carbon uptake by secondary forests in Brazilian Amazonia. *For. Ecol. Manage.* **80**, 35–46 (1996).

Johnson, C. M., Vieira, I. C. G., Zarin, D. J., Frizano, J. & Johnson, A. H. Carbon and nutrient storage in primary and secondary forests in eastern Amazonia. *For. Ecol. Manage.*, in press.

Laurance, W. F., Laurance, S. G., Ferreira, L. V., Rankin-de-Merona, J. G., Gascon, C. & Lovejoy, T. E. Biomass collapse in Amazonian forest fragments. *Science* **278**, 1117–1118 (1997).

Mesquita, R. C. G. The effect of different proportions of canopy opening on the carbon cycle of a central Amazonian secondary forest. Ph.D., University of Georgia (1995).

Moran, E. F., Brondizio, E., Mausel, P. & Wu, Y. Integrating Amazonian vegetation, land-use, and satellite data. *BioScience* **44,** 329–338 (1994).

Phillips, O. L., Mahli, Y., Higuchi, N., Laurance, W. F., Nunez, P., Vásquez, R. M., Laurance, S. G., Ferreira, L. V., Stern, M., Brown, S. & Grace, J. Changes in the carbon balance of tropical forests: evidence from long-term plots. *Science* **282,** 439–442 (1998).

Skole, D. & Tucker, C. Tropical deforestation and habitat fragmentation in the Amazon: satellite data from 1978 to 1988. *Science* **260,** 1905–1910 (1993).

Uhl, C., Buschbacker, R. & Serrao, A. S. Abandoned pastures in Eastern Amazonia. I. Patterns of plant succession. *J. Ecol.* **76,** 663–681 (1988).

Uhl, C., Nepstad, D., Buschbacher, R., Clark, K., Kauffman, B. & Subler, S. Disturbance and regeneration in Amazonia: lessons for sustainable land-use. *The Ecologist* **19,** 235–240 (1989).

Vieira, I. Forest succession after shifting cultivation in eastern Amazonia. Ph.D., University of Stirling (1996).

Tropical Ecosystems: Structure, Diversity and Human Welfare.
Proceedings of the International Conference on Tropical Ecosystems
K. N. Ganeshaiah, R. Uma Shaanker and K. S. Bawa (eds)
Published by Oxford–IBH, New Delhi. 2001. pp. 251–256.

Tree regeneration and timber extraction: Patterns, constraints, and research priorities

M. A. Pinard and M. D. Swaine*

University of Aberdeen, Aberdeen AB24 5UA, UK
*e-mail: m.a.pinard@abdn.ac.uk, m.swaine@abdn.ac.uk

One basic component of all silvicultural systems aimed at sustainable forest management is ensuring the replacement of harvested stems through seedling recruitment and sapling development. While a number of different harvesting systems might be considered compatible with 'best management practice', a common approach to timber management in tropical moist forests involves selective harvests, controlled by a minimum felling diameter, in a polycyclic system where tree replacement comes primarily from natural regeneration. In the past, tropical silviculturalists raised concerns over the adequacy of natural regeneration and frequently viewed early silvicultural systems developed for tropical forests as failing to induce prompt regeneration (Dawkins & Philip, 1998). The focus of this review is to consider patterns of tree recruitment and regeneration under selective logging systems in tropical moist forests and to identify the primary constraints and ecological research priorities for addressing management needs.

Any attempt to draw general patterns from forests that are selectively logged, across the moist tropics is difficult because the forests and the logging

Keywords: Tree regeneration, timber extraction, management needs.

systems are extremely variable (see Putz *et al.*, 2000 for discussion of variability in logging systems). We cautiously put forward this paper as an attempt to generate and focus discussion on research needs related to tree regeneration. Our 'ideal' frame of reference would be tropical moist forests where timber harvesting is conducted within the context of a management aim (e.g. harvest intensity and cutting cycles are based on estimates of sustained yield, the extraction process is planned and controlled). Unfortunately, this ideal is not typical of tropical timber exploitation. Nevertheless, the area of tropical forest coming under sustainable management, as defined by timber certification bodies, is steadily increasing, and provides some justification for directing discussion to these areas where the conditions for success of sustainable forest management (Schmidt, 1991) appear to be in place.

Our definition of tree regeneration includes both seedling recruitment and sapling development. Forest stewardship principles and criteria require maintenance of ecosystem functions, including forest regeneration. Unlogged, old growth forests are often used as the control or standard for comparison in ecological studies, however, deviations from conditions associated with old growth forests are often within management objectives and may not necessarily compromise ecosystem function (Lamprecht, 1993). Identifying the level of regeneration that one would consider 'adequate' for maintaining ecosystem function is problematic. We suggest that conceptually, it is important to consider quantity but also quality, species composition, and processes involved.

Drawing from studies of ecological impacts of selective logging, reviews, and discussion documents (particularly Horne & Hickey, 1991; Guariguata & Pinard, 1998; Putz *et al.*, 2000), it seems clear that the impacts on tree regeneration are sometimes positive, sometimes negative, and sometimes neutral. The principal patterns identified are: (1) where soil is damaged, tree establishment and growth is restricted; (2) where felling gap size exceeds a certain threshold, tree growth and survival is reduced by competition with weedy vegetation (e.g. lianas, pioneer vegetation); (3) some ecological guilds are more likely to experience declines in recruitment post-logging than others, and describing ecological groups along several axes (e.g., shade tolerance, dispersal mode, seed germination requirements, propensity to resprout) may be more informative for guiding silvicultural prescriptions than describing guilds along a single shade tolerance axis; (4) early successional species are likely to become more abundant over space and time unless cutting cycles are relatively long; (5) successful regeneration may be more likely where there was an existing seedling bank at the time of harvest; (6) variability in space and time, in terms of forest structure and composition, makes it difficult to assess logging impacts on composition and regeneration.

Factors likely to affect tree regeneration in tropical forests include parental tree fecundity, predispersal losses, availability of suitable germination sites, early survivorship (and losses due to predation, herbivory, pathogens, mechanical injury, presence of competitors), and access to resources. The relative importance of these factors for seedling demography is not well understood in tropical forests (Lieberman, 1996).

The constraints to seedling establishment and sapling development most frequently cited for selectively logged tropical forests include the following: competition from weedy invasive species and reduced fecundity, usually related to a lack of seed trees. Competition with weedy invasive species is particularly a problem in large felling gaps and heavily disturbed areas, although the relative importance of invasive species appears to be highly variable among regions. Liana proliferation in felling gaps may stall forest recovery for decades, perhaps more by reducing sapling growth and survivorship rather than by reducing seedling recruitment (Schnitzer *et al.*, 2000; Perez-Salicrup, 2000). Interactions between early successional vegetation or vegetation tangles and large browsers such as elephants, may contribute to arrested succession in large gaps (Struhsaker *et al.*, 1996 and references therein). Short-lived pioneer trees (e.g. *Macaranga, Musanga, Cecropia*) are sometimes referred to as silvicultural weeds, but under some conditions they play an important role as a nurse crop for non-pioneer species.

After competition, reductions in fecundity were frequently cited as limiting regeneration. Reduced fruit production in logged forest relative to unlogged forest has been reported from Malaysia (Johns, 1992) and Uganda (Plumptre, 1995). In Uganda, seedling densities were correlated with densities of trees above 50 cm dbh, suggesting that recruitment might be limited by seed trees numbers. Inadequate numbers of seed trees (< 12 per ha) have also been cited as a limitation to regeneration in dipterocarp forest (Meijer, 1970). Recruitment limitation due to lack of seed trees is more likely to be a problem for species with irregular size distributions than for species with size distributions represented by reverse-J curves.

Reductions in fecundity for some species of trees in selectively logged forest may be related to changes in reproductive biology. Some pollination studies and seed germinability studies in logged or fragmented forest suggest that selective logging may indirectly influence seed production and quality (Guariguata & Pinard, 1998). But, as with reductions in dispersal distance related to hunting pressures (Forget & Milleron, 1991; Guariguata *et al.*, 2000), the implications of this reduction for tree regeneration over time are unclear. Authors examining dispersal limitation through models of forest dynamics draw divergent conclusions (Liu & Ashton, 1999).

Accumulated experience with tropical silviculture suggests that inducing regeneration of particular species in tropical moist forests is difficult, even for very shade-tolerant species (Dawkins & Philip, 1998; Catinot, 1997).

However, for tropical tree species that characteristically have unpredictable recruitment and clumped distributions, failure to foster regeneration within a narrow time frame defined by management objectives may not be surprising. Fostering the establishment and development of species that originally captured the site following large-scale disturbance (e.g. hurricane, agricultural clearings) may require intensive and costly treatments, that also have large impacts on forest structure and biodiversity values (Putz *et al.*, 2000). From a management perspective it may be rationale to implement silvicultural interventions that favour species that are readily regenerating under existing, closed forest conditions.

From our brief description of patterns and constraints to natural regeneration in selectively logged forests, five research areas emerge as priorities for informing forest management decisions. These are as follows: (1) the silvics of poorly known species to support species groupings that will facilitate the application of silvicultural interventions to foster regeneration; (2) weed ecology; (3) tree fecundity issues; (4) seedling demography; and (5) setting standards for evaluating 'the maintenance of ecological functions' in terms of tree regeneration.

Silvicultural weeds – Given the apparently important role that aggressive, weedy species play in arresting tree regeneration in some selectively logged forests (*Merremia* in the Solomon Islands, *Dinochloa* in Sabah, climbers and the invasive exotic, *Chromolaena* in West Africa), it is surprising how poorly understood their ecology is. Research directed at elucidating the conditions that promote their establishment and development may be useful for setting guidelines related to maximum allowable gap sizes, for example. Could risk of weed proliferation be predicted from pre-harvest conditions?

Seed trees and fecundity issues – For species that require seed tree retention in order to maintain reproductive (and genetic) stock (see Jennings *et al.* in press), research is needed to guide the selection of trees to be maintained (Catinot, 1997). More baseline data are needed with respect to gene flow, and effective population sizes in tropical tree species before patterns can emerge to describe how tree fecundity may change with densities of mature conspecifics.

Early survivorship of seedlings – Although it is true that the future species composition of the canopy cannot be predicted on the basis of the composition of seedling populations (Swaine & Hall, 1988), it does seem to be true that seedlings that make it through the early stages of development are more likely to persist in the understorey (Lieberman, 1996), and persistence in the understorey may increase a sapling's chance of outcompeting neighbours capturing the benefits of gap formation (Jennings *et al.* in press). So we need research on seedling demography in relation to logging forest conditions.

Setting standards for evaluating sustainability – If interest and participation in timber certification is maintained or increased, ecological research is needed to provide data upon which standards (or indicators of sustainability) can be based. For many variables related to ecological functions in forests, it is difficult to measure differences between natural disturbance and that caused by logging (Horne & Hickey, 1991). Where logging intensity and extraction is planned and controlled, is disturbance frequency the primary factor to differentiate logging from natural disturbance? What are the ecological implications of an increased representation of early successional vegetation across the landscape?

Where natural forest management is considered an appropriate land use for tropical moist forest, there is a need for monitoring impacts of silvicultural interventions. Research is needed to provide a better understanding of the processes controlling tree regeneration and how changes in these processes relate to the maintenance of ecological functions in the forests.

References

Catinot, R. *The Sustainable Management of Tropical Rainforests* (Scytale Publishing, Paris, France, 1997).

Dawkins, H. C. & Philip. M. S. *Tropical Moist Forest Silviculture and Management. A History of Success and Failure* (CAB International, Oxon, UK, 1998).

Forget, P. M. & Milleron, T. Evidence for secondary seed dispersal by rodents in Panama. *Oecologia* **87**, 596–599 (1991).

Guariguata, M. R., Adame, J. J. R. & Finegan, B. Seed removal and fate in two selectively logged lowland forests with contrasting protection levels. *Conserv. Biol.* **14**, 1046–1054 (2000).

Guariguata, M. R. & Pinard, M. A. Ecological knowledge of regeneration from seed in neotropical forest trees: implications for natural forest management. *For. Ecol. Manage.* **112**, 87–99 (1998).

Horne, R. & Hickey, J. Ecological sensitivity of Australian rainforests to selective logging. *Austr. J. Ecol.* **16**, 119–129 (1991).

Jennings, S. B., Brown, N. D., Boshier, D. H., Whitmore, T. C. & J do C A Lopes. Ecology provides a pragmatic solution to the maintenance of genetic diversity in sustainably managed tropical rain forests. *For. Ecol. Manage.* (in press).

Johns, A. D. Species conservation in managed tropical forests. in *Tropical Deforestation and Species Extinctions* (eds Whitmore, T. C. & Sayer, J. A.) 15–53 (Chapman and Hall, London, UK, 1992).

Lamprecht, H. Silviculture in the tropical natural forests. in *Tropical Forestry Handbook* (ed. Pancel, L.) Volume 2, 727–810 (Springer Verlag, Germany, 1993).

Lieberman, D. Demography of tropical tree seedlings: a review. in *The Ecology of Tropical Forest Seedlings, Man and the Biosphere Series* (ed. Swaine, M. D. S.) Volume 17, 131–138 (UNESCO, Paris, and Parthenon Publishing Group, Lancs, UK, 1996).

Liu, J. & Ashton, P. S. Simulating effects of landscape context and timber harvest on tree species diversity. *Ecol. Appl.* **9**, 186–201 (1999).

Meijer, W. Regeneration of tropical lowland forest in Sabah, Malaysia forty years after logging. *Mal. For.* **33**, 204–209 (1970).

Perez-Salicrup, D. R. Effect of liana cutting on tree regeneration in a liana forest in Amazonian Bolivia. *Ecology* **82**, 389–396 (2000).

Plumptre, A. J. The importance of 'seed trees' for the natural regeneration of selectively logged forest. *Comm. For. Rev.* **74**, 253–258 (1995).

Putz, F. E., Redford, K. H., Robinson, J. G., Fimbel, R. & Blate, G. M. Biodiversity conservation in the context of tropical forest management. World Bank Environment Department, Biodiversity Series, Impact Studies. Washington, DC (2000).

Schmidt, R. C. Tropical rain forest management: a status report. in *Rain Forest Regeneration and Management*, Volume 6, *Man and the Biosphere Series* (eds Gomez-Pompa, A., Whitmore, T. C. & Hadley, M.) 181–207 (UNESCO, Paris, France, 1991).

Schnitzer, S. F., Dalling, J. W. & Carson, W. P. The impact of lianas on tree regeneration in tropical forest canopy gaps: evidence for an alternative pathway of gap-phase regeneration. *J. Ecol.* **88,** 655–666 (2000).

Struhsaker, T. T., Lwanga, J. S. & Kasenene, J. M. Elephants, selective logging and forest regeneration in the Kibale Forest, Uganda. *Biotropica* **12,** 45–64 (1996).

Swaine, M. D. & Hall, J. B. The mosaic theory of forest regeneration and the determination of forest composition in Ghana. *J. Trop. Ecol.* **4,** 253–269 (1988).

Tropical Ecosystems: Structure, Diversity and Human Welfare.
Proceedings of the International Conference on Tropical Ecosystems
K. N. Ganeshaiah, R. Uma Shaanker and K. S. Bawa (eds)
Published by Oxford–IBH, New Delhi. 2001. pp. 257–261.

Early stages of rain forest regeneration after logging and shifting agriculture in Sarawak, Malaysia

Stuart J. Davies*[†,‡], N. R. Hashim*, M. Mohamad* and H. Semui*

*Institute of Biodiversity and Environmental Conservation, Universiti Malaysia
Sarawak
[†]The Center for International Development and the Arnold Arboretum, Harvard
University, 22 Divinity Avenue, Cambridge, MA 02138, USA
[‡]e-mail: sdavies@oeb.harvard.edu

In the east Malaysian state of Sarawak, land-use change is driven by a combination of commercial timber extraction, shifting cultivation and large-scale agricultural development. Areas of economically unproductive secondary forest continue to expand. Understanding forest succession on degraded lands is essential for the protection and subsequent utilization of soil, water and forest resources. The objective of our research program has been to investigate ecological constraints on the development of secondary successional rain forest in Sarawak. In this paper we give an overview of our recent research in Sarawak, focussing particularly on factors constraining or limiting forest regeneration.

The field work for this study was conducted in adjacent Forest Reserves, Sabal and Balai Ringin in west Sarawak (c. 1°10'N, 111°E). The history of

Keywords: Borneo, community dynamics, plant competition, land-use change, succession.

land-use and forest cover was assessed for Sabal, and community dynamics studies were conducted in Balai Ringin.

Commercial logging began in the 1970s in Sabal and in the 1980s in Balai Ringin. The areas surrounding these forest reserves have long been populated by Iban villagers. The Iban conducted shifting cultivation, mostly of hill rice, on the margins of the reserves. Following the initiation of commercial logging, shifting cultivation expanded and intensified. Recently some areas of both Sabal and Balai Ringin have been converted to plantations of exotic and indigenous timber species.

Sabal includes an area of c. 7800 hectares. Aerial photographs were taken over Sabal at 1:25,000 in 1972, 1976, 1982 and 1998. We used these images along with detailed field surveys to reconstruct the history of land-use and forest cover changes at Sabal. The trajectory of land-use change involved intensive and repeated selective harvesting of the primary forest. This was followed by the expansion of shifting cultivation associated with improved access along logging roads. After 1984, when shifting cultivation was excluded from Sabal, plantations of the exotic fast-growing timber species, *Acacia mangium*, were developed. In unplanted areas, secondary forests began to regenerate. Continuous forest cover declined from more than 80% to c. 50% between 1972 and 1998. By 1998, medium-stature regenerating forest covered almost 40% of the area. In addition to the decline in area of continuous forest, almost all of the remaining forest had suffered degradation through selective harvesting by 1998. Parallel studies of soil properties in forest fragments and degraded secondary forests within Sabal revealed significantly lower soil nutrient levels in the secondary forest areas than in remaining fragments of selectively logged forest. This pattern of land-use change is typical for Sarawak and probably for much of the rest of Borneo.

Community successional patterns are influenced by a wide range of factors including: land-use history, topography, geology, soils, climate, and the distance from intact primary forest that may serve as a seed source for later successional species (Holl, 1999). To investigate community dynamics in regenerating secondary forests, and their relationships with land-use history and resource availability, we established 20 plots in Balai Ringin Forest Reserve. Plots were selected to include variation in land-use history and habitat features, focussing particularly on time since the most recent shifting agricultural cycle, and soil variation. Plots were 0.04-hectares in area and ranged from 1 to 14 years since the most recent farming season. All vegetation was surveyed within each plot and identified to species. Chemical and physical properties of the soil in each plot were measured. Plots were recensused after two years to assess successional dynamics.

Analysis of the floristic composition and structural characteristics of the 20 plots revealed a predictable increase in diversity, basal area and canopy height with age since the most recent shifting agricultural cycle. A general pattern of rain forest succession following logging and shifting cultivation was evident. The earliest stages of succession were dominated by *Ageratum conyzoides*, *Blumea balsamifera* and several grass and sedge species. This stage was followed quickly by a suite of fast-growing pioneer shrubs and trees with relatively short life-spans, including species of *Melastoma*, *Macaranga*, *Mallotus*, *Ficus* and *Trema*. Gradually these species died out and were replaced by longer-lived pioneers such as *Neolamarckia*, *Duabanga* and *Endospermum*. In some of the oldest plots shade-tolerant, later successional species had begun to establish. This pattern is consistent with reports from plot studies of succession in other parts of Malaysia (Wyatt-Smith, 1955; Kochummen & Ng, 1977; Ewel *et al.*, 1984). However, in heavily degraded sites this pattern of succession may be impeded by the dominance of lalang grass (*Imperata cylindrica*), resam fern (*Dicranopteris linearis*) or other weedy species. This vegetation may persist as a very nutrient impoverished community referred to as *Adinandra*-belukar after the dominant species, *Adinandra dumosa* (Sim *et al.*, 1992). Fire may in some cases also become an important recurrent feature of these ecosystems impeding the regeneration of tree communities.

Principal component analysis of floristic composition of the 20 plots in Balai Ringin illustrated the strong temporal axis of variation in these regenerating communities. Several woody species were restricted to the earlier successional plots, e.g. *Melastoma malabathricum* and *Macaranga costulata*. In addition to the change in floristic composition with stand age, the second principal component axis of variation among the plots was associated with variation in soil nutrients. Young plots on the richer soils appeared to be colonized by a different group of species than plots on the poorer soils.

To further investigate the impact of soil nutrient availability on community dynamics in the early stages of succession, we conducted a nutrient-addition experiment in very early successional plant communities in Balai Ringin. In this experiment, three replicate sites of one year after the most recent shifting cultivation cycle were selected. In each site, three 25 m^2 plots were treated with 10 gm^{-2} of controlled release NPK fertilizer periodically over a 12-month period. Three additional 25 m^2 plots in each of the three sites were left untreated. All woody stems were tagged and measured and herbaceous cover was estimated at the start of the experiment. After 12 months, biomass was harvested and weighed from all 18 plots. Physical and chemical properties of soil were measured at the start and end of the experiment. Nutrient addition had a dramatic effect on aboveground productivity, with approximately 30% greater biomass in the nutrient-treated plots. Besides, there also appears to be differences in community composition between treated and untreated plots.

For the primary forests of north Borneo there is strong evidence for soil resource effects on community composition and dynamics (Ashton and Hall, 1992). Pioneer *Macaranga* species growing in natural forest differ in diameter growth rates between different soils (Davies, 2001). Recent analyses of 12 other common pioneer species in a 52-hectare primary forest plot in Lambir National Park in northern Sarawak have also found significant differences in performance across a natural soil gradient. It remains unclear which particular soil characteristics are the key limiting resources affecting performance across soil gradients in Borneo. Some preliminary evidence suggests that soil phosphorus may be the primary limiting soil nutrient (Sollins, 1998).

We conducted a second experiment to investigate the details of nutrient effects on the early stages of forest succession after shifting agriculture. This involved a nursery experiment in which communities of pioneer tree seedlings were treated with different nutrients. For this experiment, seven common early successional species from the Balai Ringin and Sabal Forest Reserves were selected. Seeds of all seven species were sown in large wooden tubs in natural forest soil. Each tub consisted of 98 seedlings (14 per species) planted in a hexagonal array, such that each seedling was surrounded by six equidistant and heterospecific neighbours. Replicate tubs were treated with nitrogen, phosphorus, nitrogen and phosphorus, or left untreated. The striking result of this experiment was that nutrient addition had a relatively small impact on community dynamics. Competitive interactions among the species were extremely strong, with one species (*Melastoma malabathricum*) completely dominating the other six species, regardless of nutrient treatment. Despite this competitive outcome, overall community productivity was significantly increased by nutrient addition.

The sequence of forest loss in Sarawak involving logging, followed by shifting agriculture, and the subsequent establishment of large-scale plantations of exotic species is typical of much of the current land-use change occurring in Borneo. One of the major impacts of this change is on soil properties. While a range of factors influence the regeneration potential of secondary forest, our field studies and experiments demonstrate that the growth and performance of pioneer tree species and consequently the dynamics of regenerating secondary successional plant communities in Sarawak are strongly influenced by soil resources. Techniques need to be developed for restoring soils in degraded lands, and for ameliorating the effects of deforestation on soils.

Acknowledgements

This research was funded by grant #08-02-09-0508 from the Intensification of Research in Priority Areas program of the Malaysian Ministry of Science and Technology to SJD, and a J. & J. Ruinen Fellowship to NRH. We thank the Sarawak Forest Department for permission to work in Sabal and Balai Ringin, and for assistance in many aspects of the project.

References

Ashton, P. S. & Hall, P. Comparisons of structure among mixed dipterocarp forests of north-western Borneo. *J. Ecol.* **80,** 459–481 (1992).

Davies, S. J. Tree mortality and growth in 11 sympatric *Macaranga* species in Borneo. *Ecology* **82,** in press (2001).

Ewel, J. J., Chai, P. & Lim, M. T. Biomass and floristics of three young second-growth forests in Sarawak. *Malay. For.* **46,** 347–364 (1984).

Holl, K. D. Factors limiting tropical rain forest regeneration in abandoned pasture: seed rain, seed germination, microclimate, and soil. *Biotropica* **31,** 229–242 (1999).

Kochummen, K. M. & Ng, F. S. P. Natural plant succession after farming in Kepong. *Malay. For.* **40,** 61–78 (1977).

Sim, J. W. S., Tan, H. T. W. & Turner, I. M. *Adinandra* belukar: an anthropogenic heath forest in Singapore. *Vegetatio* **102,** 125–137 (1992).

Sollins, P. Factors influencing species composition in tropical lowland rain forest: Does soil matter? *Ecology* **79,** 23–30 (1998).

Wyatt-Smith, J. Changes in composition in early natural plant succession. *Malay. For.* **18,** 44–49 (1955).

Tropical Ecosystems: Structure, Diversity and Human Welfare.
Proceedings of the International Conference on Tropical Ecosystems
K. N. Ganeshaiah, R. Uma Shaanker and K. S. Bawa (eds)
Published by Oxford–IBH, New Delhi. 2001. pp. 262–265.

Patterns of plant and soil carbon accumulation in abandoned tropical agricultural and pasture lands

R. Ostertag*[,†], W. L. Silver*[,†,§] and A. E. Lugo[†]

*University of California, Berkeley, CA 94720, USA
[†]International Institute of Tropical Forestry, Rio Piedras, PR 00928, USA
[§]e-mail: wsilver@nature.berkeley.edu

Approximately half of the tropical biome is in some stage of recovery from past human disturbance, most of which is in secondary forests growing on abandoned agricultural lands and pastures. Reforestation of these abandoned lands, both natural and managed, has been proposed as a means to help offset increasing carbon emissions to the atmosphere. Restoration and reforestation have the potential to contribute to carbon storage directly through biomass and soil carbon accumulation (Richter *et al.*, 1999), and indirectly by providing an alternative to fossil fuels for energy generation (Fearnside, 1999), but the time period required is not well known. Forest composition and structure, land use history, and climate are all likely to affect the rate and character of carbon sequestration. In this paper we discuss the potential of tropical secondary forests to serve as sinks for atmospheric CO_2 in aboveground biomass and soils. We explore general patterns in carbon accumulation and storage with reforestation in relation to stand age, climate and land use history, and examine the potential for reforestation to provide a carbon offset alternative for tropical countries.

Keywords: Carbon offset, land use, reforestation, tropical forest.

We used data from the literature to examine patterns in aboveground biomass and soil carbon following reforestation in the tropics. For aboveground biomass we limited our analysis to secondary succession. Plantations have been reviewed elsewhere (Lugo et al., 1988; Lugo and Brown, 1992). All sites were completely cleared and the majority of them were also burned prior to forest regrowth. Life zone categorizations were determined by the authors, or were estimated from total annual rainfall reported in the studies according to the following classifications: dry (< 1000 mm/yr), moist (1000–2500 mm/yr), and wet (> 2500 mm/yr).

To examine patterns in soil carbon pools with reforestation we compiled data from the literature that reported forest age together with soil carbon content or soil carbon concentrations and bulk density values. We used data from both natural secondary succession and plantations, because these data have not been thoroughly reviewed elsewhere. We examined patterns with regard to the same land use and life zone categories as for aboveground biomass. When organic matter was reported we used a conversion factor of 0.5 to estimate soil carbon pools. Studies reported a variety of depths ranging from 7 to 50 cm, but were conducted predominantly in the top 25 cm. In an effort to roughly standardize depths across studies, we used a regression technique to determine the relationship of soil carbon pools with depth (Silver et al., 2000). Soil data were used for forests ≥ 3 years since abandonment or plantation establishment.

Aboveground biomass increased significantly with time following reforestation, similar to results reported previously (Brown and Lugo, 1990). The overall rate of aboveground biomass accumulation was 2.36 Mg/ha/yr. Most studies have examined only the first 20 years of forest regrowth; here the rate of aboveground biomass accumulation was significantly faster during the first 20 years of regrowth (6.17 Mg/ha/yr) than over the subsequent 60-year period ($P < 0.01$). There was no effect of life zone during the first 20 years of regrowth, but wet forests accumulated carbon faster than moist forests during the 20–80 year period ($P = 0.07$). Overall (80 years), wet forests accumulated biomass at a rate of 3.24 Mg/ha/yr ($n = 44$) and moist forests at a rate of 2.17 Mg/ha/yr ($n = 91$), while dry forests showed no significant pattern with time.

Land use also had a significant effect on aboveground biomass gain with reforestation ($P = 0.07$). Forests regrowing on old agricultural fields accumulated biomass at slightly faster rates than forests grown on abandoned pastures. Sites that were cleared but not cultivated showed the slowest rate of aboveground regrowth, but these sites also had the smallest sample size. During the first 20 years of regrowth, there was a significant increase in aboveground biomass following agricultural use, but not with other land uses. The rate of aboveground regrowth in abandoned agricultural fields was 6.04 Mg/ha/yr during the first 20 years.

There was a statistically significant relationship between soil carbon content and forest age during the first 100 years following establishment, although the predictive power of the relationship was low. During the first 20 years of forest establishment, soil carbon pools averaged 60 ± 4 Mg C ha^{-1} ($n = 33$); this increased significantly to 74 ± 6 Mg C ha^{-1} ($n = 24$) during the subsequent 20–100 years ($P < 0.01$). Soil carbon pools in mature forests were 72 ± 7 Mg C ha^{-1} ($n = 12$). Soil carbon accumulated at a rate of 1.30 Mg/ha/yr during the first 20 years and at a rate of 0.20 Mg/ha/yr for the subsequent 80-year period.

Overall, soil carbon accumulated at a rate of 0.41 Mg/ha/yr over a 100-year period following reforestation. There were no strong patterns in the rate of soil carbon accumulation with life zone. Moist forests accumulated soil carbon at a rate of 0.51 Mg/ha/yr, which was a slightly slower rate ($P < 0.10$) than dry forests (1.02 Mg/ha/yr). Wet forests did not show a significant increase in soil carbon pools over time. Dry forest sites followed a significant linear increase in soil carbon with time, although it is important to note the relatively small sample size for this life zone.

Past land use had an impact on the rate of soil carbon accumulated with reforestation ($P < 0.01$), although patterns are much less clear than for aboveground biomass. Sites that were deforested but not managed prior to forest re-establishment tended to accumulate soil carbon at a faster rate (1.17 Mg C ha^{-1} yr^{-1}, $n = 12$) than pasture sites (0.49 Mg/ha/yr; $n = 21$), or agricultural sites (0.25 Mg /ha/yr; $n = 12$). This effect was not apparent during the first 20 years of forest recovery, but was significant ($P < 0.01$) during the subsequent 80 years. The rate of change in soil carbon accumulation differed significantly over time (Figure 3, $P < 0.01$) with fastest rates early in forest development for all three previous land use types. There was no distinguishable effect of cover type (plantation versus natural secondary succession) on the rates of soil carbon accumulation following reforestation, although plantations had significantly ($P < 0.01$) more soil carbon (90 ± 9 Mg C ha^{-1}, $n = 10$) than secondary forests (61 ± 3 Mg C ha^{-1} $n = 47$).

Our results clearly demonstrate the importance of the previous land use on the rate of carbon sequestration in vegetation and soils. When the previous land use results in low degradation or high relative fertility such as some types of fertilized agriculture, biomass accumulation through succession is faster than when the previous land use degrades the site for aboveground regrowth, as was the case for pastures in our analysis. In addition, we observed a continuous accumulation of aboveground biomass in forests up to 80 years of age. The rate of sequestration varied with life zone as previously shown (Brown and Lugo, 1990), and slowed over time. Our results showed that soil carbon accumulated faster for approximately 20 years in the top 25 cm of mineral soil, than later in forest development. While this is likely to be the most active zone for carbon accumulation over short time periods, the dynamics of deeper soil profiles are important to consider. In summary, tropical forests growing on abandoned land have the potential to sequester carbon aboveground and belowground for at least 40–80 years, and possibly much longer, depending on conditions of climate and past land use.

Acknowledgements

This work was supported by a grant from the A.W. Mellon Foundation to W. Silver, and by grant # BSR-8811902 from the NSF to the Terrestrial Ecology Division of the University of Puerto Rico and IITF as part of the Long Term Ecological Research Program, and was done under the California Agricultural Experiment Station projects # 6363-MS.

References

Brown, S. A. & Lugo, A. E. Tropical secondary forests. *J. Trop. Ecol.* **6**, 1–32 (1990).

Fearnside, P. M. Forests and global warming mitigation in Brazil: opportunities in the Brazilian forest sector for responses to global warming under the 'clean development mechanism'. *Biomass Bioenergy* **16**, 171–189 (1999).

Lugo, A. E. The future of the forest. Ecosystem rehabilitation in the tropics. *Environment* **30**, 16–20 (1988).

Lugo, A. E. & Brown, S. Tropical forests as sinks of atmospheric carbon. *For. Ecol. Manage.* **54**, 239–255 (1992).

Richter, D. D., Markewitz, D., Trumbore, S. E. & Wells, C. G. Rapid accumulation and turnover of soil carbon in a re-establishing forest. *Nature* **400**, 56–58 (1999).

Silver, W. L., Neff, J., McGroddy, M.. Veldkamp, E. Keller, M. Patterns in soil chemical properties and root biomass along a soil texture gradient in a lowland Amazonian tropical forest. *Ecosystems* **3**, 193–201 (2000).

Tropical Ecosystems: Structure, Diversity and Human Welfare.
Proceedings of the International Conference on Tropical Ecosystems
K. N. Ganeshaiah, R. Uma Shaanker and K. S. Bawa (eds)
Published by Oxford–IBH, New Delhi. 2001. pp. 266–267.

Forest degradation and regeneration in eastern Madagascar: A landscape perspective

Joelisoa Ratsirarson*[†], John Silander**,
Roland de Gouvenain** and Jeannin Ranaivonasy*

*Département des Eaux et Forêts ESSA, BP 175 Université d'Antananarivo,
Antananarivo (101), Madagascar
**Ecology and Evolutionary Biology Dept. U 43 University of Connecticut,
Storrs CT 06269-3043, USA
[†]e-mail: j.ratsirarson@simicro.mg

Landscape degradation and forest regeneration in the eastern lowland (< 500 m) rainforests of Madagascar were examined. Analyses were based on satellite images, historical aerial photographs and maps, plus ground surveys (ecological, socio-economic and archeological). Together this information provides critical insight into land use and forest cover change over time.

Our study revealed that the eastern lowland forest landscape of Madagascar is highly degraded with only scattered fragments of forest remaining. Some larger patches of forest are found at higher elevations. Regeneration is limited outside of forest patches. Cutting and burning the landscape for upland rice cultivation and cash crop plantations (cloves, coffee beans) have been the main cause of landscape degradation in the region. The underlying

Keywords: Forest degradation, forest regeneration, Madagascar.

causes are mediated by economic, institutional and socio-cultural conditions. The recent fall in cash crop values has led villagers to redirect their efforts to slash and burn activities for upland rice cultivation. Weak capacity of local traditional authorities and distrust of government agencies (e.g. the forest service) have tended to intensify forest clearing. Rice cultivation has always played a central role in local culture. However, topographical heterogeneity in the landscape limits the location and extent of traditional irrigated paddy rice cultivation. In many places local residents are consequently forced to clear more forests for the less efficient upland rice cultivation. Demographic pressures have driven a search for more land, and clearing forest is one way for locals to acquire land tenure.

Forest fragments remain in the landscape either because of well-organized local institutions that manage and monitor the forest, or because the forests are located on rocky soils with difficult access and where upland rice cultivation is not possible.

After forest clearing and burning, the soil organic horizon is destroyed and soil fertility plummets. This encourages the invasion of aggressive alien plant species including *Lantana camara*, *Aframomum angustifolium*, *Psiadia altissima*, whose distributions depend on local topography and soil conditions. Moreover, the fallow period between slash-and-burning cycles has been reduced from ten or more years to three. Together these land-use practices have effectively eliminated natural forest regeneration from the landscape of eastern Madagascar.

Tropical Ecosystems: Structure, Diversity and Human Welfare.
Proceedings of the International Conference on Tropical Ecosystems
K. N. Ganeshaiah, R. Uma Shaanker and K. S. Bawa (eds)
Published by Oxford–IBH, New Delhi. 2001. pp. 268–269.

Seedling regeneration over 35 years in Australian tropical and subtropical rainforests

Joseph H. Connell* and Peter T. Green**

Department of Ecology, Evolution and Marine Biology, University of California,
Santa Barbara, California 93106, USA
**Centre for the Analysis and Management of Biological Invasions, Department of
Biological Sciences, Monash University, Clayton, Victoria, 3168, Australia
*e-mail: connell@lifesci.ucsb.edu

The recruitment, growth, and mortality of mapped seedlings, saplings, and trees have been measured for many years at two rainforest plots in Queensland, Australia. They are in a tropical rainforest plot at 17°S lat., and a subtropical rainforest plot at 28°S lat. We have monitored seedling regeneration on these long-term plots at intervals over 35 years, between 1965 and 2000.

Over this time, seedling numbers varied over several orders of magnitude. On the tropical plot, of the 121 species that had at least 2 adults, the number of new seedlings mapped and tagged since 1965 ranged from zero to 12,677. Also, the majority of the common species showed a pattern of markedly episodic recruitment, with few or no seedlings produced for long intervals. In some trees such barren periods lasted up to 26 years. In both the tropical and subtropical forests studied, only a very few species produced seedlings annually, in contrast to some Neotropical rainforests where the majority of species did so.

Keywords: Long-term trends, episodic seedling recruitment, Australia, rain forest.

Long-term trends in recruitment varied considerably. Over 35 years on our permanently-marked belt transects, seedling recruitment per year of some species gradually increased, others decreased, others showed a reversing trend (rising followed by falling, or vice-versa), and still others showed no significant pattern. Long-term records were essential in revealing these patterns or lack of them (Connell, 1978; Connell et al., 1984; Connell and Green, 2000).

References

Connell, J. H. Diversity in tropical rain forests and coral reefs. *Science* **199**, 1302–1310 (1978).

Connell, J. H., Tracey, J. G. and Webb, L. J. Compensatory recruitment, growth, and mortality as factors maintaining rain forest tree diversity. *Ecol. Monographs* **54**, 141–164 (1984).

Connell, J. H. and Green, P. T. Seedling dynamics over thirty-two years in a tropical rainforest tree. *Ecology* **81**, 568–584 (2000).

Tropical Ecosystems: Structure, Diversity and Human Welfare.
Proceedings of the International Conference on Tropical Ecosystems
K. N. Ganeshaiah, R. Uma Shaanker and K. S. Bawa (eds)
Published by Oxford–IBH, New Delhi. 2001. pp. 270–273.

Acacia-dominated rainforests in the Australian humid tropics: Contrasts of structure and regeneration in natural and anthropogenic forests

Andrew W. Graham*, Mike S. Hopkins and Bob Hewett

Tropical Forest Research Centre, CSIRO Sustainable Ecosystems and Rainforest
Cooperative Research Centre, PO Box 780, Atherton, Queensland, Australia 4883
*e-mail: andrew.graham@cse.csiro.au

Much of the future effort in rainforest conservation and research will be focussed on rehabilitation of cleared rainforest landscapes, secondary rainforests, and simplified, degraded or fragmented primary rainforests. Which successional processes will best assist a rapid return to the high values of species diversity and species richness that are considered to be the key characteristics of tropical rainforests, both at the community level and amongst the trees of the forest canopy? Studies of succession in rainforests where canopy floristic composition is relatively simple may provide some useful guidance.

In the humid tropics of northern Australia, the canopies of primary rainforests are rarely dominated by a single tree species. When simple canopy floristics do occur, the variant usually reflects specific edaphic or historical factors (Webb and Tracey, 1981). One tree that can dominate rainforest canopies is *Acacia celsa* Tindale (previously *Acacia aulacocarpa* Cunn.

Keywords: *Acacia*, old-field, succession, rainforest, cyclones.

ex Benth. *sensu lato*). This wattle grows to about 25 m tall with a diameter at breast height (dbh) about 1 m. It occurs as a canopy dominant in mesic rainforests in two contrasting situations. Within primary rainforest, *A. celsa* may occur on low-nutrient soils along exposed hillslopes or ridgelines subject to periodic localized destruction by cyclonic winds. It also occurs in old-field rainforest successions on oligotrophic or degraded mesotrophic soils. There are strong contrasts between these primary and secondary forests in terms of both physiognomy and regeneration dynamics.

The humid tropical region of North Queensland (15°S to 20°S, 146°E) has a rugged topography with steep gradients of rainfall (typically 2000 to 3000 mm but > 8000 mm on mountain peaks), altitude and temperature. The rainforests of the region were described by Tracey (1982) using the physiognomic classification of Webb (1968). These examples of wattle-dominated forests of the Innisfail and Atherton districts are based on more detailed descriptions and 1:25,000 scale mapping (Hopkins and Graham, 1981; Graham *et al.*, 1995, 1996).

Along the sub-coastal escarpments, the Notophyll Vine Forests and Simple Notophyll Vine Forests dominated by *Acacia celsa* on granite or metamorphic ridges are restricted to the tops of the most exposed ridges where slopes often exceed 50°. The granitic sandy soils, of variable depth, have many surface boulders while soils on the metamorphic rocks are generally very shallow. Viewed from above, these forests have a smooth to undulating canopy of interlaced crowns of wattles (*A. celsa*), with the characteristic uniform pale grey-green wattle colour and uniform crown size and canopy evenness.

Structural features include a dense, closed, very smooth canopy dominated by *A. celsa*; canopy height 18–24 m; very dense subcanopy of many species 12–18 m with large numbers of stems in the 3–20 cm dbh range, often merging into a uniform and dense understorey with large numbers of small, woody stems 1–2 cm dbh; plank buttresses and stilt roots absent; stem and root suckers are conspicuous to very common; strangling figs absent; leaf size predominantly notophyll; toothed leaves are more conspicuous than in other upland forest types but are still rare; walking stick palms absent; tree ferns and pandans rare; robust canopy lianes absent; slender and wiry lianes conspicuous in understorey; epiphytes rare on tree trunks; hanging lichens absent; *Calamus* rare to absent in canopy.

It is likely that these wattle-dominated forests are seral forms produced through extensive disturbance by cyclones, particularly as the distributions of the wattle-dominated forests and adjacent putative parent types cannot be consistently correlated with obvious environmental features such as soil depths, altitudes, or slopes. Most of the canopy species typical of these forest types before disturbance (Notophyll or Simple Mesophyll/Notophyll Vine Forests) are present in the understorey and subcanopy of the seral forests. In addition, most of the transitional stages consistent with a succession from a wattle-dominated canopy to the parent forest types have been observed, including the conspicuous remains of dead wattles in the understorey of advanced successional stages. It appears that the characteristic abundance of the later phase canopy species in the understorey facilitates the progressive phases of successional development. Seeds of *A. celsa* were present at low densities in the soil seed bank beneath the parent forest types (seed density to 4 cm depth < 3 m^{-2}). In contrast, wattle was the most abundant seed bank species below the wattle-dominated forests (seed density to 4 cm depth > 150 m^{-2}). This large seed bank suggests that, once established, these wattle forests may re-develop easily and repeatedly with future disturbances in a long-term cyclical and deflected successsional pathway (*sensu* Hopkins, 1981).

Secondary rainforests dominated by *A. celsa* occur on abandoned farmlands with infertile or degraded soils. Pasture break-up starts with isolated trees around which 'islands' of exotic weeds and native species colonize. These 'islands' expand and coalesce forming a vegetation mosaic with components of varying structures and ages (Figure 1). The oldest trees, usually wattles, have a characteristic open-grown 'paddock' habit with broad crowns and low branching, often at ground level. This distinctive form is a key structural feature for identifying secondary rainforest developed on abandoned pastures. Re-disturbance by grazing, fire or reclearing disrupts succession and may result in patches of less distinctive structure.

Despite their relatively well-developed canopy structure (Table 1), these secondary wattle forests often have a very sparse seedling and sapling understorey until the forest reaches a stage at which the canopy wattles

Figure 1. A diagrammatic representation of an advanced stage of old-field secondary succession in which the rainforest canopy is dominated by wattle trees (*Acacia celsa*). The wattles that established in the open pasture have a characteristic low-branched 'paddock' form.

Table 1. Structural data for replicated sites in primary rainforests and adjacent secondary rainforests dominated by wattle (*A. celsa*) on the Atherton Tableland, North Queensland. Stem density and basal area data were determined for all stems > 10 cm dbh on 20 m × 50 m transects. Leaf Area Index data are averages of five positions within each site and were recorded using a Licor LAI 2000.

Structural data (mean values)	Primary forest (logged)	Secondary forest (40–60 years old)
Stem density (stems ha^{-1}, $n = 5$)	1222	872
Basal area (m^2 ha^{-1}, $n = 5$)	43	47
Leaf Area Index (m^2 m^{-2}, $n = 4$)	5.9	5.1

senesce. Previous research indicates that the wattles are extremely efficient in exploration of the upper soil profile (Hopkins *et al.*, 1996) and may capture the scarce nutrient resources very effectively.

On-going studies of seedling survival in a replicated exclosure trial suggest that, over a period of three years, disturbance by birds and ground-dwelling vertebrates may be an important cause of mortality, possibly retarding the rate of successional development. Other current studies in these forests include soil physical and chemical characteristics, floristic composition, growth rates, and biomass accumulation.

References

Graham, A. W., Hopkins, M. S. & Maggs, J. Succession and disturbance in the remnant rainforest type complex Notophyll Vine Forest on Basalt (Type 5b). I. Vegetation map and explanatory notes. A report prepared for the Wet Tropics Management Authority. CSIRO Division of Wildlife and Ecology, Tropical Forest Research Centre, Atherton. Report No. VM 1/1295-6. 45 pp + map (1995).

Graham, A. W., Hopkins, M. S. & Maggs, J. Succession and disturbance in the remnant rainforest type complex Mesophyll Vine Forest on Basalt (Type 1b). I. Vegetation map and explanatory notes. A report prepared for the Wet Tropics Management Authority. CSIRO Division of Wildlife and Ecology, Tropical Forest Research Centre, Atherton. Report No. VM 1/0496-8. 31pp + map (1996).

Hopkins, M. S. Disturbance and change in rainforests and the resulting problems of functional classification. in *Vegetation Classification in Australia: Proceedings of a Workshop* (eds A. N. Gillison & D. J. Anderson) 42–52, Canberra, October, 1978 (CSIRO/ANU Press, Canberra, 1981).

Hopkins, M. S. & Graham, A. W. Structural typing of tropical rainforest using canopy characteristics in low-level aerial photographs – a case study. in *Vegetation Classification in Australia: Proceedings of a Workshop* (eds A. N. Gillison & D. J. Anderson) 53–65, Canberra, October 1978 (CSIRO/ANU Press, Canberra, 1981).

Hopkins, M. S., Reddell, P., Hewett, R. K., & Graham, A. W. Comparison of root and mycorrhizal characteristics in primary and secondary rainforest stands on a metamorphic soil in North Queensland, Australia. *J. Tropical Ecol.* **12**, 871–885 (1996).

Tracey, J. G. The vegetation of the humid tropical region of North Queensland, CSIRO Aust. Division of Plant Industry, Indooroopilly (1982).

Webb, L. J. Environmental relationships of the structural types of Australian rainforest vegetation. *Ecology* **49**, 296–311 (1968).

Webb, L. J. & Tracey, J. G.. Australian rainforests: patterns and change. in *Ecological biogeography of Australia* (ed. Keast, A.) 606–694 (Junk, W, The Hague, 1981).

Tropical Ecosystems: Structure, Diversity and Human Welfare.
Proceedings of the International Conference on Tropical Ecosystems
K. N. Ganeshaiah, R. Uma Shaanker and K. S. Bawa (eds)
Published by Oxford–IBH, New Delhi. 2001. pp. 274–277.

Forest successional stage affects survival and growth of rain forest tree species differing in shade tolerance

Marielos Peña-Claros

Department of Plant Ecology, Utrecht University, P. O. Box 800.84, 3508 TB
Utrecht, The Netherlands
e-mail: m.pena@bio.uu.nl

During the course of succession, the light availability at the forest floor decreases dramatically due to the regrowing vegetation. Because tree species differ in their shade tolerance, one can expect that species perform differently in terms of survival and growth during different phases or stages of succession. Pioneer species are expected to have a lower survival rate as the age of the secondary forest increases, while the survival of shade-tolerant species should be relatively high in all successional stages (cf. Augspurger, 1984). In terms of growth rates, at high and moderately low light levels (higher > 3% light), tree pioneer species have higher relative growth rates (RGR) than shade-tolerant species (Veneklaas and Poorter, 1998, and references therein). When the plants are growing in deep shade (< 2% light), however, irradiance falls below the light compensation point of the pioneer species. Under these conditions, shade-tolerant species have a higher RGR than pioneer ones (Walters and Reich, 1996). Therefore, in early stages of succession seedlings of pioneer species are expected to attain higher growth rates than seedlings of shade-tolerant species. In later stages of succession, the response of species will depend on the light level in the successional stage.

Keywords: Bolivia, growth analysis, secondary forests, seedlings, succession.

In this study I transplanted seedlings of nine tree species differing in shade tolerance to secondary forests in various successional stages and allowed them to grow for two years. My questions were the following: (1) are seedlings of different species restricted to a given successional stage because they can only survive under very specific light conditions? (2) if seedlings are able to survive in various successional stages, do species growth rates vary with successional stage?, (3) how does the successional stage affect their biomass allocation and morphology?, and (4) what plant traits are correlated with species survival and growth?

The study was carried out in the Bolivian Amazon in the forest reserve El Tigre (11°59'S, 65°43'W) and in nearby areas owned by local farmers (< 5 km from El Tigre). The area receives an annual rainfall of 1780 mm with a dry season (< 100 mm/month) from May to September. Nine tree species were included in the experiment: 3 pioneers (*Bellucia pentamera*, *Cecropia sciadophylla*, *Schizolobium amazonicum*), 4 long-lived pioneers (*Cedrela odorata*, *Inga thibaudiana*, *Jacaranda copaia*, *Tachigali cf. vasquezii*) and 2 shade tolerants (*Buchenavia cf. punctata* and *Aspidosperma* sp.). Seedlings of the nine tree species were planted in three successional stages: 1-year-old regrowth (2 sites), 10-year-old secondary forests (4 sites) and 20-year-old secondary forests (3 sites). In summary, the experiment contained 3 ages, 4 to 6 plots per age, 9 species per plot and 13–15 seedlings per species, making a total of 1899 seedlings.

The experiment lasted two years. Plants were measured non-destructively in March 1997 and after 3, 6, 12, 18, and 24 months. At each census plant survival and total height were recorded. Two destructive measurements were made during the course of the experiment, at the beginning and at the end of the experiment. Based on the final destructive measurement the following parameters were calculated: root mass fraction (RMF), stem mass fraction (SMF), leaf mass fraction (LMF), specific leaf area (SLA), and leaf area ratio (LAR). Additionally, using the two destructive measurements I calculated the relative growth rate (RGR) and the net assimilation rate (NAR, $g\ m^{-2}\ day^{-1}$).

Survival data were analyzed with a survival analysis. Differences in survival patterns among the different successional stages and species were determined using Cox regressions. Plant responses in terms of growth (HGR and RGR), biomass allocation (LMF, SMF, RMF), morphology (LAR and SLA) and whole-plant assimilation (NAR) were tested using a two-way ANOVA with successional stage and species as factors. I used Pearson's correlation coefficient to determine how allocation patterns (LMF, SMF, RMF), morphological characteristics (SLA, LAR) and assimilation rate (NAR) affected species mortality and growth.

At the end of the experiment 32% of the seedlings were still alive. The survival rate varied with successional stage, so that seedling survival decreased significantly as age of successional stage increased (48, 29, and 25% survival by the end of the experiment in the 1 year old regrowth, the 10 and the 20 year old secondary forest, respectively). Apparently, for most of the species, regardless of their shade tolerance, it is more advantageous to establish early in succession than later on. A positive effect of higher light levels on survival has also been reported for other tree species (e.g. Augspurger, 1984). Species differed in their survival rate; light demanding species (*Cecropia*, *Schizolobium* and *Jacaranda*) having a lower survival rate than more shade-tolerant species (*Tachigali*, *Buchenavia* and *Inga*). Other studies also demonstrated that pioneer species have low survival rates (Augspurger, 1984; Walters and Reich, 2000). This phenomenon has been related to their sensitivity to herbivory (Coley, 1988) and poor resistance to drought.

In general, HGR and RGR decreased as forest age increased. Both HGR and RGR were considerably higher in the 1-year-old regrowth than in the 10 and 20-year-old secondary forests.

This decrease in growth is probably due to a decrease in light availability, as canopy openness decreases with stand age. These results also suggest that species, regardless of their shade tolerance, would benefit from arriving early in succession. In general, pioneer and long-lived pioneer tree species had higher RGR than shade-tolerant species in young and old successional stages. Pioneer tree species are reported to have high growth rates, and to be able to out-compete other species relatively rapidly (e.g. Rose, 2000). As a consequence, in young successional stages pioneer species can dominate and form a closed canopy after a few years (< 5 years), although non-pioneer species also become established at the same time. These results support the idea that under high light conditions differences in species growth rates can play an important role in determining the species composition of the canopy of early successional stages (cf. Finegan, 1996).

All parameters but SMF varied with successional stage. Tree seedlings adjusted their allocation of biomass and morphology to the successional stage where they grow. Very young successional stages are characterised by a higher light level and lower air humidity. Consequently, seedlings tended to invest more in roots to obtain more water and have lower SLA and lower LAR to reduce transpiration (Poorter and Nagel, 2000). At the same time they attained higher NAR than seedling in older successional stages. In secondary forests of 10 or 20 years old, light was probably the limiting resource, and seedlings enhanced their light interception by allocating more biomass to leaves and by producing leaves with high SLA, which resulted in high LAR. Similar responses have been reported by other studies comparing seedlings growing under low and high light conditions (e.g. Popma and Bongers, 1988).

Survival in the 20-year-old secondary forests was negatively related to SMF and positively to LMF. The RGR in the 1-year-old regrowth increased with NAR and LAR. These results support the idea that different plant traits enhance survival at low light and growth at high light.

Pioneer tree species have a small time-window of opportunity during succession to establish due to their high light requirements for establishment. Long-lived pioneer species show a large variation of responses, implying that their recruitment is not restricted to early successional stages as has been proposed before (Finegan, 1996). Additionally, my study supports the idea that different plant traits are important for determining high survival rates in low light than high growth rates in high light, as has been proposed before (Kitajima, 1994). Consequently, differences in survival and growth rates among species at different successional stages should play an important role in determining the set of species that will be present in a given successional stage.

References

Augspurger, C. K. Light requirements of neotropical tree seedlings: a comparative study of growth and survival. *J. Ecol.* **72**, 777–795 (1984).

Coley, P. D. Effects of plant growth rate and leaf lifetime on the amount and type of anti-herbivore defense. *Oecologia* **74**, 531–536 (1988).

Finegan, B. Pattern and process in neotropical secondary rain forests: the first 100 years of succession. *TREE* **11**, 119–124 (1996).

Kitajima, K. Relative importance of photosynthetic traits and allocation patterns as correlates of seedling shade tolerance of 13 tropical trees. *Oecologia* **98**, 419–428 (1994).

Poorter, H. & Nagel, O. The role of biomass allocation in the growth response of plants to different levels of light, CO_2, nutrients and water: a quantitative review. *Austr. J. Plant Physiol.* **27**, 595–607 (2000).

Popma, J. & Bongers, F. The effect of canopy gaps on growth and morphology of seedlings of rain forest species. *Oecologia* **75**, 623–632 (1988).

Rose, S. Seeds, seedlings and gaps: size matters. A study in the tropical rain forest of Guyana. Topenbos–Guyana Series 9. Ph.D. thesis (Utrecht University, Utrecht, the Netherlands 2000).

Veneklaas, E. J. & Poorter, L. Growth and carbon partitioning strategies of tropical tree seedlings in contrasting light environments. in *Inherent variation in Plant Growth. Physiological Mechanisms and Ecological Consequences* (eds Lambers, H., Poorter, H. & van Vuuren, M. M. I.) 337–361 (Backhuys Publishers, Leiden, The Netherlands, 1998).

Walters, M. B. & Reich, P. B. Are shade tolerance, survival, and growth linked? Low light and nitrogen effects on hardwood seedlings. *Ecology* **77**, 841–853 (1996).

Walters, M. B. & Reich, P. B. Seed size, nitrogen supply, and growth rate affect tree seedling survival in deep shade. *Ecology* **81**, 1887–1901 (2000).

Tropical Ecosystems: Structure, Diversity and Human Welfare.
Proceedings of the International Conference on Tropical Ecosystems
K. N. Ganeshaiah, R. Uma Shaanker and K. S. Bawa (eds)
Published by Oxford–IBH, New Delhi. 2001. pp. 278–281.

Hedgerow intercropping for soil improvement in the upcountry of Sri Lanka

H. M. S. P. Madawala Weerasinghe* and
B. W. Bache**

*Department of Botany, University of Peradeniya, Peradeniya, Sri Lanka
**Department of Geography, University of Cambridge, Downing Site,
Cambridge, UK
e-mail: sumedha@botany.pdn.ac.lk

Sri Lanka faces serious problems of agricultural sustainability due to increasing human population pressure on the existing land resources. Peasant agriculture in Sri Lanka, characterized by small holdings less than 1 ha with low levels of income and productivity, frequently leaves farmers with no option but to cultivate unstable and infertile sloping lands, accelerating land degradation and soil erosion. Hedgerow intercropping (HI) (also known as alley cropping or avenue cropping) is a land management practice recommended to overcome these problems. In Sri Lanka it is being promoted under the Sloping Agricultural Land Technology (SALT) program.

The ecological zone, to which Doragalla belongs, covers a large part of the upper Mahaweli watershed area. This area was subject to drastic social, economical and environmental changes over a short period due to hydropower and irrigation project in 1978. Consequently, the vast extent of

Keywords: Hedgerow intercropping, soil fertility, natural terrace development, Sri Lanka.

fertile lands was submerged in the reservoirs and farmers were compelled to cultivate the sloping lands in the peripheral areas. This shift from the fertile valleys to the infertile sloping lands presented the farmers with several problems related to agricultural sustainability. Therefore, the government demarcated this area (Upper Mahaweli Watershed) as a pilot area to introduce land management techniques such as SALT-HI, to find solutions to problems faced by farmers in this region.

SALT-HI systems have a dual role to play. They prevent soil erosion and improve soil fertility. Their success in the former role can frequently be assessed visually and has been quantified in many studies (Young, 1989), although not in Sri Lanka. The focus of the present study and its general objective is to attempt to answer the question 'Can HI improve soil fertility in the marginal lands of the upcountry of Sri Lanka?' Under the main objective of this study, the nutrient distribution pattern due to within alley soil erosion (also known as natural terrace development) at a SALT model plot was investigated.

A four-year SALT-HI demonstration plot at Dorgalla (WU2-wet-upcountry) was chosen to carry out this study. The SALT model plot was a tea land abandoned more than 30 years ago. It had remained as a grassland until it was cleared in early 1992 to establish the SALT-HI model plot, where six selected hedgerow species were planted as mono-culture, double-hedgerows. The inter-hedgerow spacing was 6 m. A nearby farmer's plot without a HI system and an abandoned tea land were also selected with a similar slope as that in the SALT-HI model plot, for comparison.

A rigorous soil sampling procedure was carried out to determine the effects of HI against the 'noise' of the natural soil variability. Representative soil samples were collected from the SALT-HI model plot, from the nearby abandoned tea land and from the farmers' land. The soil was sampled in

Table 1. Soil pH distribution down the slope in the SALT-HI plot and abandoned tea lands at Doragalla, Sri Lanka. The hedgerow species used were *Calliandra calothyrsus* (Cal.), *Tithonia diversifolia* (Tit.), *Gliricidia sepium* (Gli.) and *Bana* sp. (Ban.).

	Salt-Hi plot							Abandoned tea land					
	Distance in meters down the slope							Distance in meters down the slope					
Depth (cm)	Cal 3 (m)		Tit. 9 (m)	Gli.	15 (m)	Ban	Mean (n = 21)	0	5 m	10 m	15 m	20 m	Mean (n = 15)
0–10	5.7	5.9	6.5	5.1	5.6	6.1	5.7	5.8	4.10	4.30	4.30	4.20	4.24	4.23
10–30	5.6	6.2	6.4	5.4	5.8	6.2	5.9	5.9						
30–50	5.8	6.2	6.1	5.5	5.6	5.4	5.7	5.7						
>50	5.8	6.0	6.1	5.4	5.5	5.1	5.3	5.6	4.20	4.20	4.23	4.21	4.25	4.22

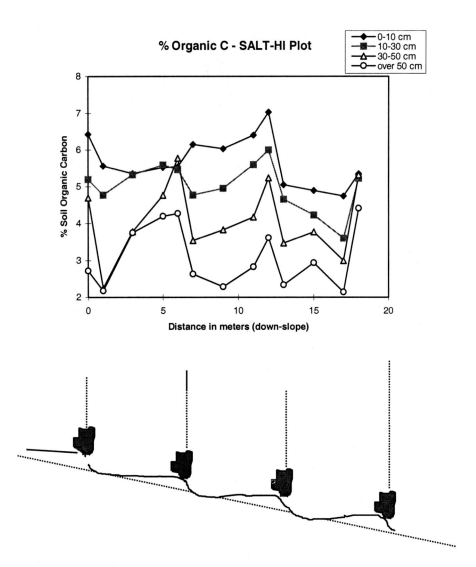

Figure 1. Variation in percentage soil organic carbon gradient within alleyways and down the landscape in SALT-HI plots in Doragalla, Sri Lanka.

three transects, aligned down the slope, covering four hedgerows and three alleyways. Soil samples were taken from (i) the hedgerow itself, (ii) 1 m and 3 m away from the hedge, along the whole transect and (iii) four depths down the soil profile. In each of the farmer's plot and in the abandoned tea land, three 20 m transects were sampled at 5 m intervals giving 5 sampling positions down the slope. The dried and sieved soil samples were analyzed for various chemical and physical parameters using the standard procedures at the laboratory of the Dept. of Geography, University of Cambridge, UK.

At Doragalla, HI plots had higher organic C and total N compared to the farmer's and the abandoned tea plots in the same vicinity. The SALT-HI plots compared to the control sites indicate that inclusion of N-fixing species has some effect at this site. The HI system appeared to be more efficient in tapping unavailable forms of P and achieved a higher P status than the control plots without hedgerows. The SALT-HI plots at Doragalla had lower acidity levels (pH 5.7) than that at the abandoned tea land nearby (pH 4.2), indicating that the mulch added to the HI plots reduced the soil acidity problem (Table 1).

The increasing trend of soil nutrient availability within alleyways down the slope was evident, but not consistent with every nutrient. At the SALT-HI plot, natural terrace development was evident and the mineralizable N and organic carbon were higher under hedges compared to that in the open alleyways. The natural terrace development on the steep slopes at Doragalla gave a 'saw-tooth' pattern for some of the measurements (Figure 1). A similar 'saw-tooth' pattern was also observed by Turkelboom et al. (1993) in an experiment carried out on distribution of soil organic matter (SOM) down an entire slope with HI. This alternating high and low soil nutrients in a SALT-HI plot can have a profound effect on crop yields, making such a system less attractive to the farmers. Therefore, the benefits of HI are hard to generalize. The selection of appropriate hedgerow tree species to a specific soil/climatic condition also seems to play a critical role in determining the success of the HI system.

In contrast, the control sites did not show any position effect down the slope, suggesting that the presence of hedgerows enhanced the within alley erosion, thereby altering the surface length, contour of the slope, soil profile depth, and soil chemical and hydrological properties. Therefore, there are still critical questions on HI under different soil-climatic conditions that need to be addressed before adopting HI as a possible sustainable farming practice.

Acknowledgements

We would like to thank the Cambridge Commonwealth Trust for providing us the financial assistance to carry out fieldwork in Sri Lanka.

References

Turkelboom, F., Ongprasert, S. & Taejajai, U. Alley cropping on steep slopes: soil fertility gradients and sustainability. Paper presented at the international workshop on sustainable agricultural development. Concepts and measures. Asian Institute of Technology, Bangkok, December 14–17 (1993).
Young, A. Agroforestry for Soil Conservation (CAB International, Wallingford, UK, 1989).

Tropical Ecosystems: Structure, Diversity and Human Welfare.
Proceedings of the International Conference on Tropical Ecosystems
K. N. Ganeshaiah, R. Uma Shaanker and K. S. Bawa (eds)
Published by Oxford–IBH, New Delhi. 2001. pp. 282–285.

Canopy gap characteristics in a logged Albertine Rift Afromontane forest island

Dennis Babaasa*, Gerald Eilu†, Aventino Kasangaki* and Robert Bitariho*

*Ecological Monitoring Programme, Institute of Tropical Forest Conservation, P.O.
Box 44 Kabale, Uganda
†Faculty of Forestry and Nature Conservation, Makerere University,
P.O. Box 29018 Kampala, Uganda
e-mail: itfc@infocom.co.ug

Although studies of gap dynamics are important in understanding processes influencing regeneration and maintenance of tree diversity, there have been few such studies in tropical rainforests of Africa. This is alarming given that the forests are exploited for timber yet knowledge of gap dynamics has not been considered in forest conservation, management and exploitation.

This study investigated gap characteristics in a logged Albertine Rift Afromontane forest island of Bwindi Impenetrable National Park, Uganda. The area was reserved by the Forest Department in 1932, then made a National Park in 1991. Pitsawing for timber was the most prevalent human activity in the forest between 1947 and 1991. This activity significantly altered the structure and composition of the forest. About 10% of Bwindi remain essentially undisturbed, 61% was intensively pitsawn and 29% selectively pitsawn (Howard, 1991). Forest clearings, characterised by a ground cover dominated by a dense tangle of herbaceous or semi-woody climbers, are

Keywords: Afromontane forest gaps, regeneration, pitsawing.

common throughout the forest. The impact of such dense ground cover on forest-wide, gap-phase regeneration is unknown.

We focus on the effect of the following variables on tree recolonisation in the gaps: past pitsawing intensity, gap size and density, dominant herbaceous ground vegetation in the gaps, slope angle and altitude. The effects of these variables on gap regeneration are poorly known and interactions between them are even less well known yet they are of importance to regeneration, stand development and wildlife conservation. Three study sites were chosen based on past timber harvesting intensity.

Gap sizes in the two heavily pitsawn sites were significantly larger and covered a significantly larger proportion of forest area than the undisturbed site. We attribute this difference in gap sizes and area of the forest they cover to indiscriminate removal of valuable timber tree species and multiple tree cuts for construction of timber pitsawing platforms and shelters. Wind-throw or snapping of surviving trees can also increase in clearings, causing an expansion of the cleared area (Kasenene and Murphy, 1991).

Large gaps ($> 20,000 \text{ m}^2$) contribute greatly to the area available for gap-phase regeneration in Bwindi. In the undisturbed site they contributed 28%, selectively pitsawn site 49% and the intensively pitsawn site 17% of the total area of gaps. The undisturbed site show a roughly negative exponential gap size-class distribution similar to that reported for montane forests under natural disturbance regimes (Arriaga, 1988).

The gap sizes in Bwindi are the largest reported anywhere in tropical rainforests. For example, the mean gap size in the undisturbed site is four times the mean gap size for a heavily logged compartment of a medium altitude forest of Kibale, Uganda. This lends support to the idea that other factors, in addition to logging, are responsible for the large gaps in the Afromontane forest.

The soils of Bwindi are weakly structured or structureless and become loose and friable as they dry and therefore are less stable. Such soils are susceptible to landslides especially where forest vegetation has been removed. Sites in Bwindi where excessive timber extraction occurred coupled with the steep slopes and deep valleys as well as unstable soils, could have generated landslide-prone zones leading to large gap formation. Although we found no significant relationship between gap size and slope angle, Denslow and Hartshorn (1994) provide evidence that the edges of gaps on steep slopes continue to expand because of unstable soils, exposure to turbulent air and structurally unbalanced trees. They also consider poor drainage and water logging, which are characteristics of Bwindi, to cause frequent treefalls during periods of heavy rainfall and high winds, thus creating large gaps.

Fire has been recognised as another factor contributing to large gap formation on the ridges of Bwindi. The impact of fire differs from other natural or man-made gap forming processes in that pre-existing seedlings and saplings are killed by fire. Bracken fern and woody climbers quickly colonize such burnt areas and overwhelm any new seedlings and saplings.

Elephants play a big role in large gap creation in Bwindi. Damage caused by elephants to forests is much greater at high altitude (especially between 2,000 m to 2,500 m) than in lowland forests (Jackson, 1956). Elephants are therefore likely to convert a montane forest into a mosaic of forest and bushland through their browsing and trampling activities. Succession in the gaps of Bwindi is partly retarded by elephants, which excavate bracken fern rhizomes. Even though they are trampled, large herb communities in the gaps are not affected, or even increase leading to failure of trees to regenerate.

We found a significant relationship between the number of sapling/pole species and gap size for the combined data of the three study sites. This implies that tree species regeneration is a function of gap size. Unlike the gaps in the two heavily pitsawn sites, the undisturbed site gaps had the same density of stems and species richness in the sapling/pole size cohort like the adjacent forest understorey. The limited number of seedling and pole species and stems in the gaps of the two heavily pitsawn sites relative to the adjacent forest understorey suggest that only a small proportion of the tree community in Bwindi regenerates well in large gaps. Also, the undisturbed site had more stems and species of young trees than the two heavily pitsawn sites. This is attributed to the larger, more abundant gaps neighbouring each other in the heavily pitsawn sites which influence regeneration through an edge effect between the gaps. A neighbourhood gap effect in Bwindi is suggested by data showing that when gap size is held constant, the number of species and stems of young trees decreased as pitsawing intensity increased.

Ordination analysis for sapling/pole species and environment parameters in the Bwindi gaps reveal that *Sericostachys* and *Pteridium aquilinium* have the greatest influence on young tree density and species composition in the gaps by exerting mechanical suppression and competition for resources. No tree species had saplings or poles that were typically found in the gaps. This is in sharp contrast to the Neotropics, which have defined secondary forest species whose saplings are typically found in the gaps (Brokaw, 1985). Secondary forest species – *Macaranga* and *Neoboutonia* – which are common in a wide range of gap sizes in Bwindi seem not to invade disturbed areas quickly. This lack of aggressive colonising tree species in the area and the potential that many tree species may require shade to grow and survive may contribute to the paucity of secondary forest trees in the big gaps and to the fact that many large canopy openings are dominated by aggressive herbs and shrubs.

284

Some of the gaps in the undisturbed site had canopy forest species. The saplings/poles of these canopy forest species were found in gaps of less than 2,000 m^2 while majority stems of the secondary forest species were in gaps greater than 1,000 m^2. This indicates that in Bwindi, the gap-switch size for secondary and canopy forest species lies between 1,000 m^2 and 2,000 m^2. This gap-switch size is much larger than that occurs in lowland rainforest, and presumably reflects the cool, cloudy lower climate of the lower montane forest zone (Whitmore, 1996).

We conclude that the high incidence of large gaps, steep slopes, poor soils, poor drainage, fires, elephant activity in large gaps, aggressive bracken fern and climbers and lack of aggressive colonising trees have combined to stall regeneration in an Afromontane forest of Bwindi.

Acknowledgements

This research was supported by a grant to the Institute of Tropical Forest Conservation from the Royal Netherlands Government through Mgahinga and Bwindi Impenetrable Forest Conservation Trust. A. McNeilage, W. Olupot, P. Mwima and W. Kakuru reviewed earlier versions of this paper. J. Rothman provided bibliographic assistance. R. Barigyira identified the plants.

References

Arriaga, L. Gap dynamics of a tropical cloud forest in northeastern Mexico. *Biotropica* **20**, 178–184 (1988).

Brokaw, N. V. Treefalls, regrowth and community structure in tropical forests. in *The Ecology of Natural Disturbance and Patch Dynamics* (eds Pickett, S. T. A. & White, P. S.) 53–69 (Academic Press, New York, 1985).

Denslow, J. S. & Hartshorn, G. S. Treefall gap environments and forest dynamic processes. in *La Selva* (eds McDade, L. A., Bawa, K. S., Hespenheide, H. A. & Hartshorn, G. S.) 120–127 (Chicago University Press, Chicago, 1994).

Howard, P. C. *Nature conservation in Uganda's forest reserves* (IUCN, Gland, Switzerland, 1991).

Jackson, J. K. The vegetation of the Imatong mountains. *J. Ecol.* **44**, 290–298 (1956).

Kasenene, J. M. & Murphy, P. G. Post-logging tree mortality and major branch losses in Kibale Forest, Uganda. *For. Ecol. Manage.* **46**, 295–307 (1991).

Whitmore, T. C. A review of some aspects of tropical rainforest seedling ecology with suggestions for further enquiry. in *The Ecology of Tropical Forest Tree Seedlings* (ed. Swaine, M. D.) 3–39 (MAB Series, vol. 17, UNESCO, 1996).

Forest Fragmentation and Community Structure

Tropical Ecosystems: Structure, Diversity and Human Welfare.
Proceedings of the International Conference on Tropical Ecosystems
K. N. Ganeshaiah, R. Uma Shaanker and K. S. Bawa (eds)
Published by Oxford–IBH, New Delhi. 2001. pp. 289–290.

A comprehensive approach towards designing sustainable large-scale industrial plantation landscapes

John Grynderup Poulsen

Center for International Forestry Research, P.O. Box 6596, JKPWB,
Jakarta 10065, Indonesia
e-mail: J.Poulsen@cgiar.org

Tropical large-scale industrial plantations are rapidly expanding as a source of industrial wood and fuel. These plantations are structurally simple and often contain only a single species. They effectively sever the inter-connectivity of the natural landscape, and thus act as barrier to species movement, dispersal, and migration. I explore how appropriate design and management of natural forest corridors and remnant patches may be used to mitigate or reduce the negative impacts of large-scale industrial plantations on native biodiversity and towards maintenance of environmental (e.g., water quality) and social (i.e., the importance of natural forests for the livelihoods of the local people) functions of the original natural forest landscape. First, plantation landscapes should be designed so that on the one hand, the landscapes are penetrable and permeable for those biodiversity components which are of conservation concern in the area under consideration, and on the other hand, impenetrable and impermeable for pests, weeds, and invasives. Second, from a human/social standpoint, the priority must be to design plantation

Keywords: Industrial plantations, forest corridors, design of plantations, criteria and indicators.

and manage landscapes in a way that minimizes the adverse impacts on the local people and communities living in and around these areas. I present work conducted in Riau, Sumatra, by the Centre for International Forestry Research.

Incentives for better landscape/spatial design and management of those areas which are set aside from production for corridors and conservation areas are identified. There are several issues which may provide disincentives and/or be constraints to better design and management of corridors in the long-term. Leases on natural (selectively, logged-over) areas classified for plantation development, are approximately 20 years, too short a period for the companies to fully commit to long-term planning, and current lease-holders have no guarantee that their lease can be extended beyond the current lease period. This uncertainty is a further impetus for maximizing short-term incomes at the expense of long-term sustainable management. Corridors have been considered as merely constraints to maximizing short-term earnings. Potential short-term economic incentives and benefits include: minimizing plantation damage caused by wind throw, fire, and insect pest attack (by functioning as windbreaks, firebreaks, and providing habitat for biological control organisms, respectively), protecting and maintaining water quality and supply to both the plantation stands and the people living there, providing resources such as beneficial plants, and other NTFPs, for those local people who live within and around the plantation landscape, and, decreasing the plantation establishment and maintenance cost (e.g. keeping 20% of their concession area as conservation area, the company will theoretically 'save' 20% of the cost in establishing the plantation).

A further immediate economic incentive could be obtained, if certification processes incorporated appropriate criteria and indicators with respect to the design and condition of corridors in terms of both their landscape connectivity and maintenance of the integrity of individual corridors (such as the vertical structure and composition). I describe improved sets of criteria and indicators for sustainable management of large-scale plantations, by (i) suggesting increasing emphasis on landscape scale and conservation and socio-economic issues, and (ii) ensuring that the criteria and indicators are sufficiently linked to practical management, specifically by establishing explicit links between the criteria and indicator sets and codes of practice, the latter used by the plantation companies for planning and management.

Tropical Ecosystems: Structure, Diversity and Human Welfare.
Proceedings of the International Conference on Tropical Ecosystems
K. N. Ganeshaiah, R. Uma Shaanker and K. S. Bawa (eds)
Published by Oxford–IBH, New Delhi. 2001. pp. 291–294.

Breeding patterns of a tropical dry forest tree species, *Enterolobium cyclocarpum*, in disturbed and undisturbed habitats

J. L. Hamrick

Departments of Botany and Genetics, University of Georgia, Athens,
GA 30602, USA
e-mail: hamrick@dogwood.botany.uga.edu

The breeding structure of a plant species influences virtually every aspect of its life history as well as levels and patterns of genetic variation within and among its populations. Yet, despite the ecological and evolutionary consequences of plant breeding patterns and the availability of genetic markers and advanced multilocus analytical procedures for its description, actual breeding patterns of most plant species are poorly understood. In particular, there is a lack of information on variation in the breeding structure of plant populations among habitats and reproductive events. Of special concern is the generalization of such limited information to develop strategies for the conservation of species and their genetic diversity. Tropical tree species provide an excellent context in which to investigate spatial and temporal variation in breeding patterns since they are long-lived, have a variety of pollination mechanisms and show considerable spatial and temporal variation in the number, density and configuration of flowering populations.

Keywords: Breeding patterns, *Enterolobium cyclocarpum*, habitat fragmentation, paternity analyses, temporal variation.

This study represents a comparative analysis of the breeding structure of populations of *Enterolobium cyclocarpum* (Mimosoideae), the Guanacaste tree, located in fragmented and relatively undisturbed neotropical, dry-forest sites in Guanacaste Province, Costa Rica. Specific objectives of this research include: (1) To determine the number of individuals contributing pollen to trees in fragmented and relatively undisturbed landscapes. Do the pollen donor pools of trees vary across reproductive events? (2) To determine the proportion of successful pollen that immigrates into multi-tree clusters (i.e. rates of pollen flow). Do pollen flow rates vary among habitats and reproductive events? (3) To estimate the spatial extent of the breeding population for each individual and cluster. Do breeding population sizes vary among trees, among habitat types and among reproductive events? (4) To determine whether single, isolated trees have more pollen donors than trees grouped in clusters. Do isolated trees experience more temporal variation in their pollen donor pools than trees within clusters?

E. cyclocarpum is widely distributed throughout the Neotropics, ranging from Central Mexico to northern South America (Janzen, 1983) where it is most commonly found in lowland deciduous and semi-deciduous dry forests. It is a common, but patchily, distributed dominant tree species in the seasonally dry forests of Guanacaste Province in northwestern Costa Rica. Clusters of adult trees range from two to a maximum of about 12 individuals. Isolated individuals (> 250 m) are relatively common and are found both in highly disturbed and relatively undisturbed landscapes.

Typically, numerous inflorescences are initiated in March when the tree is leafless or just expanding its first leaves. Pollination is by moths, hawkmoths, beetles and other small nocturnal insects (Janzen, 1982) as well as diurnal insects such as bees (O. J. Rocha, pers. obs.). Each inflorescence initiates between 0 and 3 fruits but typically only one completes development. Fruits enlarge to full size (6–15 cm in diameter) in December and mature from March through May (Janzen, 1982; Zamora, 1991). As many as 21 viable seeds per fruit have been observed.

Like many mimosoid legumes, seeds within the ear-shaped pods are full-sibs ($\approx$ 99.8% of the fruits examined) due to pollen being grouped into polyads consisting of 32 pollen grains. As a result, the multilocus genotype of the pollen donor for each fruit can be inferred from the progeny genotypic array. Mature self-pollinated fruits are rare (< 1%). We have resolved 15 polymorphic allozyme loci (34 alleles) with which we can identify more than 172 million genotypes. Based on allele frequencies in Guanacaste Province, the frequency of the most common genotype is expected to be only 0.0023.

Fruit production varies greatly from tree-to-tree. From 1994 to 2001 some adults produced no fruits while others reproduced copiously every year. The most common pattern is to fruit two of three years. Total fruit production for a given area also varies temporally with < 40% of the adults fruiting some years and 95% fruiting other years.

The study site is located in Palo Verde National Park, (10°21′N; 85°20′W) in Guanacaste Province in northwestern Costa Rica. Dominant topographic features are steep limestone hills and large freshwater marshes associated with the Tempisque River. This area was predominantly ranch land until the late 1970's when it became a national park. Much of the flatter thorn-shrub areas were heavily grazed until that time. The steep, rocky slopes were allowed to maintain much of the original dry forest vegetation but some of the more valuable tree species were removed. Guanacaste trees on this site occur primarily on flatter sites and on lower hillsides. They are commonly found in primary and secondary dry upland forests and abandoned pastures but can also be occasionally found in well-developed riparian forests.

Full-sib analyses of progeny arrays from 40 fruits per tree were used to infer the multilocus genotype of the pollen donor for each fruit. The observed pollen donor arrays were used to estimate the total number of individuals contributing pollen to each tree (Burnham and Overton, 1979; Nason et al., 1996; 1998). For trees located within clusters, pollen immigration rates (i.e. gene flow) were also determined. Completed analyses are available from the Palo Verde area over a five-year period (1995–1999) for 25 designated maternal trees representing six multi-tree clusters and for seven isolated trees (> 250 m):

Tree type	# trees	Mean estimated # pollen donors	Mean % of pollen immigration	Estimated radius of breeding population (m)
Cluster	25	50.3 ± 24.6	75.4 ± 15.0	1476 ± 531
Isolated	7	91.7 ± 13.8	100.0	1527 ± 237

These results indicate that isolated trees obtain pollen from significantly more pollen donors and that the area from which pollen is obtained is larger than that of trees located within clusters. There was also an indication that trees in primary forests had fewer pollen donors than trees in secondary forests and open pastures. These results are not surprising as cluster trees may obtain up to 50% of their pollen from other trees located within their specific cluster. During some years, 35% of the pollinations within a cluster may come from one individual.

We were also able to obtain data on variation among the five reproductive events for these population parameters. The ability to uniquely genotype each pollen donor allowed us to also calculate similarities in pollen donor arrays among trees within a cluster and among years for individual trees. The number of reproductive events varies among trees since all trees did not produce fruit every year. The results of these analyses for four trees from Palo Verde are:

Tree	Habitat type	# reproductive events	Est. # pollen donors mean (range)	Pollen donor similarity mean (range)	Rates of pollen flow mean (range)
65	CL, SF	5	38.8 (16–50)	0.68 (0.52–0.77)	0.46 (0.33–0.65)
D2	CL, OP	3	122.0 (102–139)	0.27 (0.23–0.29)	0.94 (0.92–0.94)
34	CL, PF	3	87.7 (30–120)	0.16 (0.07–0.21)	0.98 (0.95–1.00)
24	I, SF	4	57.5 (31–83)	0.37 (0.00–0.54)	1.00

I, isolated tree; CL, cluster tree; OP, open pasture; SF, secondary forest; PF, primary forest.

These results indicate that considerable variation exists in these parameters across reproductive events and trees. This is especially true for comparisons of pollen donor arrays from different reproductive events. Some trees (e.g., #65) have high pollen-donor pool similarity between years. Other trees have consistently low year to year similarity values (e.g. D2 and 34). These results indicate that to fully understand the breeding patterns of tropical trees, several populations should be analyzed over multiple reproductive events.

References

Burnham, K. P. & Overton, W. S. Robust estimation of population size when capture probabilities vary among animals. *Ecology* **60**, 927–936 (1979).

Janzen, D. H. Variation in average seed size and fruit seediness in a fruit crop of a Guanacaste tree (Leguminosae: *Enterolobium cyclocarpum*) *Am. J. Bot.* **69**, 1169–1178 (1982).

Janzen, D. H. *Enterolobium cyclocarpum*. in *Costa Rican Natural History* (ed. Janzen, D. H.) 241–243 (University of Chicago Press, Chicago, 1983).

Nason, J. D., Herre, E. A. & Hamrick, J. L. Paternity analysis of the breeding structure of strangler fig populations: Evidence for substantial long-distance wasp dispersal. *J. Biogeog.* **23**, 501–520 (1996).

Nason, J. D., Herre, E. A. & Hamrick, J. L. The breeding structure of a tropical keystone plant resource. *Nature* **391**, 685–687 (1998).

Zamora, N. Tratamiento de la famila Mimosaceae (Fabules) de Costa Rica. *Brenesia* **36**, 63–149 (1991).

Tropical Ecosystems: Structure, Diversity and Human Welfare.
Proceedings of the International Conference on Tropical Ecosystems
K. N. Ganeshaiah, R. Uma Shaanker and K. S. Bawa (eds)
Published by Oxford–IBH, New Delhi. 2001. pp. 295–296.

Regeneration of tropical dry forest communities in a fragmented landscape

Oscar J. Rocha*[‡], Marco Gutierrez** and Noel Holbrooke[†]

*Escuela de Biología, Universidad de Costa Rica, Ciudad Universitaria
'Rodrigo Facio', San José, Costa Rica
**Estación Experimental Agrícola 'Fabio Baudrit', Facultad de Agronomía,
Universidad de Costa Rica, Barrio San José, Alajuela, Costa Rica
[†]Department of Organismic and Evolutionary Biology, Harvard University,
Cambridge, MA 02138, USA
[‡]e-mail: ojrocha@cariari.ucr.ac.cr

W e studied the changes in species composition, seed rain and seed and seedling survivorship rates in successional gradient in the seasonally dry forest of Santa Rosa National Park, Guanacaste, Costa Rica. In 1998, we selected eight sites on basis of the time since their last disturbance (age). The age of the sites were 3.5, 7.5, 12, 18, 40 and 60 years. In addition, we included two sites, one that is burned every year by the park rangers and the other which is considered a mature forest. Within each of these sites, we established thirty 4×4 m quadrants distributed along 5 transects. Next to each quadrant we placed one 50×50 cm trap to sample seed rain in the site. We found major differences in species composition among the woody species in the successional gradient. However, species richness was similar in all sites. In contrast, species composition in the seed rain (over 200 sp. recorded), and their patterns of spatial and temporal distribution were similar in all sites. Finally, we found differences in the rate of survivorship of seeds and

---

Keywords: Regeneration, Costa Rica, species composition, seed rain.

seedlings of different woody species along the successional gradient. Survivorship of species that are abundant in early stages of succession is higher than that of species that are abundant in late stages. We propose that such differences could explain the observed pattern in species composition across the succession gradient.

Tropical Ecosystems: Structure, Diversity and Human Welfare.
Proceedings of the International Conference on Tropical Ecosystems
K. N. Ganeshaiah, R. Uma Shaanker and K. S. Bawa (eds)
Published by Oxford–IBH, New Delhi. 2001. pp. 297–301.

Patterns of diversity of species assemblages in islands: Testing the predictions in dung beetles

A. R. V. Kumar*, K. Chandrashekara* and K. N. Ganeshaiah[†,‡]

Departments of *Entomology, and [†]Genetics & Plant Breeding, University of Agricultural Sciences, GKVK, Bangalore 560 065, India
[‡]e-mail: kng@vsnl.com

Owing to differential rates of extinction and invasion associated with the islands of different sizes, island biogeography theory predicts that, each island shall have an optimum number of species, which increases non-linearly with island size (McArthur and Wilson, 1967). However, the theory does not offer predictions on how the composition of the species would change with size of the island. Ganeshaiah et al. (1997) predicted that at assortative equilibrium states of the islands, the similarity in the species composition among islands of similar sizes, increases with the island size. They provided evidence to this using plant species assemblages of shola forests. Chandrashekara and Kumar (1999), using dung beetles, showed that while these predictions do hold for moderate and large sized dung pats, small dung pats deviated from this prediction. In this paper, we explore the causes for the observed non-conformity of the small dung pats to the predictions derived by Ganeshaiah et al. (1997). We show that these discrepancies could partly be explained due to certain assumptions made by them, which may not always hold true. Accordingly, we propose new predictions and test them using data from dung beetles.

Keywords: Species assemblage, dung beetles, dung pats, island size.

Ganeshaiah et al. (1997) assume that all the species of the global set can occur in any of the islands, irrespective of the size. However, to be viable, each species requires a minimum population, which the smaller islands may not be able to accommodate. Similarly, physical constraints such as body size of the species may also constrain their occurrence in islands of certain sizes. In dung beetles, for instance, the large elephant dung beetles of the genus *Heliocopris* (> 30 mm long) cannot occur in small dung pats. Likewise, owing to severe competition posed by large dung beetles, certain medium and small dung beetles may be eliminated in larger pats. Thus, it is expected that only a subset of the global set of species can occur in islands of a given size, despite increase in the number of species with island size. Consequently, both the number of species that occur in an island and the proportion of the global set that can occur in an island increase with the island size as a nonlinear function. Therefore, the similarity in the species composition among the islands of similar size is a function of these two relationships. We simulated this possibility using a range of island sizes, considering the raw data and found that the monotonic increase in similarity proposed by Ganeshaiah et al. (1997) holds true for moderate to large size islands and for small islands, the relation can be highly variable.

It is generally expected that the rate of invasion increases with the island size, as greater opportunity exists in terms of space and resources. Also in larger islands, the probability of accommodating minimum viable populations of resident species increases. As a result, the extinction rate is expected to decrease with the island size. Consequently, for a given set of taxa (such as trees or dung beetles), there could exist an optimum size of the islands above which the species composition is stable but below which the species composition will be in a dynamic flux (Figure 1). Since in small islands the rate of extinction is higher than invasion, the species numbers and composition could be in a continuous flux. However, for large islands since invasion will be higher than extinction, there is stability of species richness. Thus we predict that the coefficient of variation of species richness decreases with island size.

We tested the above two predictions by using dung beetle assemblages occurring on dung pats of different sizes. Artificial dung pats of 10, 25, 50, 100, 200, 400, 800, 1000, 1500 and 2000 g were laid using fresh dung in a pasture land at the Military Dairy Farm, Hebbal, Bangalore, following the methodology adopted by Poornima (1998). All the dung beetles were extracted from each pat two days after placing the dung, by floatation technique. Beetles were then sorted into morphospecies and their numbers recorded. The procedure was repeated six times in a year to account for seasonal changes in species assemblages. Similarity between pats of the same size was estimated by the correlations for frequencies of species. Similarly, coefficient of variation was worked out for species richness of each dung pat size.

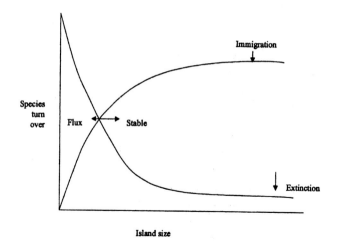

Immigration

Species
turn
over

Flux ← → Stable

Extinction

Island size

Figure 1. Species turnover as a function of island size (see text for details).

As predicted, the mean of correlation coefficients among pats of similar sizes indicated a lack of monotonic increase in all the six sampling bouts (Figure 2: means of the six bouts plotted). Although a general increasing trend was evident, from moderate to large size pats, the fluctuations were marked in pat sizes of up to 400 g. Smaller pats also showed a high variation for similarities.

In nature, mammalian dung that supports dung beetles can vary from small rodent dung pellets to large dung balls of elephants, which are thousands of times larger than the size of the former. Thus, the pat sizes we simulated do not cover the entire range of dung sizes available in nature and it would be meaningful to evaluate the natural range of dung pat sizes for dung beetle assemblages.

Further, as predicted, the coefficients of variation for species richness decreased nonlinearly with the pat size (Figure 3).

It is possible that the high coefficient of variation observed for small pats could arise due to sampling problem. To verify this, we simulated the small sample sizes in the computer by repeatedly drawing smaller samples from the pool of dung beetle data of the large pats. But the natural flux continued to persist in the simulated small samples also. We compared the variations in species richness thus created with the observed and found that the high CV observed in small pats is not due to sampling effect.

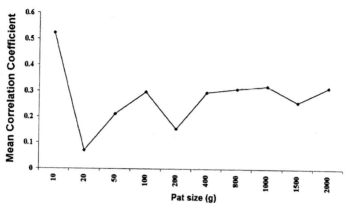

Figure 2. Mean correlation coefficients of dung beetle assemblages for pats of different sizes.

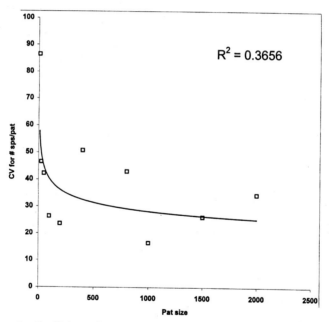

Figure 3. Coefficient of variation for number of dung beetle species found in pats of similar sizes declined with the pat size, thus indicating higher flux for the number of species in small islands.

In conclusion, these findings suggest that the smallest islands may not provide the highest diversity in species assemblages as predicted by Ganeshaiah *et al.* (1997).

References

Chandrashekara, K. & Kumar, A. R. V. *Species assemblages and community organisation of dung beetles (Coleoptera: Scarabaeidae)*, Final report of a DST, Govt. of India sponsored project (Mimeographed) (1999).

Ganeshaiah, K. N., Uma Shaanker, R. & Bawa, K. S. Diversity of species assemblages of islands: Predictions and their test using tree species composition of shola fragments. *Curr. Sci.*, **73**, 188–194 (1997).

McArthur, R. H. & Wilson, E. O. The theory of Island Biogeography. *Monogrpahs in Population Biology* (Princeton University Press, Princeton, NJ, 1967).

Poornima, G. C. *Diversity, Abundance and Seasonality of Dung Beetles (Coleoptera: Scarabaeidae: Scarabaeinae) of Bangalore*, M.Sc (Agri) thesis submitted to University of Agricultural Sciences, Bangalore (1998).

Tropical Ecosystems: Structure, Diversity and Human Welfare.
Proceedings of the International Conference on Tropical Ecosystems
K. N. Ganeshaiah, R. Uma Shaanker and K. S. Bawa (eds)
Published by Oxford–IBH, New Delhi. 2001. pp. 302–306.

Rainforest fragmentation and small carnivores in the Western Ghats, India

Divya Mudappa*, Barry Noon[†], Ajith Kumar[‡] and Ravi Chellam*

*Wildlife Institute of India, P.B. # 18, Dehradun 248 001, India
[†]Department of Fisheries and Wildlife Biology, Colorado State University, Fort Collins, CO 80523, USA
[‡]Salim Ali Centre for Ornithology and Natural History, Moongil Pallam, Anaikatti, Coimbatore 600 108, India
e-mail: shankar@ces.iisc.ernet.in;

In recent years, tropical forests and their species diversity are being threatened by extensive deforestation and forest fragmentation. Many studies have reported the consequences of such disturbances on small mammals, primates, and birds (Johns, 1988; Laurance, 1994; Turner, 1996). Few studies have been carried out on the impacts of rainforest fragmentation or other disturbances on carnivores, particularly small carnivores, which play an important role as predators, prey, and seed dispersers in natural ecosystems (Heydon and Bulloh, 1996). Changes in small carnivore community structure could affect native floral and faunal communities (Crooks and Soule, 1999). Here, we assess the distribution and relative abundance of small carnivores in a relatively undisturbed rainforest and in a fragmented landscape. We also attempt to identify the habitat correlates of small carnivore distribution and abundance.

Keywords: Tropical rainforest conservation, camera-trapping, brown palm civet, fragmented landscape, small carnivore distribution.

The study was carried out between June 1996 and December 1999 in the relatively undisturbed rainforests of Kalakad–Mundanthurai Tiger Reserve (KMTR, 895 km², 8°25′–8°53′N and 77°10′–77°35′E) and between January and May 2000 in the rainforest fragments embedded in the man-modified landscape of the Anamalai hills (958 km²). The rainforests in KMTR were surrounded by other natural habitats like dry deciduous forests, wooded and high altitude grasslands. The matrix surrounding the rainforest fragments in the Anamalais are tea, coffee, and eucalyptus plantations. Eight species of small carnivores occur in the rainforests of the Western Ghats (Mudappa, 1998). This includes the endemics – the Nilgiri marten (*Martes gwatkinsi*) and the brown palm civet (*Paradoxurus jerdoni*) – and other more widespread species such as the brown and stripe-necked mongooses (*Herpestes fuscus* and *H. vitticollis*, respectively), small Indian civet (*Viverricula indica*), small-clawed and common otters (*Ablonyx cinereus* and *Lutra lutra*, respectively), and the leopard cat (*Prionailurus bengalensis*). A combination of methods, such as track plots, camera-traps, spot-lighting (night walks and drives), and direct sightings were used to record the distribution of small carnivores. The sampling effort varied in relation to fragment sizes (Table 1).

Six habitat parameters were measured using standard methods to identify the correlates influencing small carnivore occurrences in the rainforests (tree and food tree densities, basal area, canopy cover, canopy height and shrub density). Visitation rate was calculated out of the total number of track plot or camera-trap days. Direct sightings (during drives, foot surveys) were used to calculate encounter rates for each species. Sites were categorised according to altitude, area, landscape, and disturbance levels and differences among categories in small carnivore occurrence were analysed using the chi-square test.

Table 1. List of rainforest sites surveyed for small carnivores and sampling effort in each.

Sites	Area in ha (altitude, m)	No. of camera trap nights	No. of track plot nights	No. of hours spent
KMTR	25000 (650–1300)	112	177	12240 h
ANAMALAIS				
Akkamalai–Iyerpadi complex	2500 (1250–1500)	15	60	28 h 45 m
Varagaliar	2000 (650–800)	15	20	8 h 20 m
Andiparai	200 (1250)	10	30	13 h 45 m
Manamboli	200 (800)	10	30	12 h 05 m
Karian Shola	500 (750)	10	30	16 h 15 m
Korangumudi	50 (1000)	10	30	10 h 55 m
Puthuthottam	100 (1000)	10	30	16 h 50 m
Varattuparai	8 (1100)	5	20	9 h 35 m
Tata Finley	25 (1000)	5	25	10 h 35 m
Pannimade	10 (1100)	5	20	11 h 15 m

The two species that were endemic to the Western Ghats, the brown palm civet and Nilgiri marten, were the most frequently sighted nocturnal and diurnal carnivores (encounter rates = 0.003/hr and 0.001/hr, respectively) in KMTR. The most frequently sighted species in the Anamalais was the stripe-necked mongoose (1.6/hr). The visitation rates in the track plots were significantly different between KMTR and Anamalais (48% and 32.5%, respectively; $\chi^2 = 11.73$, $df = 1$, $P < 0.001$). Similarly, the visitation rate at camera-traps was significantly higher in KMTR than in the Anamalais (37.5% and 16.8%, respectively; $\chi^2 = 10.88$, $df = 1$, $P < 0.001$). The three species that were photo-captured were the brown palm civet, small Indian civet, and the brown mongoose in both the study areas. The results indicate significantly lower occurrence of small carnivores in fragmented landscapes of Anamalais. A study of effect of logging on civets in Borneo also showed that though all the species persist, their abundances are lowered (Heydon and Bulloh, 1996).

The frequency of occurrence of both the brown palm civet and other small carnivores was significantly different between KMTR and the Anamalais ($\chi^2 = 48.44$, 14.61, respectively, $df = 1$, $P < 0.001$). In the Anamalais, the brown palm civet was the most frequently photo-trapped small carnivore, contributing to nearly 50% of the visitations, but occurred at a significantly lower rate than in KMTR (88%, $\chi^2 = 15.30$, $df = 1$ $P < 0.001$). Brown mongoose was the second most frequently photo-trapped species of small carnivore (37.5%), followed by the small Indian civet (12.5%). Although the brown mongoose was photo-trapped a greater number of times in the Anamalais than in KMTR, and the small Indian civet fewer times, these differences were not statistically significant ($\chi^2 = 2.84$ and 2.12, respectively, $df = 1$, $P > 0.05$).

Although there was no significant difference in the occurrence of small Indian civet and brown mongoose between the two study areas, a significantly lower occurrence of the brown palm civet in the Anamalais resulted in a significant change in relative abundances of small carnivores in the fragmented landscape of the Anamalais (Figure 1). Studies on primates and birds have shown frugivorous and folivorous species to benefit by moderate level of disturbance (Johns, 1988). However, a few other studies have indicated highly specialised species to be negatively affected due to habitat fragmentation (Heydon and Bulloh, 1996; Laurance, 1991, 1994). Small carnivores, being predominantly omnivorous in diet, were observed to slightly increase or even stay unchanged in abundances in some logged and disturbed sites (Johns, 1983; Oehler and Litvaitis, 1996; Travaini et al., 1997).

There was a significant difference in the success rate between fragments of varying sizes ($\chi^2 = 11.363$, $df = 3$, $P < 0.001$), with success being highest in medium-sized fragments (25%, Puthuthottam and Korangumudi) compared to small (20%), large (16.7%) and very large (10%) fragments. There were no photo-captures of any of the small carnivores in Varagaliar and Tata Finley,

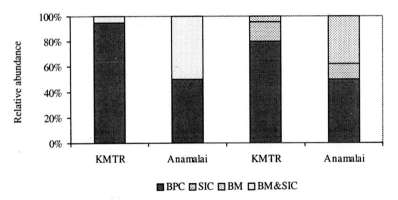

Figure 1. Relative abundances of small carnivores estimated from track plot (left) and camera-trap (right) methods. BPC, brown palm civet; SIC, small Indian civet; BM, brown mongoose.

one of the largest and one of the smallest fragments, respectively. The brown palm civet is a predominantly frugivorous species. The persistence of the brown palm civets in many of the rainforest fragments may be a result of the occurrence of food tree species in some of the fragments, which although highly disturbed, contributed to high success rates. The two medium-sized fragments had coffee in the understorey and some of the native species as canopy cover, whose fruits the civets consumed. Similar patterns of increase in abundances of herpetofauna, small mammals, and arboreal mammals in medium-sized fragments have been reported in this region (Vasudevan, 2000; Umapathy and Kumar, 2000).

The brown palm civet distribution was positively correlated with altitude ($R = 0.624$, $df = 8$, $P = 0.05$), similar to that found in the altitudinal distribution of the species in the undisturbed rainforests of KMTR (Mudappa, 1998). The presence and protection of relatively large tracts of rainforest fragments, seem to be acting in their favour in the otherwise highly disturbed landscape. The occurrence of small Indian civet was positively correlated to the size of the fragment ($R = 0.702$, $df = 8$, $P = 0.02$). The more widespread and omnivorous species exhibit no change in relative abundances, although they were more common in the fragmented landscape. The success rates were not significantly correlated with any of the habitat or site parameters as found in most other studies (Smallwood and Schonewald, 1998), although matrix characteristics are likely to play an important role in the persistence of specialised species (Laurance, 1991). Encounter rates during night walks and drives, and opportunistic sightings were lower in KMTR than in Anamalais, and these could be attributed to greater visibility in the fragmented landscape than in the undisturbed forest in KMTR. Long-term survival of the small carnivores, particularly the endemics like the brown palm civet and the Nilgiri marten can be ensured only by protection of large tracts of relatively undisturbed mid-elevation rainforests, and with prudent planning and management of protected areas, particularly in highly threatened areas like the Anamalais.

Acknowledgements

We would like to thank Wildlife Institute of India, United States Fish and Wildlife Service, and Wildlife Conservation Society for providing funds to carry out this project. We thank T. R. Shankar Raman who provided helpful suggestions on earlier drafts of this paper. Tamil Nadu Forest Department provided the necessary permits to carry out this research. Poovan, Doraipandi, Sivakumar, Rajamani, Ganesan, and P. Jeganathan helped with field work. Narendra Babu and P. R. Shankar assisted in vegetation data collection.

References

Crooks, K. R. & Soulé, M. E. Mesopredator release and avifaunal extinctions in a fragmented system. *Nature* **400**, 563–566 (1999).

Heydon, M. J. & Bulloh, P. The impact of selective logging on sympatric civet species in Borneo. *Oryx* **30**, 31–36 (1996).

Johns, A. D. Wildlife can live with logging. *New Sci.* **99**, 206–211 (1983).

Johns, A. D. Effect of 'selective' timber extraction on rainforest structure and composition and some consequences for folivores and frugivores. *Biotropica* **20**, 31–37 (1988).

Laurance, W. F. Ecological correlates of extinction proneness in Australian tropical rain forest mammals. *Conserv. Biol.* **5**, 79–89 (1991).

Laurance, W. F. Rainforest fragmentation and the structure of small mammal communities in tropical Queensland. *Biol. Conserv.* **69**, 23–32 (1994).

Mudappa, D. Use of camera-traps to survey small carnivores in the tropical rainforest of Kalakad–Mundanthurai Tiger Reserve, India. *Small Carn. Conserv.* **11**, 9–11 (1998).

Oehler, J. D. & Litvaitis, J. A. The role of spatial scale in understanding responses of medium-sized carnivores to forest fragmentation. *Can. J. Zool.* **74**, 2070–2079 (1996).

Smallwood, K. S. & Schonewald, C. Study design and interpretation of mammalian carnivore density estimates. *Oecologia* **113**, 474–491 (1998).

Travaini, A., Delibes, M., Ferreras, P. & Palomares, F. Diversity, abundance or rare species as a target for the conservation of mammalian carnivores: a case study in Southern Spain. *Biodiv. Conserv.* **6**, 529–535 (1997).

Turner, I. M. Species loss in fragments of tropical rain forest: a review of the evidence. *J. Appl. Ecol.* **33**, 200–219 (1996).

Umapathy, G. & Kumar, A. The occurrence of arboreal mammals in the rain forest fragments in the Anamalai Hills, south India. *Biol. Conserv.* **92**, 311–319 (2000).

Vasudevan, K. Amphibian species assemblages of the wet evergreen forests of southern Western Ghats of India and the effect of forest fragmentation on their diversity. Ph.D. thesis, Utkal University, Orissa, India (2000).

Tropical Ecosystems: Structure, Diversity and Human Welfare.
Proceedings of the International Conference on Tropical Ecosystems
K. N. Ganeshaiah, R. Uma Shaanker and K. S. Bawa (eds)
Published by Oxford–IBH, New Delhi. 2001. pp. 307–309.

Tropical dry forest detection in Costa Rica: Its impact on tropical biology research

Pablo Arroyo-Mora* and G. Arturo Sanchez-Azofeifa*,†

*Earth Observation Systems Laboratory (EOSL), Department of Earth and
Atmospheric Sciences, University of Alberta, Edmonton, Alberta, Canada T6G 2H3
†e-mail: arturo.sanchez@ualberta.ca

With their unique biodiversity and highly fragmented landscape, the tropical dry forests (TDF) are considered one of the most endangered ecosystems in the lowland tropics (Janzen, 1988). Dry forests are characterized climatically by six effectively dry months per year (Hartshorn, 1983). It is estimated that less than 2% of the tropical dry forest remains as relatively undisturbed wildlands in Mesoamerica, and only 0.9% of it lies within national parks or other kinds of protected areas, mainly in the province of Guanacaste, Costa Rica (Janzen, 1986). In Costa Rica, the deciduous forest is located in the northwestern province of Guanacaste, being areas mainly in the transitional zones to Moist Forest. However, it is possible to find some areas with pure dry forest. Furthermore, this ecosystem has been understudied, and can prove to be one of the most important field laboratories in the tropics, encompassing broad research topics, ranging from endangered species protection to landscape structure studies.

Keywords: GIS, dry forest, remote sensing, expert knowledge classifiers, Costa Rica.

The first objective of this paper is to evaluate the usefulness of remote sensing classification.techniques on tropical dry forest ecosystems as a tool to extract their extent at a regional level. These techniques involve the integration of base knowledge classifiers, ground truthing and ancillary data published by on-the-ground projects aimed at the characterization of the tropical dry forest ecosystem (Pacheco, 1998; Herrera, 1998). A second objective is to assess the detection of different successional stages using single imagery and ancillary data. We are focusing this study on a Tropical Dry Forest ecosystem in northwest Costa Rica. The geographical center of the study area is 10°41'50"N and 85°35'38"W. Two radiometrically and atmospherically Landsat Enhanced Thematic Mapper (ETM+) scenes, acquired on January 2000 (transition season) and April 2001 (dry season), were used to evaluate the role of different classification algorithms as a function of phenology.

A supervised classification method was applied, selecting training areas using color aerial photography from 1997 and ground truthing data collected in June 2000. Once the forest regions were classified, the amount of underestimated dry forest area was obtained using a change matrix analysis that counts areas classified as forest in January 2000 and classified as a different class in April 2001.

A characterization of the different succession stages of the forest in terms of number of trees per hectare, basal area, frequency, abundance, species, photosynthetic active radiation and diametrical classes was developed to correlate this data with that from the forest areas of the January image. Furthermore, Principal Component Analysis (PCA) and Band Ratios (BR) were applied to a subset of the image to separate the different forest classes.

Our results show that by using a dry season image, the percentage of underestimated dry forest is around 30, which also encompasses the classification error (15%). In terms of successional differentiation, early secondary growth (0–3 years) is hard to observe because it is confused with some pasture lands and agriculture fields. It was possible to obtain a secondary succession between 3 and 15 years for tropical dry forest. After this age it is not possible to separate between primary forest and intermediate dry forest.

The main impact of our results in terms of tropical biology research is the fact that most of the tropical dry forest and moist forest mapped for the Guanacaste Region, is secondary growth. This secondary growth, in addition to the remnants of the tropical dry forest provides an opportunity to study and discover new ecological relationships.

Acknowledgements

This work was supported by the Earth Observatory Laboratory Systems at the University of Alberta and the R.E. Train Education Program of the World Wildlife Fund (WWF). We would like to thank the generous support of the Canada Foundation for Innovation (Grant No. 2041 to Sanchez-Azofeifa), and the U.S. National Science Foundation (Grant No. BCS 9980252). We are grateful to the Regional Fellowship Program for Graduate Studies in the Social Sciences that represents the John D. and Catherine T. MacArthur Foundation, the Ford Foundation and the William and Flora Hewlett Foundations for helping Central America students to achieve their goals in terms of conservation. We extend our gratitude to the Tropical Science Center, the Organization for Tropical Studies, the National Institute of Biodiversity (INBio), the National Fund of Forestry Financing (FONAFIFO) and Costa Rican Ministry of the Environment (MINAE) for providing useful information to complete this study.

References

Hartshorn, G. Chapter 7: Plants. in *Costa Rican Natural History* (OTS. San José, Costa Rica), p. 118–323 (1983).

Herrera, F. Inventario Florístico de fragmentación del bosque seco en diferentes estados de sucesión durante la época lluviosa. Informe de Práctica de Especialidad. ITCR-Departamento de Ingeniería Forestal (1998).

Janzen, D. H. Guanacaste National Park: Tropical ecological and cultural restoration. Editorial UNED. San José, Costa Rica. 40 p (1986).

Janzen, D. H. Tropical dry forest: The most endangered major tropical ecosystem. in *Biodiversity* (ed. Wilson, E. O.) 130–137 (Natural Academy Press, Washington DC, 1988).

Pacheco, A. Inventario Florístico durante la sucesión del bosque tropical seco, Parque Nacional Santa Rosa, Guanacaste. Informe de Práctica de Especialidad. ITCR-Departamento de Ingeniería Forestal. 114 p (1998).

Tropical Ecosystems: Structure, Diversity and Human Welfare.
Proceedings of the International Conference on Tropical Ecosystems
K. N. Ganeshaiah, R. Uma Shaanker and K. S. Bawa (eds)
Published by Oxford–IBH, New Delhi. 2001. pp. 310–313.

Effects of rainforest fragmentation on the amphibian diversity in the Western Ghats, Southern India

Karthikeyan Vasudevan*[‡], Ravi Chellam*, Ajith Kumar**
and Barry Noon[†]

*Wildlife Institute of India, P.O. Box 18, Dehradun 248 001, India
**Salim Ali Centre for Ornithology and Natural History, Anaikatti P.O.
Coimbatore 641 108, India
[†]Department of Fisheries and Wildlife Biology, Colorado State University, Fort
Collins, CO 80523, USA
[‡]e-mail: karthik@wii.gov.in

The fragmented rainforests of the Western Ghats and the biological diversity it holds present a challenge to forest managers, biologists and conservationists. This is particularly true in the case of amphibians, among which species richness (nearly 130 species) and endemism (nearly 75%) are very high in the Western Ghats. In this paper, we examine the species richness and density of forest floor amphibians in remnant rainforest fragments in relation to contiguous rainforests. We also attempt to identify the major factors that influence the amphibian species richness in rainforest fragments.

The data come from the contiguous and undisturbed rainforests (nearly 400 km^2) in the Kalakkad-Mundanthurai Tiger Reserve (KMTR) in the

Keywords: Rainforest, fragmentation, Western Ghats, amphibians, species richness, density.

southern end of the Western Ghats, sampled in 1996–97, and 14 rainforest fragments in the Anamalai Hills, sampled in 1998–99. In KMTR, amphibians were sampled at three sites (Kannikatti, Sengaltheri and Kakkachi), at an altitudinal range of 700 m to 1300 m. The forest fragments in the Anamalai Hills ranged in area from < 0.5 ha to about 2000 ha, at an altitudinal range of 800 m to 1400 m. We sampled amphibians using adaptive cluster sampling, which gives an unbiased estimate of density (Thompson *et al.*, 1992). In total 530 quadrats (5 m × 5 m) were sampled in KMTR, and 638 quadrats in the forest fragments, with seasonal replicates in both the areas. The number of quadrats sampled in forest fragments varied from 13 to 110. Several habitat and landscape level variables were also measured, which were used to estimate a measure of habitat heterogeneity using discriminant function analysis. Species richness was measured as the number of species recorded for a fragment and the mean number of species per quadrat, while density was estimated as the number of amphibians per quadrat.

The number of species in a fragment varied from 0 to 15 (Figure 1). Species richness increased as a function of fragment area ($r = 0.786$, $n = 13$, $P < 0.001$). A partial correlation of species richness and fragment area, after controlling for the number of frogs, was considerably weaker, but still significant ($r = 0.576$, df = 11, $P < 0.05$). Area, degree of isolation (a function of distance from the nearest fragment and time since isolation) and habitat heterogeneity explained 56% of the variance in the number of species in the rainforest fragments. Degree of isolation (regression slope = –0.524) and habitat heterogeneity (–0.478) were better predictors of species than fragment area (0.268). The parameters contributing to habitat heterogeneity were shrub density, canopy height, number of large rocks in the plot, soil moisture and air temperature. Thus, fragment area seems to be a major, but not the only, determinant of species richness.

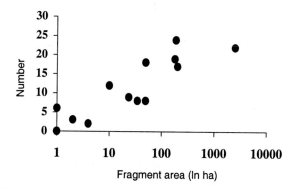

Figure 1. Number of amphibian species vs fragment area.

A total of 30 species was recorded from KMTR and 40 from the Anamalai Hills, even though the area of forest in the former was considerably greater. The turnover of species with drainage and altitude (Vasudevan *et al.*, 1998) makes the comparison of the species richness between these two hill-ranges difficult. We, therefore, used the mean number of species per quadrat (MSPQ) as a surrogate of species richness for comparison between KMTR and the Anamalai Hills. We grouped forest fragments in the Anamalai Hills as large (150–2500 ha), medium (10–149 ha), and small (< 10 ha). Since the large fragments were in the 1100–1400 m altitude range, we compared MSPQ of these fragments with MSPQ estimated from quadrats in the Kakkachi area in KMTR which had similar altitudes. Since medium-sized fragments occupied a wide altitudinal range (700–1400 m), we used the entire sample set from KMTR. For small fragments, we selected the sample set from Sengaltheri, of similar altitude range, for comparison. The MSPQ for different size classes of fragments was lower than that obtained from comparable areas in KMTR (Figure 2, Mann–Whitney U test: large fragments, $Z = 4.174$, $n_1 = 189$, $n_2 = 102$, $P < 0.001$; medium-sized fragments, $Z = 4.599$, $n_1 = 160$, $n_2 = 24$, $P < 0.001$; small fragments, $Z = 4.017$, $n_1 = 82$, $n_2 = 24$, $P < 0.001$). Thus, forest fragments have a lower species density compared with contiguous and undisturbed forest. Moreover, the difference between them seemed to increase with decreasing fragment area (Figure 2). Among the forest fragments, however, the MSPQ was only weakly correlated with area (0.50, $n = 14$, $P = 0.066$), and not correlated with any of the other habitat parameters.

The density of amphibians in forest fragments ranged from 0 to 1.12 animals/quadrat, with a mean of 0.29 (SE = 0.078, $n = 14$). In contrast, KMTR had a much higher density (0.81 animals/quadrat). However, the density of amphibians was neither related to fragment area ($r = 0.42$, $n = 14$, $P = 0.136$) nor to any of the other habitat parameters.

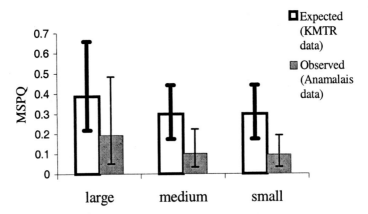

Figure 2. A comparison of mean species richness per quadrat (MSPQ) in three fragment classes and KMTR.

Thus, compared to the contiguous rainforest in KMTR, the forest fragments have considerably fewer species, species per unit area, as well as considerably lower density of all amphibians together. Among the fragments, however, species richness seems to increase only marginally with area, while decreasing more strongly with degree of isolation and habitat heterogeneity. The absence of streams, with which amphibians are strongly associated, might be one of the reasons for the marked decline in density. The turnover of species with drainage might be a reason for the overall greater species richness in the fragmented landscape in the Anamalai Hills, which has a greater number of drainages than KMTR.

Acknowledgements

We thank Wildlife Institute of India, and US Fish & Wildlife Service for funding this project, Mr. B. C. Choudhury for co-ordinating it, Dr. S. K. Dutta for assistance in taxonomy, Tamil Nadu Forest Department for necessary research permits, Poovan, Doraipandi, Sivakumar, Rajamani, and Ganesan for field assistance, and P. Jeganathan, Narendra Babu and P. R. Shankar for assistance in data collection.

References

Vasudevan, K., Kumar, A. & Ravi Chellam. The distribution of stream amphibians in the rainforests of the Western Ghats. Proc. National Wildlife Seminar, Wildlife Institute of India, Dehra Dun (1998).
Thompson, S. K., Ramsey, F. L., Seber, G. A. F. An adaptive procedure for sampling animal populations. *Biometrics* **48**, 1195–1199 (1992).

Tropical Ecosystems: Structure, Diversity and Human Welfare.
Proceedings of the International Conference on Tropical Ecosystems
K. N. Ganeshaiah, R. Uma Shaanker and K. S. Bawa (eds)
Published by Oxford–IBH, New Delhi. 2001. pp. 314–318.

Fragment sizes and diversity of species assemblages in sholas and sacred groves: Are small fragments any worth?

B. S. Tambat*, V. Channamallikarjuna**,
G. Rajanikanth*, G. Ravikanth*, C. G. Kushalappa[+],
R. Uma Shaanker* and K. N. Ganeshaiah**

*Department of Crop Physiology and **Department of Genetics and Plant Breeding,
University of Agricultural Sciences, Bangalore 560 065, India
[+]College of Forestry, Ponnampet, Kodagu
e-mail: kng@vsnl.com

Island biogeography theory predicts that islands attain equilibrium with respect to the number of species they can harbor and that this number increases nonlinearly with the size of islands (Mac Arthur and Wilson, 1967; Kohn and Walsh, 1994); but it does not suggest the composition of the species in these islands. Given a global set of N number of species, if two islands of similar size harbor n number of species each, then there can be $^{N}C_{n}$ combination of species in these islands. Accordingly, Ganeshaiah et al. (1997) showed that similarity among islands of similar size increases at the rate of $1/N$. In other words, species composition of the larger islands would be more similar among themselves than the smaller islands (Ganeshaiah et al., 1997) (Figure 1). Ganeshaiah et al. (1997) also predicted that smaller islands would exhibit high variation (CV) for their similarity than the larger islands. We

Keywords: Forest fragments, sholas, sacred groves, island biogeography.

tested these two predictions using two forest fragment systems, namely sacred groves and shola forests.

Sacred groves are unique forest fragments of south India maintained by local rural communities amidst their agricultural landscape. Sholas are relatively undisturbed natural fragments of forests available at higher altitude. Sholas of Brahmagiri Wildlife Sanctuary (11°57′E, 75°58′N) and sacred grooves of Ponnampet range (12°E, 75°N) both in Coorg district, Karnataka, India were chosen for the present study. Both the shola and sacred grooves were categorized into three size class, viz. for shola into: small (< 1 ha), medium (1–3 ha) and, large (> 3 ha) and for sacred groves: small (< 2 ha), medium (2–4 ha) and large (> 4 ha). In sholas, all tree species (> 1 cm gbh) were enumerated and in sacred grove all the species were enumerated in plots of 10 m × 10 m laid randomly within each fragment. The proportion of species shared and the correlation of the frequency of species were computed for all possible combination of fragments within each size class.

As predicted, the proportion of species shared and the correlation for the frequency of species increased with the size of shola (Figure 2a) and sacred grooves (Figure 2b), suggesting that larger fragments harbor similar composition of species compared to smaller fragments.

Further, as predicted by Ganeshaiah *et al.* (1997), the CV for the correlation and proportion of species shared decreased with the fragment size (Ganeshaiah *et al.*, 1997) (Figure 3a and b). In other words, the diversity of species assemblages is high among smaller fragments than larger fragments.

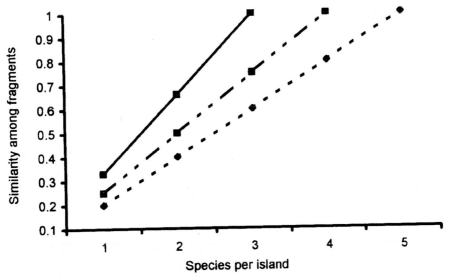

Figure 1. Similarity among islands of the same size containing combination of the species from the global set of 3 (solid line), 4 (dashed line) and 5 (dotted line) speices.

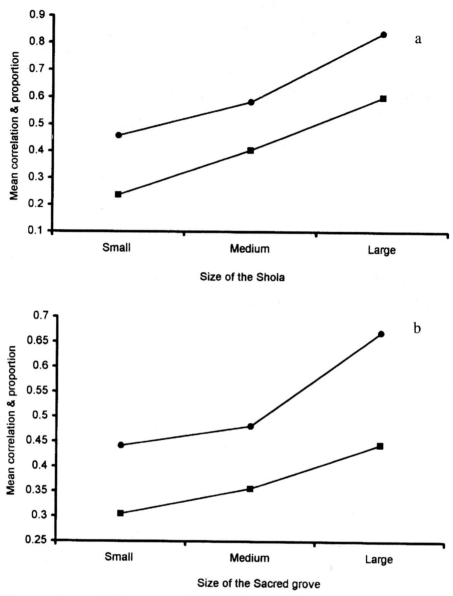

Figure 2. Average similarities among fragments of similar size of shoals (a) and sacred groves (b), average correlation (filled circles) and average proportion of species shared (filled squares).

Thus, our results indicate that, smaller fragments of forests are more diverse among themselves with respect to their species composition than the larger fragments and this may have important implications for designing the size of the protected areas. Since functional diversity of an ecosystem could depend heavily on diversity of the assemblage of species, it might be profitable

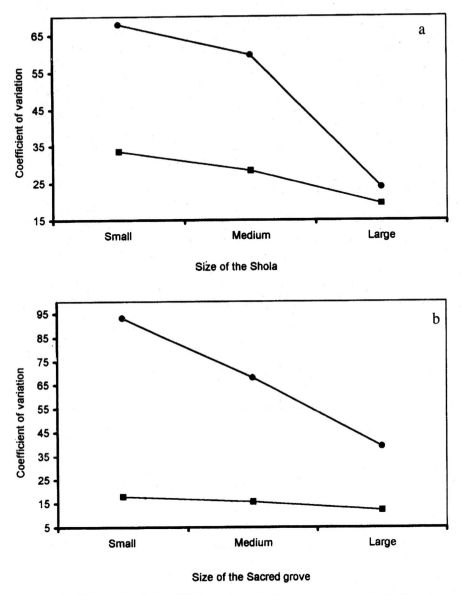

Figure 3. Coefficient of variation (CV) for similarity indices among fragments of different size in shola (a) and sacred groves (b). The CV has been computed for correlation values (filled circles) and proportion of species shared (filled squares) between fragments.

to increase the diversity of the assemblages by designing appropriate sizes of the protected areas. Ganeshaiah *et al.* (1997) have shown that, the species diversity of assemblage is maximum at medium-sized fragments than at very small and very large fragments.

Acknowledgements

The work reported here is supported by grants from IPGRI-Rome Forest Genetic Resources program. We acknowledge the cooperation of the Karnataka Forest Department for permission to conduct field studies at the sholas and the sacred groves.

References

Dickerson, Jr. J. E. & Robinson, J. V. Microcosms as islands: A test of the Mac-Arthur Wilson equilibrium theory. *Ecology* **66,** 966 (1985).

Ganeshaiah, K. N., Uma Shaanker, R. & Bawa, K. S. Diversity of species assemblages of isalnds: Predictions and their test using tree species composition of shola fragments. *Curr. Sci.* **73,** 188–194 (1997).

Kohn, D. D. & Walsh, M. Plant species richness – the effect of islands size and habitat diversity. *J. Ecol.* **82,** 367 (1994).

Mac Arthur, R. H. & Wilson, E. O. The Theory of Island Biogeography. *Monographs in Population Biology* (Princeton University Press, Princeton, NJ, 1967).

Fire Ecology

Tropical Ecosystems: Structure, Diversity and Human Welfare.
Proceedings of the International Conference on Tropical Ecosystems
K. N. Ganeshaiah, R. Uma Shaanker and K. S. Bawa (eds)
Published by Oxford–IBH, New Delhi. 2001. pp. 321–324.

Late Quaternary history of three Brazilian ecosystems: The impact of climate, fire and human interference on vegetation

Paulo E. De Oliveira*[†] and Paul A. Colinvaux**

*Department of Botany, The Field Museum of Natural History
Chicago, IL 60605, USA
*Department of Sedimentary and Environmental, Geology, University of São Paulo,
São Paulo, Brazil
**Marine Biological Laboratory, Woods Hole, MA 02543, USA
[†]e-mail: poliveira@ufu.br

We review pollen and charred particle data to reconstruct the impact of climate, fire histories and human interference on three Brazilian ecosystems: the Amazon Rain Forest, the Cerrados of the central plateau and the Caatinga of northeastern Brazil during the last 50,000 years. Our results are presented below:

The Amazon: We find that plant communities of the neotropics do produce copious amounts of pollen that allow the recognition of community composition variations through time. Three main palynological studies in the Amazon clearly indicate vegetational histories characterized by remarkable constancy and stability. By comparing them with deposits of equal antiquity from the Cerrado region of central Brazil, it becomes clear that the Amazon

Keywords: Brazil, human interference, climate, fire, vegetation, late Quaternary history.

lowlands remained forested throughout the last glacial cycle (Colinvaux *et al.*, 1996; Colinvaux *et al.*, 2000, Colinvaux and De Oliveira, 2001, in press). There has been no evidence for forest fragmentation as suggested by Haffer (1969) and Prance (1982) and other refugialists. Instead, the intact forest of glacial times included significant populations of plants that are now restricted to high elevations, suggesting that the global warming of the Holocene resulted in the expulsion of heat-intolerant plants from the lowland forest. The expansion of *Podocarpus*, *Rapanea* (*Myrsine*), *Ilex*, *Hedyosmum*, *Weinmannia*, Myrtaceae, Melastomataceae, in association with various other rainforest trees can only be explained by the existence of former cooler climates conducive to wet forests, where plant composition was marked by the sympatry of taxa that are now disjunct by the effects of the global Holocene warming.

The Pata pollen record (northwestern Brazilian Amazonia) (Colinvaux *et al.*, 1996) suggests that around 5,500 years BP, the Upper Rio Negro region must have been impacted by human activities because charred particles content in the Lake Pata sediments increase dramatically in the profile during this period. The size of these particles indicates the occurrence of intense local fires. Palynological spectra of the charred particles indicate that there could have been a substantial increase in the *Mauritia* palm. There has been some speculation in the literature that certain increases in the pollen of this taxon could also reflect human interference since this palm is a primary source of food and housing materials for local Amazonian populations.

The Cerrado: Palynological analyses in the central Brazilian cerrados have flourished in the literature in the last decade (De Oliveira, 1992; Ledru, 1993; Ferraz-Vicentini and Salgado-Labouriau, 1996; Ferraz-Vicentini (1999), Barberi (2001). The Serra Negra (De Oliveira, 1992) and the Salitre Record (Ledru, 1993) are the oldest sedimentary sequences available so far for the Brazilian cerrado. The Serra Negra record shows the existence of cool and moist *Araucaria* forests throughout the last glacial cycle in a region presently covered by cerrado vegetation. The modern cerrado has been established locally only during the last few millennia. Charred particle analysis were found throughout the > 100,000 year old sedimentary sequence of Serra Negra and can be attributed to local and natural fires, since there is no evidence of human interference on that landscape until the last 1,000 years. At Salitre, higher resolution of the sediments permitted the reconstruction of *Araucaria* forests extending into the early Holocene (up to 8,000 yrs BP) in the region presently covered by typical cerrado vegetation. Additional palynological analyses in central Brazil by Barberi (2001) indicate the occurrence of cold and moist forest with abundant conifer pollen (*Podocarpus*) in association with other cold and moist forest indicators throughout the last glacial maximum (LGM). Ferraz-Vicentini (1999) shows that forest and cerrado remained stable since the last glacial maximum, contrary to all predictions of the Refuge Hypothesis.

Charred particle contents during the LGM of central Brazilian cerrado region are now interpreted as evidence of natural fires of the forests in the drier seasons. This interpretation is supported by the presence, in these charred particles, samples of palynological spectra rich in forest elements, many of which are now restricted to gallery and montane forests where orographic rain is high. Increase in charred particle analysis attributed to human impact appears only in the late Holocene, with increase of human interference brought about by the arrival of the first Portuguese settlers.

Caatinga: The only palynological analysis combined with charred particle studies available so far from the semi-arid caatinga vegetation of central Brazil is that of the Icatu River Valley peat bog (De Oliveira *et al.*, 1999). Pollen spectra and charcoal profile of these peat sediments revealed very moist conditions at the Pleistocene/Holocene transition (i.e. 11,000 to 10,000 yr BP). The high humidity permitted the expansion of dense tropical forest in the watershed of the Icatu River. Floristic composition revealed by the different pollen taxa clearly indicates that these forests had both Amazonian and Atlantic forest trees. This result is of fundamental importance for it supports an important biogeographical hypothesis for which there was until now no direct botanical evidence: the former connection of Amazonian and Atlantic forests. The palynological results from the Icatu River implying very moist Late Pleistocene conditions in a presently semi-arid region are supported by the palaentological work conducted by Hartwig and Cartelle (1996) and Cartelle and Hartwig (1996) who have found in a cave in the same region as that of the Icatu River, a complete skeleton of a large-bodied *Protopithecus* primate of Pleistocene age. Their findings on this and other caatinga fossils provide strong support for the former existence of tropical forests in the present semi-arid region of northeastern Brazil. Charred particle analysis along the Icatu River peat bog reveals an increase in fire intensity at 4240 years BP, which is interpreted by us as consequence of increasingly dry climates that prevailed since. This augmentation in aridity appears to be synchronous to the reported intense ENSO events of the mid-Holocene onwards in Brazil (Martin & Suguio, 1992).

All palynological investigations conducted so far in Brazil reveal no evidence of disturbance or impact that can be unequivocally attributed to human activity, despite the archeological evidence, specially that provided by Roosevelt *et al.* (1991) for the central Amazon where human presence is now believed to go back as far as 11,300 years BP. The only direct evidence of agricultural practices in the mid Holocene of the Brazilian amazon is provided by Bush *et al.* (2000). According to these authors, a single and undoubtedly *Zea* pollen grain (114 µm) occurred at 3,350 yrs BP in a sample rich in charred particles, thus implying intensification of burning and evidence of agriculture and changes in the land management practices of these early human occupants of the Hilea.

We believe that the lack of palynological support for widespread early and mid Holocene human impact in the three ecosystems might be a reflection of the fact that early human groups were rather small and widely separated in the vast immensity of the landscape.

References

Barberi, M. Mudanças Paleoambientais no Cerrado do Planalto Central durante o Quaternário Tardio: O Estudo da Lagoa Bonita/DF. Ph.D. Dissertation. Insitute of Geosciences. University of Sao Paulo, Sao Paulo, Brazil (2001).

Bush, M. B., Miller, M. C., De Oliveira, P. E. & Colinvaux, P. A. Two histories of environmental change and human disturbance in eastern lowland Amazonia. *The Holocene* 10, 543–553 (2000).

Cartelle, C. & Hartwig, W. C. Macacos sul-americanos: ossos que são um verdadeiro tesouro. *Ciência Hoje* 21, 31–36 (1996).

Colinvaux, P. A. & De Oliveira, P. E. Palaeoecology and climate of the Amazon basin during the last glacial cycle. *J. Quarter. Sci.* 15. (2001) in press.

Colinvaux, P. A., De Oliveira, P. E., Moreno, J. E., Miller, M. C. & Bush, M. B. A Long Pollen Record from Lowland Amazônia: Forest and Cooling in Glacial Times. *Science* 274, 85–88 (1996).

Colinvaux, P. A., De Oliveira, P. E. & Bush, M. B. Amazonian and neotropical plant communities on glacial time-scales: The failure of the aridity and refuge hypothesis. *Quarter. Sci. Rev.* 19, 141–169 (2000).

De Oliveira, P. E. A Palynological Record of Late Quaternary Vegetation and Climatic Change in Southeastern Brazil. Ph.D. Dissertation. Ohio State University, Columbus. USA (1992).

De Oliveira, P. E., Barreto, A. M. F. & Suguio, K. Late Pleistocene/Holocene climatic and vegetational history of the Brazilian Caatinga: the fossil dunes of the middle São Francisco River. *Palaeogeogr. Palaeoclimatol. Palaeoecol.* 152, 319–337 (1999).

Ferraz-Vicentini, K. Historia do Fogo no Cerrado: Uma Analise Palinologica. Ph.D. Dissertation. University of Brasilia. Brasilia, Brazil (1999).

Ferraz-Vicentini, K. & Salgado-Labouriau, M. L. Palynological Analysis of a Palm Swamp in Central Brazil. *J. South Am. Earth Sci.* 9, 207–219 (1996).

Haffer, J. Speciation in Amazonian forest birds. *Science* 165, 131–137 (1969).

Hartwig, W. C. & Cartelli, C. A complete skeleton of the giant South American primate *Protopithecus. Nature* 381, 307–311 (1996).

Ledru, M. P. Late Quaternary environmental and climatic changes in Central Brazil. *Quarter. Res.* 39, 90–98 (1993).

Martin, L. & Suguio, K. Variations of coastal dynamics during the last 7,000 years recorded in beach-ridge plains associated with river mouths: example from the central Brazilian coast. *Palaeogeogr. Palaeoclimatol. Palaeoecol.* 99, 119–160 (1992).

Prance, G. T. (ed.). *Biological Diversification in the Tropics* (Columbia University Press. New York, 1982).

Roosevelt, A. C., Housley, R. A., Lima Da Costa, M., Lopes Machado, C., Michab, M., Mercier, N., Valladas, H., Feathers, J., Barnett, W., Imazio Da Silva, M., Henderson, A., Sliva, J., Chernoff, B., Reese, D. S., Holman, J. A., Toth, N. & Schick, K. Paleoindian cave dwellers in the Amazon: The peopling of the Americas. *Science* 272, 373–384 (1996).

Systematics and Evolution of Tropical Plants

Tropical Ecosystems: Structure, Diversity and Human Welfare.
Proceedings of the International Conference on Tropical Ecosystems
K. N. Ganeshaiah, R. Uma Shaanker and K. S. Bawa (eds)
Published by Oxford–IBH, New Delhi. 2001. pp. 327–331.

Modeling herbivore dynamics on a Marantaceae: An empirical test of a metapopulation approach

Derek M. Johnson

Department of Biology, University of Miami, P.O. Box 249118, Coral Gables,
FL 33124, USA
e-mail: derek_delgado@yahoo.com

Ecologists are increasingly utilizing metapopulation models to generate predictions about the dynamics and viability of spatially structured populations (Hanski, 1994). One such model, the incidence function model (IFM), is promising as a predictive tool because it can be parameterized from a relatively modest amount of spatial data and is a reasonable predictor of patch occupancy (Hanski, 1997). The IFM treats within-patch populations in a binomial manner, either as 'occupied' or 'empty', and focuses on inter-population dynamics. The metapopulation concept, however, has come under criticism for being largely theory-based and lacking empirical support. Critics contend that stringent model assumptions limit their applicability to most systems (Harrison and Taylor, 1997). Because of its spatial structure and dispersal rate, *C. fenestrata* is an ideal species to empirically test the predictive value of the IFM and its robustness to assumption violations.

The hispine beetle *C. fenestrata* is a specialist herbivore on *Pleiostachya pruinosa* (Marantaceae) in the lowland humid to wet forests of Costa Rica. *C. fenestrata* spends its entire lifetime on *P. pruinosa*, laying eggs in the concavity

---

Keywords: Hispinae, Marantaceae, metapopulation, assumption violations, predictive model, herbivore.

of the leaf petioles where the larvae subsequently feed and pupate. Adults feed primarily in the immature rolled leaves of the plant, but may also feed in the petioles. Seventy-five patches of the host plant *P. pruinosa* are distributed along a 1.5 km long area, primarily near and in the flood zone (20 per cent of ramets in the flood zone) of the Puerto Viejo River at La Selva Biological Station. The river floods at an estimated 50 per cent probability every six months. Patch sizes range from 1 to 735 ramets. Between March 1999 and January 2001 *C. fenestrata* occupied from 45 to 64 per cent of the patches. While beetles move among ramets many times throughout their lives, inter-patch movement was detected in only 20% of recaptures. Thus, dispersal rates of *C. fenestrata* are low enough that there is some separation of populations, yet extinctions and recolonizations are frequent enough that population turnover is observable over the course of a two-year study.

The incidence function model is based on a linear first-order Markovian chain model that estimates the probability that a specific patch within a metapopulation is occupied (J_i) based on the probabilities of colonization (C_i) and extinction (E_i) over a specific time step (Hanski, 1994). The probabilities of occupancy are based on two patch characteristics, size and connectivity to other patches. Extinction probability is a function of patch size (A_i). Colonization probability is a function of the number of migrants into patch i (M_i) over one time step. The number of migrants is a function of the dispersal curves (a power function of inter-patch distances d_{ij}), the product of beetle density and emigration probability (β), occupancy pattern (p_j's equal 1 if donor patch j is occupied, 0 if empty), and sizes (A_j's) of the donor patches.

$$J_i = C_i / (C_i + E_i - C_i E_i), \tag{1}$$

$$E_i = \min(\mu / A_i^x, 1), \tag{2}$$

$$C_i = M_i^2 / (M_i^2 + y^2), \tag{3}$$

$$M_i = \Sigma_j (p_j d_{is}^{-\alpha} \beta A_j) \; j \neq i. \tag{4}$$

Two assumptions of the IFM are,

1) Emigration rate is independent of patch size.
2) Extinction is asynchronous among populations.

I censused the 75-patch network of *P. pruinosa* every six months from March 1999 to December 2000 to collect patch occupancy data. All life stages were recorded for each patch. A patch was considered 'occupied' only if immature stages were present, to exclude transitional adults. In addition, I individually marked over 1,200 adults, of which 40% were recaptured at least once, to estimate density and dispersal parameters.

Analyses of the census and mark/recapture data indicate that both model assumptions are violated. First, emigration rate is negatively correlated with patch size, meaning that beetles are more likely to emigrate from a small patch than a larger one. Secondly, patches had semi-synchronous extinctions (meaning only a fraction of the patches were affected) caused by a flooding event between the 3rd and 4th presence/absence censuses. I modified the migration equation to relax the assumption of patch size-independent migration, where

$$M_i = \Sigma_j(p_j e^{-\alpha d_{ij}} \beta A_j^\zeta),$$ (5)

and where $j \neq i$ and ζ is the patch size-dependent migration parameter.

I modified the extinction equation, adding the power function w, to relax the assumption of asynchronous extinction, so when a flood occurs probabilistically,

$$E_i = \min(\mu/A_i^w, 1),$$ (6)

for patches located in the flood zone.

Parameters were estimated for four model types [the basic model (Basic), the model with patch size-dependent migration (PDM), with semi-synchronous extinction (SSE), and with both modifications (PDM & SSE)] using a maximum likelihood technique. Model fit is quantified as the error term, negative log (likelihood), where better model fit is indicated by lower numbers. Adding SSE extinction to the models significantly improved the model fit to the La Selva data, but PDM did not (Figure 1).

Simulations of the four model types were run for 200 time steps and statistics were calculated only for the final 100 time steps to allow the simulated metapopulation to reach steady state. One time step is 6 months and, in the SSE models, flooding probability was set at 50% each time step to simulate observed patterns at La Selva. The models with PDM had higher proportions of occupied patches and the models with semi-synchronous extinction had lower proportions of occupancy. The differences were not striking (Basic = 0.48, SSE = 0.44, PDM = 0.53, SSE and PDM = 0.52) which might suggest that the IFM is robust to the assumption violations. However, when patches were removed from the simulation, thus changing the structure of the patch network, patch persistence times varied up to an order of magnitude among the models, depending on which patches were removed. These results are not surprising because simulations of the different models will converge on the same patch occupancy pattern when the same patch network structure is used to estimate the parameters. The more rigorous test of the predictability of a model is to change the structure of the patch network and again compare predictions of the models. In this case the model predictions diverge greatly.

Lastly, the predictive value of each model was tested by applying each of the models to one 'snapshot' of occupancy data of another 41-patch metapopulation of *C. fenestrata* on *P. pruinosa*. The comparison metapopulation is in Hacienda Baru on the Pacific coast in Dominical, Costa Rica. Hacienda Baru is a humid forest with a more distinct dry season than La Selva. The mean nearest neighbor at Hacienda Baru is 32.87 ± 21.96 m compared to 39.69 ± 49.49 m at La Selva, showing that patches at Hacienda Baru are more contiguous than at La Selva. Hacienda Baru has a slightly greater median patch size than La Selva. Model fit was calculated by running 100 simulations to 100 time steps and then calculating the –log (likelihood) transition probability from the occupancy pattern of the last simulation step to the observed occupancy pattern at Hacienda Baru (see Moilanen, 1999). The patches at Hacienda Baru do not experience flooding events, so flooding probability was set at 0 in the SSE. Note that while simulations of both the Basic and the SSE models had no flooding events, thus had no semi-synchronous extinction, differences in other parameter values gave different model fits. The model with patch size-dependent migration (PDM) was a significantly better fit ($p < 0.05$) with the occupancy pattern at Hacienda Baru, while adding semi-synchronous extinction (SSE) did not improve the predictive value of the model (Figure 1).

This study demonstrates that the IFM is highly sensitive to violations of the model assumptions of patch size-independent migration and asynchronous extinction. The potential magnitude of persistence time sensitivity is only

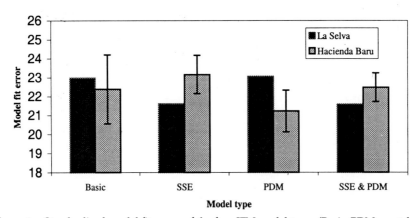

Figure 1. Standardized model fit errors of the four IFM model types (Basic, PDM = patch size-dependent migration, SSE = semi-synchronous extinction, and PDM & SSE). The lesser value indicates a better model fit.

observable when the structure of the patch network is altered in model simulations. While the models with SSE were significantly better fits to the La Selva data than were those with PDM, the models with PDM were better fits to the Hacienda Baru data. This result is perplexing. One possible explanation is that the ecology of *C. fenestrata* at Hacienda Baru is different from that at La Selva, thus none of the models sufficiently describe the occupancy pattern. The model is being fitted to more *C. fenestrata* metapopulations to test whether this result withstands replication.

References

Hanski, I. A practical model of metapopulation dynamics. *J. Anim. Ecol.* **63**, 151–162 (1994).
Hanski, I. Metapopulation dynamics: From concepts and observations to predictive models. *in Metapopulation Biology: Ecology, Genetics, and Evolution* (eds Hanski, I. & Gilpin, M. E.) 69–91 (Academic Press, San Diego, California, 1997).
Harrison, S. & Taylor, A. D. Empirical evidence for metapopulation dynamics. in *Metapopulation Biology: Ecology, Genetics, and Evolution* (eds Hanski, I. A. & Gilpin, M. E.) 27–42 (Academic Press, San Diego, California, USA, 1997).
Moilanen, A. Patch occupancy models of metapopulation dynamics: Efficient parameter estimation using implicit statistical inference. *Ecology* **80**, 1031–1043 (1999).

Tropical Ecosystems: Structure, Diversity and Human Welfare.
Proceedings of the International Conference on Tropical Ecosystems
K. N. Ganeshaiah, R. Uma Shaanker and K. S. Bawa (eds)
Published by Oxford–IBH, New Delhi. 2001. pp. 332–336.

Impact of rain forest fragmentation on the growth and reproductive potential of an Amazonian understorey herb (*Heliconia acuminata*)

Emilio M. Bruna*·†, Olavo Nardy†,‡, Sharon Y. Strauss*
and Susan Harrison*

*Center for Population Biology, University of California, 1 Shields Ave.,
Davis, CA 95616, USA
†Biological Dynamics of Forest Fragments, PDBFF-INPA, CP 478, Manaus,
AM 69011-970, Brazil
‡Departamento de Ecologia, Universidade Estadual Paulista Rio Claro,
SP 13506-900, Brazil
†e-mail: embruna@ucdavis.edu

The impact of fragmentation on abiotic conditions in habitat fragments has consistently been shown to be widespread and severe, particularly in forest ecosystems (Didham *et al.*, 1998; Laurance *et al.*, 1998; Harrison and Bruna, 1999). Fragments of rain forest, for example, often have increased air and soil temperatures, reduced relative humidity, and reduced soil moisture levels (Kapos, 1989; Kapos *et al.*, 1997). Abiotic modifications such as these appear to drive many of the negative community-wide impacts seen in fragmented landscapes, such as reductions in species diversity and changes in plant community dynamics (Leach and Givnish, 1996; Carvalho and Vasconcelos, 1999). It has recently been suggested that

Keywords: Fragmentation, *Heliconia*, Amazon, growth, abiotic conditions.

in addition to community-level effects, the modification of abiotic conditions could cause physiological and morphological changes in individuals that survive in habitat isolates (Weishampel *et al.*, 1997; Sumner *et al.*, 1999). These changes could occur because altered environmental conditions directly impact development and growth, or because they influence resources upon which individuals are dependent. While a few studies have found consistent morphological differences between individuals found in fragments and continuous forests (Weishampel *et al.*, 1997; Sumner *et al.*, 1999), the lack of any manipulative experiments makes it difficult to determine whether this differentiation followed fragmentation or reflects pre-isolation variation. Furthermore, the long-term consequences of these differences are unclear, as the link between the characters measured and fitness is often unknown.

Understorey plants may be particularly susceptible to altered environmental conditions in rainforest fragments, as they grow in areas with lower temperatures, high humidity, and reduced light availability. We show that individuals of the Amazonian understorey herb *Heliconia acuminata* transplanted into rainforest fragments suffered dramatic reductions in size when compared to those transplanted to continuous forest, and that these changes could disrupt reproduction by greatly reducing the number of flowering individuals found in forest fragments.

We conducted this study at the Biological Dynamics of Forest Fragments Project (BDFFP), located 80 km north of Manaus, Brazil. Four one-hectare fragments of lowland rainforest, all isolated between 1980 and 1986 by the creation of cattle pastures, were paired with four nearby areas of continuous forest. We then reciprocally transplanted a total of 160 *H. acuminata* individuals between the forest fragments and the nearest continuous forest sites. An additional 160 plants were removed from the ground and re-planted within the same sites to serve as controls for the effects of travel, transplanting, and local adaptation. To test for potential edge effects within fragments, control and experimental plants were transplanted at varying distances from the edges of fragments, ranging from the fragment edge to 40 m into the fragment interior.

Plants were initially transplanted during the early part of the 1999 rainy season, at which time we counted how many vegetative shoots each plant had and calculated its total leaf area. Seven months after transplanting, following the completion of the rainy and dry seasons, we re-measured all plants. A third measurement was made fourteen months after the initial planting, at the end of the second rainy season.

During the 1999 dry season, plants transplanted to continuous forest sites lost approximately 5% of their total leaf area. During the same period, plants transplanted to forest fragments lost almost four times this amount

(18.7%). They also lost 14% of their vegetative shoots, while plants moved into continuous forest grew new ones. This asymmetrical loss in both leaf area and shoot number suggests that the normal dry season temperature stressses to which these understorey plants are exposed were greatly exacerbated in forest fragments. While by the end of the second rainy season plants in both habitat types had recovered some of the lost leaf area by growing new leaves, net loss in fragments was significantly greater than in continuous forest. Plants in continuous forest also had almost 10% more shoots than when the experiment began, whereas plants in fragments had yet to recover completely from dry season losses (Figure 1).

The reductions in plant size observed in forest fragments could have substantial demographic impacts. At the individual level, the probability of flowering in H. acuminata increases with shoot number (Bruna and Kress, unpublished data), therefore shoot loss could have significant consequences for individual plant fitness. Furthermore, reducing the average size of individuals also has population-level demographic consequences. The reductions in shoot numbers observed between the initial planting and the second measurement, at which time plants would generally initiate flowering, result in a 27–35% decrease in the expected number of reproductive individuals in forest fragments. This is in sharp contrast to the population of plants transplanted to continuous forest, in which the number of flowering plants expected actually increased by 25–37%. This disparity in the expected number of reproductive individuals closely mirrors the 50% difference in flowering documented in continuous forest and fragment Heliconia demography plots at the BDFFP (Bruna and Kress, unpublished data).

Our results suggest that changes in flowering frequency resulting from shifts in plant size, alone or in concert with other mechanisms reducing

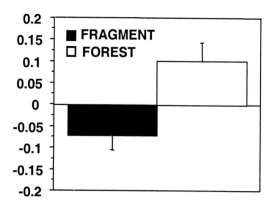

Figure 1. Proportional change in shoot number after 14 months.

reproductive success such as lower pollination or seed germination (Aizen and Feinsinger, 1994; Bruna, 1999; Jules and Rathcke, 1999), could substantially reduce the recruitment of seedlings into rainforest fragments. Furthermore, these reduced numbers of flowering individuals could lead to a substantial increase in inbreeding in fragmented populations. Environmentally induced changes in morphology, if severe enough, may help explain why populations of plants in habitat fragments often fail to persist over the long term (Turner *et al.*, 1995; Corlett and Turner, 1997; Jules, 1998).

Acknowledgements

We thank F. Marques, O. F. da Silva, and J. Ribamar for assistance in the field and W. J. Kress for many helpful discussions. We would also like to thank the BDFFP for providing logistical support and the Manaus Free Trade Zone Authority (SUFRAMA) for permission to conduct the research. This work was supported by the US National Science Foundation (Dissertation Improvement Grant INT 98-0635 and a Minority Predoctoral Fellowship to EMB), The University of California-Davis (Office of Graduate Studies and the Center for Population Biology), and the Smithsonian Institution (The Graduate Student Fellowship Program).

References

Aizen, M. A. & Feinsinger, P. Forest fragmentation, pollination, and plant reproduction in a chaco dry forest, Argentina. *Ecology* **75**, 330–351 (1994).
Bruna, E. M. Seed germination in rainforest fragments. *Nature* **402**, 139 (1999).
Carvalho, K. S. & Vasconcelos, H. L. Forest fragmentation in central Amazonia and its effect on litter-dwelling ants. *Biol. Conserv.* **91**, 151–158 (1999).
Corlett, R. T. & Turner, I. M. Long-term survival in tropical forest remnants in Singapore and Hong Kong. in *Tropical Forest Remnants: Ecology, Management, and Conservation of Fragmented Communities* (eds Laurance, W. F. & Bierregaard, Jr. R. O.) 333–345 (University of Chicago Press, Chicago, IL, USA, 1997).
Didham, R. K., Hammond, P. M., Lawton, J. H., Eggleton, P. & Stork, N. E. Beetle species responses to tropical forest fragmentation. *Ecol. Monogr.* **68**, 295–323 (1998).
Harrison, S. & Bruna, E. Habitat fragmentation and large-scale conservation: what do we know for sure? *Ecography* **22**, 225–232 (1999).
Jules, E. S. Habitat fragmentation and demographic change for a common plant: *Trillium* in old-growth forest. *Ecology* **79**, 1645–1656 (1998).
Jules, E. S. & Rathcke, B. J. Mechanisms of reduced *Trillium* recruitment along edges of old-growth forest fragments. *Conserv. Biol.* **13**, 784–793 (1999).
Kapos, V. Effects of isolation on the water status of forest patches in the Brazilian Amazon. *J. Trop. Ecol.* **5**, 173–185 (1989).
Kapos, V., Wandelli, E., Camargo, J. L. & Ganade, G. Edge-related changes in environment and plant responses due to forest fragmentation in Central Amazonia. in *Tropical Forest Remnants: Ecology, Management, and Conservation of Fragmented Communities* (eds Laurance, W. F. & Bierregaard, Jr. R. O.) 33–44 (University of Chicago Press, Chicago, IL, USA, 1997).
Laurance, W. F., Ferreira, L. V., Rankin-De-Merona, J. M. and Laurance, S. G. Rain forest fragmentation and the dynamics of Amazonian tree communities. *Ecology* **79**, 2032–2040 (1998).
Leach, M. K. & Givnish, T. J. Ecological determinants of species loss in remnant prairies. *Science* **273**, 1555–1558 (1996).
Sumner, J., Moritz, C. & Shine, R. Shrinking forest shrinks skink: Morphological change in response to rainforest fragmentation in the prickly forest skink (*Gnypetoscincus queenslandiae*). *Biol. Conserv.* **91**, 159–167 (1999).

Turner, I. M., Chua, K. S., Ong, J. S. Y., Soong, B. C. & Tan, H. T. W. A century of plant species loss from an isolated fragment of lowland tropical rain forest. *Conserv. Biol.* **10,** 1229–1244 (1995).

Weishampel, J. F., Shugart, H. H. & Westman, W. E. Phenetic variation in insular populations of a rainforest centipede. in *Tropical Forest Remnants: Ecology, Management, and Conservation of Fragmented Communities* 111–123 (University of Chicago Press, Chicago,IL, USA, 1997).

Tropical Ecosystems: Structure, Diversity and Human Welfare.
Proceedings of the International Conference on Tropical Ecosystems
K. N. Ganeshaiah, R. Uma Shaanker and K. S. Bawa (eds)
Published by Oxford–IBH, New Delhi. 2001. pp. 337–341.

Study on the flexistyly pollination mechanism in *Alpinia* plants (Zingiberaceae)

Qing-Jun Li*, W. John Kress[†,1], Zai-Fu Xu*,
Yong-Mei Xia*, Ling Zhang*, Xiao-Bao Deng*
and Jiang-Yun Gao*

*Xishuangbanna Tropical Botanical Garden, The Chinese Academy of Sciences,
Mengla, Yunnan 666303, China
[†]Department of Botany, National Museum of Natural History, Smithsonian
Institution, Washington, DC 20560-0166 USA
[1]e-mail: kress.john@nmnh.si.edu

For the avoidance of inbreeding depression that follows repeated selfing (Darwin, 1916; Antonovics, 1968), plants have evolved diverse mechanisms to promote outcrossing, e.g. dioecy, dichogamy (Lloyd and Webb, 1986), herkogamy (Webb and Lloyd 1986), enantiostyly (Ornduff and Dulberger, 1978), heterostyly (Barrett, 1992), and self-incompatibility (Haring *et al.*, 1990). The phenomenon in some plants in which various floral parts move during anthesis has been recognized by several researchers (Stout, 1927; Fetscher and Kohn, 1999; Endress, 1984), but it has been either ignored as an outcrossing mechanism (Richards, 1997) or considered as a type of dichogamy or herkogamy (Lloyd and Webb, 1986; Endress, 1994; Fægri and Pijl, 1979). Recently, we found an unambiguous example of temporal dioecy in which a unique floral mechanism involving extreme stigmatic movement,

---

Keywords: *Alpinia*, flexistyly, hyperflexistyly, cataflexistyly, pollination biology.

which we call flexistyly, enhances the outcrossing rate in plants (Li *et al.*, 2001).

Alpinia, a member of the Ginger family (Zingiberaceae), is composed of more than 250 species found in Southeast Asia extending to Japan in the north and Australia in the south (Smith, 1990). They are perennial herbs, usually 1.5– 4 m tall with leafy, many-bladed and frond-like shoots. Terminal inflorescences on the leafy-shoots are made up of congested bracts with each bract subtending a single hermaphroditic flower or a cincinnus with two to several flowers. The flowers are composed of relatively inconspicuous sepals and petals, but with a petaloid, showy labellum derived from two fused staminodes that is usually fused with the single fertile stamen. The free part of the labellum is expanded as a landing platform for visiting insects. During flowering, each inflorescence produces two to ten flowers per day, each flower lasting only one day, with the blooming season over a two-month period. We made observations on the flowering behaviour of nine native and introduced species of *Alpinia* (*A. blepharocalyx, A. bractea, A. conchigera, A. galanga, A. katsumadai, A. kwangsiensis, A. maclurei, A. platychilus* and *A. zerumbet*) (Table 1) and conducted controlled experiments for large populations of two of them (*A. kwangsiensis* and *A. galanga*) in a natural reserve of tropical seasonal rain forest in Xishuangbanna (21°41′N, 101°25′E, 580 m elev.), Southwest China.

Each species of *Alpinia* has two phenotypes present in all populations that differ in floral behaviour. We have named the two phenotypes

Table 1. The blooming season, flowering behaviour and the pollinators of nine species of *Alpinia* distributed and cultivated in Xishuangbanna, Yunnan, Southwest China

Species	Blooming season	Flowering behaviour (Time)				
		Stigma receptivity[1]		Pollen dispersal[2]		Pollinators
		Hyper-	Cata-	Cata-	Hyper-	
A. blepharocalyx	April–May	06:30–11:20	14:00–18:30	07:00	13:55	*Xylocopa* spp.
A. bractea	May–April	06:30–12:10	13:50–19:30	06:40	13:30	*Xylocopa* spp.
A. conchigera	May–July	06:45–11:40	13:45–19:00	06:50	13:40	*Bombus* spp.
A. galanga	May–July	06:30–11:30	14:00–19:30	06:45	13:40	X. spp., B. spp. and *Nomia* spp.
A. katsumadai	April–Jun	06:35–12:00	14:30–20:00	06:50	14:20	*Xylocopa* spp.
A. kwangsiensis	March–May	06:30–11:30	14:30–19:30	06:45	14:25	*Xylocopa* spp.
A. maclurei	March–July	06:30–12:00	14:45–19:50	06:40	14:40	*Xylocopa* spp.
A. platychilus	April–May	06:30–11:30	14:00–19:00	06:40	13:45	*Xylocopa* spp.
A. zerumbet	April–Jun	06:40–12:00	13:50–19:00	06:45	13:40	*Xylocopa* spp.

The data are three days' observation for each species under the normal weather condition at that season.
[1]The time when the stigma at the receptive position of about 170° with the anthers' ventral face.
[2]The time when the anthers begin to dehisce.

hyperflexistyly and cataflexistyly (Li *et al.* 2001). The duration of anthesis of both phenotypes is 24 h and begins before dawn. When the flowers are fully open (6:00 to 6:30), cataflexistyled forms have the stigma exserted above the anther and the pollen sacs are dehiscent (Figure 1). At the same time, the receptive stigma of hyperflexistyled flowers is curved downward below the

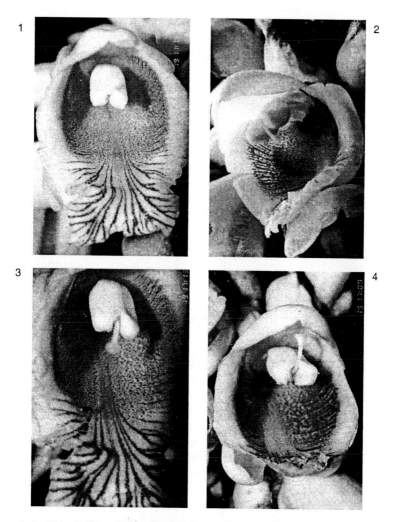

Figure 1–4. The position of the stigmas of the two flower phenotypes at different stages of anthesis in *Alpinia zerumbet*. **1,** Cataflexistyled flower in its male phase (before noon) in which the stigma is reflexed above the dehiscent anther. **2,** Hyperflexistyled flower in its female phase (before noon) in which the stigma is deflexed below the indehiscent anther. **3,** The same cataflexistyled flower as in **1** during its female phase (afternoon) in which the stigma moves down towards the labellum; note the pollen has been removed from the anther by visitors (mainly *Xyclocapa* bees). **4,** The same hyperflexistyled flower as in **2** in its male phase (afternoon) in which the stigma has become erect above the anther and the pollen sacs are dehisced after the stigma has moved out of any visitor's way. The quartz data show the time of day.

undehisced anther sacs from which pollen has not yet been shed (Figure 2). Flowers of both types retain these respective positions until about mid-day (11:00–12:00). At that time the stigmas of hyperflexistyled flowers begin to elongate and become erect above the anther. As the stigma moves upward and out of direct contact with any floral visitor (170° between the stigma and the anther's ventral face; 11:45–13:30; Figure 4), the anther dehisces and the pollen is presented (14:30 to 15:00). The movement of the stigma of the cataflexistyled flower is slower: here the stigma begins to move downward and enters the receptive position (about 170° from the anther ventral face; Figure 3) between 14:40 and 15:00 (several minutes after anther dehiscence in hyperflexistyled flowers). Anthesis in both forms ends after dark (20:30 to 21:00) when the anthers collapse and the corolla limply hangs down. The speed of stigmatic movement varies depending on different weather conditions, but all the flowers of the same phenotype that open on the same day are strictly synchronous and the hyperflexistyled anthers never dehisce before all of the same phenotype's stigmas have moved out of the receptive position. Non-fertilized flowers abscise in two or three days while the ovaries of fertilized flowers will enlarge in the next several days; bracts and corollas of fertilized flowers will remain attached almost until fruit maturity.

Different controlled pollination combinations within and between phenotypes of *A. kwangsiensis* were experimentally conducted in the field. The results indicate that plants of the genus *Alpinia* are self-compatible, that insects are the natural pollinators (e.g., *Xylocopa magnifica* and *X. tenuiscapa*), and that the selfing rate is minimal.

It has been suggested that species of plants that are hyperdispersed in forested habitats and produce only a few flowers per day are adapted to pollination by traplining animals (Linhart 1973; Janzen 1971; Kress and Beach 1994). Pollination by trapliners is an effective means of outcrossing, and traplined plants show a 'steady-state' flowering pattern (Gentry, 1974; Endress, 1994). The flowering traits of *A. kwangsiensis* and the behaviour of its visitors appear to be an example of traplining pollination that may in association with the floral movement promote out-crossing in this species.

This floral mechanism prevents not only self-pollination within a flower (autogamy) and within the same individual (geitonogamy), but also among individuals of the same phenotype. We conclude that this elaborate floral mechanism results in a very low level of inbreeding despite the presence of self-compatibility in the populations.

Recent molecular analyses of the phylogenetic relationships of seven of the nine species of *Alpinia* (Rangsiruji *et al.*, 2000), showed that they are distributed in three distinct paraphyletic clades in the Tribe *Alpineae* (Kress *et al.*, unpublished data). These results suggest that flexistyly has either evolved independently several times in the *Alpineae* or is widespread in the Tribe but as yet unrecorded in other species and genera.

Acknowledgements

We thank the constructive discussions or comments from Peter H. Raven, Xiao-Long Cui, Jin Chen, Xing-Guo Han, Ai-Zhong Liu, Ken Marr, Linda Prince, and Ruby Marr; we thank Dr. Dan Nicolson for discussion on nomenclature; figures were scanned and modified by Hong-Mei Li and Si-Hai Wang; Prof. Da-Yong Yang helped to identify the insects. This is part of the work of CAS Innovation Projects, supported by the National Natural Science Foundation of China Grant 39700019 and the Smithsonian Scholarly Studies Program.

References

Antonovics, J. Evolution in closely adjacent plant populations. V. Evolution of self-fertility. *Heredity* **23**, 219–238 (1968).

Barrett, C. C. H. (ed.). *Evolution and Function of Heterostyly* (Springer, Berlin, 1992).

Darwin, C. *The effects of cross and self fertilization in the vegetable kingdom* (2nd edn) (John Murray, London, 1916).

Endress, P. K. *Diversity and Evolutionary Biology of Tropical Flowers* (Cambridge University Press, 1994).

Endress, P. K. The flowering process in the Eupomatiaceae (Magnoliales). *Bot. Jahrb. Syst.* **104**, 297–319 (1984).

Fægri, K. & Pijl, L. van der. *The Principles of Pollination Ecology* (3rd revised), Pergamon Press, New York (1979).

Fetscher, E. A. & Kohn, J. R. Stigma behavior in *Mimulus aurantiacus* (Scrophulariaceae). *Am. J. Bot.* **86**, 1130–1135 (1999).

Gentry, A. H. Flowering phenology and diversity in tropical Bignoniaceae. *Biotropica* **6**, 64–68 (1974).

Haring, V. *et al.* Self-incompatibility: a self-recognition system in plants. *Science* **250**, 937–941 (1990).

Janzen, D. H. Euglossine bees as long distance pollinators of tropical plants. *Science* **171**, 203–205 (1971).

Kress, W. J. & Beach, J. H. Flowering plant reproductive systems at La Selva Biological Station. in *La Selva: Ecology and Natural History of a Neotropical Rain Forest* (eds McDade, L. A., Bawa, K. S., Hespenheide, H. & Hartshorn, G.) 161–182 (University of Chicago Press, 1994).

Li, Q.-J. *et al.* Flexistyly promoting plants' outcrossing. *Nature* **410** (2001).

Linhart, Y. B. Ecological and behavioral determinants of pollen dispersal in hummingbird pollinated. *Heliconia. Am. Nat.* **107**, 511–523 (1973).

Lloyd, D. G. & Webb, C. J. The avoidance of interference between the presentation of pollen and stigmas in angiosperms. I. Dichogamy. *N. Z. J. Bot.* **24**, 135–162 (1986).

Webb, C. J. & Lloyd, D. G. The avoidance of interference between the presentation of pollen and stigmas in angiosperms. II. Herkogamy. *N. Z. J. Bot.* **24**, 133–178 (1986).

Ornduff, R. & Dulberger, R. Floral enantiomorphy and the reproductive system of *Wachendorfia paniculata* (Haemodoraceae). *New Phytol.* **80**, 427–434 (1978).

Rangsiruji, A., Newman, M. F. & Cronk, Q. C. B. A study of the infrageneric classification of *Alpina* (Zingiberaceae) based on the ITS region of nuclear rDNA and the trnL-F spacer of chloroplast DNA. in *Monocots – Systematics and Evolution* (eds Wilson, K. L. & Morrison, D. A.) 695–709 (CSIRO Publishing, Collingwood, Australia, 2000).

Richards, A. J. *Plant Breeding Systems*. 2nd edn (Chapman & Hall, London, 1997).

Smith, R. M. Alpinia (Zingiberaceae): a proposed new infrageneric classification. *Edinb. J. Bot.* **47**, 1–75 (1990).

Stout, A. B. The flower behavior of avocados. *Mem. New York Bot. Gard.* **7**, 154–203 (1927).

Tropical Ecosystems: Structure, Diversity and Human Welfare.
Proceedings of the International Conference on Tropical Ecosystems
K. N. Ganeshaiah, R. Uma Shaanker and K. S. Bawa (eds)
Published by Oxford–IBH, New Delhi. 2001. pp. 342–346.

Pollination guilds and the evolution of floral characters in Bornean Zingiberaceae and Costaceae

Shoko Sakai*,#, Hidetoshi Nagamasu†, Kazuyuki Ooi‡, Makoto Kato* and Tamiji Inoue

*Graduate School of Human and Environmental Studies, Kyoto University, Kyoto 606-8501, Japan
†The Kyoto University Museum, Kyoto 606-8501, Japan
‡Department of Biology, Kyusyu University, Fukuoka 812-8581, Japan
#e-mail: sakai@sinfo.net

There is no doubt that floral morphology and flowering phenology are important in plant–pollinator interactions. Convergence in floral morphology among plants with the same pollination systems has been indicated repeatedly. On the other hand, character displacement among plants sharing pollinators has been suggested in floral morphology and flowering phenology. In some plant groups variation in the site of pollen placement on the pollinator body is thought to promote coexistence of species sharing common pollinators. Temporal segregation in flowering periods to avoid competition for pollinators has often been suggested, although it is rarely demonstrated statistically (Ollerton and Lack, 1992). To evaluate convergence and divergence in a group of plants, we need to know phylogenetic relationships among the plants.

Keywords: Character displacement, dipterocarp forest, floral morphology, reproductive phenology, phylogeny.

We studied floral morphology and flowering phenology of the Zingiberaceae and Costaceae (Zingiberales) in a mixed dipterocarp forest in Lambir Hills National Park (hereafter Lambir), Sarawak, Malaysia. The Zingiberaceae is a family that contains about 50 genera and 1000 species. Most species of Zingiberaceae occur in the Indomalesian region; approximately 160 of which are found in Borneo. Although Zingiberaceae is the most diverse herbaceous plant family found in Bornean lowland forests (Poulsen, 1997), with members that are both economically and ecologically important, the ecology and systematics of Zingiberaceae have received little study until recently. The Costaceae, which is thought to be closely related to the Zingiberaceae, has a distribution that is clearly centered in Neotropics. Only three species of Costaceae are recorded from Borneo. At Lambir, 45 species of Zingiberaceae and three species of Costaceae have been recorded. Among them 18 species and one genus of Zingiberaceae were undescribed when we started the study in 1994.

Three groups of pollinators were found to visit the 27 species of Zingiberaceae and two species of Costaceae. We classified the plant into three pollination guilds corresponding to the three pollinator groups. These guilds included spiderhunter-pollinated plant (*Arachnothera longirostra*, Nectariniidae, 8 plant species), medium-sized bee-pollinated plant (2 spp. of *Amegilla*, Anthophoridae, 11 species), and small bee-pollinated plant (4 species of Halictidae, 10 species) guilds. Spiderhunters are birds with long bills (36 mm) used to catch arthropods and suck floral nectar. The *Amegilla* and halictid bees are solitary bees with long prosbosces (11–12.2 mm and 3.9–8.7 mm, respectively) foraging in the understory. All of the pollinators that we observed were classical trapliners that traveled long distances between small, scattered resource patches. Each of the plant species studied was visited by pollinators of only one of the three pollinator groups. Other flower visitors were rarely observed.

Canonical discriminant analysis of seven characters of floral morphology revealed significant correlation between floral morphology and pollination guilds. Most of the species in the three guilds were separated on the plot by the first and second canonical variables, suggesting the importance of floral morphology in plant-pollinator interactions. Spiderhunter-pollinated flowers had the longest floral tubes, while medium-sized bee-pollinated flowers had the widest lips, which function as a platform for the pollinators. The pistils and stamens of halictid-pollinated flowers were generally smaller than those of spiderhunter or bee pollinated plants. In addition to the floral morphology, floral color and nectar production were also correlated with the pollination guilds (Sakai *et al.*, 1999a). Segregation in the pollen deposition site on the pollinator body among plants sharing pollinators was not supported in the present study.

In Borneo annual cycle in rainfall is weak, although rainfall does change greatly. Forests of the region have strong supra-annual fluctuation in plant reproductive activities at the community level, which may be related with El Niño Southern Oscillation (ENSO) or other climatic phenomena at multi-year intervals (Sakai et al., 1999b). Mass flowering synchronized among many plant species at multi-year intervals found on Borneo is quite different from the marked annual seasonality found at many Neotropical locations.

Our study of the reproductive phenology of 20 species of the Zingiberaceae and Costaceae at Lambir (Sakai, 2000) showed that most of the species reproduce more than once a year, or flower continuously with short interruptions. While mass flowering of emergent and canopy trees was observed during the study period (Sakai et al., 1999b), no significant changes in flowering intensity of our focal species were observed. Significant but weak synchronization in flowering events among conspecific individuals was detected for only two species out of the five examined. The low levels of synchronization within the populations, the overlap of flowering periods among species sharing common pollinators, and the high flowering frequency observed in Borneo contrast markedly with the reproductive phenology of hummingbird-pollinated and large bee-pollinated plants of related taxa in the Neotropics. Although the plants studied do not include all members of each pollination guild, at least one species within a guild was observed to be flowering at any time except in the spiderhunter-pollinated guild. Some bee-pollinated species with long flowering periods may serve as keystone species for survival of the traplining bees by providing floral nectar and pollen when resources are scarce.

A molecular phylogeny was estimated based on DNA sequence data from the internal transcribed spacer (ITS) region of nrDNA and the chloroplast gene matK. Our data suggest Tamijia, a genus recently described from Lambir had diverged from the rest of the species in the family very early on (Sakai and Ooi, unpublished data). This may explain deviation of the genus from the criteria of any tribes in the family (Sakai and Nagamasu, 2000). The phylogeny also supports that Amomum is polyphyletic.

Within the Zingiberaceae we detected seven changes in pollination systems. The data also showed medium-sized bee pollination to be ancestral in the family, from which spiderhunter and small bee pollination were derived. Pollinator shifts between small bees and birds were unlikely. Floral morphology was under strong phylogenetic constraints. Species of different lineages sharing pollinators often have quite different floral shapes, and morphological convergence within a pollination guild was not supported. However, this does not necessarily mean the absence of morphological adaptation to the pollinators. It more likely means that floral parts do not always function in the same way in different lineages. This demonstrates that shifts in pollination systems may contribute morphological diversity within a pollination guild.

The 27 species of Zingiberaceae and two species of Costaceae that we studied at Lambir belong to only three pollination guilds. Phylogenetic relationships of the focal plants suggested that speciation associated with a shift of pollination systems probably did not occur frequently. The pollination guilds that we found in Zingiberaceae and Costaceae in Borneo are comparable to the hummingbird-, and euglossine-bee-pollinated guilds of neotropical Zingiberales (Stiles, 1975; Kennedy, 1978; Schemske, 1983). However, the number of pollinator species that we found to be involved in each pollination guild ranged from one to four, many fewer than those in Neotropics (Sakai et al., 1999a). Plant species within a guild have similar floral morphology, but the similarity may be due to a high degree of relatedness between species in the same guild rather than to convergence. Flowering periods of different species within a guild overlapped considerably, and there was no evidence for temporal segregation. This result is in striking contrast to reproductive phenology of hummingbird-pollinated and large bee-pollinated species, which mostly flower annually with temporal segregation among species within a pollination guild to reduce competition for pollinators (Stiles, 1975; 1978; Frankie et al., 1983).

Acknowledgements

The authors thank Dr H. S. Lee and Dr A. A. Hamid, and other staff of the Forest Department Sarawak for logistical help with the study. This project was supported in part by Grants-in-Aid from the Japanese Ministry of Education, Science and Culture (numbers 04041067, 06041013, 09NP1501 and 10041169) and by JSPS Research Fellowship for Young Scientist for SS.

References

Frankie, G. W., Haber, W. A., Opler, P. A. & Bawa, K. S. Characteristics and organization of the large bee pollination system in the Costa Rican dry forest. in *Handbook of Experimental Pollination Biology* (eds Jones, C. E. & Little, R. J.) 441–447 (Van Nostrand Reinhord, New York, 1983).

Kennedy, H. Systematics and pollination of the 'closed flowered' species of *Calanthea* (Marantaceae). *Univ. California Publ. Bot.* **71**, 1–90 (1978).

Ollerton, J. & Lack, A. J. Flowering phenology: an example of relaxation of natural selection? *Trends Ecol. Evol.* **7**, 274–276 (1992).

Poulsen, A. D. The herbaceous ground flora of the Batu Apoi Forest Reserve Brunei Darussalam. in *Tropical Rainforest Research–Current Issues* (eds Edwards, S., Booth, W. E. & Choy, S. C.) 43–51 (Kluwer Academic Publishers, Dordrecht, 1997).

Sakai, S. Reproductive phenology of gingers in a lowland mixed dipterocarp forest in Borneo. *J. Trop. Ecol.* **16**, 337–354 (2000).

Sakai, S., Kato, M. & Inoue, T. Three pollination guilds and variation in floral characteristics of Bornean gingers (Zingiberaceae and Costaceae). *Am. J. Bot.* **86**, 646–658 (1999a).

Sakai, S., Momose, K., Yumoto, T., Nagamitsu, T., Nagamasu, H., Hamid, A. A., Nakashizuka, T. & Inoue, T. Plant reproductive phenology over four years including an episode of general flowering in a lowland dipterocarp forest, Sarawak, Malaysia. *Am. J. Bot.* **86**, 1414–1436 (1999b).

Sakai, S. & Nagamasu. H. Systematic studies on Bornean Zingiberaceae III. *Tamijia*: a new genus. *Edinb. J. Bot.* **57**, 245–255 (2000).

Schemske, D. W. Breeding system and habitat effects on fitness components in three Neotropical *Costus* (Zingiberaceae). *Evolution* **37**, 523–539 (1983).

Stiles, F. G. Ecology, flowering phenology and hummingbird pollination of some Costa Rican *Heliconia* species. *Ecology* **56**, 285–301 (1975).

Stiles, F. G. Temporal organization of flowering among the humming bird food plants of a tropical wet forest. *Biotropica* **10**, 194–210 (1978).

Tropical Ecosystems: Structure, Diversity and Human Welfare.
Proceedings of the International Conference on Tropical Ecosystems
K. N. Ganeshaiah, R. Uma Shaanker and K. S. Bawa (eds)
Published by Oxford–IBH, New Delhi. 2001. pp. 347–350.

The evolutionary history of the Zingiberales

W. John Kress

Department of Systematic Biology, Botany, MRC-166, National Museum of Natural
History, Smithsonian Institution, Washington, DC 20560-0166, USA
e-mail: kress.john@nmnh.si.edu

The Zingiberales are a group of entirely tropical monocotyledons that include the bananas, gingers and their relatives. Advances in the understanding of the diversity, systematics, phylogeny, morphology, evolution, and ecology of these plants over the last twenty years have placed the Zingiberales as one of the better known plant groups in the tropics. This increase in the biological information available for these plants makes them a prime candidate as a 'model' group for focused investigations on the evolution and ecology of tropical plants.

Eight families are recognized in the order Zingiberales (Kress, 1990; Kubitzky, 1998). These families can be divided into the 'banana group' that includes the Musaceae (the bananas, three genera), Strelitziaceae (the bird-of-paradise, three genera), Lowiaceae (one genus), and Heliconiaceae (lobster claws, one genus); and the 'ginger group' that includes the Costaceae (four genera), Zingiberaceae (gingers, 51 genera), Cannaceae (one genus), and Marantaceae (prayer plants, 31 genera). The largest family is the Zingiberaceae with over 1000 species followed by the second largest family the Marantaceae with about 500 species. The smallest families are Lowiaceae and Cannaceae, each with one genus of approximately 10 species. Six of the eight families (Musaceae, Heliconiaceae, Costaceae, Zingiberaceae, Maran-

---

Keywords: Monocotyledons, phylogeny, DNA, fossils, pollination.

taceae, Cannaceae) have species of significant economic importance as starch sources, spices and condiments, herbal remedies, fiber plants and ornamentals.

The Zingiberales are found in tropical and sub-tropical habitats around the world. The usually large petiolate leaves with a central midrib and transverse venation, and the colourful, bracteate inflorescences characterizing many members of this order make them readily identifiable in the field. Members of three of the families (Zingiberaceae, Costaceae and Marantaceae) are distributed pantropically in Africa, Asia and the Americas. The Strelitziaceae (Africa and South America), Musaceae (Africa and Asia with fossil representatives in the Americas), and Heliconiaceae (American tropics and South Pacific islands) are each found on two continents. Only the Lowiaceae (Southeast Asia) and Cannaceae (neotropics) are restricted primarily to a single land mass. This wide pantropical distribution suggests a Gondwanan origin for the Zingiberales in the mid-Cretaceous over 70 million years ago (mya).

Zingiberales are a dominant component of forest understorey vegetation in the humid tropics around the world. Many species, such as some heliconias, exploit forest gaps and margins, whereas others, such as *Canna* and *Thalia*, inhabit marsh edges and swamps. Although much remains to be studied about fruit and seed dispersal (Horvitz and Schemske, 1986; Bruna, 1999), a remarkable number of pollination mechanisms have evolved in the order, including insect, bird, bat, and lemur pollination (e.g., Schemske, 1981; Kress and Stone, 1993; Kress *et al.*, 1994; Sakai and Inoue, 1999; Sakai *et al.*, 1999; Li *et al.*, 2001). Hummingbirds are particularly dependent on heliconia flowers for nectar in the American tropics (Stiles, 1975). This dependency is exemplified in the Lesser Antilles where sex-specific specialization by hummingbirds on individual species of *Heliconia* has evolved (Temeles *et al.*, 2000). The flowers of the Marantaceae exhibit a very specialized and unique trip mechanism triggered by visiting bees in both the Old and New World tropics (Schemske and Horvitz, 1984; Kennedy, 2000). In contrast, dependence on pollination for reproduction has been lost in a few species now known only from cultivation, such as *Musella lasiocarpa*, a diminutive member of Musaceae grown for animal fodder and used as a medicinal in southern China (Kress and Liu, unpubl.).

Although fossil pollen is unknown for Zingiberales due to the very reduced exine layer, the macrofossil record for gingers is relatively abundant (Rodríguez-de la Rosa and Cevallos-Ferriz,. 1994). The earliest fossils are Zingiberaceae leaves of the Santonian of Wyoming (Cretaceous, > 80 mya) and Musaceae fruits from the Cretaceous of Mexico. Fossil leaves of Cannaceae, Marantaceae, Musaceae, and Heliconiaceae are widely distributed from Texas to Greenland and England and date from the Eocene through the Miocene. Fossil fruits attributed to *Musa* have been collected in

Eocene deposits of India and Oregon whereas fossil gingers are known in Eocene and Miocene deposits of Denmark. These fossil records provide insights into both the current and past distribution of the taxa.

The phylogenetic relationships among the families have been investigated using morphological and molecular characters (Kress, 1990; Kress *et al.*, 2001). There is universal support for the monophyly of the eight families as defined above and the combined morphology and DNA data sets provide a well-supported estimate of phylogenetic relationships among these families with the following topology: Musaceae ((Strelitziaceae, Lowiaceae) (Heliconiaceae ((Zingiberaceae, Costaceae) (Cannaceae, Marantaceae))))). Evidence from branch lengths in the molecular analyses and from the fossil record suggest that the Zingiberales underwent a rapid and extensive radiation in the early- to mid-Cretaceous at which time most extant family lineages had diverged. Since that time the most extensive radiations at the generic and species level have occurred in the two highly specialized families Zingiberaceae and Marantaceae.

A reasonable estimate of the evolutionary relationships among genera within families is now also available, particularly in the large families Zingiberaceae (Searle and Hedderson, 2000; Kress *et al.*, unpubl.) and Marantaceae (Prince and Kress, unpubl.). In addition, phylogenetic relationships within some genera have been recently investigated (e.g. Wood *et al.*, 2000). The results of these studies now allow us to formulate testable hypotheses on the patterns of character evolution and ecological relationships at the family, generic and species levels in the Zingiberales. For example, through a combined study of present day ecology, phylogenetic relationships, and fossil history, an ancient relationship between herbivorous hispine beetles and their ginger hosts has been shown to date back to the late Cretaceous (66 mya) which not only tells us about the history of these two groups of organisms, but about the evolution of plant–animal interactions in general (Wilf *et al.*, 2000).

The exceptional information available on taxonomy, phylogeny, fossil history, and geographic distribution of the Zingiberales paves the way for a coordinated approach to understanding the evolution, ecology, and conservation of this conspicuous group of tropical plants.

References

Bruna, E. M. Seed germination in rainforest fragments. *Nature* **402**, 139 (1999).
Horvitz, C. & Schemske, D. Seed dispersal and environmental heterogeneity in a neotropical herb: A model of population and patch dynamics. in *Frugivores and Seed Dispersal* (eds Estrada, A. & Fleming, T. H.) (W. Junk, Dordrecht, 1986).
Kennedy, H. Diversification in pollination mechanisms in the Marantaceae. in *Monocots – Systematics and Evolution* (eds Wilson, K. L. & Morrison, D. A.) 335–343 (CSIRO Publishing, Collingwood, Australia, 2000).
Kress, W. J. The phylogeny and classification of the Zingiberales. *Ann. Missouri Bot. Gard.* **77**, 698–721 (1990).

Kress, W. J. & Stone, D. E. The morphology and floral biology of *Phenakospermum* (Strelitziaceae), an arborescent herb of the neotropics. *Biotropica* **25,** 290–300 (1993).

Kress, W. J., Schatz, G. E., Andrianifahanana, M. & Morland H. S. Pollination of *Ravenala madagascariensis* (Strelitziaceae) by lemurs: evidence for an archaic coevolutionary system? *Am. J. Bot.* **81,** 542–551 (1994).

Kress, W. J., Linda M. Prince, William J. Hahn & Elizabeth A. Zimmer. Unraveling the evolutionary radiation of the families of the Zingiberales using morphological, molecular, and fossil evidence. *Syst. Biol.* (in press) (2001).

Kubitzki, K. (ed.). *The Families and Genera of Vascular Plants. Vol. IV – Monocotyledons: Alismatanae and Commelinanae (except Gramineae)* 278–293 (Springer, Berlin, 1998).

Li, Q.-J., Xu, Z.-F., Kress, W. J., Xia, Y.-M., Zhang, L., Deng, X.-B., Gao, J.-Y. & Bai, Z.-L. Flexible style that encourages outcrossing. *Nature* **410,** 432 (2001).

Rodríguez-De La Rosa, R. A. & Cevallos-Ferriz, S. R. S. Upper Cretaceous Zingiberalean fruits with *in situ* seeds from southeastern Coahuila, Mexico. *Int. J. Plant Sci.* **155,** 786–805 (1994).

Sakai, S. & Inoue, T. A new pollination system: dung-beetle pollination discovered in *Orchidantha inouei* (Lowiaceae, Zingiberales). *Am. J. Bot.* **86,** 56–61 (1999).

Sakai, S., Kato, M. & Inoue, T. Three pollination guilds and variation in floral characteristics of Bornean gingers (Zingiberaceae and Costaceae). *Am. J. Bot.* 86: 646–658 (1999).

Searle, R. J. & Hedderson, T. A. J. A preliminary phylogeny of the Hedychieae tribe (Zingiberaceae) based on ITS sequences of the nuclear rRNA cistron. in *Monocots – Systematics and Evolution* (eds Wilson, K. L. & Morrison, D. A.) 710–718 (CSIRO Publishing, Collingwood, Australia, 2000).

Schemske, D. W. Floral convergence and pollinator sharing in two bee-pollinated tropical herbs. *Ecology* **62,** 946–954 (1981).

Schemske, D.W. & Horvitz, C. C. Variation among floral visitors in pollination ability: a precondition for mutualism specialization. *Science* **225,** 519–521 (1984).

Stiles, F. G. Ecology, flowering phenology and hummingbird pollination of some Costa Rican *Heliconia* species. *Ecology* **56,** 285–301 (1975).

Temeles, E. J., Pan, I. L., Brennan, J. L. & Horwitt, J. N. Evidence for ecological causation of sexual dimorphism in a hummingbird. *Science* **289,** 441–443 (2000).

Wilf, P., Labandeira, C. C., Kress, W. J., Staines, C. L., Windsor, D. M., Allen, A. L. & Johnson, K. R. Timing the radiations of leaf beetles: Hispines on gingers from latest Cretaceous to Recent. *Science* **289,** 205–348 (2000).

Wood, T. H., Whitten, W. M. & Willams, N. H. Phylogeny of Hedychium and related genera (Zingiberaceae) based on ITS sequence data. *Edinb. J. Bot.* **57,** 261–270 (2000).

Tropical Ecosystems: Structure, Diversity and Human Welfare.
Proceedings of the International Conference on Tropical Ecosystems
K. N. Ganeshaiah, R. Uma Shaanker and K. S. Bawa (eds)
Published by Oxford–IBH, New Delhi. 2001. pp. 351–352.

Specialization on birds vs. ants as dispersal agents for Marantaceae: Why go far if near works?

Carol C. Horvitz

Department of Biology, University of Miami, Coral Gables, FL 33124, USA
e-mail: chorvitz@umiami.ir.miami.edu

The context of this study is an hypothesis that ant-dispersed species are less gap-dependent than bird-dispersed species. I studied six bird-dispersed species and one ant-dispersed species in the family Marantaceae at two wet tropical forests, in the Atlantic and Pacific lowlands of Costa Rica. Four features of seed dispersal and seed ecology were investigated comparatively across plant species and sites: (1) the attractiveness of seeds to ants and birds, (2) the distance of dispersal, (3) disperser assemblages, and (4) the effects of gaps on seeds and seedlings. To address issues 1–3, a set number of seeds (the number naturally available to dispersers) was observed as follows. All ant-seed interactions in the leaf-litter were recorded during 524 trials (total of 1440 seeds) and all bird–seed interactions on the plants were recorded during 495 trials (total of 2279 seeds). To address issue 4, seeds were planted in 204 wire mesh germination boxes (total of 1936 seeds) into gap, understorey and intermediate light environments in the field.

To determine how seed and seedling fitness contributed to overall fitness, population growth rate, and population expansion rate, overall demography was studied by tagging individual plants in study plots and following their

Keywords: Ant-dispersal, bird-dispersal, treefall gaps, demography.

fates and reproductive output at two censuses per year for three years. To determine whether demographic rates of plants were sensitive to light and shade, hemispherical fish-eye photographs were taken over each tagged plant at each census.

Ant–seed and bird–seed studies both showed that: plant species varied in their attractiveness to dispersers; and that disperser assemblage variation across sites led to differences in dispersal success, with a higher probability of dispersal in the Atlantic forest. A difference between ant- and bird-studies was that the distance of seed removal by ants was affected by plant species. In contrast, the distance of seed removal by birds was independent of plant species. Germination and seedling survival varied among plant species and sites in gap-dependence. The bird-dispersed plant species with the most conspicuous display had the highest probability of dispersal by birds and show enhanced germination in gaps compared to other bird-dispersed species, as predicted. However, the ant-dispersed species also had very enhanced germination in gaps, not predicted. Surprisingly, seedling survival of these two species was not very much enhanced in gaps compared to other species. Further analyses combining demography and dispersal are presented to investigate sensitivity of both population growth rate and of expansion rate to particular life history transitions.

Tropical Ecosystems: Structure, Diversity and Human Welfare.
Proceedings of the International Conference on Tropical Ecosystems
K. N. Ganeshaiah, R. Uma Shaanker and K. S. Bawa (eds)
Published by Oxford–IBH, New Delhi. 2001. pp. 353–355.

Natural history of *Zamia portoricensis* Urban (Cycadophyta: Zamiaceae)

Eva N. Dávila* and Ana Arguello-López

Universidad Metropolitana, San Juan, Puerto Rico
*e-mail: mrnorat@coqui.net

Puerto Rico is the smallest and easternmost Greater Antille of the Caribbean. It is a subtropical island with rainy, wet, moist and dry forests. *Zamia portoricensis* Urban is the only endemic cycad of Puerto Rico (Stevenson, 1987). It is a small understorey plant with a subterranean stem. Leaves are compound with 17 to 125 leaflets, which are long and narrow with entire margins, different from all other *Zamia* in Puerto Rico. Female cones are reddish, ovoid, with an acuminate apical tail; megasporophylls are hexagonal with a flat center. They measure up to 13.3 cm long. Male cones are cylindrical and reddish and measure up to 15.7 cm long. Seeds are ovoid and elongated, with a bright red sarcotesta (Eckenwalder, 1980; Jones, 1994).

No information about the ecology, biology or life history of *Z. portoricensis* was known although it was described in 1899. This species is cited as living in serpentine soils at the Susúa Forest, a subtropical moist premontane forest. However, in 1998 we discovered two populations living in limestone soils at the Subtropical Dry forest. Susúa moist forest is located in the central southwestern part of the island at an elevation of 80 to 473 m. The average annual precipitation is 1339 mm and the average annual temperature is 25°C. The dry forest at Guánica is located in the south part of the island at an elevation from 0 to 228 m. The average annual precipitation is 791 mm and

Keywords: *Zamia portoricensis*, Pureto Rico, natural history, cone phenology, cyads.

the average annual temperature is 26°C. Because these populations were found in two different life zones with different soil types, we decided to study and compare the populations at the Susúa moist forest (serpentine soils) and the Guánica dry forest (limestone soils). One hundred adults, in one hectare, were randomly selected in each site.

A morphometric study showed no differences between the populations, although sexual dimorphism was evident: female plants produced longer leaves, more leaves per plant, more leaflets per leaf, and a smaller number of strobili per plant than males. Although males of both populations produced an average of three cones per plant, the cones phenology varied between these populations. Guánica male cones emerged from underground between July and September, during the rainy season. Their average life span was 21 (1999) to 23 (2000) weeks. Anthesis started in October (1999) and September (2000). These cones dehisced between December and February, during the dry season. Similarly, Susúa males cones emerged from underground between July and October, during the rainy season. Their average life span was 19 (1999) to 25 (2000) weeks. Anthesis started in November (1999) and September (2000). These cones dehisced between December and February, during the dry season. Shedding the pollen in the dry season would permit it to be dry and easier to be carried by the pollinator.

There was no difference in female cone phenology when the two populations were compared: between 1999 and 2000 the Susúa females produced an average of 1.5 cones per plant. The cones emerged between August and October, during the rainy season. Seed production and abortions were between January and February, during the dry season. The Guánica females produced an average of 2 cones per plant. They emerged between August and October, as the Susua population. Seed setting and abortions were between December and January, during the dry season and the average life span was 69 weeks.

Only five plantlets were observed at the Guánica dry forest. If a high abortion rate and a low seed setting were found, then this would explain – in part – the low number of plantlets observed. A very interesting fact was that the Guánica population had only 1% of seed setting, and an abortion rate of 99%. In comparison, the Susúa population had 45% seed setting and an abortion rate of 55%, about half of the ovule are aborted. The average number of seeds produced per cone in Guánica was 0.56 while in Susúa it was 21. Female at Guánica produced an average of 2 cones per plant with more megasporophylls per cone. Female at Susúa produced only 1.5 cones per plant. Based on these results we can explain the low number of plantlets observed at the Guánica dry forest due to the high abortion rate and low seed setting in this population. The Guánica *Zamia* population is practically not leaving behind offspring even when the females are producing more cones per plant and more megasporophylls per cone (Eckenwalder, 1980; Tang, 1988).

354

Are these plants at Guánica resource limited? Both soils were physically and chemically analyzed. Both were rather poor in nutrients and the pH was close to neutral, the best pH for nutrient absorption. Therefore, nutrient availability does not seem to be a limiting factor. Cianobionts that live symbiotically in the coralloid roots of *Zamia* fix nitrogen; therefore, these cianobionts might be limiting factors. Guánica is an open canopy forest, and the plants receive plenty of light. Although Guánica is a dry forest, *Zamia* is adapted to these conditions, and then water availability should not be the limiting factor. Is the pollinator present and efficient? We have not found the pollinator. Is a genetic factor affecting the reproduction in this population? Has the population reached its carrying capacity? We discuss some of these questions.

Acknowledgements

We thank Martha Norat, Melody Taveras, Katty Jimenez, Leida Rohena, Xiomara Davila and Michelle Romero for all their help and support. We also thank the MIE project (Model Institute of Excellence) of the NSF at Universidad Metropolitana for the financial support.

References

Stevenson, D. W. Again the West Indian Zamias. *Fairchild Trop. Garden Bull.* **7**, 23–27 (1987).
Jones, D. L.. *Cycads of the World* (Smithsonian Institution Press, 1994).
Tang, W. Seed dispersal in the Cycad *Zamia pumila* in Florida. *Can. J. Bot.* 67 (1988).
Eckenwalder, J. E. Taxonomy of the West Indian Cycads. *J. Arnold Arboretum* **61**, 701–722 (1980).
Eckenwalder, J. E. Dispersal of the West Indian Cycad, *Zamia pumila* L. *Biotropica* **12**, 79–80 (1980).

Ecological and Bio-Geographical Contrasts among Tropical Rainforests with Conservation Implications

Tropical Ecosystems: Structure, Diversity and Human Welfare.
Proceedings of the International Conference on Tropical Ecosystems
K. N. Ganeshaiah, R. Uma Shaanker and K. S. Bawa (eds)
Published by Oxford–IBH, New Delhi. 2001. pp. 359–362.

A comparison of flowering phenology and pollination systems in the tropical forests of southeast Asia and the Neotropics

Shoko Sakai

Graduate School of Human and Environmental Studies, Kyoto University,
Kyoto 606-8501, Japan
e-mail: sakai@sinfo.net

One of the principal characteristics of flowering phenology in tropical forests may be the high diversity partly due to weaker physical constraints on the schedules of biological activities. Gentry (1974) was among the first to indicate the existence of high diversity in phenology among species *within* a tropical forest, and to discuss its significance in relation to pollination systems. Recently, Sakai (in press) also suggested diversity in phenology *among* tropical forest communities. Since favourable flowering patterns for pollinators are different depending on pollinator groups, flowering phenology and pollination systems are thought to be correlated with each other. In this paper pollination systems of neotropical and SE Asian forests are compared in terms of the strikingly contrasting flowering phenology of the two regions based on limited available data.

One prominent theme in tropical community studies is the periodicity or regularity of biological activities, because in the tropics annual changes in mean temperature and photoperiod are very small. Pioneer studies mostly conducted in Neotropics suggested that in tropics periodic change in rainfall caused by movements of the intertropical convergence zone

Keywords: Bee, beetle, general flowering, Lambir, La Selva, phenology.

often plays an important role as proximate and ultimate factors for plant phenology. Dry seasons with annual cycles occur in most tropical regions, and many studies have shown the existence of a correlation between tropical plant phenology and rainfall. Most seasonal tropical forest communities studied so far show flowering and fruiting peaks near the end of dry season (reviewed in van Schaik *et al.*, 1993). The pattern may be due to high insolation and photosynthesis in dry seasons, and/or to enhance germination and seedling survival by adjusting fruiting to precede the beginning of the wet season.

In contrast with neotropical forests, annual cycle in rainfall is weak in a large portion of Asian tropical forests from Sumatra to the Philippines. This weakness of seasonality is largely due to the fact that both the northeast monsoon in the northern hemisphere summer and the southwest monsoon in winter bring predominantly warm, humid air masses and precipitation to this region. Interestingly, the phenomenon with multiyear intervals, called general flowering (GF, or mass flowering), is known from lowland dipterocarp forests in this region (Ashton *et al.*, 1988). During GF, many plant species including most of dipterocarp species flower sequentially for several months. Few plant species flower during non-GF periods (Sakai *et al.*, 1999).

Dipterocarp forests with GF phenomenon show much lower intensity of plant reproduction than neotropical forests. In wet neotropical forests, around 15% of trees on average are flowering throughout the year. In contrast, the proportion of plants flowering during non-GF periods in dipterocarp forests is very low, from zero to 3%. The maximum is about

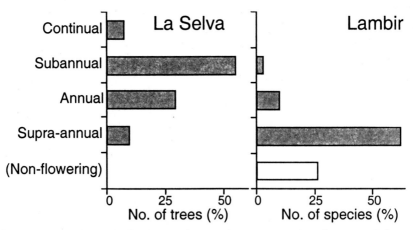

Figure 1. The proportion of sub-annual, annual, supra-annual and continual flowering types among trees at La Selva, Costa Rica based on 12-year observation (254 trees, Newstrom *et al.* 1994) and Lambir based on 43 months observation (187 tree species, Sakai *et al.* 1999). Plants at Lambir which did not flower during the study period are tentatively categorized into 'non-flowering'. From Sakai (in press).

20% even in a GF, which occurs quite irregularly only once in several years. More than half of plants at La Selva flower twice or more in a year, while flowering intervals of most trees are much longer than one year at Lambir (Figure 1). The different flowering phenology may be related to differences in pollination biology between the two sites.

Pollination biology at the community level in neotropical forests has been most extensively studied at La Selva, Costa Rica (Kress *et al.*, 1994) and in Asian tropics at Lambir, Borneo, Malaysia (Momose *et al.*, 1998). The discussion below is mostly based on studies from the two sites.

At La Selva, most canopy and emergent trees are pollinated by solitary and eusocial bees of many different sizes. Although bees are also important pollinators at Lambir, most large trees are pollinated by small, social bees (honeybees and stingless bees) (Momose *et al.*, 1998). At Lambir, few large trees are pollinated by large bees. Nomadic giant honeybees immigrate into the forest at the start of GF. Resident honeybees and stingless bees store nectar and pollen, and stabilize fluctuation in resource availability. Highly social system may be important to survive in forests with unpredictable changes in resource level.

The other abundant pollination system in the upper strata at Lambir is beetle pollination. Many dipterocarp trees are pollinated by weevils and leaf beetles. Interestingly, some pollinator beetles are herbivores feeding on new leaves of dipterocarp trees during non-GF periods without dipterocarp flowers. In GF, however, they were observed foraging on petals of the nocturnal flowers of many *Shorea* species (Dipterocarpaceae), and contribute to pollination. An increase of floral resources may cause their feeding niche to shift. At La Selva beetle pollination was found for only one species in the canopy layer.

In the Neotropics, many species of humming-birds and euglossine bees coexist by partitioning floral resources, and serve as important long-distance pollinators. The Asian counterparts of the American hummingbirds and euglossine bees are spiderhunters, sunbirds, and *Amegilla* bees. Interestingly, the diversity of nectarivorous birds and *Amegilla* bees found in the Asian tropics is much lower than the diversity of hummingbirds and euglossine bees found in the Neotropics. The proportion of plants they pollinate is smaller in Asian tropics than in Neotropics. Other categories of long-distance pollinators including mammals, lepidopterans, and large bees are also rare at Lambir.

Apparently, long-distance and specific pollinators have less important roles in the lowland dipterocarp forest of Lambir than in the forest of La Selva, in spite of higher plant species diversity of the former. Long-distance pollinators may require a continuous supply of rich resources, because their costs for body maintenance and foraging are high. Irregular and ephemeral floral resources in lowland dipterocarp forests are inadequate for their survival.

To conclude the comparison of phenology and pollinators systems found at La Selva and Lambir suggests that at Lambir pollinators are dominated by various generalist flower visitors. Leaf beetles pollinating *Shorea* spp. feed both leaves and petals depending on their availability. Giant honeybees have wide range of potential foraging area and migrate to change their nesting site depending on flowering intensity. Long-tongued solitary bees and long-billed bird, moths and bats play limited roles. The differences can be explained by different flowering phenology in the two forests; general flowering phenomenon at Lambir and dominance of sub-annual flowering trees at La Selva. Out-breeding in the species-rich forest of Lambir may be mostly achieved by attracting large amount of flower visitors with massive display at long intervals.

Acknowledgement

We thank Dr H. S. Lee, Dr A. A. Hamid, and the staff of the Forest Department Sarawak for logistical help with the study. I also thank the members of the Canopy Biology Sarawak for valuable discussions and suggestions. This project was partly supported in part by Grants-in-Aid of the Japanese Ministry of Education, Science and Culture (numbers 04041067, 06041013, 09NP1501 and 10041169) and by JSPS Research Fellowship for Young Scientist for SS.

References

Ashton, P. S., Givnish, T. J. & Appanah, S. Staggered flowering in the Dipterocarpaceae: new insights into floral induction and the evolution of mast fruiting in the aseasonal tropics. *Am. Nat.* **132,** 44–66 (1988).

Gentry, A. H. Flowering phenology and diversity in tropical Bignoniaceae. *Biotropica* **6,** 64–68 (1974).

Kress W. J. & Beach, H. Flowering plant reproductive systems. in *La Selva: Ecology and Natural History of a Neotropical Rain Forest* (eds McDade, L. A., Bawa, K. S., Hespenheide, H. A. & Hartshorn, G. S.) 161–182 (The University of Chicago Press, Chicago and London, 1994).

Momose, K., Yumoto, T., Nagamitsu, T., Kato, M., Nagamasu, H., Sakai, S., Harrison, R. D., Itioka, T., Hamid, A. A. & Inoue, T. Pollination biology in a lowland dipterocarp forest in Sarawak, Malaysia I: characteristics of the plant-pollinator community in a lowland dipterocarp forest. *Am. J. Bot.* **85,** 1477–1501 (1998).

Newstrom, L. E, Frankie, G. W., Baker, H. G. & Colwell, R. K. Diversity of long-term flowering patterns. in *La Selva: Elogy and Natural History of a Neotropical Rain Forest* (eds McDade, L. A., Bawa, K. S., Hespenheide, H. A. & Hartshorn, G. S.) 142–160 (The University of Chicago Press, Chicago and London, 1994).

Sakai, S. Phenological diversity in tropical forests. Population Ecology (in press).

Sakai, S., Momose, K., Yumoto, T., Nagamitsu, T., Nagamasu, H., Hamid, A. A., Nakashizuka, T. & Inoue, T. Plant reproductive phenology over four years including an episode of general flowering in a lowland dipterocarp forest, Sarawak, Malaysia. *Am. J. Bot.* **86,** 1414–1436 (1999).

van Schaik, C. P., Terborgh, J. W. & Wright, S. J. The phenology of tropical forests: adaptive significance and consequences for primary consumers. *Annu. Rev. Ecol. Syst.* **24,** 353–377 (1993).

Tropical Ecosystems: Structure, Diversity and Human Welfare.
Proceedings of the International Conference on Tropical Ecosystems
K. N. Ganeshaiah, R. Uma Shaanker and K. S. Bawa (eds)
Published by Oxford–IBH, New Delhi. 2001. pp. 363–365.

Pollination and fruit dispersal in the wet forests of the southern Western Ghats

T. Ganesh*[‡], M. S. Devy* and P. Davidar[†]

*Ashoka Trust for Research in Ecology and the Environment, 659, 5th A Main Road,
Hebbal, Bangalore 560 024, India
[†]Salim Ali School of Ecology, Pondicherry University, Pondicherry 605 014, India
[‡]e-mail: tganesh@atree.org

Pollination and seed dispersal can play an important role in the functioning of tropical ecosystems and the maintenance of diversity in them. Wet forests in tropics are known to have exceptionally high diversity of pollination and dispersal mechanisms that are complex and can vary across continents. Identifying modes of pollination and dispersal can, to some extent, delineate this complexity and help in better understanding of these systems (Ibarra-Manriquez and Oyama, 1992). Only recently a detailed analysis of the pollination and dispersal modes of the wet forests of Western Ghats, a biodiversity hotspot, has been carried out (Ganesh and Davidar, 2001; Devy, 1998). This study documents the pollination and seed dispersal modes of tree species in a mid-elevation wet forest at Kakachi in the southern Western Ghats. A total of 89 flowering species and 82 fruiting species were observed. Flowers were watched for visitors from the commencement of anthesis. Similarly fruiting trees were watched for visitors during ripe fruit stage and also frugivore identity was inferred from fruits and seeds in droppings and scats. Nocturnal observations were also done for visitors of several species. The data were obtained over a span of seven years and are based on direct observations of visitors to flowers and fruits. Many of these observations were recorded by ascending the canopy of trees using tree ladders.

Keywords: Pollination, fruit dispersal, Western Ghats, wet forest.

A total of 12 pollination guilds were identified. This ranged from social bees, beetles, carrion flies and mammals. Social bees and beetles pollinated 17 and 18% of the plant species respectively and diverse small and big insects accounted for 25% of the species. Uniquely there were no bird-pollinated species. About 75% of the species were specialized to a single order of pollinators such as bees, beetles, or moths. Pollinators from diverse orders pollinate the other 25% of the tree species. Majority of the tree species are pollinated by less than 4 species of pollinators and a high proportion are pollinated by just one species (Figure 1). Six species of birds and 5 species of mammals were involved in dispersal or predation of seeds. Bird-dispersed species were the most common ones (60%) and 9% were dispersed exclusively by large birds such as the Imperial pigeons (*Ducula badia*) followed by mammal-dispersed species (26%); primates were less important than bats and civets. Fifteen per cent of the species had no apparent adaptation for biotic dispersal (mechanically dispersed) except one wind-dispersed species (Table 1).

Kakachi was characterised by a high diversity of pollination modes but with a very low number of pollinator species within each guild compared to other lowland sites. Global comparison with other wet forest sites showed that diversity and specialised pollination modes observed in Kakachi, bore closer resemblance to other lowland than montane sites described so far. However, the number of pollinators involved in pollination was comparable with the montane sites. We examined the consequences that might have led to selection of the observed pollination modes in Kakachi. Multiple year observations on the pollinator fauna at Kakachi showed no perceptible shift in pollinator order or species between years.

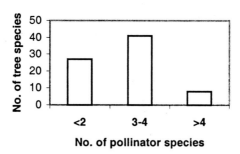

Figure 1. Frequency distribution of pollinator species.

Table 1. Number of species under the three dispersal categories at Kakachi. Species under closed canopy and edge/gap habitats are also shown. Percentages in parentheses.

Species	All birds	Mammal	Mechanically dispersed	Large birds
Total	49 (60)	21 (26)	12 (15)	7 (9)
Closed canopy	30 (61)	12 (57)	9 (75)	6 (86)
Edge/gap	19 (32)	9 (35)	3 (21)	1 (12)

The disperser assemblage was less diverse and species poor at Kakachi when compared with sites in Central America and South-east Asia. There could be many reasons for this, but overall paucity of avian dispersers might be due to lack of year-round availability of fruit resources which could eliminate sedentary frugivores from the area, low density of species producing fleshy fruits, and low diversity of bird dispersed *Ficus* species in the area which are more common at lower elevations in the same reserve and elsewhere where avian frugivore richness is high. It therefore appears that seed dispersal of tree species in Kakachi, though not highly specialized, is dependent on very few vectors. This is particularly important for large bird-dispersed species as elimination of these frugivores could affect the dispersal and regeneration of the dependent tree species. We discuss the conservation implications of these results.

References

Devy, M. S. Pollination of canopy and sub-canopy trees by social bees in a wet forest of South Western Ghats. Doctoral dissertation work, Madras University, India (1998).
Ganesh, T. & Davidar, P. Dispersal modes of tree species in the wet forests of southern Western Ghats. *Curr. Sci.* **80**, 394–399 (2001).
Ibarra-Maríquez, I. & Oyama, K. Ecological correlates of reproductive traits of the Mexican forests trees. *Am. J. Bot.* **79**, 383–394 (1992).

Tropical Ecosystems: Structure, Diversity and Human Welfare.
Proceedings of the International Conference on Tropical Ecosystems
K. N. Ganeshaiah, R. Uma Shaanker and K. S. Bawa (eds)
Published by Oxford–IBH, New Delhi. 2001. pp. 366–369.

Seed dispersal in tropical rainforests: Inter-regional differences and their implications for conservation

Richard T. Corlett

Department of Ecology & Biodiversity, University of Hong Kong, Pokfulam Road,
Hong Kong, China
e-mail: corlett@hkucc.hku.hk

Most seeds in Neotropical, African and Asian rain forests are dispersed by birds, bats or primates, with small carnivores, large terrestrial herbivores and rodents also significant for some species. This superficial similarity at the ordinal level, however, masks large differences at the family level in the major dispersal agents, particularly between New and Old World rain forests, which show almost no overlap. The other major rainforest areas, in Madagascar and New Guinea, differ also at the ordinal level, with both areas lacking large terrestrial herbivores and New Guinea also lacking primates and carnivores. Fruits are thus eaten and seeds dispersed in each major rain forest region by animals which often differ in size, behaviour, sensory capabilities and/or fruit acquisition and processing techniques from those in other regions.

The major groups of fruit-eating birds in Asia and Africa usually take fruits from a perch and swallow them whole. Fruit consumption and seed dispersal by these birds are therefore limited principally by their maximum gape width, which is generally larger in bigger birds. Seed-dispersing birds in Asian rain forests range in size from 5 g, flowerpeckers (*Dicaeum*) to 3 kg,

Keywords: Degraded landscapes, frugivory, seed dispersal, succession, tropical forests.

hornbills (Bucerotidae: the largest flying frugivores), and the largest fruits that can be swallowed whole range from < 8 mm to > 30 mm in diameter (Corlett, 1998). African rain forests have a similar disperser avifauna, but a somewhat smaller range of bird sizes. The New Guinea avifauna includes flowerpeckers and a single hornbill, but also the world's largest frugivorous birds, the flightless cassowaries (*Casuarius*), the biggest of which can swallow whole fruits up to 7 cm in diameter (Stocker and Irvine, 1983). Some birds of paradise (*Paradisaeini*), another endemic group, have an apparently unique ability to break open the woody capsules of some Myristicaceae and Meliaceae to get at the arillate seeds inside (Beehler and Dumbacher, 1996).

Many seeds in Neotropical rain forests are dispersed by endemic bird families with fruit acquisition and processing behaviours that are rarely seen in the Old World. Most fruit-eating suboscines, including manakins, cotingas and tyrant flycatchers, have wide gapes, take fruits on the wing, and swallow them whole (Levey *et al.*, 1994). Tanagers and related groups of nine-primaried oscines have narrow gapes, take fruits while perching, and crush them in the bill, squeezing out all but the smallest seeds, before swallowing the pulp. These 'mashers' drop most large seeds under the parent plant, but are important dispersal agents for species with seeds < 2 mm in length, which they cannot separate from the pulp.

Frugivorous bats of Old and New World rain forests belong to different suborders with separate evolutionary histories. Old World fruit bats (Megachiroptera) locate fruits by a combination of smell and vision, usually land at least briefly when harvesting them, and swallow only the juice and the smallest seeds, dropping larger seeds under a nearby feeding roost (Corlett, 1998). *Pteropus* flying foxes in Madagascar, Asia and New Guinea are the largest flying mammals (< 1500 g) and can carry > 200 g, but are more likely than smaller species to process fruits in the fruiting tree and drop the seed underneath. New World fruit bats (Microchiroptera) are more varied in everything but size, the largest species being < 100 g. They locate fruits by a combination of smell, echolocation and vision, often hover to take fruits, and include both seed swallowing and seed spitting species (Fleming, 1986).

Frugivorous Old World monkeys (Cercopithecinae) are medium to large (1–30 kg) primates, with human-like trichromatic colour vision. They usually swallow only small seeds and drop or spit most larger ones, often after transport in their extensible cheek pouches. Many Old World monkeys are more or less terrestrial and most frugivorous species will cross open areas on the ground. Old World rain forests also support apes, including the largest primates (gorilla, < 200 kg) and largest arboreal primates (orangutan, < 85 kg) as well as the medium-sized and exclusively Asian gibbons. Apes are trichromatic, like Old World Monkeys, but swallow the seeds in most ripe fruits they eat. New World monkeys are a diverse group of small to medium-sized (0.1–10 kg) primates, mostly with dichromatic colour vision, that typically swallow even large seeds, and are unwilling to cross open areas. Prosimians are present in Asian and African rain forests but not important in seed dispersal. In Madagascar, the prosimian lemurs are the only primates and, although the largest and most frugivorous species extant weighs only 3–4 kg, far larger species (< 200 kg) existed until recently. Frugivorous lemurs swallow and defecate many seeds (Birkinshaw, 1999).

The most frugivorous Carnivora in Old World and New World forests – the Viverridae (civets) and Procyonidae, respectively – are at least superficially very similar in sizes (< 10 kg), diets and behaviour. Old World rain forests support megaherbivores (elephants in Asia and Africa, rhinoceroses in Asia) capable of dispersing seeds from the largest fruits in the forest. In Neotropical rain forests, the smaller (< 400 kg) tapirs are now the only large terrestrial herbivores which disperse seeds, although there were megaherbivores in the Pleistocene. Seed dispersal by rodents has received very little attention in the Old World, but there is now evidence that both murid rodents (in Asian and Australian rain forests) and ground squirrels (in Asian and African rain forests) do scatter-hoard seeds (Forget and Vander Wall, 2001). Scatter hoarding by New World rodents in the endemic caviomorph families Dasyproctidae and Echimyidae has been shown to be very important for secondary seed dispersal in Neotropical rain forests.

What are the ecological and evolutionary consequences of these differences in disperser fauna between rain forest regions? Unfortunately, the necessary inter-regional comparisons with standardised methodology have not yet been made. There is, however, evidence that Neotropical fruits are generally smaller than elsewhere, which is consistent with the smaller body sizes of the disperser fauna (Mack, 1993). There are also striking differences in the seed sizes of the common woody pioneers, with those in the Neotropics typically having much smaller seeds than those in Africa and Asia (Corlett, 2001). This may reflect the tendency of the major bird and bat dispersers of pioneers in the Neotropics to swallow only very small seeds, while most Old World pioneers are dispersed by birds that swallow fruits whole. The data needed to compare other fruit characteristics, such as colour, presentation and chemical composition, are not yet available, except in anecdotal form.

Seed dispersal is a key process in the persistence of plant species in disturbed and fragmented forests, and in forest recovery on logged or cultivated sites (Corlett, 2001), so inter-regional differences in the disperser fauna may have major implications for conservation in degraded tropical landscapes. Dispersers of large seeds seem, in general, more willing to enter disturbed and open habitats in Asian and African rain forest regions than in the Neotropics, which would be expected to affect the rate at which large-seeded species invade secondary vegetation. Conversely, the large size of many fruits and seeds in Old World rain forests may make them particularly vulnerable to hunting pressures which selectively remove the largest dispersal agents (Bennett and Robinson, 2000). Although the evidence for significant conservation consequences of differences in the disperser fauna is, at best, anecdotal, it is sufficient to warn against the uncritical extrapolation of results from one region to another. The same warning should probably apply to all biologically-mediated processes in tropical rainforests.

Acknowledgements

This paper has benefitted greatly from discussions with Nate Dominy, Michael Leven, Doug Levey, Peter Lucas, Richard Primack and many other people.

References

Beehler, B. M. & Dumbacher, J. P. More examples of fruiting trees visited predominantly by birds of paradise. *Emu* **96**, 81–88 (1996).

Bennett, E. L. & Robinson, J. G. Hunting for sustainability: the start of a synthesis. in *Hunting for Sustainability in Tropical Forests* (eds Robinson, J. G. & Bennett, E. L.) 499–520 (Columbia University Press, New York, 2000).

Birkinshaw, C. R. The importance of the Black Lemur (*Eulemur macaco*) for seed dispersal in Lokobe Forest, Nosy Be. in *New Directions in Lemur Studies* (eds Rakotosamimanana, B., Rasamimanana, H., Ganzhorn, J. U. & Goodman, S. M.) 189–199 (Kluwer Academic, New York, 1999).

Corlett, R. T. Frugivory and seed dispersal by vertebrates in the Oriental (Indomalayan) Region. *Biol. Rev.* **73**, 413–448 (1998).

Corlett, R. T. Frugivory and seed dispersal in degraded tropical East Asian landscapes. in *Seed Dispersal and Frugivory: Ecology, Evolution and Conservation* (eds Levey, D. J., Silva, W. R. & Galetti, M.) (CABI Publishing, Wallingford, Oxfordshire, UK, 2001).

Fleming, T. H. Opportunism versus specialization: the evolution of feeding strategies in frugivorous bats. in *Frugivores and Seed Dispersal* (eds Estrada, A. & Fleming, T. H.) 105–118 (Dr. W. Junk Publishers, Dordrecht, Netherlands, 1986).

Forget, P. M. & Vander Wall, S. B. Scatter-hoarding rodents and marsupials: convergent evolution on diverging continents. *TREE* **16**, 65–67 (2001).

Levey, D. J., Moermond, T. C. & Denslow, J. S. Frugivory: an overview. in *La Selva: Ecology and Natural History of a Neotropical Rain Forest* (eds McDade, L. A., Bawa, K. S., Hespenheide, H. A. & Hartshorn, G. S.) 287–294 (University of Chicago Press, Chicago, 1994).

Mack, A. L. The sizes of vertebrate-dispersed fruits: A neotropical-palaeotropical comparison. *Am. Nat.* **142**, 840–856 (1993).

Stocker, G. C. & Irvine, A. K. Seed dispersal by cassowaries (*Casuarius casuarius*) in North Queensland's rainforests. *Biotropica* **15**, 170–176 (1983).

Tropical Ecosystems: Structure, Diversity and Human Welfare.
Proceedings of the International Conference on Tropical Ecosystems
K. N. Ganeshaiah, R. Uma Shaanker and K. S. Bawa (eds)
Published by Oxford–IBH, New Delhi. 2001. pp. 370–371.

Many tropical forests: An ecological, biogeographical and conservation comparison

Richard B. Primack[*,†] *and Richard T. Corlett***

*Department of Biology, Boston University, Boston, MA, USA
**University of Hong Kong
[†]e-mail: primack@bio.bu.edu

There are five large regions of tropical rain forests: three large ones in the Neotropics, Africa and Southeast Asia, and two smaller areas in Madagascar and New Guinea. Each of these regions has a distinct biogeographic history, unique features of its flora and fauna, special features to its climate, and a special structure to its community organization. Further, each region has had a different relationship to human impact, and faces different threats from human activities at the present time. In order to better study and conserve tropical rain forests, it is important to appreciate these differences. Even though the differences among these forests are known to specialists in particular groups, few attempts have been made to synthesize these differences into a comprehensive view of tropical forests. Such a synthesis can lead to valuable cross-continental comparative studies that highlight unique features of each area in the world. As one example, the plant communities of each region have both similarities in terms of many shared families, such as the Lauraceae, Myristicaceae and Moraceae, but also major differences as well. For example, Asian forests are dominated by numerous

Keywords: Rain forests, biogeography, South America, Africa, Southeast Asia, dipterocarps, bromeliads, primates, bird communities.

tree species in the dipterocarp family, which flower and fruit on three to seven year cycles. As a result, these forests have relatively little fruit during other periods leading to lower populations of insects, birds and other animals, and selection for large size in bird and mammal species to be able to survive for long periods with low food supply. American forests have large numbers of bromeliad epiphytes, and epiphytes generally, resulting in an abundance of water and food sources in the forest canopy, contributing to the abundance of insect, bird and other animal life in the tree canopy. African forests are typically lower in rainfall than American and Asian rain forests, and there is a much lower diversity of plant species, particularly epiphytes, apparently because past episodes of drying have caused the extinction of vulnerable species. Among the animal communities, South American forests are richest in species, due to a large extent because the American rain forest is largest in area. The African and Asian rain forest are most similar in species due to their periodic connections via land bridges. Madagascar and New Guinea have been isolated from the main rain forest regions and have developed their own unique rain forest faunas, lemurs most notably in Madagascar and tree kangaroos, cassowaries, birds of paradise and bower birds in New Guinea. Conservation policies informed by these large scale differences will be more effective in protecting what remains of these valuable biodiversity and exciting public opinion. Tropical biologists could contribute to these efforts by undertaking comparisons among rain forest regions. Does careful comparative study truly support the impression of field biologists that the abundance of animals in general and insects, birds and herps in particular is lower in Asian forests than Neotropical forests? It would be of particular value to examine the impact of human activities and invasive species on rain forest communities. What will be the impact of introduced honey bees on the native insect communities of the Amazonian rain forest? What will be the impact on New Guiana birds of the introduced macaque? Such questions are of both scientific value and immediate practical concern.

Tropical Ecosystems: Structure, Diversity and Human Welfare.
Proceedings of the International Conference on Tropical Ecosystems
K. N. Ganeshaiah, R. Uma Shaanker and K. S. Bawa (eds)
Published by Oxford–IBH, New Delhi. 2001. pp. 372–375.

Biodiversity survey of arthropods from Australia to Borneo: Patterns and processes

R. L. Kitching

Australian School of Environmental Studies, Griffith University, Brisbane,
Qld 4111, Australia.
e-mail: r.kitching@mailbox.gu.edu.au

The establishment of a series of fully mensurated 50 ha plots in the rainforest regions of the world has contributed to our understanding of vegetation dynamics in these most complex of all terrestrial ecosystems. We have paralleled this in recent years by establishing a set of one-hectare plots in which we have carried out intensive survey of arthropods on a strictly comparative basis (Kitching *et al.*, 2000). Our work, encompassing nine sites along a latitudinal transect from south-east Queensland to northern Borneo is now being mirrored in many additional sites in the Pacific Rim 'green belt' from Siberia to New Zealand as a part of the IBOY program.

At each focal plot we have made full inventories, measurements and maps of all trees > 5 cm dbh. This serves to link our work to the botanical work of others as well as providing us with an underpinning tool for the subsequent explanation of arthropod diversity. Our arthropod surveys are based on a set of trapping methods each designed to remove any human aptitude dimension from the size and composition of our catches. Recognizing that there are many habitat components supporting the arthropod assemblage in any forest site and, further, that no trap is without bias, we have used seven

Keywords: Arthropods, biodiversity, rainforests, Australia, Borneo.

survey techniques. We extracted leaf-litter (focussing on the interstitial and leaf faunae of the forest floor) and established pitfall traps of two different dimensions (targeting the epigeic fauna). We used yellow pan traps (aimed at the aerial plankton and the cursorial fraction of the forest floor fauna) and Malaise traps at both ground level and suspended in frames in the canopy (sampling the free flying or wind-borne faunae in each location). The middle and high canopy faunae were targeted using canopy pyrethrum knockdown. We used pyrethrum aerosols to spray the bark of selected trees (targeting the specialized fauna of tree trunks) and, finally we light trapped at both ground level and in the canopy (aimed at free-flying Lepidoptera and Coleoptera). We sampled in the late wet season in each case (except in the perhumid climate of Brunei) working, in each case over a two to five week periods.

Our study sites (Figure 1) are located in reserved rain forest areas from the subtropical rain forest of south-east Queensland to Papua New Guinea to coastal northern Borneo. Their altitudinal range spans sea level to about 1800 m.

Initially all arthropods sampled were sorted to Order. Selected taxa were then sorted to family and species, generally reflecting available expertise, time and money. We have variously targeted macro-Lepidoptera (particularly in studies of the impact of disturbance), the Coleoptera (especially for guild analyses), the Collembola (as key decomposers in ground and perched litter), the ants (as a taxonomically amenable ubiquitous group) and the Diptera (as a major non-phytophagous group).

We demonstrated (unsurprisingly) clear differences in the ordinal profiles generated by different sampling methods. In an analysis of the results

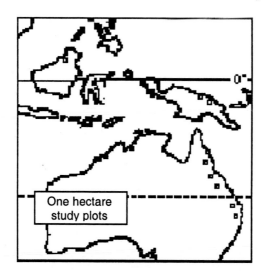

Figure 1. Location of focal one-hectare study sites.

from our first four sites we showed that pyrethrum knockdown, bark sprays and litter extractions gave maximal between-site discrimination and that focus on Collembola, hemipteroids and ants also optimized the identification of differences (Kitching *et al.*, in press 1). These results will assist in the design of future 'rapid biodiversity assessment' (RBA) protocols.

We replicated each of our sampling methods *within* each one hectare focal plot. In testing hypotheses about the impact of geographical location (latitude, altitude) on the composition of our catches our approach may or may not be pseudoreplicated depending on what an 'appropriate' scale of sampling is determined to be for arthropod assemblages. At six of our locations we established two additional quarter hectare plots and, using litter samples and yellow pan catches we ran parallel analyses on samples within the focal hectare, samples from each quarter ha plus a quarter hectare nested within the focal hectare, and randomly assembled sub-sets of 5 trapping locations selected from among the total set from all plots at a location. We showed clearly that, except in one instance, where we had *a priori* reasons to expect differences associated with a particular quarter hectare, there was consistent identification of significant location to location differences, regardless of the scale of sampling – leaving us to conclude that sampling only within the one-hectare plots provides an acceptable measure of geographical change at the continental scale.

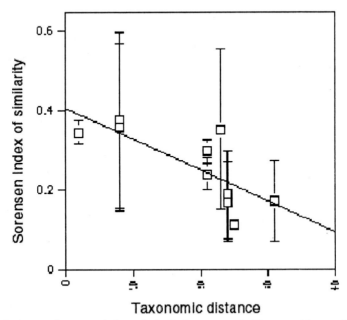

Figure 2. Sorensen index of similarity among bettle assemblages plotted against the taxonomic distance between the pairs of tree species (see Kitching *et al.*, in press 2).

In work derived from these baseline surveys we have shown for beetle assemblages that the degree of similarity between assemblages from different tree crowns reflects the taxonomic distance between the species of trees being sampled (Figure 2).

In a separate study based on these surveys which focussed on Lepidoptera, we have demonstrated dramatically different faunas exist in uncleared, regrowth and freshly cleared forest patches. Differences are reflected very clearly at the family/sub-family level. A small subset of readily recognized families make an excellent 'predictor set' of the level of disturbance.

We anticipate that the data derived from our focal surveys would be useful in not only investigating the pattern of diversity across the landscape but also in understanding the underlying processes. In addition to the work already referred to on the role of disturbance and basic animal/plant interactions, we are currently addressing the impact of flowering cycles in the canopy; on the relationships between gap dynamics and arthropod diversity; and the role of fungi in determining arthropod diversity.

References

Kitching, R. L., Vickerman, G., Laidlaw, M. & Hurley K. *The Comparative Assessment of Arthropod and Tree Biodiversity in Old-World Rainforests: The Rainforest CRC/Earthwatch Protocol Manual*, CRC for Tropical Rainforest Ecology & Management, Technical Report, Cairns, pp. 70 (2000).

Kitching, R. L., Li, D. & Stork, N. E. Assessing biodiversity 'sampling packages': How similar are arthropod assemblages in different tropical rainforests? *Biodiv. Conserv.* (in press 1).

Kitching, R. L., Hurley, K. M. & Thalib, L. Tree relatedness and the similarity of canopy insect assemblages: pushing the limits? in *Arthropods of Tropical Forests* (eds Y. Basset *et al.*), (Cambridge University Press, Cambridge, in press 2).

Tropical Ecosystems: Structure, Diversity and Human Welfare.
Proceedings of the International Conference on Tropical Ecosystems
K. N. Ganeshaiah, R. Uma Shaanker and K. S. Bawa (eds)
Published by Oxford–IBH, New Delhi. 2001. pp. 376–377.

Role of seed-dispersal by frugivorous bats in the landscape-level population dynamics of a large-seeded timber tree of the Neotropics

Rachel T. King

Department of Biology, University of Miami, PO Box 249118, Coral Gables,
FL 33124, USA
e-mail: rachel@bio.miami.edu

In lowland neotropical rainforests, river meanders generate a spatially explicit dynamic landscape of forest patches of differing successional states and elevations (Kalliola *et al.*, 1991), in contrast to the more well-known, time-dependent dynamic landscape of forest patches generated by treefall gap-phase regeneration (Horvitz and Schemske, 1986). Bats may be one of the most important modes of dispersal for plants requiring specific environments in the river meander habitat mosaic (Foster *et al.*, 1986). Many small-seeded successional species are dispersed by frugivorous phyllostomid bats (Charles-Dominique, 1986) that have been found to evenly distribute ingested seeds in open habitats and gap centers (Charles-Dominique, 1986; Gorchov *et al.*, 1993). For large seeds that are carried by bats to feeding roosts (Morrison, 1980) rather than swallowed, the dispersal pattern is likely to be very different. Large seeds that are not ingested are dropped under these roosts (Fleming and Heithaus, 1981) which may be especially suitable sites for recruitment for shade- and flood-tolerant tree species. For these species,

Keywords: Seed dispersal, demography, *Calophyllum brasiliense*, frugivorous bats, neotropics.

the benefit of dispersal may be more of directed dispersal (*sensu*: Howe and Smallwood, 1982) than colonization of open areas.

To investigate the dispersal benefit provided by bats, I compared the pattern of feeding roost use to sites of better recruitment, growth and survival of a large-seeded bat dispersed tree at Cocha Cashu biological station in Manu National Park, Peru. *Calophyllum brasiliense* is a timber tree found mostly in seasonally flooded depressions of mid-successional floodplain forest of Central and South America. At Cocha Cashu, *C. brasiliense* seeds are dispersed by large frugivorous bats (*Artibeus lituratus*, *A. jamaicensis* and *Phyllostomus elongatus*) that carry the fruits to feeding roosts away from the parent tree. *C. brasiliense* seedling survival and growth, as well as growth of sapling and juvenile stages were highest in early successional habitat, while trees large enough to produce fruit were only found in mid-successional and mature forest. During the fruiting season of *C. brasiliense*, (Jan 1–Mar 20, 2000), I found over 450 feeding roosts containing up to five species of large seeds, including *C. brasiliense*. Feeding roost density was higher in mid-successional than in early successional forest, reflecting higher fruit availability in that habitat. However, roost densities were similar between early succession and mature forest, even though fruiting trees are not found in the early successional habitat, implying that dispersal may be biased away from mature forest and toward earlier succession. If bats do tend to fly towards the river (and therefore, towards earlier successional forest) then they may be directionally dispersing seeds to the habitat where they do best.

References

Charles-Dominique, P. Inter-relations between frugivorous vertebrates and pioneer plants: Cecropia, birds and bats in French Guyana. in *Frugivores and Seed Dispersal* (eds Estrada, A. & Fleming, T. H.) 119–135 (Dr W. Junk Publishers, Dordrecht, 1986).

Fleming, T. H. & Heithaus, E. R. Frugivorous bats, seed shadows, and the structure of tropical forests. Reproductive Botany, *Biotropica 13 (Supplement)* 45–53 (1981).

Foster, R., Arce, B. J. & Wachter, T. S. Dispersal and the sequential plant communities in Amazonian Peru floodplain. in *Frugivores and Seed Dispersal* (eds Estrada, A. & Fleming, T. H.) 357–370 (Dr W. Junk Publishers, Dordrecht, 1986).

Gorchov, D. L., Cornejo, F., Ascorra, C. & Jaramillo, M. The role of seed dispersal in the natural regeneration of rain forest after strip-cutting in the Peruvian Amazon. in *Frugivory and Seed Dispersal: Ecological and Evolutionary Aspects* (eds Fleming, T. H. & Estrada, A.) 339–349 (Kluwer Academic Publishers, Belgium, 1993).

Horvitz, C. C. & Schemske, D. W. Dispersal and environmental heterogeneity in a neotropical herb: a model of population and patch dynamics. in *Frugivores and Seed Dispersal* (eds Estrada, A. & Fleming, T. H.) 169–186 (Dr W. Junk Publishers, Dordrecht, 1986).

Howe, H. F. & Smallwood, J. Ecology of seed dispersal. *Annu. Rev. Ecol. Syst.* **13**, 201–228 (1982).

Kalliola, R., Salo, J., Puhakka, M. & Rajasilta, M. New site formation and colonizing vegetation in primary succession on the western amazon floodplains. *J. Ecol.* **79**, 877–901 (1991).

Morrison, D. W. Foraging and day roosting dynamics of canopy fruit bats in Panama. *J. Mammal.* **61**, 20–29 (1980).

Tropical Ecosystems: Structure, Diversity and Human Welfare.
Proceedings of the International Conference on Tropical Ecosystems
K. N. Ganeshaiah, R. Uma Shaanker and K. S. Bawa (eds)
Published by Oxford–IBH, New Delhi. 2001. pp. 378–381.

Conservation of riverine biodiversity: Is there hope for monsoonal Asia?

David Dudgeon

Department of Ecology and Biodiversity, The University of Hong Kong, Pokfulam
Road, Hong Kong SAR, China
e-mail: ddudgeon@hkucc.hku.hk

Human impacts threaten the biodiversity of rivers and their associated wetlands at global, regional and local scales, and impair or significantly reduce ecosystem. For example, analysis of imperilment of freshwater fishes, amphibians, crayfishes and unionid mussels in North America showed predicted extinction rates of 4% per decade, five times higher than the average rate for terrestrial fauna. Recent data on qualitative global trends in freshwater wetland vertebrates point to a general decrease during the last 40 years (Groombridge and Jenkins, 2000). Quantitative time-series data for a subset of 70 species (mostly waterfowl) showed declines of ~ 50% between 1970 and 1999, with more severe dips in tropical latitudes (Loh, 2000). The limited data on the status of freshwater biodiversity in Asia also support the global trend of imperilment (Kottelat and Whitten, 1996). However, perhaps nowhere in the world are the effects of human activities on aquatic biodiversity more apparent than in Asia. Here, I evaluate the proposition that Asian rivers are among the most threatened ecosystems in the world, and describe present and projected trends in riverine biodiversity. I draw upon recent reviews (Dudgeon, 1999, 2000a, 2000b, 2000c; Dudgeon *et al.*, 2000) that give further details and primary sources.

Keywords: Rivers, dams, pollution, overharvest, fishes.

Rivers in the Oriental region (i.e. monsoonal Asia south of 30°N) support a significant proportion of global biodiversity, yet these habitats are under severe and increasing threat from human activities. Flow regulation, habitat degradation, and conversion of riverine wetlands to agriculture have been going on for centuries, while pollution and over-harvest of fishes, crocodiles, turtles and frogs are threats that have become important in recent decades. Concomitant species loss seems inevitable. A matter of special concern is the fact that fish biodiversity in Asian rivers is high, yet we do not know the extent of this species richness. What is known is that the region is host to at least 3,500 freshwater fish species. The countries of the region, and the rivers they contain, are among the richest in the world and make up half of the global 'hotspots' for fish biodiversity. In terms of contribution to the global total species of fishes, and species per unit area, Asia seems disproportionately rich. Among other animals, crocodilians are an important and endangered part of riverine biodiversity in Asia: eight out of 23 species known globally occur in the region. Three of the five species of 'true' river dolphins (which never enter the sea) are found only in Asia. All are highly endangered – the Yangtze dolphin (*Lipotes vexillifer*) numbering fewer than 200 individuals. An array of nominally terrestrial and semi-aquatic mammals, many of them globally endangered, are associated with riparian corridors and wetlands (e.g. swamp forest and grassy floodplain) along Asian rivers. Among these 'charismatic megafauna' are rhinoceroses, buffalo, an array of swamp deer (*Ruvicervus* spp.), tapirs, proboscis monkeys and (in swamp forest) orangutan.

Turtles are emblematic of the grave threats faced by riverine biodiversity in Asia. At least 90 species of non-marine chelonians occur in southern Asia. Thirty-seven were categorized in the 1996 IUCN Red List as vulnerable (VU), endangered (EN) or critically endangered (CE); a further 18 were data deficient (DD) but thought to be under threat (Van Dijk, 2000). Four years later, the 2000 IUCN Red List contained twice as many threatened species (18 CR; 27 EN; 21 VU; 6 DD) – i.e. almost three-quarters of Asian turtles were threatened. Habitat degradation has contributed to the decline of turtles, but the greatest threat is collection for trade. The meat and shells are thought to have medicinal value, and the main consumers are in East Asia (China, Japan, Korea) and Chinese ethnic communities around the world. Turtles in India, Burma, Laos, Vietnam, Cambodia, Malaysia and Indonesia are subject to intensive collection pressures, and are shipped (or transshipped) live to China. A minimum estimate of the annual trade is 13,000 tons, including at least 5,000 tons of wild-collected animals from Indonesia, but the actual figures may be substantially higher (Van Dijk, 2000).

A detailed understanding of the ecology of tropical Asian rivers is lacking. In general, however, it is clear that the life histories and life cycles of fishes and other inhabitants of Oriental rivers reflect the seasonal pattern of flow variation caused by monsoonal rains. For example, breeding by fishes (and many invertebrates) is stimulated by rising water levels during the wet season, and may involve longitudinal (upstream and downstream) or lateral migrations (onto inundated floodplain or into swamp forest) by different groups of fish species. Unfortunately, accurate data on species composition and fish biodiversity are not available for most countries – even for major rivers such as the Mekong and Kapuas. This creates the problem that the effects of environmental degradation on the riverine biota may go undetected, at least in the initial stages, because of inadequate knowledge of what is present in the systems. Nor do we know the extent of temporal variation in communities of fishes and other organisms, and this makes it difficult to separate the anthropogenic 'signal' of declining abundance or species richness from the background 'noise' of natural variation. Even fisheries statistics are lacking for most rivers, and the available data do not distinguish between aquaculture and capture fisheries or provide species-level information.

Despite inadequacies in the data, it is clear that over-harvesting has had severe effects on fish population (especially large species) in many Asian rivers, but this impact rarely occurs in isolation. Other threats include pollution by untreated urban and industrial wastes, and run-off of nutrients and pesticides from cultivated land. Flow regulation, which includes dam-building for hydroelectricity and impoundment of rivers to control floods and provide water for irrigation, can have severe effects ranging from changes to the natural flow regime (changes in current speed, flow volume, water temperature and oxygen levels) to blockage of fish breeding migrations by dams. Conversion of floodplains to agriculture, deforestation and drainage-basin degradation are also widespread, and lead to changes in river ecology. The combined and long-term effects of these impacts cannot be predicted accurately, but the prognosis for riverine biodiversity is grim. Some rivers are already so degraded that fisheries stocks have collapsed entirely. The effects of large-scale flow regulation (such as the array of dams planned for the Mekong mainstream, or the Three Gorges Project in China) present a challenge as they highlight a conflict between biodiversity conservation and human welfare. Although damaging to river ecosystems, the construction of dams and impoundments provide immediate benefits for human populations through relief from floods and droughts. Hydroelectricity is also seen as a 'green' alternative to use of fossil fuels.

A wider appreciation of the value of freshwater biodiversity is essential to ensure its long-term preservation. Realistic economic valuations of the ecosystem benefits provided by rivers and other inland waters will be an essential first step. The promotion of flagship species as conservation icons will also be needed. While more data on river ecology and biodiversity will increase our understanding of these complex systems, information alone is of limited utility. Indeed, it is by no means certain that the acquisition of more data will do anything to serve the ends of conservation of Asian rivers. Applying information effectively remains a major challenge; perhaps *the* major challenge. Without this conversion of knowledge to action, we will be unable to secure any commitment from politicians and decision-makers to the conservation of freshwater ecosystems.

References

Dudgeon, D. *Tropical Asian Streams: Zoobenthos, Ecology and Conservation* pp 830 (Hong Kong University Press, Hong Kong, 1999).

Dudgeon, D. Large-scale hydrological alterations in tropical Asia: prospects for riverine biodiversity. *BioScience* **50**, 793–806 (2000a).

Dudgeon, D. The ecology of tropical Asian streams in relation to biodiversity conservation. *Annu. Rev. Ecol. Syst.* **31**, 239–263 (2000b).

Dudgeon, D. Riverine wetlands and biodiversity conservation in tropical Asia. in *Biodiversity in Wetlands: Assessment, Function and Conservation* (eds Gopal, B., Junk, W. J. & Davis, J. A.) 35–60 (Backhuys Publishers, The Hague, 2000c).

Dudgeon, D., Choowaew, S. & Ho, S. C. River conservation in Southeast Asia. in *Global Perspectives on River Conservation: Science, Policy and Practice* (eds Boon, P. J., Petts, G. E. & Davies, B. R.) 279–308 (John Wiley & Sons, Chichester, 2000).

Groombridge, B. & Jenkins, M. *Global Biodiversity. Earth's Living Resources in the 21st Century.* UNEP World Conservation Monitoring Centre, pp 246 (World Conservation Press, Cambridge, 2000).

Kottelat, M. & Whitten, T. Freshwater biodiversity in Asia with special reference to fish. *World Bank Tech. Pap.* **343**, 1–59 (1996).

Loh, J. Living Planet Report 2000. *World Wide Fund for Nature, Gland*, pp 32 (eds Ricciardi, A. & Rasmussen, J. B.) 1999; Extinction rates of North American freshwater fauna. *Conserv. Biol.* **13**, 220–222 (2000).

Van Dijk, P. P. The Status of Turtles in Asia. in *Asian Turtle Trade: Proceedings of a Workshop on Conservation and Trade of Freshwater Turtles and Tortoises in Asia* (eds Van Dijk, P. P., Stuart, B. L. & Rhodin, A. G. J.) 15–23 (Chelonian Research Monographs No. 2, Chelonian Research Foundation, Lunenberg, 2000).

Tropical Ecosystems: Structure, Diversity and Human Welfare.
Proceedings of the International Conference on Tropical Ecosystems
K. N. Ganeshaiah, R. Uma Shaanker and K. S. Bawa (eds)
Published by Oxford–IBH, New Delhi. 2001. pp. 382–385.

Biodiversity hotspots in West Africa

Lourens Poorter*[†] and Jan Wieringa[‡]

*Ecosyn Project, Silviculture and Forest Ecology Research Group, Wageningen
University, P.O. Box 342, 6700 AH Wageningen, The Netherlands
[‡]Ecosyn Project, National Herbarium, the Netherlands, Wageningen Branch,
Wageningen University, Wageningen, The Netherlands
[†]e-mail: Lourens.Poorter@btbo.bosb.wau.nl

The rain forests of West Africa have been earmarked as one of the world's hotspots of biodiversity (Myers *et al.*, 2000). These forests extend from Ghana to Senegal, and are referred to as the Upper Guinean forests (White, 1983). The Upper Guinean forests are separated from the rest of the African rain forests by the Dahomey gap; a woodland savanna which extends in Ghana from the Sahel to the Gulf of Guinea.

Because of their isolated position the Upper Guinean forests still harbor a large number of endemic animal and plant species. At the same time they are disappearing rapidly because of logging activities, shifting cultivation, and conversion into plantations. For an effective conservation policy, information is needed on the distribution of rare and endemic species in the region, and the regions in which they are concentrated. A problem of many tropical countries is that such botanical background information is scarce, or highly fragmented. To rapidly generate the necessary information one may carry out botanical surveys in which areas are screened for their species composition. An example of such an approach is given by Hawthorne and Abu-Juam (1995) for Ghana. Another option is to use existing herbarium collections as a

Keywords: Biodiversity, tropical rain forest, West Africa, endemic species.

data source. This paper explores how, and to what extent herbarium collections can be used to define hotspots of biodiversity in West Africa.

Based on the Flora of West Tropical Africa (Hutchinson *et al.*, 1954), inventories in Ghana (Hall and Swaine, 1981; Hawthorne, 1995) and taxonomic revisions, a compilation was made of 1000 species which are rare or endemic for the closed forests of Upper Guinea. Emphasis was on woody tree, shrub and liana species, but other life forms such as herbs, grasses, and epiphytes were included as well. From this compilation 600 species were selected for a shortlist. Care was taken to include species from different families, and to include species with different distribution patterns. For the species belonging to this shortlist, all herbarium specimens collected from Senegal to Togo were entered in a data-base. To this end, specimens from herbaria in Wageningen, Brussels, Kew, Geneve, and Legon were checked and entered. These data were complemented by distribution data from taxonomic revisions. In total the database contained 48,000 records.

Botanical collections have several advantages; they provide an existing source of information, are correctly identified, and provide permanent records which can always be rechecked. A disadvantage is that specimen collection is not carried out in a stratified or random way. As a consequence, sampling efforts are not equally distributed, and areas of high biodiversity often coincide with areas of high collection intensity (Nelson *et al.*, 1990). To allow for comparison between regions, one may construct a species collection curve, in a similar way as a species-area curves is made. For each grid cell of half a degree latitude by half a degree longitude, we randomly drew collections, and constructed 100 species-collection curves. The number of species found in a given area increases with the number of collections made, until an asymptote is reached (Steege *et al.*, 1998; Colwell and Coddington, 1995). Therefore we fitted per cell an asymptotic curve through the data, according to the equation $Sn = (S_{max} * n)/(c + n)$, where S is the number of species in a sample of n collections, S_{max} the estimated maximum number of species in a given area, and c is a constant.

Apart from the S_{max}, a weighted species score was also calculated. The weighted species score is a weighted measure of species diversity, which takes the rarity of the species into account. Each species is weighted for the number of half degree grid cells in which it occurs. A common species, which occurs in 100 half degree grid-cells receives thus a weighing score of 1/100 for each cell in which it is occurs. A rare species which occurs in 4 half degree grid-cells receives a weighing score of 1/4.

Both measures of species diversity were related to environmental variables that shape the structure and composition of plant communities. Environmental variables used were rainfall (mm/year), soil water holding capacity (mm water/m soil), altitude (m), and soil fertility (mmol cations per g soil).

A rainfall map was created based on a compilation of 600 weather stations in the region. Data on soil fertility and water holding capacity were derived from the FAO soil map of Africa. All geostatistical analysis were carried out with ArcView (Esri).

Most collections in the database are confined to the southern part of West Africa and closely follow the forest zone. Collection efforts have been particularly high near the capitals (Abidjan, Monrovia) and near botanical research stations. If we divide the observed species number by the predicted maximal species number, we get a measure of sampling intensity. Sampling intensity is particularly low (< 20%) in southeast Liberia. This area contains one of the largest remaining forest blocks in West Africa, and receives also a high amount of rainfall. It might therefore harbor a rich and unexplored flora, which definitely merits further attention.

Species collection curves show that regions differ in their richness in rare and endemic species. Areas of a similar size, with a similar number of collections, may differ considerably in their species number. In general, there is a latitudinal gradient in S_{max}, which parallels the rainfall gradient; the species number increases from the Sahel towards the coast. Areas of particularly high richness are found in the very wet regions near the coast. Examples are the coastal fog zone near Freetown in Sierra Leone, the high rainfall zone near Monrovia in Liberia, and Ankasa forest in Ghana. In addition, montane areas have a very high S_{max}. Examples are the Mount Nimba at the frontier of Ivory Coast, and Mount Ziama. Some of these centers of diversity (Mount Nimba, Cape Palmas, and Cape Three Points) have been postulated to be Pleistocenic refuges (Sosef, 1994).

The weighted species score shows the same pattern as the species richness. The speciality of the montane habitat is accentuated, however, as many montane species have a very localized distribution. Montane habitats are therefore characterized by a high degree of endemic species. S_{max}, and the weighted species score will be related to environmental variables, to give insight into which factors give rise to a high diversity.

References

Colwell, R. K. & Coddington, J. A. Estimating terrestrial biodiversity through extrapolation. in *Biodiversity Measurement and Estimation* (ed. Hawksworth, D. L.) 101–118 (Chapmann & Hall, London, United Kingdom, 1995).

Hall, J. B. & Swaine, M. D. *Distribution and Ecology of Vascular Plants in a Tropical Forest Vegetation in Ghana.* (Dr W. Junk Publishers, The Hague, 1981).

Hawthorne, W. D. *Ecological profiles of Ghanaian Forest Trees* (Tropical Forestry Papers 29, Oxford Forestry Institue, Oxford, 1995).

Hawthorne, W. D. & Abu-Juam, M. Forest protection in Ghana. IUCN, Gland (1995).

Hutchinson, J. J. M., Dalziel. & Keay, R. W. J. *Flora of West Tropical Africa* (The Whitefriars Press Ltd., London, 1954).

Myers, N., Mittermeier, R. A., Mittermeier, C. G., Da Fonseca, G. A. B. & Kent, J. Biodiversity hotspots for conservation priorities. *Nature* **403**, 853–858 (2000).

Nelson, B. W., Ferreira, C. A., Da Silva, M. F. & Kawasaki, M. L. Endemism centres, refugia, and botanical collection density in the Brazilian Amazon. *Nature* **345**, 714–716 (1990).

Sosef, M. S. M. Refuge Begonias. Taxonomy, phylogeny and historical biogeography of Begonia sect. Loasibegonia and sect. Scutobegonia in relation to glacial rain forest refuges in Africa. Ph.D. Thesis, Department of Plant Taxonomy, Wageningen Agricultural University (1994).

Steege, H., T., Jansen-Jacobs. M. J. & Datadin, V. Can botanical collections assist in a National Protected Area Strategy in Guyana? *Biodiv. Conserv.* **9**, 215–240 (1998).

White, F. The vegetation of Africa. A descriptive memoir to accompany the UNESCO/AETFAT/UNSO map of Africa. UNESCO, Paris (1983).

Tropical Ecosystems: Structure, Diversity and Human Welfare.
Proceedings of the International Conference on Tropical Ecosystems
K. N. Ganeshaiah, R. Uma Shaanker and K. S. Bawa (eds)
Published by Oxford–IBH, New Delhi. 2001. pp. 386–391.

The Philippine biodiversity hotspot: Opportunities and future directions

Perry S. Ong

Conservation International-Philippines, No. 5, South Lawin, Philam Homes,
Quezon City, Philippines
e-mail: ongperry@yahoo.com

The Philippines is an archipelago of more than 7,100 islands. Its complex geological history and long periods of isolation from the rest of the world are primary reasons for its high levels of biological diversity and endemism (Heaney *et al.*, 1999; Punongbayan *et al.*, 1998). In fact, it has several centers of diversity and endemism and thus has been described as Galapagos times ten (Heaney and Regalado, 1998).

The Philippines has lost more than 97% of its original forest cover in the last 500 years (Figure 1; Environment Science for Social Change, 1999; Philippine Department of Environment and Natural Resources, 1999), yet more new species are still being discovered on these islands than any other areas on earth in recent times (e.g., see Brown *et al.*, 1999).

Of the more than 1130 terrestrial wildlife species so far recorded, more than half are found nowhere else in the world (Table 1). The floral diversity is equally extraordinary, with between 10,000 and 13,000 species of plants, of which more than half are endemic to the Philippines (Merrill, 1923–1926).

Keywords: Biodiversity hotspot, Philippines, wildlife, marine biodiversity.

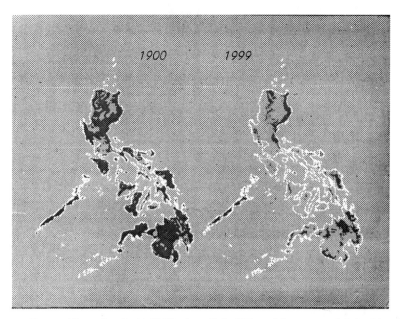

Figure 1. Extent of forest loss in the Philippines. Only 18% total forest cover is left with only 3% representing the original forests (ESSC 1999; Philippine DENR, 1998).

Table 1. Diversity, endemism and conservation status of the Philippine wildlife (Alcala and Brown, 1998; Diesmos and Herpetofauna Working Group, 2000; Collar *et al.*, 1994; 1999; Dickenson *et al.*, 1991; Heaney *et al.*, 1998; Mallari and the Bird Working Group, 2000; Ong, 1998; Wildlife Conservation Society of the Philippines, 1997; Tan, 1996)

	No. of species	No. of endemic species	% endemics	No. of threatened species	No. of threatened endemic species
Amphibians	105[+]	82[+]	78%	24	24
Reptiles	254[+]	209[+]	82%	8	4
Birds	576[*]	192[+*]	33%	74	59
Mammals	202+[*#]	110[+]	54%	51	41
Total	1137[+]	592[+*]	52%	157	128

[+]Includes new species (38 sp. for amphibians, 35 sp. for reptiles); [*]includes rediscovered species; [#]22 species of dolphins, whales and dugong.

The country's marine biodiversity is equally exceptional. With a coastline of 22,450 km and an estimated 27,000 km^2 of coral reefs, it has nearly 500 coral species of the more than 800 known coral species worldwide (Nañola *et al.*, 2000; Gomez *et al.*, 1994; Conservation International, in press). It also contains more than 2,000 species of fish (Herre, 1953; Dela Paz and Gomez, 1995) and more than 40 species of mangrove plants making the country one of the richest concentrations of marine life in the world (Zamora, 1995).

Unfortunately, mangroves and seagrass beds had been reduced to 120,000 ha from 500,000 ha while only 5% of coral reefs are in excellent condition (Calumpong, 1994; Aliñio and the Marine Working Group, 2000).

The Philippines' remaining biodiversity and the ecosystems that support it are under tremendous threats. Extractive industries such as logging and mining have destroyed most of the forests. High human population density and growth rate (Cincotta et al., 2000) have further put enormous pressure. In turn, rainforests have been converted to agriculture and plantations to augment the lack of land to support a growing population. In addition, cyanide and dynamite fishing, along with rapid development in coastal areas, have destroyed coral reefs and mangroves. Conservative land and resources-use trend projections indicate that profound degradation of the country's biogeographic regions will occur in approximately 10 to 15 years. Because of the dire conditions of Philippine biodiversity, several authors have written off the country as being damaged beyond repair (Terborgh, 1999; Linden, 1998).

These indeed are gloomy pictures of the country's biodiversity. However, the combination of an active community of civil society organizations, progressive legislation associated with integrated protected areas management, legal precedents recognizing the rights of indigenous peoples to manage resource use within their ancestral domains, private sector interest to be involved, provides a glimmer of hope that provide a small window of opportunity for positive action to be undertaken before a point of no return is reached.

The Philippines' National Biodiversity Strategy and Action Plan (NBSAP) was formulated and published in 1997. A re-assessment and the second iteration of the NBSAP was undertaken through the recently concluded Conservation Priority-setting Workshop (CPW), co-convened by Conservation International with the PAWB-DENR and the Biodiversity Conservation Program of the University of the Philippines' Center for Integrative and Development Studies (BCP-UP CIDS).

The CPW was undertaken based on the principle that no single individual or organization can save Philippine Biodiversity. Alliance and partnership building became an integral part of the whole process, from the data collection, compilation and analysis that culminated in the consensus-building workshop. In the end, more than 300 scientists and more than 100 institutions shared their expertise and resources to arrive at the sets of priority areas and strategic actions to make conservation happen.

On a per unit area basis, the Philippines is the Hottest of the Biodiversity Hotspots (Heaney et al., 1999). More than 50% of Philippine biodiversity is found nowhere else in the world, and once lost, the whole world losses. Thus

Philippine biodiversity forms part of global patrimony and as such its conservation should be a global priority, despite its gloomy circumstances. Causes of biodiversity loss in the country are complex and often inter-linked. These can be broadly divided into ultimate and proximate drivers. Below are some of these drivers (not an exhaustive list):

a) Reckless pursuit of economic development at the expense of the environment, particularly biodiversity.
b) A rapidly expanding population that requires more natural resources to meet its needs and demands.
c) Lack of a national constituency that advocates for biodiversity conservation.
d) These factors in turn lead to increasing fragmentation of the remaining habitats and ecosystems.
e) These factors also stem from the under-valuation and under-appreciation of biodiversity.

The recently concluded second iteration of the NBSAP, through the CPW, provides the country with an opportunity to seize the small window of opportunity that has emerged, but this window is rapidly closing.

A number of innovations were introduced in the Philippine CPW. One is the conduct of a series of regional consultations, as a means to communicate the CPW as a national exercise and at the same time collect or validate data that were collected; another is the inclusion of inland waters and marine as an integral part of the CPW; while another is the establishment of a mechanism that will ensure that the outputs of the CPW will be implemented. This mechanism is tentatively called the Network for Nature (N4N) and has the potential of being the mechanism as well forming the core group for a Hotspots Monitoring Program which will tell us if we are winning the biodiversity war or not.

Two scenarios are predicted to happen within the next 10–50 years. These are:

Scenario A: If present trends continue, it is predicted that the remaining forests will be down to 6.6% in ten years (ESSC, 1999) and wiped out completely in 50 years (Linden, 1998).

Scenario B: If drastic steps are undertaken now, forest cover would remain at about 19% in 10 years (ESSC, 1999) while damaged habitats and ecosystems would commenced to be restored and rehabilitated in 25 years.

For *Scenario B* to occur the following needs to happen. This will involve the development of a national constituency for biodiversity conservation through the establishment of alliances and partnerships at the local, regional and national level and building upon the success of CPW as well as other initiatives in attaining conservation goals. Among the different steps and activities that needs to be done are:

i)   A road show to promote the results of the CPW
ii)   Creation of the Network for Nature
iii) Development of a hotspots monitoring system at various scales and levels
iv)   Scientific and policy studies to support national and local level actions
v)   Innovative approaches for national, regional and local conservation actions such as creation of biodiversity corridors.

These proposed actions are not either or options, but actions that need to be undertaken simultaneously, if we are to seize that small window of opportunity to turn a gloomy situation around. There is no other way.

References

Alcala, A. C. & Brown. W. C. *Philippines Amphibians: An Illustrated Field Guide* (Bookmark, Makati City, Philippines, 1999).
Aliño. & the Marine Working Group. Marine Biodiversity Conservation Priority Concerns. Paper presented at the National Biodiversity Conservation Priority Setting Workshop, White Rock Hotel, Subic, Philippines (2000).
Brown, W. C., Alcala, A. C. Ong, P. S. & Diesmos, A. C. A new species of *Platymantis* (Amphibia: Ranidae) from the Sierra Madre Mountains, Luzon Island, Philippines. *Proceedings of the Biological Society of Washington.* **112,** 510–514 (1999).
Calumpong, H. P. Status of mangrove resources in the Philippines. in *Proceedings of the Third ASEAN-Australia Symposium on Living Coastal Resources* (eds Wilkinson, C. R., Sudara, S. & Ming, C. L.) 215–228 (Australian Institute of Marine Sciences, Australia, 1994).
Cincotta, R. P., Wisnewski, J. & Engelman, R. Human population in the biodiversity hotspots. *Nature* **404,** 990–992 (2000).
Collar, N. J., Crosby, M. J. & Stattersfield. A. J. *Birds to Watch 2: The World List of Threatened Birds.* Conservation Series No. 4. Birdlife International, Cambridge, United Kingdom (1994).
Collar, N. J., Mallari, N. A. D. & Tabaranza, B. R. *Threatened Birds of the Philippines: Haribon Foundation-Birdlife International's Red Data Book.* Bookmark, Makati City, Philippines (1999).
Conservation International. Preliminary report on the 1998 Marine Rapid Appraisal Program in the Calamianes, Palawan, Philippines (in press).
Dela Paz, R. & Gomez, E. D. Faunal Diversity in the Marine Coastal Zone. Biodiversity Conservation Report No. 2. University of the Philippines Center for Integrative and Development Studies, Diliman, Quezon City, Philippines (1995).
Dickinson, E. C., Kennedy, R. S. & Parkes, K. C. The birds of the Philippines, an annotated checklist. *British Ornithological Union Checklist* **12,** 1–507 (1991).
Diesmos, A. & the Herpetofauna Working Group. Philippine Amphibians and Reptiles: An Overview of Diversity, Biogeography and Conservation. Paper presented at the National Biodiversity Conservation Priority Setting Workshop, White Rock Hotel, Subic, Philippines (2000).
Environmental Science for Social Change. *Decline of Philippine Forests* (ESSC Inc. and Bookmark, Makati, Philippines, 1999).
Gomez, E. D., Aliño, P. M., Licuanan, W. Y. & Yap, H. P. Status report of the coral reef of the Philippines. In *Proceedings of the 3rd ASEAN-Australia Symposium on Living Coastal Resources, May 16–20, 1994* (eds Wilkinson, C. R., Sudara, S. & Chow, L. M.) 57–76 (Chulalongkorn University, Bangkok, Thailand, 1994).

Heaney, L. R. & Regalado, J. *Vanishing Treasures of the Philippine Rainforest* (The Field Museum, University of Chicago Press, Chicago, USA, 1998).

Heaney, L. R., Balete, D. S., Dolar, M. L., Alcala, A. C., Dans, A. T. L., Gonzales, P. C., Ingle, N. R., Lepiten, M. V., Oliver, W. L. R., Ong, P. S., Rickart, E. A., Tabaranza, B. R. & Utzurrum, R. C. B. A synopsis of the mammalian fauna of the Philippine Islands. *Fieldiana Zoology, New Series* **88**, 1–61 (1998).

Heaney, L. R., Ong, P. S., Mittermeier, R. A. & Mittermeier, C. G. The Philippines. in *Hotspots: Earth's Biologically Richest and Most Endangered Terrestrial Ecoregions* (eds Mittermeier, R., Myers, N. & Mittermeier, C.) 308–315 (Conservation International and Cemex, Mexico City, 1999).

Herre, A. H. *Checklist of Philippine Fishes*. Research Report 20. Fish and Wildlife Service, US Dept. of Interior, Government Printing Office, Washington, DC (1953). pp. 977.

Linden, E. *The Future in Plain Sight: Nine Clues to the Coming Instability* (Simon and Schuster, New York, USA, 1998).

Mallari, N. A. D. & the Bird Working Group. Philippine Birds: Setting an Agenda for Conservation. Paper presented at the National Biodiversity Conservation Priority Setting Workshop, White Rock Hotel, Subic, Philippines (2000).

Merrill, E. D. *An Enumeration of Philippine Flowering Plants*. vol. 1–IV (Bureau of Printing, Manila, Philippines, 1923–1926).

Nañola, Cleto Jr. L., Dantis, A. L., Hilomen, V., Ochavillo, D. G. & Aliño, P. M. Philippine Reef Fish Diversity: conservation significance and concerns. Paper presented at the National Biodiversity Conservation Priority Setting Workshop, White Rock Hotel, Subic, Philippines (2000).

Ong, P. S. The Philippine menagerie. in *The Philippine Archipelago* Vol. 1 of *Kasaysayan: A History of the Filipino People* (eds Punongbayan, R. S., Zamora, P. M. & Ong, P. S.) 227–255 (Asia Publishing Co. Ltd., Makati, Philippines, 1998).

Philippine Department of Environment and Natural Resources. Forestry Statistics. Quezon City, Philippines (1998).

Punongbayan, R. S., Zamora, P. M. & Ong, P. S. *Philippine Archipelago*. Volume 1: *Kasaysayan: A History of the Filipino People* (Asia Publishing Co. Ltd., Makati, Philippines, 1998).

Wildlife Conservation Society of the Philippines. Philippine Red Data Book. Wildlife Conservation Society of the Philippines and Bookmark, Makati City, Philippines (1997).

Tan, J. M. L. *A Field Guide to the Whales and Dolphins of the Philippines* (Bookmark, Makati City, Philippines, 1996).

Terborgh, J. *Requiem for Nature*. Island Press, Washington, DC, USA (1999).

Zamora, P. M. Diversity of flora in the Philippine mangrove ecosystem. University of the Philippines Center for Integrative and Development Studies, Diliman, Quezon City, Philippines. *Biodiv. Conserv. Rep.* 1–92.1 (1995).

Tropical Forest Canopies

Tropical Ecosystems: Structure, Diversity and Human Welfare.
Proceedings of the International Conference on Tropical Ecosystems
K. N. Ganeshaiah, R. Uma Shaanker and K. S. Bawa (eds)
Published by Oxford–IBH, New Delhi. 2001. pp. 395–398.

Do specific arthropods forage in the upper canopy layer of rainforests? Examples from the closed forests of Africa

Yves Basset

Smithsonian Tropical Research Institute, Apartado 2072, Balboa,
Ancon, Panamá.
Research associate, Programa Centroamericano de Maestría en Entomología,
Universidad de Panamá
e-mail: bassety@tivoli.si.edu

It is probable that most of the variance in the distribution of insect herbivores in tropical rainforest is accounted for by the following: (1) host plant effect (2) local and regional effects, including historical factors (3) successional gradients (4) altitudinal gradients (5) rainfall gradients (6) vertical gradients (7) seasonal gradients and (8) diel activity.

With reference to vertical gradients of arthropod distribution in tropical rainforests, the literature is replete with studies analysing samples obtained from the 'canopy', often meaning samples obtained 15 m or higher above the ground. More precisely, the 'canopy' is defined as the aggregate of every tree crown in the forest, including foliage, twigs, fine branches and epiphytes (Nadkarni, 1995). In botany, the *canopée* or 'canopy surface' is also defined as the interface between the uppermost leaf layer and the atmosphere (Hallé and Blanc, 1990; Bell *et al.*, 1999). Because

Keywords: *Agrilus*, insect herbivores, species richness, understorey, upper canopy.

entomological samples are difficult to obtain from such vegetation surface which, further, has no depth by definition, the term 'upper canopy' is used hereafter to denote the uppermost leaf layer, which is often 1–2 m deep in closed tropical rainforests (Hallé and Blanc, 1990).

The arthropod fauna of the upper canopy has been rarely sampled and studied. Most entomological studies, either with insecticidal fogging (e.g., Erwin, 1983), light traps (e.g. Sutton, 1983) or by felling trees (Basset et al., 1999) cannot sample the upper canopy selectively. The origin of the material collected by fogging cannot be ascertained with precision and it is probable that specimens from the canopy and upper canopy are mixed in the samples. Whether fogging performed at ground level is able to kill the fauna of the upper canopy efficiently and whether this fauna eventually falls in the collecting trays at ground level are also doubtful.

Many abiotic and biotic characteristics of the upper canopy of closed tropical rainforests are different from other forest layers below, especially from the understorey. The implications for the distribution of insect herbivores along vertical gradients in tropical rainforests may be significant. Insect herbivores foraging and feeding in the upper canopy encounter a serious hygrothermal stress during the day, and water condensation at night. Further, the high level of plant defenses in the upper canopy may pressure them to specialize on leaves from the upper canopy of particular tree species. Conversely, the supply of young leaves available to them is greater in the upper canopy than in the understorey. This suggests several strategies in order to overcome this apparently conflicting situation: (1) a specialized, distinct and well-adapted fauna to the extreme microclimatic conditions of the upper canopy; (2) interchanges of fauna between the upper canopy and lower layers, such as individuals resting in lower layers at day and moving up in the upper canopy to feed at night, perhaps taking advantage of air movements (e.g., Haddow and Corbet, 1961; Sutton, 1983); or (3) both of the above. This contribution discusses the results of three sampling programmes that were performed to assess hypotheses (1) and (2) mentioned above, during the mission of the Canopy Raft in Gabon in 1999 (Hallé, 2000).

The abundance, activity and species richness of arthropods, particularly of insect herbivores, were investigated in the upper canopy and understorey of a lowland rainforest at La Makandé, Gabon. In total 14,161 arthropods were collected with beating, flight interception and sticky traps, from 6 canopy sites, during day and night, and from mid-January to mid-March 1999. Access to the canopy was provided by the canopy raft, canopy sledge and treetop bubble.

The effects of stratum were most important, representing between 40 and 70% of the explained variance in arthropod distribution. Site effects were significant for many arthropod taxa and guilds. Sticky traps showed best these effects, followed by beating and flight interception traps. Site effects represented 19, 41 and 29% of the variance explained by environmental variables for beating, flight interception and sticky trap data, respectively. This emphasizes the need for spatial replicates, but also the problems of obtaining them in the upper canopy. The use of fixed structures, such as canopy cranes (e.g., Wright and Colley, 1994), may generate interesting data with regard to temporal replication but cannot be used easily to study the important aspects of spatial variability of arthropod distribution in highly heterogeneous rainforests. Time effects (diel activity) explained a much lower percentage of variance (6–9%).

The density and abundance of many arthropod taxa and species were significantly higher in the upper canopy than in the understorey. Arthropod activity was also higher during day than night. In particular, insect herbivores were 2.5 times more abundant and twice as speciose in the upper canopy than in the understorey, a probable response to the greater and more diverse food resources in the former stratum. Faunal overlap between the upper canopy and understorey was low. The most dissimilar herbivore communities foraged in the understorey at night and the upper canopy during day. Further, a taxonomic study of a species-rich genus of herbivore collected there (*Agrilus*, Coleoptera Buprestidae) confirmed that the fauna of the upper canopy was different, diverse and very poorly known in comparison to that of the understorey. Herbivore turnover between day and night was rather high in the upper canopy and no strong influx of insect herbivores from lower foliage to the upper canopy was detected during night.

With reference to the questions (a)–(c) posed in the introduction, the results of the three sampling methods suggest that decrease in the abundance activity and species richness from day to night may comparatively be higher in the upper canopy than in the understorey. Since few compensatory effects occur, the data do not indicate a strong influx of insect herbivores from lower foliage to the upper canopy at night. This suggests that insect herbivores of the upper canopy may be resident and well adapted to environmental conditions there.

Since faunal stratification in tropical rainforests may depend on slope (e.g., Sutton, 1983), it may be optimum and may lead to a diverse fauna in the upper canopy of closed and wet lowland forests (in contrast with montane forests), which also represent the most endangered type of rainforest. Whether the fauna collected in the upper canopy is very specialised and whether it may be different from that foraging a few metres below in the canopy represents the next question to explore. Since the upper canopy may well be distinguished from the canopy only in closed and undisturbed rainforests, the implications for the conservation of tropical rainforest arthropods may also be important.

Acknowledgements

My sincere thanks to members of Océan Vert, Pro-Natura International, les Accro-Branchés and the many colleagues in France, Italy and Panama that helped me to achieve this study.

References

Basset, Y., Charles, E. C. & Novotny. V. Insect herbivores on parent trees and conspecific seedlings in a Guyana rain forest. *Selbyana* **20**, 146–158 (1999).

Bell, A. D, Bell, A. & Dines, T. D. Branch construction and bud defence status at the canopy surface of a West African rainforest. *Biol. J. Linn. Soc.* **66**, 481–499 (1999).

Erwin, T. L. Beetles and other insects of tropical forest canopies at Manaus, Brazil, sampled by insecticidal fogging. in *Tropical Rain Forest: Ecology and Management* (eds Sutton, S. L., Whitmore, T. C. & Chadwick, A. C.) 59–76 (Blackwell, Oxford, 1983).

Haddow, A. J. & Corbet, P. S. Entomological studies from a high tower in Mpanga Forest, Uganda. V. Swarming activity above the forest. *Trans. R. Entomol. Soc. London* **113**, 284–300 (1961).

Hallé F. (ed.) Biologie d'une canopée de forêt équatoriale – IV. Rapport de Mission: Radeau des Cimes Janvier Mars 1999, La Makandé, Gabon. Pro-Natura International and Opération Canopée, Paris (2000).

Hallé, F. & Blanc, P. (eds). Biologie d'une canopée de forêt équatoriale. Rapport de Mission Radeau des Cimes Octobre–Novembre 1989, Petit Saut – Guyane Française. Montpellier II et CNRS-Paris VI, Montpellier/Paris (1990).

Nadkarni, N. M. Good-bye, Tarzan. *The Sciences*, January/February 1995. 28–33 (1995).

Sutton, S. L. The spatial distribution of flying insects. in *Tropical Rain Forest Ecosystems. Bigeographical and Ecological Studies* (eds Lieth, H. & Werger, M. J. A.) 427–436 (Elsevier, Amsterdam, 1983).

Wright, S. J. & Colley, M. (eds). Accessing the Canopy. *Assessment of Biological Diversity and Microclimate of the Tropical Forest Canopy: Phase I.* United Nations Environment Programme, Nairobi, Kenya (1994).

Tropical Ecosystems: Structure, Diversity and Human Welfare.
Proceedings of the International Conference on Tropical Ecosystems
K. N. Ganeshaiah, R. Uma Shaanker and K. S. Bawa (eds)
Published by Oxford–IBH, New Delhi. 2001. pp. 399–404.

Bromeliads and habitat fragmentation in the Atlantic rainforest of North-eastern Brazil: The remaining species

T. Fontoura

Universidade Estadual de Santa Cruz, Depto. de Biologia, Universidade Estadual de
Campinas, IB-Depto. de Ecologia/PPGE, CP6109 CEP13083-970, SP Brazil
e-mail: talita@jacaranda.uescba.com.br,

Habitat reduction and fragmentation, edge effects, and selective logging can act diversely on arboreal populations modifying juvenile recruitment, mortality rates (Primack and Hall, 1992), fruit set and pollination processes (Aizen and Feizinger, 1994). Nevertheless, none of these local scale processes have been investigated in the epiphytic community. At a regional scale, Koopowitz (1992) and Koopowitz *et al.* (1993, 1994a,b) have studied the extinction probability rates due to deforestation in tropical areas. Based on the geographic distribution of orchid species and deforestation rates in tropical areas, Koopowitz (1992) predicted that slightly more than 1 in 5 orchid species may have already disappeared from the wild or are at the point where extinction is inevitable.

Brazil has one of the highest absolute extinction rates with an estimated 52 orchid species disappearing each year (Koopowitz *et al.*, 1994a). Epiphytes seem to be especially highly vulnerable to extinction because most of epiphytic Orchidaceae and Bromeliaceae representatives are restricted to one or few localities (Koopowitz *et al.*, 1994b). The processes governing epiphyte

Keywords: Bromeliads, epiphytes, habitat fragmentation, Atlantic rainforest, cocoa plantation.

spatial distribution seem to be very different from those acting on essentially terrestrial species. More recently, Andersen *et al.* (1997) re-evaluated Koopowitz (1992) and Koopowitz *et al.* (1993, 1994a, b) data and reported that – among others – Bromelioideae and Pitcairnioideae (Bromeliaceae) are the sub-families which are most prone to extinction.

Twenty-six bromeliads species occur at Reserva Biológica de Una (Bahia, northeastern Brazil, Thomas *et al.*, 1998). Ten of these are endemic and the remaining 16 belong to the sub-family Bromelioideae. This conservation unit is an environmental mosaic where undisturbed forest and disturbed forest areas are embedded in a complex matrix amidst cattle pasture, rubber, piaçava, and cocoa plantations. Such systems can be envisaged as simplified primary ecosystems contributing significantly to biodiversity protection as far as they are patches of managed forests. This is the case with cocoa plantations which are established after understorey logging and the remaining canopy trees shadow the cocoa trees. This work aims to characterize habitat fragmentation effects on epiphytic bromeliads, focusing on the species remaining on different habitats after environmental disturbance.

The Reserva Biológica de Una (RBU) is about 40 km south Ilhéus, BA (15°10'S, 39°03'W). The vegetation of this region is classified as southern Bahian wet forest ('mata higrófila sul-Bahiana') to inland regions, and restinga vegetation over sand near the coast. This region is included in the Af climatic region of Köppen with occasional 1 to 3 months being rainfree. Annual rainfall is 1800 mm (Mori *et al.*, 1981; Thomas *et al.*, 1998). Cocoa is a common plantation system in the surrounding area of the RBU. In it, all understorey is logged and the remaining canopy trees shadow the cocoa trees (*Theobroma cacao* L.). In this region, bromeliad species were collected in order to establish a reference collection.

Beyond the cocoa plantation, two habitats were selected in forested areas over 800 ha: forest interior and forest edge. Three sample blocks of 5 × 5 km were selected at the Una region in such a way to contain these three habitats. In each block 2 sample plots of 110 × 20 m were settled within each habitat. Thus, each sample block contained 6 sample plots and 18 sample plots were settled in the 3 sample blocks. Perimeter of all trees (> 20 cm) was measured, and tagged. Bromeliad species and their abundance were recorded for each tree. Data presented here refer to 16 of the 18 sample plots.

In total, 34 species were found. Approximately one-third of species are not found in the cocoa plantation (Table 1), and eleven species occurred in all the three habitats. Multivariate analysis was performed to assess habitat fragmentation effects on species and sample plots (habitats). Data on species abundance per sample plot were included in a matrix in which simple Euclidean distance coefficient was used to calculate distance matrix. Cluster analysis was run using Ward's method on 11 species and 16 sample plots. The Principal Component Analysis (PCA) was used to investigate species gradient among all habitats.

In a total of 34 species occurring in the three habitats (Table 1), 11 (32.35%) share all 3 environments, 12 (35.29%) species are not found in the cocoa plantation and 10 (29.41%) occur in the interior and edge forests. Almost all of the endangered species (11) inhabit the understorey and/or subcanopy in the forested areas. The remaining 11 species on the three habitats belong to the canopy and/or subcanopy.

Table 1. Epiphytic bromeliads in Una region. I, forest interior; E, forest edge; CP, cocoa plantation; U, understorey; SC, subcanopy; C, canopy.

Species	I	E	CP
Aechmea aquilega – C	X	X	X
A. blanchetiana –	X		X
A. conifera – C	X	X	X
A. lingulata var. 1			X
A. lingulata var. 2 – C/SC	X	X	X
A. mollis – U	X	X	
Aechmea sp.	X	X	X
Areococcus parviflorus – U	X	X	
Billbergia sp. – SC/U	X	X	X
Bromeliaceae indet.	X	X	X
Guzmania monostachia – U		X	
Hohenbergia brachycephala – U			X
H. hatschbachi – C	X	X	X
Lymania alvimii – D	X	X	
L. azurea – U	X	X	
L. smithii – U	X	X	
Neoregelia longifolia – C	X	X	X
Neoregelia sp1 – C/SC	X	X	X
Nidularium amorimii – U	X	X	
Streptocalyx curranii – U/SC	X	X	
Tillandsia bulbosa – C	X	X	X
T. kautskii – C	X	X	X
Tillandsia sp.	X	X	X
T. spiculosa	X		X
T. stricta		X	X
Tillandsia sp 2 – C	X		
Vriesea drepanocarpa – U	X	X	
V. duvaliana – U	X	X	
V. ensiformis – SC	X	X	
V. platynema – SC	X	X	X
V. procera			X
Vriesea sp 1			X
V. sp 2 – SC/C	X	X	X
Vriesea sp.	X	X	X

The cluster analysis of sample plots (Figure 1) identifies three distinct groups: (a) the cocoa plantation group including 1 sample plot of forest interior; (b) forest edge group including 1 forest interior and 1 cocoa plantation sample plots; (c) forest interior group, including 2 edge sample plots. The comparison between dendrogram and original matrix showed that the first group comprises the low abundance species ($N \le 29$ individuals). The forest interior sample plot placed in this group showed low epiphytes. The second group comprised the intermediate abundance sample plots, with 40 to 65 individuals. Included in this group is the high abundance cocoa plantation with 65 individuals distributed among 9 species. The third group consisted of more than 61 individuals per sample plot. Groups (b) and (c) can also be recognized by the presence and abundance of *Vriesea platynema*: it is absent or present at low abundance ($N < 27$ individuals) in the forest edge group. Its presence in the forest interior group is characterized by high abundance ($N > 58$ individuals).

The cluster analysis of species revealed 4 groups: (a) species with low abundance in all sample plots or low abundance at the forest interior ($N > 6$ individuals): *Aechmea aquilega, Neoregelia longifolia*; (b) species with high abundances in all sample plots or high abundance in the forest interior ($N > 14$ individuals); groups (c) and (d) with the most abundant species in the area: *Vriesea* sp. 2 (com 134 individuals) and *V. platynema* (411 individuals). All cluster analysis resulted in high values of cophenetic correlation (cc): sample plots cc = 0.83, species cc = 0.98.

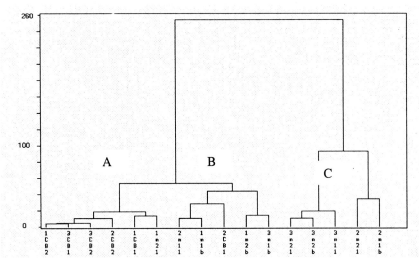

Figure 1. Cluster analysis of sample plots in Una region using Ward's method on simple Euclidean distance coefficient. Codes of sample plots: first number refers to sample block number; CB, cocoa plantation; m, forest habitats; b, forest edges. All sample plots ending with 1 refers to forest interior. A, cocoa plantation group; B, forest edge group; C, forest interior group.

Ordination analysis of sample plots (Figure 1) confirmed the results obtained by cluster analysis showing the cocoa plantation-edge-interior gradient. The cocoa plantation sample plots were positioned on the most negative values of the first component and skewed to zero values of the second component. The same samples of forest edge group from cluster analysis were represented on negative values of first component and intermediary values of the second. The remaining forest interior sample plots had the highest positive values on the first component and intermediary values to the second component. Species ordination delimited species with highest abundance in the surveyed area: *V. platynema* e*V. sp2*. The accumulated variance to the first axis in the sample plot analysis was 76.2% and to the species analysis was 84.76%. The broken stick analysis revealed a low probability of aleatory results (27.45 for sample plot analysis and 21.13 for species analysis).

In the context of the results presented, three interconnected points deserve attention: the meaning of the groups found here is based on species abundances, the fragmentation process, and the canopy importance in managed forests. One of the problems concerning habitat fragmentation is how to determine areas affected by fragmentation. The analyses employed here were useful to assess abundance modifications on species populations. The inclusion of forest interior sample plots in the cocoa plantation group or in the forest edge group indicates that such areas were probably adversely affected by fragmentation as far as all forest interior sample plots were positioned more than 200 m away from edge. Some forest interior sample plots were depauperated on its bromeliad abundance to levels either compared to the managed areas of cocoa plantation or to forest edge sample plots. The decrease in abundance of bromeliads apparently depends on how long grass areas were opened close to the surveyed forest areas. Temporal scale must also be concerned because some of the forest edge and forest interior sample plots were positioned in the same group revealing equivalent high bromeliad abundances. Thus, beyond the clear extinction of one third of the 34 species found in the area, fragmentation process acts indirectly on edges and interior of forests.

Although the plantation system impedes survival of 11 bromeliad species, canopy has an obvious importance in the maintenance of some bromeliads. Considering that these systems can present high abundance and considerable species richness, these areas can act as important species corridor between large forest remnants.

References

Andersen, M., Thornhill, A. & Koopowitz, H. Tropical forest disruption and stochastic biodiversity losses. in *Tropical Forest Remnants. Ecology, Management and Conservation of Fragmented Communities* (eds Laurence, W. F. & Bierregard, R. O.) 281–291 (The University of Chicago Press, Chicago, 1997).

Aizen, M. A. & Feizinger, P. Forest fragmentation, pollination, and plant reproduction in a chaco dry forest, Argentina. *Ecology* 75, 330–350 (1994).

Koopowitz, H. A stochastic model for the extinction of tropical orchids. *Selbyana* **13**, 115–122 (1992).

Koopowitz, H., Thornhill, A. & Andersen, M. Species distribution profile of the Neotropical orchids *Masdevallia* and *Dracula* (Pleurothalidinae): implications for conservation. *Biodiversity Conserv.* **2**, 681–690 (1993).

Koopowitz, H., Andersen, M. & Thornhill, A. Comparison of distribution of terrestrial and epiphytic African orchids: implications for conservation. in *Proceedings of the Fourteenth World Orchid Conference* (1994a).

Koopowitz, H., Thornhill, A. & Andersen, M. A general stochastic model for the prediction of biodiversity losses based on habitat conversion. *Conserv. Biol.* **8**, 425–438 (1994b).

Mori, S., Boom, B. M. & Prance, G. T. Distribution patterns and conservation of eastern Brazilian coastal forest tree species. *Brittania* **33**, 233–245 (1981).

Primack, R. B. & Hall, P. Biodiversity and forest change in Malaysian Borneo. *BioScience* **42**, 829–837 (1992).

Thomas, W. W., Carvalho, A. M. V., Amorim, A. M. A., Garrison, J. & Arbeláez, A. L. Plant endemism in two forests in southern Bahia, Brazil. *Biodiversity Conserv.* **7**, 311–332 (1998).

Tropical Ecosystems: Structure, Diversity and Human Welfare.
Proceedings of the International Conference on Tropical Ecosystems
K. N. Ganeshaiah, R. Uma Shaanker and K. S. Bawa (eds)
Published by Oxford–IBH, New Delhi. 2001. pp. 405–409.

Bromeliad invertebrate ecology in the Luquillo Experimental Forest, Puerto Rico

Barbara A. Richardson

165 Braid Road, Edinburgh EH10 6JE, UK
e-mail: mjrichardson@clara.net

The Bromeliaceae dominate many neotropical epiphytic communities, and are specialised morphologically for the interception and utilisation of non-soil nutrients. Anemophilous types intercept wind-blown nutrients deposited in rain or clouds. Tank bromeliads, with rosettes of broad-based leaves, intercept canopy litter and throughfall water, and provide a microhabitat for largely detritivorous animal communities. Specialised trichomes at the base of the leaves absorb nutrients from decomposition products of the litter and animal excreta (animal-assisted nutrition, Benzing, 1990). They form true microcosms, providing a gradation of microhabitats for their animal communities, from newly intercepted litter to the truly aquatic habitat of the phytotelmata at the base of the leaves. The biological, physical and chemical parameters of these microcosms can be sampled in their entirety and they can, therefore, be used to examine patterns of α and β animal diversity, and to answer fundamental ecological questions such as relationships between resource productivity, animal abundance and species richness, along productivity gradients (Richardson, 1999; Richardson *et al.*, 2000a). Such questions are difficult to answer in a more diffuse and complex whole-forest ecosystem.

Keywords: Communities, diversity, elevation, nutrients, phytotelmata.

The Luquillo Experimental Forest (LEF), in the eastern, windward, part of Puerto Rico, is the wettest part of the island. Steep elevations, and the consequent changes in climate, particularly rainfall and the associated waterlogging of soils, have resulted in different vegetation types. Four easily distinguished forest areas are dominated by different tree species. Tabonuco (*Dacryodes excelsa*) forest occupies areas < 600 m a.s.l., the mid-elevation palo colorado forest (*Cyrilla racemiflora*) forest occurs in areas above the average cloud condensation level (600–900 m a.s.l.), and the dwarf forest system (dominant tree *Tabebuia rigida*), with stunted vegetation and waterlogged anoxic soils, is located on the highest peaks > 900 m a.s.l. Palm forest (*Prestoea montana*) occurs at all elevations, principally on windward slopes and in wet gullies and valleys. Annual rainfall increases with elevation, from *ca* 3.5 m in the tabonuco, through 4.2 m in the palo colorado, to 4.8 m in the dwarf forest, and the number of rainless days per year decreases with elevation, from 97 (tabonuco), 69 (palo colorado) to 53 (dwarf forest). Net primary productivity (NPP) is related to rainfall, and declines with elevation. Mean annual temperature declines from 23°C to 19°C over the same gradient.

In the LEF, epiphytic and saxicolous tank bromeliads (mainly *Vriesia* and *Guzmania* spp.) are a noticeable component of lower canopy and understorey plant communities. Plant density was measured to determine the relative importance of bromeliads in the forest ecosystems. They increase in density with elevation, such that in the dwarf forest they form patches of closely packed ground cover and extend into the canopy of the small trees, forming a notable component of their epiphytic mass (density 32000 plants ha^{-1}). As these plants are intolerant of deep shade, they occur only along watercourses in the tabonuco forest, particularly on branches overhanging the water and on large rocks in the river bed (45 plants ha^{-1}). Whole plants, and their impounded water, were collected from each forest over a period of three years. Plants were dissected, leaf-by-leaf, to extract all organisms and organic debris. Organisms were counted and identified to morphospecies, and their biomass estimated. Chemical analyses were carried out on debris, impounded water and bromeliad plants.

Within forests, bromeliads behaved as islands, in that impounded organic matter, animal abundance and species richness all increased with plant size. In the dwarf forest, where plants were significantly smaller than in the other two forests, α diversity was lowest in terms of both abundance and species richness and was unrelated to the large total plant area available for colonisation. Community dominance increased with elevation (Mc Naughton index 37, 54, and 73 for tabonuco, palo colorado and dwarf forest respectively). Typical litter detritivores, such as isopods, millipedes and cockroaches, were reduced in abundance in the dwarf forest, as were tipulid larvae *Trentepohlia dominicana*, detritivores in the fine organic soil at the base of the phytotelmata. Scirtid beetle (*Scirtes* sp.) larvae, the most

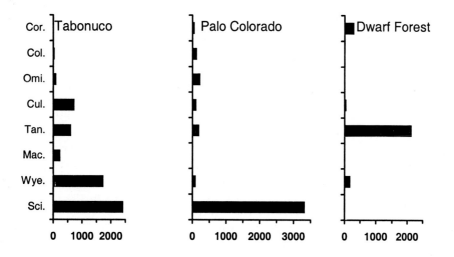

Figure 1. Total number of animals of selected species recovered from 60 plants from each forest type (1993–1996). Cor. = *Corethrella*; Col. = *Coleophoridae*; Omi. = *Omicrus*; Cul. = *Culex*; Tan. = *Tanytarsus*; Mac. = *Macrochernes*; Wye. = *Wyeomyia*; Sci. = *Scirtes*.

abundant species in the lower two forests, were absent from the dwarf forest, as were hydrophilid beetles, *Omicrus ingens* (Hansen and Richardson, 1998), the naidid worm *Aulophorus superterrenus*, and the large predatory larvae of *Platycrepidius* beetles. In general, predators were absent, or their populations reduced, in the dwarf forest bromeliads, which is consistent with the inability of the food chain to support them. An exception was the predatory larvae of the chaoborid midge, *Corethrella* sp., in the plant pools (Richardson, 1999). Communities could be distinguished among forests (Sørensen's similarity coefficient), and could be characterised by the relative abundance, or presence/absence, of certain indicator species (Figure 1).

Nutrient studies of inputs (canopy-derived organic matter and throughfall water) in the bromeliads showed that P, K and Ca in debris declined with elevation, as did N and P in phytotelm water. Annual nutrient budgets showed that microcosms accumulated only 5% of most nutrients passing through them, exceptions being P and K in the dwarf forest, where accumulation was 25% of inputs. These microcosms are, therefore, nutrient abundant, but the nutrients are not necessarily available to the fauna, as the detrital food source is difficult to digest, because of its high lignin content.

In a three-year study it was observed that animal abundance declined with elevation, along with the decline in nutrients in the microcosm, decreasing decay rates (C/N ratios) and forest NPP. The species richness pattern was, however, unimodal, peaking at mid-elevation, and was independent of nutrient levels and animal abundance (167, 198 and 97 species per forest type [P = 0.009] in the tabonuco, palo colorado and dwarf forest respectively). Further evidence for the independence of species richness and abundance is given in Richardson et al. (2000b). Body size and total animal biomass also peaked at mid-elevation, and the smallest individuals and biomass were in the dwarf forest. Explanations for higher species richness in the intermediate forest may lie in the optimal conditions provided for animal and plant growth and survival (fewer extinctions), or in its structural and physical heterogeneity. It has the highest density of vines and epiphytes, soils are covered by a thick network of roots with bryophytes, and herbaceous ground cover is greater than in the tabonuco forest (Weaver, 1994). Many of the bromeliad invertebrates spend only part of their life history in bromeliads and, as adults, come under influences of the wider forest ecosystem.

The bromeliad habitat provides stable long-term conditions in which many successive generations of invertebrates can breed. Species that breed there, but leave as adults to pass into the wider forest ecosystem (e.g., larval Diptera and Coleoptera), made up approx. 80% of the fauna. In areas of high bromeliad density in the palo colorado and dwarf forests their contribution to food chains could be considerable, and in all forests they provide a habitat for many specialised species, endemic to bromeliads and endemic to Puerto Rico.

Acknowledgements

I am grateful to collaborators, F. N. Scatena, M. Hansen, W. H. McDowell and M. J. Richardson, for giving generously of their time and ideas, and to staff of the El Verde Field Station of the University of Puerto Rico and the plant analysis laboratory of the International Institute of Tropical Forestry. This research was conducted as part of the Long-Term Ecological Research Program in the Luquillo Experimental Forest, with additional funds from the USDA Forest Service, the Carnegie Trust for the Universities of Scotland, Napier University, and The Royal Society of Edinburgh.

References

Benzing, D. H. Vascular epiphytes. Cambridge University Press, Cambridge (1990).
Hansen, M. & Richardson, B. A. A new species of Omicrus Sharp (Coleoptera: hydrophilidae) from Puerto Rico and its larva, the first known larva of Omicrini. Syst. Entomol. 23, 1–8 (1998).
Richardson, B. A. The bromeliad microcosm and the assessment of faunal diversity in a neotropical forest. Biotropica 31, 321–336 (1999).
Richardson, B. A., Richardson, M. J., Scatena, F. N. & Mcdowell. W. H. Effects of nutrient availability on bromeliad populations and their invertebrate communities in a humid tropical forest in Puerto Rico. J. Trop. Ecol. 16, 167–188 (2000a).

Richardson, B. A., Rogers, C. & Richardson, M. J. Nutrients, diversity and community structure of two phytotelm systems in a lower montane forest, Puerto Rico. *Ecol. Entomol.* **25,** 348–356 (2000b).

Weaver, P. L. Baño do Oro Natural Area Luquillo Mountains, Puerto Rico. General Technical Report SO-111. USDA Forest Service, Southern Forest Experiment Station, New Orleans, LA, p. 55 (1994).

Tropical Ecosystems: Structure, Diversity and Human Welfare.
Proceedings of the International Conference on Tropical Ecosystems
K. N. Ganeshaiah, R. Uma Shaanker and K. S. Bawa (eds)
Published by Oxford–IBH, New Delhi. 2001. pp. 410–411.

Time will tell: Assessment of long-term species diversity in vertical, horizontal and temporal dimensions in a Neotropical fruit-feeding butterfly community

P. J. DeVries,** and T. R. Walla[†]*

*Center for Biodiversity Studies, Milwaukee Public Museum, Milwaukee,
Wisconsin 53233 [†]Department of Biology, University of Oregon, Eugene,
Oregon 97404, USA
**e-mail: pjd@mpm.edu

Butterflies are frequently cited as a focal group in studies that estimate tropical insect community diversity, and they are also important in studies supporting conservation policy. However, most often butterfly studies do not directly address the effects of time and space on community diversity. Rather, most are limited by short sampling periods, use of non-comparable sampling methods, small sample sizes, and contain little information on spatial and temporal distribution. Thus, it is usually difficult or impossible to assess community patterns accurately, or compare diversity studies among areas.

Two recent studies sampled a fruit-feeding guild of Ecuadorian nymphalid butterflies at 12 consecutive monthly intervals in the canopy and understorey across a range of microhabitats, including intact forest, hi-graded forest, old

Keywords: Species abundance distributions, tropical butterfly communities, spatial effects, temporal effects, conservation.

secondary forest and forest edges (DeVries *et al.*, 1997; 1999). The results concluded that butterfly communities in lowland Neotropical rainforests showed significant β-diversity in all spatial and temporal dimensions. These studies further suggested that species richness of fruit-feeding butterflies was similar in both canopy and understorey, that significant differences in species richness may occur among horizontal habitats, and that only a small fraction of total sample species richness was represented within any given month.

To test the diversity patterns recovered from our one-year studies, we used the same sampling protocols established previously to sample fruit-feeding butterflies at monthly intervals for five consecutive years in the canopy and understory of five contiguous forest plots within the same rainforest (DeVries and Walla, 2001). Five years of sampling accrued a total of 11,861 individuals in 128 species. Total species diversity (γ-diversity) partitioned into additive components within and among community subdivisions (α-diversity and β-diversity) confirmed that there was significant β-diversity in vertical, horizontal and temporal dimensions. Species abundance distributions showed that significant number of species belonged to either the canopy or understorey fauna, confirming that samples from one vertical position seriously underestimate total diversity. Significant monthly variation showed that intermittent or short-term sampling would underestimate diversity; further variation in diversity among years and areas showed a strong influence of sampling year, and that even when the underlying communities were the same, temporal interactions affected species diversity in both horizontal and vertical dimensions.

The results of our five-year study suggest that to advance our understanding of tropical diversity, long-term studies among many sites are required to verify the generality of the spatial and temporal patterns of diversity community patterns described here. Finally, our five-year study shows that quick surveys are extremely limited for evaluating tropical insect diversity, or for promoting a broad understanding of tropical communities and how to conserve them.

References

Devries, P. J., Murray, D. & Lande, R. Species diversity in vertical, horizontal, and temporal dimensions of a fruit-feeding butterfly community in an Ecuadorian ranforest. *Biol. J. Lin. Soc.* **62**, 343–364 (1997).

Devries, P. J., Walla, T. & Greeney, H. Species diversity in spatial and temporal dimensions of fruit-feeding butterflies from two Ecuadorian rainforests. *Biol. J. Lin. Soc.* **68**, 333–353 (1999).

Devries, P. J. & Walla T. R. Long-term spatial and temporal species diversity in a neotropical fruit-feeding butterfly community. *Biol. J. Lin. Soc.* (in review) (2001).

Tropical Ecosystems: Structure, Diversity and Human Welfare.
Proceedings of the International Conference on Tropical Ecosystems
K. N. Ganeshaiah, R. Uma Shaanker and K. S. Bawa (eds)
Published by Oxford–IBH, New Delhi. 2001. pp. 412–415.

Use of stable isotopes in canopy research

Peter Hietz[*][†] *and Wolfgang Wanek*[‡]

*Institute of Botany, University of Agricultural Sciences, G. Mendel-Str. 33, A-1180 Vienna, Austria
[‡]Institute of Ecology and Conservation Biology, University of Vienna, Althanstr. 14, A-1091 Vienna, Austria
[†]e-mail: hietz@edv1.boku.ac.at

As many physical, chemical or biochemical processes discriminate against the heavier or the lighter version of a chemical element or compound, stable isotopes techniques have produced important insights into physiological and ecosystem processes (Handley and Raven, 1992; Högberg, 1997). So far the method has rarely been applied to investigate tropical forest canopies and their biota, and we here present recent examples from Mexico and Costa Rica using C and N isotopes.

The fact that Rubisco discriminates stronger against $^{13}CO_2$ than PEP-carboxylase has long been used to distinguish between crassulacean acid metabolism (CAM) and C3 plants. CAM may be quite common among vascular epiphytes and the proportion of epiphytes performing CAM is higher in exposed than in shaded canopy positions (Zotz and Ziegler, 1997) and higher in the more arid forests (Earnshaw et al., 1987; Hietz et al., 1999). However, it is important to be aware that the C-isotope ratio can only show CAM activity, not the potential. Recent studies have found epiphytes with C3-like isotope ratios but nocturnal carboxylation activity, which, though

---

Keywords: Canopy, epiphyte, nitrogen, humid montane forest, stable isotope.

perhaps not resulting in net CO_2 uptake, significantly improves the total carbon balance (Holtum and Winter, 1999).

In plants under drought stress with partially closed stomata, $^{13}CO_2$ in the intercellulars is relatively enriched, which results in a lower discrimination during CO_2 uptake than in unstressed plants. This allows to detect the long-term effect of drought from tissue carbon isotope ratios. In several epiphytes drought stress was found to decrease with plant size (Zotz and Ziegler, 1999; Hietz and Wanek, 2001). Small plants either have a higher surface/volume ratio or, in bromeliads, a very inefficient tank and therefore store less water per unit leaf area than large plants. With an irregular water supply, the water storage of small plants is soon depleted and this size-related water stress appears to have substantial consequences for many aspects of epiphyte ecophysiology. One important caveat for the interpretation of carbon isotope ratios, particularly from tall rainforests is: Owing to the high soil respiration rates from the decomposition of ^{13}C-depleted plant matter and the limited exchange of air between the understorey and the upper canopy, there may be a vertical gradient in carbon isotope ratios in the air resulting in different isotope ratios in leaves, although these are not related to water stress or water use efficiency (Medina *et al.*, 1986).

The nitrogen cycle involves different forms of organic and inorganic N, many pools and fluxes. Nitrogen isotope ratios can help to analyze many of these processes, but the possible interactions of different steps involving isotopic effects also pose problems for the interpretation of the observed isotope ratios. ^{15}N in ammonium and nitrate of rainwater is generally strongly depleted relative to N in soil water. Since epiphytes rely partly on nutrients from rainwater, this is reflected in lower $\Delta^{15}N$ in epiphytes than in the leaves of their host trees (Stewart *et al.*, 1995). Among epiphytes, bromeliads appear to obtain a higher proportion of N from the rain via their water-absorbing leaf scales than other groups, and atmospheric bromeliads more than bromeliads with water and litter impounding tanks (Hietz *et al.*, 1999).

In epiphyte-rich forests and particularly in montane cloud forests, thick layers of canopy soil develop on large branches, providing anchorage, water and nutrients for epiphytes. A comparison of $\Delta^{15}N$ signatures of terrestrial and canopy soil, and tree and epiphyte leaves as well as litter studies (Nadkarni and Matelson, 1991) suggests that the canopy soil is mainly derived from epiphytes, i.e. that the nutrient cycle of epiphytes and canopy soil is largely detached from that of terrestrial soil and trees. Epiphytes rooting in canopy soil on thick branches had lower $\Delta^{15}N$ signals than individuals on thin branches with little substrate (Figure 1), suggesting that on small branches they rely more heavily on direct rain as a nitrogen source. Epiphytes rooting in canopy soil had higher N concentrations and lower $\Delta^{13}C$ values, indicating that canopy soil substantially improves N supply and reduces drought stress.

Apart from studying natural abundances, tracer studies with enriched ^{15}N can be used to follow nitrogen fluxes. Epiphylls are organisms that colonize leaf surfaces (the phyllosphere) and belong to bacteria, cyanobacteria, fungi, lichens and bryophytes. While heterotroph microorganisms largely depend on sugars and nutrients leached from plant leaves and minerals from wet and dry deposition, N_2-fixation by cyanobacteria appears to be an important N source for tropical rainforests as estimated by [15]N_2 uptake (Bentley, 1987).

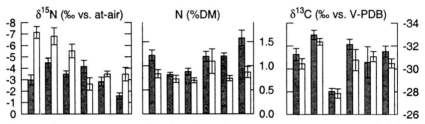

Figure 1. $\Delta^{15}N$, N content and $\Delta^{13}C$ of epiphytes from six systematic groups sampled from thin branches (empty bars) and thick branches (filled bars) on nine trees in the Monteverde cloud forest, Costa Rica.

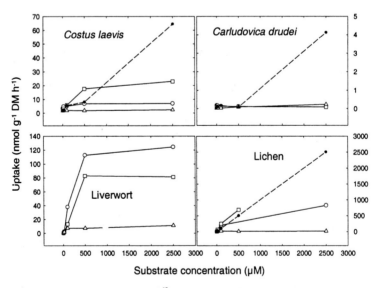

Figure 2. Tracer study on the uptake of ^{15}N compounds (circles: glycin, triangles: ammonium, squares: nitrate) by epiphylls and understorey host leaves in the Corcovado lowland rainforest, Costa Rica. Open symbols: uptake through the upper leaf surface, closed symbols: through the lower surface.

Foliicolous lichens and liverworts may form thick mats on older leaves and are thought to be carbon autotroph and highly efficient in nutrient uptake from rain. A study in a lowland rainforest (Corcovado National Park, Costa Rica) exposing leaf surfaces and epiphyllous lichens and liverworts to various ^{15}N-labelled compounds at different concentrations showed that epiphylls scavenge nutrients from artificial rain and thus compete with foliar uptake by the phorophytes. Foliar uptake by understorey plants at common rain water concentrations of ammonium and nitrate (1 to 50 μM) was low through the upper leaf surface, but significant uptake of ammonium was found through the lower surface with stomata (Figure 2). Nutrient scavenging and N_2-fixation of epiphyllous cryptogams may substantially alter nutrient fluxes in tropical rainforest ecosystems.

Acknowledgements

Travel funds made available by the FWF (Project 12241-BIO), the University of Agricultural Sciences and the University of Vienna are gratefully acknowledged.

References

Bentley, B. L. Nitrogen fixation by epiphylls in a tropical rainforest. *Ann. Missouri Bot. Gard.* **74**, 234–241 (1987).

Earnshaw, M. J., Winter, K. Ziegler, H., Stichler, W., Cruttwell, N. E. G., Kerenga, K., Cribb, P. J., Wood, J., Croft, J. R., Carver, K. A. & Gunn, T. C. Altitudinal changes in the incidence of crassulacean acid metabolism in vascular epiphytes and related life forms in Papua New Guinea. *Oecologia* **73**, 566–572 (1987).

Handley, L. L. & Raven, J. A. The use of natural abundance of nitrogen isotopes in plant physiology and ecology. *Plant Cell Environ.* **15**, 965–985 (1992).

Hietz, P, & Wanek, W. Size-dependent variation of carbon and nitrogen isotopes in epiphytic bromeliads. *Funct. Ecol.*, submitted (2001).

Hietz, P., Wanek, W. & Popp, M. Stable isotopic composition of carbon and nitrogen and nitrogen content in vascular epiphytes along an altitudinal transect. *Plant Cell Environ.* **22**, 1435–1443 (1999).

Holtum, J. A. M. & Winter, K. Degrees of crassulacean acid metabolism in tropical epiphytic and lithophytic ferns. *Aust. J. Plant Physiol.* **26**, 749–757 (1999).

Högberg, P. N-15 natural abundance in soil-plant systems. *New Phytol.* **137**, 179–203 (1997).

Medina, E., Montes, G., Cuevas, E. & Rokzandic. Z. Profiles of CO_2 concentration and delta-13C values in tropical rainforests of the upper Rio Negro Basin, Venezuela. *J. Trop. Ecol.* **2**, 207–217 (1986).

Nadkarni, N. M. & Matelson. T. J. Fine litter dynamics within the tree canopy of a tropical cloud forest. *Ecology* **72**, 2071–2082 (1991).

Stewart, G. R., Schmidt, S., Handley, L. L., Turnbull, M. H., Erskine, P. D. & Joly, C. A. [15]N natural abundance of vascular rainforest epiphytes: implications for nitrogen source and acquisition. *Plant Cell Environ.* **18**, 85–90 (1995).

Zotz, G. & Ziegler, H. The occurrence of crassulacean acid metabolism among vascular epiphytes from central Panama. *New Phytol.* **137**, 223–229 (1997).

Zotz, G. & Ziegler, H. Size-related differences in carbon isotope discrimination in the epiphytic orchid, *Dimerandra emarginata*. *Naturwissenschaften* **86**, 39–40 (1999).

Tropical Ecosystems: Structure, Diversity and Human Welfare.
Proceedings of the International Conference on Tropical Ecosystems
K. N. Ganeshaiah, R. Uma Shaanker and K. S. Bawa (eds)
Published by Oxford–IBH, New Delhi. 2001. pp. 416–419.

Arthropod diversity in epiphytic bryophytes of a Neotropical cloud forest

S. P. Yanoviak and N. M. Nadkarni

The Evergreen State College; Olympia, WA 98505, USA
e-mail: Yanoviak@racsa.co.cr, NadkarnN@evergreen.edu

Tropical montane forests typically support high diversity and abundance of epiphytic plants. For example, the forests in and around Monteverde, Costa Rica, contain several hundred species of epiphytic orchids alone (Nadkarni and Wheelwright, 2000). Bryophytes (mosses and leafy liverworts) are the most conspicuous epiphytes in the Monteverde forests, and form a nearly continuous covering on the woody portions of trees from the ground to the tips of the highest branches. Few ecological studies have specifically focused on these plants (Clark *et al.*, 1998; Nadkarni *et al.*, 2000). Moreover, unlike the fauna of temperate bryophytes (Usher, 1983), almost nothing is known of the arthropods and other invertebrates that live in association with epiphytic bryophytes in tropical forests (Gerson, 1982).

Here we present preliminary results from the first half of a two-year study of arthropods living in epiphytic bryophytes of the Monteverde Cloud Forest Preserve, Costa Rica. The Preserve is mostly primary forest with a few small (2–5 ha) secondary forest inclusions (see Nadkarni and Wheelwright 2000 for a site description). The results presented here address the following subset of questions from the larger project: 1) What kinds of arthropods live in the bryophytes of the Preserve?; 2) Does arthropod diversity and abundance in

---

Keywords: Arthropods, canopy, diversity, epiphytes, tropical cloud forest.

these epiphytes differ between secondary (~ 40 year old) and primary (> 200 year old) forest types?; and 3) What is the role of dead organic matter in structuring arthropod assemblages in this system?

To answer the first two questions, we collected small patches of bryophytes (each ca 300 ml in volume) from crowns (15–30 m) of various tree species in primary and secondary forests. Arthropods were extracted from epiphyte samples using Tullgren funnels, counted, and assigned to morphospecies. When possible (e.g. most Coleoptera), individuals were identified to family level. After arthropod extraction, bryophyte samples were dried at 60°C for 24 h, then weighed to the nearest 0.01 g. Means for arthropod abundance, arthropod morphospecies richness, and bryophyte dry mass were compared between the two forest types with repeated-measures ANOVA. Abundance data were log-transformed before analysis.

Approximately 200 arthropod species were found in association with epiphytic bryophytes in Monteverde. Mites (Acarina), Collembola, and Coleoptera (especially Curculionidae and Staphylinidae) were consistently the most diverse and most abundant groups collected. Ants (Hymenoptera: Formicidae) were very numerous in many samples, and were proportionally more abundant in secondary forest (Figure 1). In general, arthropod morphospecies composition in epiphytic bryophytes was very similar between forest types.

The average number of arthropod morphospecies collected was significantly higher in primary forest bryophytes than in secondary forest ($F_{1,68} = 6.76$, $P = 0.0114$). This difference was strongly influenced by data collected in January and May 2000 (Figure 2 a). Unlike morphospecies richness, arthropod abundance was greater in secondary forest than in primary forest on all collection dates ($F_{1,68} = 27.68$, $P = 0.0001$; Figure 2 b). This is attributed to the

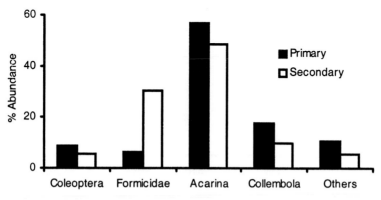

Figure 1. Relative abundance of selected arthropod groups in epiphytic bryophyte collections from primary and secondary forests in Monteverde. Values were obtained from abundance data pooled within a forest type across all collection dates.

large number of ants in the secondary forest samples; mean arthropod abundance did not differ between forest types when ants were removed. Changes in average morphospecies richness and abundance over time (Figure 2) suggest that the structure of arthropod assemblages in epiphytic bryophytes varies seasonally.

Although we standardized the approximate volume of bryophyte material collected, the average dry mass of samples was greater in primary forest on all collection dates ($F_{1,68} = 208.0$, $P = 0.0001$). The overall mean sample mass (pooled across sample dates; $\pm$ SD) was 6.7 g ($\pm$ 2.83) in primary forest and 4.1 g ($\pm$ 1.51) in secondary forest. We attribute this difference in dry mass to differences in epiphyte substrate composition and microhabitat quality between forests. Given sufficient time, epiphytes create dense mats via vegetative growth, accumulation of detritus, and aggregation of epiphyte roots. The vegetative portion of bryophytes collected from primary forest was more dense than that collected from secondary forest, and primary forest samples contained a greater fraction of detritus/root (hereafter, 'brown') material.

We hypothesized that these differences in microhabitat quality and composition contributed to the observed differences in arthropod mor-

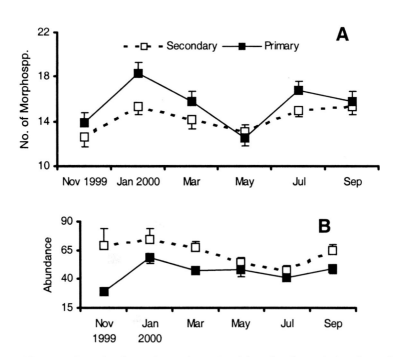

Figure 2. Mean number of arthropod morphospecies (A) and arthropod abundance (B) in bryophyte samples from primary and secondary forests in Monteverde. Bars represent $\pm$ 1 SE, $n = 35$ for each mean.

phospecies richness between forest types (Figure 2a; question 3 above). Focusing on the brown microhabitat, we predicted that the number of morphospecies in a sample would increase with the proportion of detritus material present. To test this prediction, a total of 62 bryophyte patches (10 × 10 cm) were collected from five primary forest tree crowns. Each patch was divided into vegetative and brown fractions, and arthropods were extracted and dry weights obtained (for each fraction separately) as described above.

The mean (± SE) number of morphospecies was higher in the vegetative portions (6.7 ± 0.63) than the brown portions (4.9 ± 0.38) of the samples (paired $t = 3.11$, $P = 0.003$), and the total number of morphospecies in a sample significantly decreased with increasing proportion of brown material present (linear regression: $F_{1,60} = 8.09$, $P = 0.006$, $R^2 = 0.12$). These results suggest that arthropod diversity (and perhaps observed differences in diversity between forest types) is more strongly linked to the quality of the vegetative portion than the brown portion of epiphyte mats. This hypothesis remains to be tested.

Our research on the fauna of epiphytic bryophytes in Monteverde shows that these habitats support diverse arthropod assemblages, that the structure of these assemblages differs between primary and secondary forest types, and that microhabitat characteristics may play an important role in determining those structural differences. This work has recently been expanded to include other cloud forests in the region, as well as experimental approaches to understanding the factors that regulate arthropod diversity in cloud forest bryophytes.

Acknowledgements

We extend special thanks to Rodrigo Solano, Chad Smith, and Hannah Smith for their patient assistance in the field and lab. The Monteverde Cloud Forest Preserve provided logistical support. This research was supported by a grant from the National Science Foundation (RUI #9974035) to NMN.

References

Clark, K. L., Nadkarni, N. M. & Gholz, H. L. Growth, net production, litter decomposition, and net nitrogen accumulation by epiphytic bryophytes in a tropical montane forest. *Biotropica* **30**, 12–23 (1998).

Gerson, U. Bryophytes and invertebrates. in *Bryophyte Ecology* (ed. Smith, A. J. E.) 291–332 (London: Chapman and Hall, 1982).

Nadkarni, N. M., Cobb, A. R. & Solano, R. Interception and retention of macroscopic bryophyte fragments by branch substrates in a tropical cloud forest: an experimental and demographic approach. *Oecologia* **122**, 60–65 (2000).

Nadkarni, N. M. & Wheelwright, N. T. (eds) Monteverde: ecology and conservation of a tropical cloud forest. New York: Oxford University Press (2000).

Usher, M. B. Pattern in the simple moss-turf communities of the sub-Antarctic and maritime Antarctic. *J. Ecol.* **71**, 945–958 (1983).

Tropical Ecosystems: Structure, Diversity and Human Welfare.
Proceedings of the International Conference on Tropical Ecosystems
K. N. Ganeshaiah, R. Uma Shaanker and K. S. Bawa (eds)
Published by Oxford–IBH, New Delhi. 2001. pp. 420–422.

Interactions between social bees and their foodplants in a rainforest canopy of Western Ghats, India

M. S. Devy*[,†] and C. Livingstone[‡]

*Ashoka Trust for Research in Ecology and the Environment, 659, 5th A Main,
Hebbal, Bangalore 560 024, India
‡Department of Botany, Madras Christian College, Tambaram,
Chennai 600 059, India
†e-mail: soubadra@atree.org

Integrating the phenology of pollinators and their foodplants is crucial to understand pollination guilds. In the tropics, studies on the phenology of the trees and of pollinating vectors such as bees have mostly been carried out independent of each other. This study, carried out at a wet evergreen forest of Kakachi, in Southern Western Ghats, integrates the phenology of two social bee species *Apis dorsata* and *Apis cerana* with their canopy and sub-canopy food plants. The two bee species are the only ones that are active on the canopy of this forest and visit the understorey plants occasionally (Davidar *et al.*, unpublished). Unlike other tropical forests, solitary bees are distinctly absent in this forest canopy, and therefore the two bee species play a major role in the pollination of a number of tree species in the forest.

Hive density and the foragers' abundance of the two species were quantified across different seasons. Floral resource abundance was determined by

Keywords: Canopy pollination, social bees, rainforests, Western Ghats, India.

Table 1. Food plant species of *Apis cerana* and *Apis dorsata* and their distributional range in the Western Ghats

Nectar and pollen source	Species
Sapotaceae	*Palaquium ellipticum* (Dalz.) Baillon
	Palaquium bourdillonii Brandis*
	Isonandra lanceolata Wight
Lauraceae	*Litsea wightiana* (Nees) Hook,f**
	*Actinodaphne bourdillonii** Gamble
	Litsea insignis Gamble**
	Litsea glabrata (Wall.ex Nees)**
	Neolitsea fischeri Gamble**
	Neolitsea cassia (L.) Kosterm
Elaeocarpaceae	*Elaeocarpus munronii* (Wight) Mast.
	Elaeocarpus vensustus Bedd.**
Guttiferae	*Calophyllum austroindicum* Kosterm. ex. Stevens**
Anacardiaceae	*Holigarna nigra* Bourd.
Papilionaceae	*Ormosia travancorica* Bedd.
Rosaceae	*Prunus ceylanica* (Wight) Miq.
Symplocaceae	*Symplocos cochinchinensis* (Lour.) Moore
Myrtaceae	*Eugenia thwaitesii* Duthie
Rubiaceae	*Tricalysia apiocarpa* (Dalz.) Gamble
Only pollen source	
Ebenaceae	*Diospyros malabarica* (Desr.) Kostel.
Flacourtiaceae	*Scolopia crenata* (Wight & arn.) Clos
Euphorbiaceae	*Mallotus tetracoccus* (Roxb.) Kurz
	Epiprinus mallotiformis (Muell.-Arg) Croizat
	Macaranga peltata (Roxb.) Muell.-Arg
	Acronychia pedunculata (L.) Miq
Rutaceae	

Bee-visited species = 24.
Total tree species recorded in the site = 90 (Ganesh *et al.*, 1996).
*Endemic to Agasthyamalai (Nayar, 1996).
**Endemic to Southern Western Ghats (Nayar, 1996).

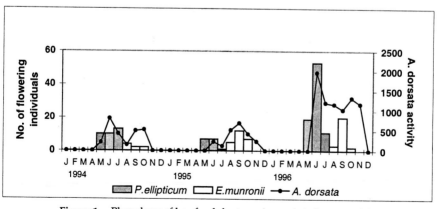

Figure 1. Phenology of key foodplant species vs *Apis dorsata* activity.

establishing transects and marking out trees which were visited intensely by both bees. *Apis cerana* and *Apis dorsata* were found to be involved in the pollination of 12% of the canopy tree species (see Table 1). The choice of food plants is related to their life history traits and energy requirements. *Apis dorsata* is a local migrant, and its arrival to the site coincides with the flowering of *Palaquium ellipticum*, which acts as a 'flag' species. *P. ellipticum* is the most common tree in the site and is a reliable resource to track as they flower annually and offer copious nectar and pollen. It probably meets the energy demands of large open hive nesting *A. dorsata*. Remarkably only two tree species namely *Palaquium ellipticum* and *Elaeocarpus munronii* determined the *Apis dorsata* presence in the site (Figure 1). *Apis cerana*, which is a resident species, exhibits low colony density and occurs in smaller colony size in the site. They visit tree species that occur in low density or exhibit non-annual flowering cycles in addition to *Palaquium ellipticum* and *Elaeocarpus munronii*. Bee foodplant composition and phenology in the adjoining secondary forest were compared with those of primary forest. The secondary forest cannot sustain the bee populations independently of primary forests. Management practices such as selection felling of foodplants, particularly *Palaquium ellipticum*, could interrupt social bee–foodplant interactions in these forests.

References

Ganesh, T., Ganesan, R., Devy, M. S., Davidar, P. & Bawa, K. S. Assessment of plant biodiversity at a mid-elevation evergreen forest of Kalakad–Mundanthurai Tiger Reserve, Western Ghats, India. *Curr. Sci.* **71**, 379–392 (1996).

Nayar, M. P. *Hotspots of Endemic Plants of India, Nepal and Bhutan* (Tropical Botanical Garden and Research Institute, Thiruvanthapuram, India, 1996).

Tropical Ecosystems: Structure, Diversity and Human Welfare.
Proceedings of the International Conference on Tropical Ecosystems
K. N. Ganeshaiah, R. Uma Shaanker and K. S. Bawa (eds)
Published by Oxford–IBH, New Delhi. 2001. pp. 423–427.

Why does the world need a global canopy programme?

Andrew W. Mitchell

ICAN-Global Canopy Programme, Halifax House, University of Oxford,
6-8 South Parks Road, Oxford OX1 3UB, United Kingdom
e-mail: a.mitchell@globalcanopy.org

The forest canopy is often popularly described as nature's last biological frontier. The amount of biodiversity contained there represents one of the greatest concentrations of life on earth. In addition, the canopy is where 'life meets the atmosphere'. The fluxes of gasses which occur at this interface are increasingly thought to play a significant role in maintaining the earth's climate. A challenge for science here is that little is known of the structure–function relationships that occur in forest canopies. We do not even know the true extent of the biodiversity there. Further, the precise mechanism and extent of the exchange of gasses across the leaf atmosphere boundary including the uptake of carbon in the canopy and its transfer to the soil, is also largely unknown.

Why is this important? First, current human-induced biodiversity loss is the sixth extinction event which will dramatically alter future evolution on the planet. Knowledge of biodiversity alone cannot save it but changes in the way we act regarding the maintenance of biodiversity do need to be backed by sound understanding. Second, the extent of our ignorance about biodiversity in forest canopies is both an economic threat through lost commercial opportunity and its value is therefore underestimated by

Keywords: Forest, canopy, carbon, biodiversity.

governments, who cannot plan on an evolutionary time scale, and discount the value of biodiversity. Third, our almost complete lack of understanding of structure–function relationships in forest canopies and their impact on maintaining global environmental conditions is a major potential threat to human welfare in many countries at the local and regional levels. And fourth, a lack of understanding at all levels of the role of forests in climate change, especially as sequesters of carbon, has led to a major breakdown in the implementation of the one treaty which is designed to fix the problem – namely the Kyoto protocol.

Why is our understanding so poor? First we lack the scientific data on which to make sound predictions and which enable us to scale up from a molecular level to a global level or from space-based observations to a leaf epiphyll. Essentially this is because of the difficulty of access in forest canopies which has delayed the implementation of experiments other than those which ask: 'what is there?' as opposed to manipulative investigations asking 'how does it work?' or 'how will it change?'. Canopy science has now emerged from the pioneering days of 30 years ago using monkeys or climbing ropes to access the crowns of tall trees to the large scale operations of today using canopy cranes, dirigibles, towers and other techniques which enable repeatable access with ease but often at considerable expense.

Second, the natural sciences have not been able corner the market in major funding for projects in its field in the way that the physical sciences have. Projects on forest canopies have not attracted the support of governmental science funding bodies at a level commensurate with either the scale or urgency of the problem to be addressed.

These funding comparisons for major physical science projects under consideration in 1998 within the European Science Foundation's Large Facilities Program (Mitchell, 2001) indicate some recent priorities:

European Spallation Source	ECU 950 million
(Project investigating the nature of neutrinos)	
X-Ray Vision Project	ECU 150 million
(A bio-molecular structure synchrotron)	
100 Telsa High magnetic field magnet	ECU 50 million
ESF Tropical Canopy Research	ECU 327,240
(Total budget estimate 1994-1998/9)	

Part of the reason for this may be that no such programme has been presented with international support which addresses the scientific questions at the right scale and that can dramatically capture the attention of the political aspects of the fund-raising process which could benefit from the results and from the positive public attention the programme could generate in terms of education and conservation.

Following the 'Tropical Rainforest Canopies: Ecology and Management' International Conference at Oxford in 1998, a number of participants, including the author, convened a workshop under the auspices of the International Canopy Network (ICAN) and funded by European Science Foundation and the National Science Foundation, to address these issues. The outcome of that meeting was a Forest Canopy Planning Workshop Report (Nadkarni et al., 2000) at which the following resolution was agreed by 29 experts present from 9 countries: 'We propose an integrated, coordinated study of canopies across major environmental and management gradients to investigate the role of forest canopies in maintaining global biodiversity, global environmental conditions, and the sustainability of forests'.

A year later in November 2000, with funding from two UK foundations the Global Canopy Programme (GCP) Secretariat Office, working in close partnership with ICAN was established in Oxford. Its mission is as follows:

- To seek to create the international framework for implementing the central research vision and goals of the Global Canopy Programme as conceived at the Oxford canopy workshop
- To determine the resources needed and to forge the international partnerships in research, education, training and conservation who wish to have a stake in the programme
- To combine these groups into an alliance of organisations who support and expect to benefit from a proposal, which seeks funding at a level necessary for the task on a global scale, for submission to major national and international donors to implement the program.

This initiative will be formally launched at the 3rd International Conference on Canopy Science in Cairns, Australia in June 2002.

Since November 2000 an international Steering Committee and Science Advisory Committee has been set up besides also establishing a network of in-country representatives. Research has been carried out into the scientific projects currently in progress around the world related to the issues mentioned above. Representatives of these groups have been contacted and all have expressed enthusiasm at the prospect of collaborating in a global programme of research provided funding is available and a mechanism for such collaboration exists. Institutions and projects currently contacted have been divided into three potential networks:

- Canopy Biodiversity and Ecosystem Function Network
- Canopy Climate Change Network
- Canopy Conservation, Policy and Education Network

It is clear that those in the climate change network are already quite well organised internationally often using standard protocols, shared databases and an extensive network of canopy access towers to obtain measurements of gas fluxes in the canopy. These include such organisations as CarboEurope, AmeriFlux and the Large Scale Biosphere–Atmosphere Exchange Experiment (LBA) and the emerging AsiaFlux. Globally, funding for these programs is already at the $100–$150 million level (P. Jarvis, pers. commun.).

Those projects concerned with biodiversity and ecosystem studies are growing rapidly but are less well organised internationally as a network with few common protocols or shared databases. These include some ten canopy access jib cranes now situated in both temperate and tropical forests plus numerous towers, walkways, and other individual canopy access systems such as the Canopy Operation Permanent Access System (COPAS) under development and Operation Canopée.

Extensive forest-related education and conservation organisations are in existence but few specifically focused on forest canopies or the science being conducted within them. The flow of information at the policy level from canopy science is yet to be investigated but may benefit from better international co-ordination and it is the intention of GCP to enable this information to be easily accessed by scientists, the public, educators, and decision makers and to support governments in their efforts to implement the Conventions on Biodiversity and on Climate Change. GCP has proposed a world training centre for capacity building in canopy science which will manage courses on canopy access techniques and canopy ecology.

One of the questions now being tackled by the GCP is how to link these three areas together into one programme and what benefits would arise to each group from such a collaboration. We welcome input from interested collaborators on this question and also welcome contacts from researchers or institutions interested in linking existing projects or contributing new ones for potential funding within the framework of the proposed GCP. A three-year grant from NSF in 2000 will enable ICAN to design the 'Big Canopy Database' which will provide a principal respository of shared information for the proposed GCP.

There never has been a better time to scale up and co-ordinate forest canopy research activities into a much better organised and integrated research effort across the world which will address two of the greatest scientific and conservation challenges facing humans over the next fifty years – namely the loss of genetic diversity and the impact of climate change.

Acknowledgements

The Global Canopy Programme is supported by the Rufford Foundation, the Maurice Laing Foundation and Earthwatch Institute. Contacts and support from The International Canopy Network (ICAN) have also been invaluable in starting the programme.

References

Mitchell, A. W. Canopy science – time to shape up. in *Tropical Rainforest Canopies: Ecology and Management*. Proceedings of the 3rd International Conference at Oxford' (eds Eduard Linsenmair, Andrew Davis, Martin Speight, & Brigit Fiala) (Kluwer Academic Publishers, Dordrecht, 2001) (in press).

Nadkarni, N., Mitchell, A. W. & Rentmeester, S. Forest Canopy Planning Workshop: Final Report. Evergreen State College, Olympia Washington, USA. Unpublished Report (2000).

Tropical Ecosystems: Structure, Diversity and Human Welfare.
Proceedings of the International Conference on Tropical Ecosystems
K. N. Ganeshaiah, R. Uma Shaanker and K. S. Bawa (eds)
Published by Oxford–IBH, New Delhi. 2001. pp. 428–432.

Bottom-up and top-down effects on canopy ants

Diane W. Davidson

University of Utah, Department of Biology, 257 South, 1400 East, Salt Lake City,
UT 84112-0840, USA
e-mail: Davidson@biology.utah.edu

Ants often dominate the rainforest canopy in numbers and biomass, and despite being viewed as predators and scavengers, they are often more abundant than herbivores. Tobin (1994) suggested that the dominant canopy ants feed mainly as herbivores (using EFN and homopteran honeydew), and stable isotope work in Panama and now Peru shows they obtain significant fractions of their nitrogen from these exudates. Tested against a wide-range of rainforest ants, stable isotopes give an intuitively satisfying picture of trophic specializations among and within major taxa (sub-families and genera). Mean delta ^{15}N values are highest for specialized predators, higher for generalized predators and scavengers than for exudate-foragers, higher for homopteran-tenders that may eat some of their associates than for exudate-feeders not tending Homoptera, and lowest of all for species feeding on pollen, fungal spores or (perhaps, see below) some as yet to be defined resource. We can therefore use these values, ranging continuously from < 2.0 to almost 11, as an aid in interpreting patterns in the effects of plant-foraging on the ecology and evolution of ants (family Formicidae).

The relationship between delta ^{15}N and the exchange ratio (ER = the degree of CHO : N limitation) generally demonstrates greater nitrogen limitation in

Keywords: Ants, avian predators, exudate-feeding, nitrogen isotopes, tempos.

ants with lower delta ^{15}N, i.e. more herbivorous diets (Figure 1). ER is computed as the ratio of the lowest CHO concentration (wt/vol of sucrose in water) for which > 50% of workers will forage, divided by the threshold concentration of amino acids (AAs) which they find acceptable as food (Kay, 2000). (Effectively, this ratio puts CHOs and AAs in the same currency.) Mean ER values for individual species are generally higher for exudate-dependent ant species, since exudates are high in CHOs but low in protein. Therefore, such species will forage only for relatively high concentrations of CHOs. For ants (e.g., *Pheidole*) with relatively balanced diets, ER = 1 (empirically, Figure 1), although this value probably has no special significance, there is no reason to assume that ants require sugars and amino acids in identical wt/vol concentrations. There are no predatory ants with high ER values. ER is lowest for ants with relatively high delta ^{15}N, and then rises steeply at delta ^{15}N values < 4 (Figure 2). Nevertheless, some species with delta ^{15}N < 3 have quite low measured ERs. These species include two myrmicines (genus *Cephalotes*) thought to forage on pollen and fungal spores, which would provide a more balanced diet than is typical for exudate-feeders, as well as several of the smaller, non-territorial *Camponotus* species. Most *Dolichoderus* species also have comparatively low ERs for their delta ^{15}N values (ranging from 3.1 to 4.6), almost certainly due in part to consumption of some of the Homoptera they tend. However, we are currently investigating mechanisms of N conservation and recycling in *Dolichoderus*.

The low ERs measured for the three small *Camponotus* and two *Cephalotes* species may reflect use of a previously unrecognized ant resource with

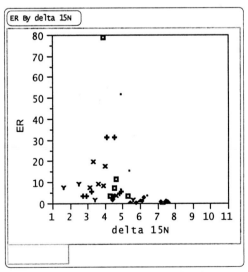

Figure 1. Exchange ratios (ER) for sucrose vs amino acid solutions vs. delta^{15}N (see text for details). Symbols: Formicinae (+: *Camponotus* and *Giganteops destructor*, Y: *Camponotus* sub-genus *Dendromyrmex*); Myrmicinae (♦: *Crematogaster*, Δ: *Pheidole*, Z: *Cephalotes*); Dolichoderinae (X: *Dolichoderus*, tiny •: *Azteca*); □: Ponerinae. To facilitate viewing of clustered points, *Cremato-gaster acuta* (body size = 4.0; ER = 79.5) was omitted.

relatively high N content but low delta^{15}N. Although this resource could be pollen, delta ^{15}N of pollen (a highly processed plant product) could be higher than isotopic ratios of plants producing the pollen. Another possibility would be N-fixing epiphylls such as cyanobacteria. All of these ants tend to be active on leaf surfaces during the day, frequently stopping to (apparently) feed at points on the leaves where there are no visible signs of exudates from either the plant or inconspicuous scale insects. We are currently searching microscopically for the nature of this resource.

Large diurnal foragers in the sub-families Formicinae and the Dolichoderinae differ consistently in crop capacities and dramatically in exudate uptake rates, with *Camponotus* spp. exceeding *Dolichoderus* spp. in both the former and the latter (Figure 2A & B). (Two species of *Camponotus* sub-genus *Dendromyrmex* and a bizarre bamboo specialist, *Camponotus depressus*, perform more poorly than do their congeners by both measures.) Among ants in the sub-family Myrmicinae, *Cephalotes* resemble *Camponotus* species in their crop capacities but *Dolichoderus* in their exudate-uptake rates. Finally, ants of the sub-family Ponerine transport exudates externally in their mandibles, rather than internally in crops, and their workers have the same exudate-transporting capacities and loading rates as do *Dolichoderus* species. An exception is *Paraponera clavata*, which more closely resembles *Camponotus* spp. in these capacities, perhaps because of its disproportionately large head size.

Exudate-feeding ants in the dominant canopy taxa have digestive adaptations permitting them to concentrate N from N-poor exudates efficiently (Eisner, 1957), as well as (except in cephalotines, 'turtle ants') having weakly developed, protein-costly exoskeleton (Davidson and Patrell-Kim, 1996, and recent data). Formicines are unique in having long, finger-like extensions of the proventriculus surrounding the crop. We believe that finger-like extensions of the formicine proventriculus alternately squeeze and release the crop to permit rapid filling.

The high CHO:N balance in exudate diets provides excess CHOs that can potentially be used as 'gasoline' to fund 'high tempo' activity and defense of spatial territories (Davidson, 1997). We find higher foraging tempos, i.e. faster worker velocities during natural foraging on horizontal branches, in more protein-starved *Camponotus* spp. than in species of similar body size but lower ER (*Dolichoderus* spp. and *Camponotus* sub-genus *Dendromyrmex* spp.). Similarly *Pheidole biconstricta* has a higher measured foraging tempo in comparison to a diversity of less exudate-dependent congeners.

Differences in crop capacities and feeding rates appear to have been driven principally by susceptibility to avian predators, because worker responses to artificial predators are correlated with types of defensive compounds. Volatile compounds (formic acid in formicines and keytones of species in *Dolichoderus* subgenus *bispinosus*) are depleted after being sprayed, and may be less likely than are persistent exocrine products (viscous defensive secretions of many *Dolichoderus* spp.) and armored exoskeleton (cephalotines) to deter predation. Reactions to an artificial avian predator were strongest among formicines (with *Camponotus* sub-genus *Dendromyrmex* spp. again exceptional for their genus and comparatively unresponsive), intermediate among cephalotines, and generally lowest in *Dolichoderus* spp. However, two

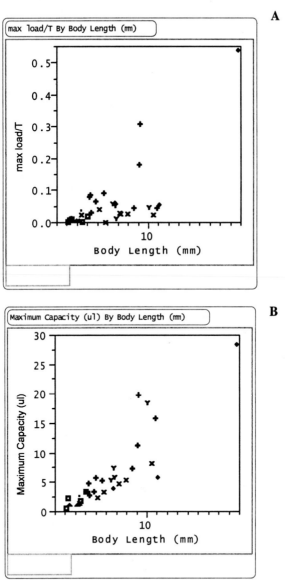

Figure 2. Maximum measured capacities of crops (or mandibles in Ponerinae) in µl (A) and maximum loading rates (B) versus body length (mm). Symbols as in Figure 1. To facilitate viewing of clustered points, *Paraponera clavata* (body length 18.75 mm; max. capacity = 28.5; max load/T = 0.54) was omitted.

species in the *D. bispinosus* sub-genus were exceptional in exhibiting formicine-like responses to predators. Although top-down forces appear to determine differences in foraging modes and performances of these relatively large-bodied ants, smaller-bodied species of, e.g. *Crematogaster* and *Azteca*, do not react strongly to artificial predators and may generally not be attractive targets of avian predators.

Differences between formicines and *Dolichoderus* spp. in the 'sociality' of foraging accompany disparities in reactions to predators. Formicines (including one of the two *Camponotus* [*Dendromyrmex*] spp.), almost always foraged alone at exudate baits, whereas *Dolichoderus* spp., including those with volatile defenses, nearly always foraged together. Rates of interruptions of single workers feeding at baits were therefore much higher in the latter than in the former ants, with *Dolichoderus* workers often or usually leaving to recruit nest-mates before feeding to satiation themselves. The two species of the *Dolichoderus bispinosus* subgroup did not behave like formicines. Part of their strategy for predator avoidance appears to be rapid and confusing running in all directions (like the *Azteca* species with similar or identical chemical defenses). Such a strategy may predispose these ants to forage together despite their lack of persistent defensive compounds. To a considerable extent, the differences in sociality of foraging are consistent with the hypothesis that unpalatable prey should forage together so that each benefits from the learning experiences of naïve predators.

Finally, unpalatability may be requisite for most diurnal and large-bodied ant species to tend Homoptera for long time periods in exposed sites. Although many *Camponotus* species do this occasionally, and such activity is characteristic of at least one small-to-medium-sized species (*C. femoratus* studied here), *Dolichoderus* species and at least one of our *Camponotus* (*Dendromyrmex*) species are consummate homopteran tenders. Neither of our *C. (Dendromyrmex)* species produced gastral secretions smelling of the formic acid typical of formicines, but rather had more fetid odors of compounds that may have offered better protection against predators. These compounds are currently under study.

References

Davidson, D. W. The role of resource imbalances in the evolutionary ecology of tropical arboreal ants. *Biol. J. Linn. Soc.* **61**, 153–181 (1997).

Davidson, D. W. & Patrell-Kim, L. Tropical arboreal ants: why so abundant? in *Neotropical Biodiversity and Conservation* (ed. Gibson, A. C.) 127–140 (UCLA Botanical Garden, Publ. No. 1, Los Angeles, CA, 1996).

Eisner, T. A comparative morphological study of the proventriculus of ants (Hymenoptera: Formicidae). *Bull. Mus. Comp. Zool.* **116**, 439–490 (1957).

Kay, A. D. Nutrient balancing in ants. Ph.D. Dissertation, Univ. of Utah, UT, USA (2000).

Tobin, J. E. Ants as primary consumers: diet and abundance in the Formicidae. in *Nourishment and Evolution in Insect Societies* (eds Hunt, J. H. & Nalepa, C.A.) 279–308 (Westview Press, Boulder CO., 1994)

Tropical Ecosystems: Structure, Diversity and Human Welfare.
Proceedings of the International Conference on Tropical Ecosystems
K. N. Ganeshaiah, R. Uma Shaanker and K. S. Bawa (eds)
Published by Oxford–IBH, New Delhi. 2001. p. 433.

Canopy herbivory and soil processes in temperate and tropical forests

Margaret D. Lowman*,[§] H. Bruce Rinker**,
Mark D. Hunter[†] and Timothy D. Schowalter[‡]

*Marie Selby Botanical Gardens, Sarasota, FL 34236, USA
**Marie Selby Botanical Gardens, USA
[†]Institute of Ecology, University of Georgia, Athens, Georgia 30602, USA
[‡]Department of Entomology, Oregon State University, Corvallis,
Oregon 97331-2907, USA
[§]e-mail: mlowman@selby.org

Insect herbivores in forest canopies have the potential to influence soil processes by introducing materials from the treetops to the forest floor. Laboratory work has shown that the products of defoliation influence soil respiration and nutrient cycling. It is now important to assess whether natural defoliation under field conditions can also modify soil processes significantly.

Using one temperate and one tropical forest site (LTER sites: Coweeta Hydrological Laboratory, North Carolina, and Luquillo Experimental Forest, Puerto Rico), we will test two hypotheses: (1) herbivore-derived inputs from canopy to forest floor influence decomposition processes and (2) the timing of inputs and subsequent floor responses vary between temperate and tropical forests. This project will link canopy herbivory and soil processes in forest systems for the first time in one study. It will also address intrinsic differences between temperate and tropical forests, particularly their seasonality and biodiversity.

Keywords: Biodiversity, canopy herbivory, tropical rain forest, temperate forest, soil processes.

Tropical Ecosystems: Structure, Diversity and Human Welfare.
Proceedings of the International Conference on Tropical Ecosystems
K. N. Ganeshaiah, R. Uma Shaanker and K. S. Bawa (eds)
Published by Oxford–IBH, New Delhi. 2001. pp. 434–435.

Importance of canopy continuity in conservation of slender loris (*Loris tardigradus*)

Kaberi Kar Gupta

Department of Anthropology, Arizona State University, Tempe, AZ 85287–2402,
USA
e-mail: kaberi@asu.edu

Canopies play an important role in the distribution of and habitat use by arboreal animals. Slender loris is a nocturnal arboreal prosimian primate endemic to southern India and Sri Lanka. An extensive study on habitat use by slender loris in Kalakad–Mundanthurai Tiger Reserve (KMTR) suggests that canopy plays important role in their life history. Connection between canopies is produced by higher tree and climber densities. Slender lorises use these connections as bridges between gaps in the canopy for movement. Dense canopy thickets produced by thorny climbers are also used as sleeping sites during the day. Another use of canopy is as cover from predators. Human disturbances such as large-scale plantations change habitat structure. A relatively smaller scale and more short-term disturbance such as lopping of plants for fuel wood collection reduces canopy continuity and cover. I conducted a study on the distribution of slender loris in different habitats of KMTR. The ongoing project focuses on the influence of loss of canopy and other microhabitat characters on individual slender lorises in disturbed and undisturbed habitats. In the preliminary phase, I measured microhabitats parameters within home ranges of four radio-collared slender lorises.

Keywords: Slender loris, canopy, conservation, primate.

Table 1. Microhabitat parameters of vegetation structures in the home ranges of four radio-collared slender lorises in KMTR. Values are mean ± S.E.

Area (sampled plots in each)	Tree density**	Liana density	Canopy continuity	Canopy cover	Stump density	Lop density
PP (8)	891 ± 218	414 ± 95	3.9 ± 0.04	95.63 ± 2.67	302.39 ± 86.55	334.22 ± 101.91
SNK (6)	531 ± 145	424 ±126	3.58 ± 0.27	77.33 ± 6.21	466.85 ± 125.90	763.99 ± 267.11
SR90 (8)	748 ± 131	621 ± 135	3.90 ± 0.04	92.13 ± 3.41	413.80 ± 133.43	302.39 ± 53.47
SR95 (10)	1528 ± 194	255 ± 63	3.93 ± 0.03	81.30 ± 9.28	331.50 ± 93.37	267.38 ± 79.29

** = $p < 0.005$.

The abundance of loris was higher in mixed forests and relatively undisturbed teak forest compared with other forest types (Kar-Gupta, in press). Vegetation parameters did not show any correlation with abundance when compared between the habitats. However, when compared within habitats between sighting and non-sighting plots, canopy continuity was significantly higher in sighting plots than in non-sighting plots. Vegetation data from the home range of four radio-collared lorises indicate that the most disturbed habitat has low canopy continuity (Table 1). Focal animal observations showed that foraging and movement are mainly in the canopy (99% of all sightings). All the sleeping sites are in the canopy at an average height of five meters. Mothers station their infants in the canopy while foraging. Although lorises may use the ground for movement where a gap in the canopy is too big, the evolutionary adaptation of their anatomy is for climbing. Reduction in canopy continuity may isolate a population. Over time these isolated populations may lead to extinction of the local population. A long-term study on ecology is required to understand the impact of loss of canopy on slender loris populations.

References

Kar-Gupta, K. A study on distribution and habitat use by slender loris in Kalakad–Mundanthurai Tiger Reserve, India. *Am. J. Primatol.* (in press).

Tropical Ecosystems: Structure, Diversity and Human Welfare.
Proceedings of the International Conference on Tropical Ecosystems
K. N. Ganeshaiah, R. Uma Shaanker and K. S. Bawa (eds)
Published by Oxford–IBH, New Delhi. 2001. pp. 436–439.

Monitoring changes in woody species composition and abundance of two tropical dry evergreen forests on the Coromandel coast of South India

R. Venkateswaran and N. Parthasarathy*

Salim Ali School of Ecology and Environmental Sciences, Pondicherry University,
Pondicherry 605 014, India
*e-mail: partha@pu.pon.nic.in

The installation and repeat inventory of permanent plots in major tropical forests are increasing day by day (Bellingham *et al.*, 1999, Phillips, 1998). Many studies on forest dynamics have been carried out in species-rich forests and data on species-poor forests are inadequate. Tropical dry evergreen forests are relatively less known forest types (Parthasarathy and Karthikeyan, 1997), and occur on the Coromandel coast of peninsular India. Though baseline plant biodiversity inventories have been carried out in the dry evergreen forests, there are very few which have monitored them for changes in diversity and population density of the woody species. These forest fragments are shrinking in their extent due to a number of anthropogenic activities. Monitoring can aid in collecting reliable scientific information on habitat composition, structure, dynamics and in evaluating existing management approaches, and their impacts on forest ecosystem. Available literature reveals that data on dynamics of dry evergreen forests are limited,

Keywords: Biomonitoring, dry evergreen forest, permanent plot, stand dynamics, species diversity.

as compared to other tropical forests. The objective of this study is to assess the changes in diversity and density of woody species in two one-hectare permanent plots of tropical dry evergreen forests on Coromandel coast of peninsular India.

This study was conducted in two tropical dry evergreen forests, namely Kuzhanthaikuppam (KK, Lat.11°45N and Long. 79°38E) and Thirumanikkuzhi (TM, Lat. 11°N and Long. 79°41E) on the Coromandel coast of Tamil Nadu, south India. These forests are sacred groves or temple forests.

Two 1 ha (100 m × 100 m) square plots were established, distributed one each in site KK and site TM. All woody species ≥ 10 cm gbh were inventoried by subdividing the plot into 10 m × 10 m quadrats as workable units. Species were identified and their girth measured at 1.3 m from ground level. For multi-stemmed trees, bole girths were measured separately basal area calculated and summed. Recensusing of both the plots was made by re-measurement of stems during the year 1999, and without re-measurement of tree girths, but only counts of stems during 2000 in site KK. For data analysis 4-year changes were analysed for site TM, while 5-year changes for site KK.

Changes in species richness, stand density and basal area of all woody species ≥ 10 cm gbh between 1995 inventories and 1999 (and 2000) recensuses of KK and TM tropical dry evergreen forest sites are summarized in Table 1. During the 1995 enumeration 42 and 38 species were recorded in sites KK and TM respectively. During 1999 and 2000 recensuses 43 and 49 species were respectively encountered in site KK, while site TM maintained the total 38 species.

When compared to the other girth classes, species richness is greater in lower girth classes in the study plots in all the years, except in 30–60 cm gbh class at KK during the year 1995 (Table 2).

In KK, tree density increased by 19.3% during 2000, over the 1995 enumeration. Changes in stem density were greater in site TM (33%) than site KK (5.53%) during the period 1995–1999. In TM, woody species density

Table 1. Changes in species richness, density and basal area of all woody plants (≥ 10 cm gbh) between 1995 and 1999 (and also 2000) in the tropical dry evergreen forest sites (KK and TM) of south India.

Variable	KK			TM	
	1995	1999	2000	1995	1999
Species richness	42	43	49	38	38
Stand density (No. ha^{-1})	1367	1447	1694	974	1295
Stand basal area (m^2 ha^{-1})	15.44	16.73	–	29.48	30.085

Table 2. Changes in species richness, stand density and basal area of all woody species ≥ 10 cm gbh between 1995 inventory and 1999 recensus (2000 also in site KK) and by period for total and by girth classes in sites KK and TM.

| Girth class (cm) | Species richness | | | | | Density (no. of individuals) | | | | | Basal area (m²) | | | |
| | KK | | | TM | | KK | | | TM | | KK | | TM | |
	1995	1999	2000	1995	1999	1995	1999	2000	1995	1999	1995	1999	1995	1999
10–30	27	39	46	30	33	744	1019	1265	591	1018	2.25	4.42	1.39	2.87
30–60	28	24	24	19	16	412	320	320	172	103	4.43	5.27	1.74	1.64
60–90	16	12	14	17	10	135	85	86	83	63	3.39	4.23	3.15	3.15
90–120	12	4	4	11	12	39	18	18	40	52	1.71	1.62	3.12	5.26
120–150	8	3	3	7	7	21	4	4	25	25	1.04	0.53	3.25	4.40
150–180	5	0	0	8	6	7	0	0	22	17	0.67	0.00	3.98	4.00
180–210	3	0	0	5	4	5	0	0	18	11	0.45	0.00	4.51	3.75
> 210	3	1	1	6	4	4	1	1	23	6	1.53	0.65	8.35	5.15

increased by 42% (427 individuals) during the period between 1995 and 1999. Species richness and stem density decreased with increasing tree girth classes. Stem basal area increased moderately in site KK (7.7%) and in site TM (2.1%). The basal area of site TM is comparable with that of a seasonal dry evergreen forest of Thailand (Bunyavejchewin, 1999) and is also well within the range reported (23.0–65.4 $m^2 ha^{-1}$, Swaine *et al.*, 1997) in many other permanent study plots. The mean annual stand basal area increment was 0.32 $cm^2 yr^{-1}$ and 0.15 $cm^2 yr^{-1}$ for sites KK and TM respectively, during the year 1995–1999. It is lesser than that of Mpanga research forest reserve, Uganda (Taylor *et al.*, 1996) but greater than that of tropical deciduous forests reported by Sukumar *et al.* (1998). Basal area nearly doubled in the lower girth classes of 10–30 cm gbh during the period between 1995 and 1999 in both the study sites, while for the most of the middle girth classes it varied marginally, but in higher girth classes it decreased considerably.

References

Bellingham, P. J., Stewart, G. H. & Allen, R. B.. Tree species richness throughout New Zealand forests. *J. Veg. Sci.* **10**, 825–832 (1999).

Bunyavejchewin, S. Structure and dynamics in seasonal dry evergreen forest in northeastern Thailand. *J. Veg. Sci.* **10**, 787–792 (1999).

Parthasarathy, N. & Karthikeyan, R. Plant biodiversity inventory and conservation of two tropical dry evergreen forests on the Coromandel coast, South India. *Biodivers. Conserv.* **6**, 1063–1083 (1997).

Phillips, O. L. Increasing tree turnover in tropical forests as measured in permanent plots. in *Forest Biodiversity Research, Monitoring and Modelling: Conceptual Background and Old World Case Studies* (eds Dallmeier, F. & Comiskey, J. A.) 221–245 (Parthenon Publishing, Paris, 1998).

Sukumar, R., Suresh, H., Dattaraja, H. S. & Joshi, N. V. Dynamics of a tropical deciduous forest: population changes (1988 through 1993) in a 50-hectare plot at Mudumalai, southern India. in *Forest Biodiversity Research, Monitoring and Modelling: Conceptual Background and Old World Case Studies* (eds Dallmeier, F. & Comiskey, J. A.) 529–540 (Parthenon Publishing, Paris, 1998).

Swaine, M. D., Lieberman, D. & Putz, F. E. The dynamics of tree populations in tropical forests: A review. *J. Trop. Ecol.* **3**, 359–366 (1997).

Taylor, D. M., Hamilton, A. C., Whyatt, J. D., Mucunguzi, P. & Bukenya-Ziraba, R. Stand dynamics in Mpanga research forest reserve, Uganda, 1968–1993. *J. Trop. Ecol.* **12**, 583–597 (1996).

Woody Invasive Species in
Tropical Ecosystems

Tropical Ecosystems: Structure, Diversity and Human Welfare.
Proceedings of the International Conference on Tropical Ecosystems
K. N. Ganeshaiah, R. Uma Shaanker and K. S. Bawa (eds)
Published by Oxford–IBH, New Delhi. 2001. pp. 443–445.

Ecology of weeds in tropical and warm temperate forests

J. S. Denslow[t], S. J. DeWalt[‡] and L. L. Battaglia[‡]*

*USDA Forest Service, 23 E. Kawili St., Hilo, HI, 96720, USA
[‡]Louisiana State University, Baton Rouge, LA 70803, USA
[t]e-amil: julie.denslow@gte.net

Among plant communities, intact tropical and warm temperate forests often are seen to be less vulnerable to invasion by exotic weeds than more open communities (Rejmánek, 1996). However both the changing ecology of forests and the increased diversity of non-indigenous species to which they are exposed may be shifting this scenario. Invasive species control is now frequently a central focus for forest management, conservation and restoration. Several forest characteristics reduce the likelihood of successful seedling establishment in either native or exotic species: Light levels under a closed tree canopy or dense shrub layer are low, especially in moist forests. Shade-grown seedlings grow slowly, have short life expectancies, and are vulnerable to insect and pathogen damage (Denslow, 1987). The effects of low light availability may be compounded by litter accumulation which smothers seedlings and prolongs the period of vulnerability before new roots reach adequate moisture supplies, a source of mycorrhizal innoculum and available nutrients. As a result, densities of established seedlings are low, especially in tropical moist forests where diversity, productivity and competition are high. Under such circumstances patches of light produced by treefalls, landslides, or other sources of canopy opening and soil disturbance become important sites for establishment and growth of native species as well as serving the most likely point of invasion by exotics.

Keywords: Invasive species, tropical forest, horticulture, weeds, ecology.

Adaptive life history traits among forest-regenerating species include large seed size, shade tolerance, animal dispersal, and vegetative reproduction. These traits confer an ability to place seeds in newly formed gaps, an ability to persist in the understorey in intervals between gaps, and an ability to produce a large, vigorous seedlings with deep roots even in the shade. Weeds associated with agricultural or disturbed areas are characterized more commonly by high light requirements, small seeds and extended dormancy. While such species are poorly adapted for life in the forest, they spread rapidly through agricultural fields, pastures, road sides, vacant lots, and other disturbed areas. For these species, intact native forests present a formidable barrier to their establishment, persistence, and spread.

In recent years, forests have begun to appear increasingly vulnerable to invasions by exotic species. Among factors contributing to the appearance of exotic species in native forests are increased forest perimeter-to-edge ratios due to habitat fragmentation and increased transport of seed and propagules by farm, logging, and road-building equipment, by hikers and by the expansion of human settlements into forest zones, and increasing impacts of alien animal pests, which alter disturbance regimes, demography, and nutrient processes. Habitat alteration, including changes in nutrient, moisture, and disturbance regimes, provide more frequent opportunities for seedling establishment. Tropical forests, once distant from centers of population and commerce, are no longer protected by their remoteness.

Forests are also exposed to exotic species with life history traits suited to establishment and growth in shaded understorey habitats. Growth in the global horticultural trade has increased the numbers of novel species introduced into the vicinity of conservation areas. Plants introduced for their landscaping potential often are selected for their foliage, their ability to quickly cover bare areas, or to grow in the shade. Networks of plant explorers, horticulturalists, nurseries, and distributors provide highly effective routes by which new species can be introduced (Reichard and White, 2000). Reichard (1994) estimates that 85% of 235 woody invasives in North America were originally introduced for landscaping purposes.

In this study we examined reports of invasive weeds in tropical and warm temperate forests published in the scientific literature, reported by knowledgeable observers on web pages and list servers, and reported by resource managers in Hawaii. We compared characteristics of species reported to be invasive in forests with those reported for invasive species world wide. In addition, we examined the kinds of forests invaded and the nature of the impact of the weed on the forest community and ecosystem.

Initial incursions of exotic species into intact forest ecosystems are often reported in treefall clearings or following large canopy-opening events such as hurricanes (Horvitz et al., 1998). In the bottomland hardwood forests along the coast of the Gulf of Mexico, Chinese tallow (*Sapium sebiferum*, Euphorbiaceae) invades old growth forest through treefall gaps and swales where open canopies are associated with high flood frequencies. This pattern is sometimes taken as evidence that eventual canopy closure will eliminate the weed with little lasting impact on the forest. However, where alien seedling densities and rates of growth are high, the natural regeneration processes of native species can be suppressed. In Hawai'i the Himalayan raspberry (*Rubus ellipticus*, Rosaeeae) invades gaps in montane rainforest where its dense canopy suppresses native seedlings and shrub. If such a pattern becomes chronic either because the exotics are able to replace natives in the shrub or tree canopy or because exotic seed rain from forest edges successfully populates natural gaps, the natural regeneration processes will be interrupted and there will be a gradual erosion of forest structure and composition.

A second common pattern of forest invasion is the spread and dominance of the understorey by shade-tolerant clonal species. In Hawaii this pattern is seen in the rhizomatous Kahili ginger (*Hedychium gardnerianum*, Zingiberaceae), the clonal strawberry guava (*Psidium cattlianum*, Myrtaceae), and the aggressive shrub *Clidemia hirta* (Melastomataceae). While these species do not appear dependent on canopy opening for their establishment, their spread may be hastened (but not limited) by the soil disturbance and seed-dispersal activities of wild pigs. The clones of Kahili ginger and palm grass (*Setaria palmifolia*, Poaceae) form a dense mat on the forest floor whereas clones of guava and *Clidemia* create a dense shade under the forest canopy. The impacts of these species on the forest are similar in many ways to those of the gap colonizers in that they reduce establishment opportunities of native species, contributing to the gradual degradation of the forest canopy structure and composition.

Exotic shade-tolerant trees may establish in gaps or in the forest understorey. Their ability to overtop and eventually kill native trees, not only alters canopy structure and composition, but also changes ecosystem processes as to affect the survival of all species. *Miconia calvescens* (Melastomataceae) casts such a dense shade in wet forest habitats of Tahiti and Hawaii that few other species are able to coexist with it. *Paraserianthes falcataria* (= *Albizzia falcataria*, Mimosaceae) trees kill native ohia (*Metrosideros polymorpha*, Myrtaceae) where it is overtopped in Hawaii. *Paraserianthes* also facilitates the establishment of other exotics such as *Clidemia* and *Psidium* because soil nitrogen supply is improved under their crowns. Similarly exotic vines and lianas may kill or deform native trees they cover. While vines often require the high light levels of treefall openings for establishment, their impacts may extend far beyond the initial site of establishment once they have reached the canopy.

Seasonally dry tropical forests are subject to the double threat of alien pests and altered fire regime. Competition from alien trees and shrubs such as Christmas berry (*Schinus terebinthefolius*, Anacardiaceae) and lantana (*Lantana camara*, Verbenaceae) for water and light depresses the regeneration potential of native species. Alien grasses such as fountain grass (*Pennisetum setaceum*, Poaceae), broomsedge (*Andropogon virginicus*, Poaceae), and molasses grass (*Melinis minutiflora*, Poaceae) enhance the sizes and frequency of fires which erode dryland forests and eliminate regeneration of fire-sensitive species (D'Antonio and Vitousek, 1992).

While both natural and anthropogenic disturbances enhance the invasibility of tropical forests as they do of other ecosystems, exotic species with appropriate trait combinations are able to establish and spread into intact communities as well. We should not assume that inherent ecological characteristics of these communities confer a special resistance to novel species. Increasing opportunities will bring more frequent incursions and unfortunately ever greater challenges to resource managers.

References

D'antonio, C. M. & Vitousek, P. M. Biological invasions by exotic grasses, the grass/fire cycle, and global change. *Annu. Rev. Ecol. Syst.* **23,** 63–87 (1992).

Denslow, J. S. Tropical rainforest gaps and tree species diversity. *Annu. Rev. Ecol. Syst.* **18,** 431–451 (1987).

Horvitz, C. C., Pascarella, J. B., Mcmann, S., Freedman, A. & Hofstetter, R. H. Functional roles of invasive non-indigenous plants in hurricane-affected subtropical hardwood forests. *Ecol. Appl.* **8,** 947–974 (1998).

Reichard, S. E. Assessing the potential of invasiveness in woody plants introduced in North America. Ph.D. Thesis. University of Washington, Seattle (1994).

Reichard, S. & White, P. Horticulture as a pathway of invasive plant introductions in the United States. *Bioscience* (2000).

Rejmánek, R. Species richness and resistance to invasions. in *Biodiversity and Ecosystem Processes in Tropical Forests* (eds Orians, G., Dirzo, R. & Cushman, J. H.) 153–172 (Springer, New York, 1996).

Tropical Ecosystems: Structure, Diversity and Human Welfare.
Proceedings of the International Conference on Tropical Ecosystems
K. N. Ganeshaiah, R. Uma Shaanker and K. S. Bawa (eds)
Published by Oxford–IBH, New Delhi. 2001. pp. 447–450.

Invasional 'meltdown' in island rain forest

Dennis J. O'Dowd*, Peter T. Green and P. S. Lake

Centre for Analysis and Management of Biological Invasions, School of Biological
Sciences, P.O. Box 18, Monash University, Victoria 3800, Australia
*e-mail: dennis.odowd@sci.monash.edu.au

Accelerating rates of species introductions have fueled increasing impacts of invasive species on native biodiversity and ecosystem processes. This is especially evident on many tropical oceanic islands where alien species that interact with native species and each other now comprise large fractions of island biotas. Although the impact of island invaders on native species can be large, the tempo, magnitude and variety of indirect effects that modify invaded ecosystems are unclear and, even when examined, are usually inferred retrospectively. The few, thorough analyses of island invasion by alien species have emphasized 'bottom-up' effects of invasive plants on ecosystem properties (e.g. Vitousek and Walker, 1989). We show here that a 'top-down' change following invasion by an alien tramp ant precipitates a rapid state change in the rain forest ecosystem on Christmas Island, a 134 km^2 high island in the northeastern Indian Ocean.

The yellow crazy ant (*Anoplolepis gracilipes*), unintentionally introduced between 1915 and 1934, has formed expansive 'supercolonies' over the last decade and rapidly extirpated the dominant native consumer, the red land crab (*Gecarcoidea natalis*), from ~15 km^2 of invaded rain forest (15% of total island rain forest). In these areas, activity of this omnivorous ant is high

Keywords: Ecosystem processes, invasion, islands, mutualism, tropical rain forest.

(Table 1A) and it forages continuously and three-dimensionally. In uninvaded forest, omnivorous red crabs typically occur at densities exceeding one crab m^{-2} and average 1440 kg ha^{-1} (Green, 1997). As supercolonies spread, crazy ants occupy red crab burrows, kill and consume resident crabs, and use burrows as nest sites. As such, this key native consumer is deleted from ant-invaded forest (Table 1B). Furthermore, during the annual migration of the red crab, large numbers of crabs are killed in transit when a migratory pathway intercepts an ant supercolony (Table 1B). This 'effect-at-a-distance' depletes crab populations in areas not (yet) invaded by the crazy ant.

Table 1. Comparison of impact variables on the forest floor and in the canopy within six paired 1-ha rain forest sites that were either invaded or uninvaded by the yellow crazy ant, *Anoplolepis gracilipes*. In all cases, block effects (paired sites) were not significant (*P*'s > 0.05). We believe that ants spread to all invaded sites 1–2 years previously. At each site five 4 m × 4 m quadrats were used to measure ant activity at sugar baits, land crab mortality and burrow density, per cent litter cover, and seedling density and species richness on the forest floor. For canopy impacts, ant densities on the trunks of canopy trees and scale insect abundance, sooty mould cover on leaves and stems, and canopy condition were determined for five trees at each site. Ant densities on canopy tree boles were determined by applying a 16.5 cm wide strip of sticky tape around one-half of the bole of each tree at breast height and then counting and identifying captured ants. Large branches were downed from each tree canopy from which five shoots were haphazardly selected and scale insects counted under magnification (10–15 ×) for 20 cm stem sections and on a randomly chosen, fully expanded leaf from each shoot. Sooty mould cover on these stems and leaves was rated as 0–20 (0), 21–40 (1), 41–60 (2), 61–80 (3), or 81–100 (4) per cent cover. Canopy condition was assessed by collecting 30 shoots from each branch and determining the proportion of shoots with new leaf growth. For each variable, values were pooled or averaged over all quadrats or trees sampled at each site to produce a site value (*N* = 6 for invaded and uninvaded sites).

Variable	Invaded	Uninvaded	$F_{1,5}$	P
Forest floor				
(A) Ant activity				
Anoplolepis activity index	6.96 (1.42)	0.06 (0.06)	45.40	0.001
Other ant activity index	1.10 (0.39)	2.91 (0.78)	7.33	0.042
(B) Land crabs				
Crab burrows/80 m^2	2.3 (1.6)	95.7 (24.5)	44.82	0.001
Dead crabs/80 m^2	51.8 (17.5)	0.0 (–)	18.81	0.007
(C) Litter cover (%)	87 (3)	43 (13)	21.43	0.006
(D) Seedlings				
No. seedlings/80 m^2	1375.8 (166.2)	44.7 (16.7)	123.59	0.000
No. spp./80 m^2	22.2 (2.6)	6.3 (1.3)	34.06	0.002
Forest canopy				
(E) Ants/100 cm^2 bole	4.5 (0.4)	0.3 (0.1)	137.65	0.000
(F) Scale insects				
Stem (no./20 cm)	114.7 (35.6)	8.2 (4.9)	17.85	0.008
Leaf (no./leaf)	122.5 (45.0)	7.1 (3.0)	59.29	0.001
(G) Sooty mould rating				
Stem	2.2 (0.4)	0.4 (0.2)	43.39	0.001
Leaf	2.0 (0.6)	0.2 (0.1)	26.11	0.004
(H) Per cent expanding shoots	72.7 (8.9)	96.0 (0.9)	10.14	0.024

Over the past decade, we have conducted experiments at a small spatial scale (25 m²) which show that removal of red crabs deregulates seedling recruitment, seedling species composition, litter breakdown, and the density of litter invertebrates (Green *et al.*, 1997; 1999). Elimination of this key native consumer by invasive ants recapitulates these effects but now at a landscape scale. Forest processes in ant-invaded areas are indirectly affected: on average, litter cover was double, seedling densities 30-fold higher, seedling species richness 3.5-fold higher, and forest understorey structure markedly different (Table 1C, D; Figure 1A, B). The relative species composition of seedlings depended significantly on ant invasion (non-metric multi-dimensional scaling (NMDS) followed by analysis of similarities (ANOSIM), $P = 0.009$), whereas we found no relationship between canopy tree composition and invasion status of sites (NMDS followed by ANOSIM, $P = 0.835$). Clearly, the direct impact of an introduced ant on the dominant native consumer has long-term implications for forest structure and composition through indirect effects on the 'advanced regeneration' and changes in habitat structure.

In addition to this dramatic, indirect fallout from alien invasion, mutualism between this invasive ant and introduced/cryptogenic scale insects has amplified and diversified rain forest impacts. This viewpoint is supported by coincidence of high, sustained densities of ants foraging on honeydew on canopy trees, population outbreaks of scale insects, and spread of honeydew-dependent sooty moulds (Capnodiaceae) on canopy stems and leaves in invaded areas (Table 1E–G). Outbreaks of the lac scale insect *Tachardina aurantiaca* (Kerridae), native to Southeast Asia, in particular, were associated with *Anoplolepis*-invaded sites. This host plant generalist is associated with at least 21 tree and vine species on the island, including most canopy dominants.

Figure 1. Impacts of invasion of island rain forest by the yellow crazy ant, *Anoplolepis gracilipes*. **A.** Uninvaded site with open understorey maintained largely by the foraging activities of the red land crab, *Gecarcoidea natalis*. **B.** Site 1–2 years after ant invasion with a dense and diverse seedling cover and thick litter layer.

This positive interaction between *Anoplolepis* and scale insect species helps explain the lower proportion of expanding shoots on canopy trees at ant-invaded sites (Table 1H). Furthermore, the frequency of tree dieback depended on ant invasion (at invaded sites, 51% of trees [146/286] showed evidence of dieback; at uninvaded sites just 18% of trees [53/294], $\chi^2 = 77.7$, $P = 0.000$, log-linear analysis). Dieback effects can be differential and affect both seedling and canopy species composition. Both trees and saplings of *Inocarpus fagifer*, a forest dominant, were heavily infested and affected by lac scale insects in areas of *Anoplolepis* invasion: *Inocarpus* comprised 24% of all seedlings we censused on uninvaded plots, but we encountered only 10 *Inocarpus* seedlings (0.1% of all seedlings) on *Anoplolepis*-invaded plots. Deaths of *Inocarpus* trees also occurred disproportionately in ant-invaded sites. At two invaded sites where we counted dead standing canopy trees, *Inocarpus* comprised 83% (15/18) and all (6/6) dead trees whereas they comprised only 28% (28/100) and 22% (22/100) of live canopy trees sampled at random at the respective sites ($\chi^2 = 17.85$, $P < 0.001$, Chi-square test, and $\chi^2 = 17.71$, $P < 0.001$, Fisher exact test).

Tropical islands provide relatively simple and tractable systems for analyses of biological invasions. Our results show that reconfiguration of species interactions following the deletion of a native dominant by an alien invader rapidly transforms the island rain forest ecosystem. Positive interactions between this dominant invader and other introduced species further amplify and diversify impacts. Together this illustrates how changes in the web of interactions following multiple alien species introductions can lead to synergism in invasive impact to precipitate invasional 'meltdown' (Simberloff and Von Holle, 1999).

Acknowledgements

The Australian Research Council and Environment Australia supported this research.

References

Green, P. T. Red crabs in rain forest on Christmas Island, Indian Ocean – Activity patterns, density and biomass. *J. Trop. Ecol.* **13**, 17–38 (1997).

Green, P. T., O'Dowd, D. J. & Lake, P. S. Control of seedling recruitment by land crabs in rain forest on a remote oceanic island. *Ecology* **78**, 2474–2486 (1997).

Green, P. T., Lake, P. S. & O'Dowd, D. J. Monopolization of litter processing by a dominant land crab on a tropical oceanic island. *Oecologia* **119**, 235–244 (1999).

Simberloff, D. & von Holle, B. Positive interactions of nonindigenous species: Invasional meltdown? *Invasion Biol.* **1**, 21–32 (1999).

Vitousek, P. M. & Walker, L. R. Biological invasion by *Myrica faya* in Hawai'i: plant demography, nitrogen fixation, ecosystem effects. *Ecol. Monogr.* **59**, 247–265 (1989).

Tropical Ecosystems: Structure, Diversity and Human Welfare.
Proceedings of the International Conference on Tropical Ecosystems
K. N. Ganeshaiah, R. Uma Shaanker and K. S. Bawa (eds)
Published by Oxford–IBH, New Delhi. 2001. pp. 451–455.

From resistance to meltdown: Secondary invasion of an island rain forest

Peter T. Green[†,‡], Dennis J. O'Dowd* and P. S. Lake**

*Centre for the Analysis and Management of Biological Invasions, School of
Biological Sciences, P.O. Box 18, Monash University, Victoria 3800, Australia
[†]For correspondence: c/o Parks Australia North, P.O. Box 867, Christmas Island,
Indian Ocean 6798
[‡]e-mail: peter.green@ea.gov.au

Nowhere is the success of alien invaders better illustrated than in the forests of tropical oceanic islands; a large fraction of island biota now constitute introduced alien species and case studies of catastrophic impacts abound (e.g., the brown tree snake on Guam, *Miconia calvescens* in Tahiti, and avian malaria in Hawaii). The success of invasions is two-sided, depending both on the attributes of potential invaders and the characteristics of the recipient community. One view for the high success of invaders on islands is that biotic 'resistance' on islands is inherently low: islands are typically species-poor, and lack the retinue of competitors, predators, and parasites that may exclude many would-be invaders from continental communities. However, empirical evidence for low biotic resistance on islands is limited (e.g., Simberloff, 1989; Simberloff and Boecklen, 1991). We have argued that biotic resistance to invaders can be high on some islands. For example, in rain forest on Christmas Island, a 134 km^2 'high' island in the northeastern Indian

Keywords: Invasion, secondary invasion, red land crabs, giant African land snails, woody weeds.

Ocean, omnivorous native land crabs dominate the forest floor and appear to present an effective barrier to many invaders (O'Dowd and Lake, 1989; Lake and O'Dowd, 1991). Here we describe our changing perspective on biotic resistance to rain forest invasion on this island over the last 15 years. Observations and experiments over this period show a rapid shift in biotic resistance to rain forest invasion – from high resistance to rapid breakdown of resistance to invasion meltdown, with an accelerated rate of secondary invasion (Simberloff and Von Holle, 1999).

From 1986 to 1996, biotic resistance to invasions appeared high in undisturbed forest on Christmas Island. We based this viewpoint on the presence of extraordinary numbers of red land crabs (*Gecarcoidea natalis*) in rain forest (0.4–1.8 crabs m^{-2}, and 758–1519 kg ha^{-1}; Green, 1997). These omnivorous crabs include leaf litter, seeds and seedlings, and larger invertebrates in their diet and are a key regulator of seedling recruitment, seedling species composition and litter breakdown in the forest (Green et al., 1997; 1999a). Three lines of evidence indicated that this dominant consumer conferred a high degree of biotic resistance to invasion by a range of alien species:

(1) Giant African land snails (GALS, *Achatina fulica*), introduced to Christmas Island nearly 60 years ago, had failed to invade primary rain forest (Lake and O'Dowd, 1991). This snail is highly invasive on many tropical islands. Experiments showed that red crabs are major predators of *Achatina*, killing almost all (97%) of snails tethered in rain forest within 48 h whereas they killed only 23% in disturbed forest edges. Further, the density of red crabs and snails showed inverse patterns across a gradient from the disturbed forest edge to primary rain forest. Thus, the biotic resistance provided by this native omnivore appears to explain the failure of GALS to invade primary rain forest.

(2) Seedling surveys, crab exclusion and seedling caging experiments suggested that red crabs play a key role in preventing the widespread establishment of alien weeds in intact rain forest. First, seedling counts on sixteen 100 m × 4 m transects established across rain forest failed to detect the presence of a single alien plant. Second, woody weed seedlings transplanted into rain forest were rapidly consumed by red crabs (O'Dowd and Lake, 1989). Third, in two separate crab exclusion experiments, a total of 7 individuals of 3 weed species (*Capsicum frutescens*, *Turnera ulmifolia*, and an unidentified Asteraceae) germinated and established on fenced exclusion plots, whereas no weeds germinated on unfenced control plots (Green et al., 1997).

(3) *Muntingia calabura*, an exotic pioneer tree species from Central America, is abundant and widespread along roadsides and abandoned mined fields across the island. Its tiny seeds are widely dispersed into rain forest by fruit bats and pigeons where they form an extraordinarily abundant seed bank (c. 19,000 seeds m^{-2}; Green et al., 1999b). Despite the

introduction of *Muntingia* to the island over 50 years ago, the presence of a large, viable seed bank, and the occurrence of large natural light gaps (>150 m^2) sufficient for its recruitment, we had never seen established *Muntingia* in natural disturbances in rain forest. Given the large impact of red crabs on the abundance and diversity of seedlings in light gaps, we strongly suspected that this native consumer played a key role in the failure of *Muntingia* to establish in natural forest disturbances.

Our discovery of the invasion of intact rain forest by an alien tramp ant, *Anoplolepis gracilipes*, has forced a radical shift in our thinking on strong biotic resistance. Since the mid-1990s, yellow crazy ants have formed extensive supercolonies with high, sustained densities of foraging ants in rain forest across the island. These supercolonies have rapidly expanded and now occupy at least 15% of island rain forest. Red crabs offer no biotic resistance to this invasion and are themselves highly vulnerable to crazy ants; red crabs are killed by invading ants and crab densities are rapidly reduced to nil. The extirpation of the resident land crab population within supercolonies has not only had manifold consequences for the dynamics and structure of native forest, but is also facilitating a variety of secondary invasions:

(1) A recent tethering experiment has shown that while *Achatina* mortality due to red crabs is still high in forest uninvaded by crazy ants (92% after 48 h), there was nil snail mortality in plots invaded by crazy ants after 6 days. Crazy ants were not observed attacking GALS in this experiment, and these snails can evidently coexist with crazy ants; GALS have been observed in crazy ant-invaded forest more than 100 m from the nearest forest edge. By extirpating red crabs, crazy ants appear to be facilitating the invasion of rain forest on Christmas Island by GALS.

(2) Crazy ants are facilitating the entry and range expansion of alien woody weeds in rain forest on the island. In the absence of red crabs, woody weeds such as chillies (*Capsicum*) and papaya (*Carica papaya*) are becoming common components of the understorey in some areas of crazy-ant invaded forest. *Muntingia* is now firmly established in crazy ant-infested forest, in an area we know to have been ant-free, and *Muntingia*-free, just several years ago. Crazy ants have also facilitated secondary invasions of alien cockroaches in the forest, including *Blatella germanica* and *Periplaneta americana*. These cockroaches occur in much greater abundance in forest dominated by crazy ants than forest with an intact population of land crabs.

(3) Crazy ants are probably accelerating the rate of spread by woody weeds that show partial resistance to herbivory by land crabs. *Clausena excavata* (Rutaceae), a small understorey tree, is an alien plant on Christmas Island with the ability to invade undisturbed primary rain forest. We conducted an experiment in which we transplanted *Clausena* seedlings into high (roadside) and low (adjacent forest interior) light environments, and caged half of the seedlings in each light treatment to protect them from

grazing by land crabs. *Clausena* seedlings were long lived, with 84% survivorship after 9 months. Land crabs had no significant impact on survival in high light (95% survival of caged seedlings vs 80% for unprotected seedlings), but significantly lowered seedling survival in the forest interior (100% survivorship of caged seedlings vs 60% of unprotected seedlings). At two forest sites, *Clausena* accounted for 24% and 67% of all stems ≤ 2 m tall, and significant areas of invasion are known from at least four widely-spaced locations around the island. These data indicate that while *Clausena* can invade crab-dominated rain forest, crazy ants are likely to indirectly facilitate and accelerate the spread of this invasive weed, by extirpating red crabs and thereby increasing seedling longevity.

For a decade we held the view that natural communities on Christmas Island were largely intact, because abundant, omnivorous land crabs conferred a high degree of biotic resistance to alien invasions. Following the advent of supercolony formations by invasive crazy ants, we now hold a radically different view; not only is rain forest on Christmas Island susceptible to invasion, but the rate of ecosystem change following in the wake of crazy ants has been breathtakingly rapid. Secondary invasions by a range of alien plants and animals, facilitated in large part by the extirpation of the principal agent of biotic resistance on the island, the red land crab, are contributing to these changes. Christmas Island has had a relatively short period of human occupation, and it is clear that while red crabs have afforded a level of biotic resistance to invasion somewhat unique among oceanic islands, it was, perhaps, just a matter of time before exotic species with superior competitive abilities were introduced to the island and shifted the balance from invasional resistance to invasional meltdown.

Acknowledgement

The Australian Research Council and Environment Australia supported this research.

References

Green, P. T. Red crabs in rain forest on Christmas Island, Indian Ocean – Activity patterns, density and biomass. *J. Trop. Ecol.* **13,** 17–38 (1997).

Green, P. T., O'Dowd, D. J. & Lake, P. S. Control of seedling recruitment by land crabs in rain forest on a remote oceanic island. *Ecology* **78,** 2474–2486 (1997).

Green, P. T., Lake, P. S. & O'Dowd, D. J. Monopolization of litter processing by a dominant land crab on a tropical oceanic island. *Oecologia* **119,** 235–244 (1999a).

Green, P. T., Hart, R., Jamil Bin Jantan, Metcalfe, D. J., O'Dowd, D. J. & Lake, P. S. Red crabs in rain forest on Christmas Island, Indian Ocean: No effect on the soil seed bank. *Austr. J. Ecol.* **24,** 90–94 (1999b).

Lake, P. S. & O'Dowd, D. J. Red crabs in rain forest, Christmas Island: biotic resistance to invasion by an exotic snail. *Oikos* **62,** 25–29 (1991).

O'Dowd, D. J. & Lake, P. S. Red crabs in rain forest: Differential herbivory of seedlings. *Oikos* **58,** 289–292 (1989).

Simberloff, D. Which insect introductions succeed and which fail? in *Biological Invasions: A Global Perspective* (eds Drake, J. A., Mooney, H. A., Castri, F. D., Groves, R. H., Kruger, F. J., Rejmanek, M. & Williamson, M.) 61–75 (Wiley and Sons, New York, 1989).

Simberloff, D. & Boecklen, W. Patterns of extinctions in the introduced Hawaiian avifauna: A reexamination of the role of competition. *Am. Nat.* **138**, 300–327 (1991).

Simberloff, D. & Von Holle, B. Positive interactions of nonindigenous species: Invasional meltdown? *Invasion Biol.* **1**, 21–32 (1999).

Tropical Ecosystems: Structure, Diversity and Human Welfare.
Proceedings of the International Conference on Tropical Ecosystems
K. N. Ganeshaiah, R. Uma Shaanker and K. S. Bawa (eds)
Published by Oxford–IBH, New Delhi. 2001. pp. 456–458.

The role of forest structure in plant invasions on tropical oceanic islands

Jean-Yves Meyer and Christophe Lavergne

Conservatoire Botanique National de Mascarin, RD 12 Domaine des Colimaçons, F-97436 Saint-Leu, Ile De La Reunion, France
e-mail: paulam@wanadoo.fr

Invasibility of oceanic islands, i.e. their greater susceptibility to invasion by alien species, is generally explained by (1) the lack of specific natural enemies for the newly introduced species; (2) poor competitive abilities of insular species which evolve in isolation; (3) low number of species on small islands; (4) disharmonic biota, i.e. the absence of taxonomic groups or ecological guilds, due to remoteness; (5) natural or human-related disturbances (Carlquist, 1974; Loope and Mueller-Dombois, 1989; Vitousek, 1988). Some of these hypotheses are difficult to test if not controversial (Simberloff, 1995). Another poorly explored hypothesis is whether island forest structure may facilitate invasion success.

Two small tree species, *Miconia calvescens* and *Ligustrum robustum* subsp. *walkeri*, are uncommon in the rainforests of their native country (respectively tropical America and Sri Lanka), but are highly invasive in undisturbed forest on tropical oceanic high islands (respectively in French Polynesia and Hawai'i in the Pacific Ocean, and Réunion Islands and Mauritius Islands in the Indian Ocean). A comparative study of these two species in both native and introduced areas shows that their recruitment, growth and reproduction is enhanced in islands where the native (or primary) forest physiognomy is

Keywords: Forest structure, invasive plants, *Ligustrum robustum*, *Miconia calvescens*, oceanic islands.

456

characterized by a low canopy (generally between 12 and 15 m in height), simplified vertical structure, low number of canopy species, trees with small crowns and small leaves, and difference in life forms/ecological groups (Cadet, 1977; Strasberg, 1996; for Réunion Islands, Fosberg, 1992; Florence, 1993; for French Polynesia; Mueller-Dombois, 1987; Gagné and Cuddihy, 1990; for Hawai'i). The particular vegetation structure found on tropical oceanic high islands may be explained by their young geological age, a steep relief with deep valleys and high peaks, a thin soil layer, the occurrence of strong winds and cyclones.

– In their native range of tropical America, most *Miconia* species tends to produce larger crops, have more extended fruiting episodes, and fruits more frequently in gaps than under closed canopy. Some species require a canopy opening to reproduce. In the island of Tahiti (French Polynesia), the small tree (10–12 (–18) m in height) *Miconia calvescens* is able to attain or overtop the canopy of the primary wet forests, and flowers and fruits profusely (Meyer, 1998). It covers now more than 70,000 ha, and forms dense monospecific stands which exclude all the native plants (Meyer and Florence, 1996). This species is also naturalized in the rainforests of Sri Lanka where it thrives only on forest edges and gaps (C. Lavergne, pers. obs.).

– *Ligustrum robustum* subsp. *walkeri* is a tall shrub to small tree (up to 10 m in height) only found along streams, forest margins and roadsides in the submontane rain forests of its native range of Sri Lanka. It forms dense thickets in all parts of the montane wet forests of Mauritius Islands (Lorence and Sussman, 1986), and in Réunion Islands where 3,000 ha are infested, including 500 ha in undisturbed primary vegetation (Lavergne *et al.*, 1999).

Both species can be considered as late secondary successional small trees. They are characterized by a high seed germination rate, rapid growth, and an early reproductive maturity. They are adapted to low light levels for germination and growth (i.e., shade-tolerant species) but benefits greatly from additional light for flowering and fruiting, such as in open vegetation (treefall gaps, forest edges, riverbanks), and in island forests where the absence of large emergent trees and a reduced canopy cover allow a greater light availability in the understorey.

We suggest that the forest structure of oceanic islands may play an important role in explaining the success of invasion by introduced plant species, in relation to their life-history traits. This hypothesis has also been suggested to explain the penetration of the introduced Red-whiskered bulbul *Pycnonotus jocosus*, a bird more commonly found in open and anthropogenic habitats, into native forests of Réunion Islands (Mandon-Dalger *et al.*, 1999), and the predation success of the introduced snake *Boiga irregularis* in Guam (Pacific

Ocean) where the low forest canopy may facilitate its access to prey species (Savidge, 1987).

Acknowledgements

The first author is grateful to Julie Denslow (Institute of Pacific Island Forestry, USDA Forest Service, Hawaii, USA) for her invitation to attend the ATB conference at Bangalore, and the International Program of the US Forest Service for financial support.

References

Cadet, T. La végétation de l'île de la Réunion : étude phytoécologique et phytosociologique. I. Texte. Thèse de Doctorat, Université d'Aix-Marseille III (1977).

Carlquist. *Island Biology* (Columbia University Press, New York, 1974).

Florence, J. La végétation de quelques îles de Polynésie française. *In* Atlas de la Polynésie française, planches 54–55. Editions de l'ORSTOM, Paris (1993).

Fosberg, F. R. Vegetation of the Society Islands. *Pacific Sci.* **46**, 232–250 (1992).

Gagne, W. C. & Cuddihy, L. W. Vegetation. in *Manual of the Flowering Plants of Hawaii Islands* (eds Wagner, W. L., Herbst, D. R. & Sohmer, S. H.) 45–113 (Bishop Museum Press, Honolulu, 1990).

Lavergne, C., Rameau, J.-C. & Figier, J. The invasive woody weed *Ligustrum robustum* subsp. *walkeri* threatens native forests on La Réunion. *Biol. Inv.* **1**, 377–392 (1999).

Loope, L. L. & Mueller-Dombois, D. Characteristics of invaded islands, with special reference to Hawaii. in *Biological Invasions: A Global Perspective* (eds Drake, J. A., Mooney, H. A., Di Castri, F., Groves, R. H., Kruger, F. J., Rejmanek, M. & Williamson, M.) 257–280 (John Wiley & Sons Ltd, Chichester, 1989).

Lorence, D. H. & Sussman, R. W. Exotic species invasion into Mauritius wet forests remnants. *J. Trop. Ecol.* **2**, 147–162 (1986).

Mandon-Dalger, I., Le Corre, M., Clergeau, P., Probst, J.-M. & Besnard, N. Modalités de la colonisation de l'île de la Réunion par le Bulbul orphée (*Pycnonotus jocosus*). *Revue d'Ecologie (Terre Vie)* **54**, 283–295 (1999).

Meyer, J.-Y. Observations on the reproductive biology of *Miconia calvescens* DC (Melastomataceae), an alien invasive tree on the island of Tahiti (South Pacific Ocean). *Biotropica* **30**, 609–624 (1998).

Meyer, J.-Y. & Florence, J. Tahiti's native flora endangered by the invasion of *Miconia calvescens* DC. (Melastomataceae). *J. Biogeogr.* **23**, 775–781 (1996).

Mueller-Dombois, D. Forest dynamics in Hawaii. *Tree* **2**, 216–220 (1987).

Savidge, J. A. Extinction of an island forest avifauna by an introduced snake. *Ecology* **68**, 660–668 (1987).

Simberloff, D. Why do introduced species appear to devastate islands more than mainland areas? *Pacific Sci.* **49**, 87–97 (1995).

Strasberg, D. Diversity, size composition and spatial aggregation among trees on a 1-ha rain forest plot at La Réunion. *Biodiver. Conserv.* **5**, 825–840 (1996).

Vitousek, P. M. Diversity and biological invasions of oceanic islands. in *Biodiversity* (ed. Wilson, E. O.) 181–189 (National Academic Press, Washington DC, 1988).

Tropical Ecosystems: Structure, Diversity and Human Welfare.
Proceedings of the International Conference on Tropical Ecosystems
K. N. Ganeshaiah, R. Uma Shaanker and K. S. Bawa (eds)
Published by Oxford–IBH, New Delhi. 2001. pp. 459–462.

Biological control of invasive plants in South India

R. Muniappan

Agricultural Experiment Station, University of Guam, Mangilao, Guam 96923, USA
e-mail: rmuni@uog9.uog.edu

The world became aware of the invasion of alien species in 1982 when the General Assembly of the Scientific Committee on Problems of the Environment rang the wake up call (Drake *et al.*, 1989). The seriousness of the invasive alien plants has been recognized in India in the early 1800s. India was one of the pioneering countries in utilizing beneficial insects for control of invasive plants. In 1836 biological control programme for the invasive weed, *Opuntia vulgaris* Miller was implemented in South India. The Government of India detailed Ramachandra Rao in 1916 to survey the invasive plants and natural enemies of *Lantana camara* L. in India and Burma. A decade later, Tadulingam and Venkatanarayana (1932) published *A Handbook of Some South Indian Weeds* with descriptions of several weeds. The 1989 Nainital workshop reviewed ecology of some invasive species in India (Ramakrishnan, 1991). Muniappan and Viraktamath (1993) highlighted the problem of alien invasive weeds in the Western Ghats region of southern India.

The first biological control of a weed in the world was serendipitously achieved in India in 1795 by introducing *Dactylopius ceylonicus* Green from Brazil. The introduction was made with the mistaken identity for *D. cacti* (L.), which was the cochineal insect of commerce. Efforts to culture it on

Keywords: Invasive plants, biological control, south India.

O. vulgaris were given up, as the dye produced by it was inferior to *D. coccus* Costa. However, it established on *O. vulgaris* in the field and effectively controlled it (Rao *et al.*, 1971). Redistribution of this insect in South India in 1836 marked the first attempt at biological control of a weed (Julian and Griffiths, 1998). Efforts to control *O. dillenii* (Ker-Gawler) Haworth and *O. elatior* Miller utilizing *D. ceylonicus* in the late 19th century in South India were not successful, as it did not feed on them (Rao *et al.*, 1971). In 1836 *D. confusus* (Cockerell) from South America via Germany and South Africa was introduced to India to control *O. vulgaris* but it did not establish (Julian and Griffiths, 1998). In 1926, the North American *D. opuntiae* (Cockerell) was introduced to South India from Sri Lanka where it was imported from Australia in 1924. *D. opuntiae* has spread throughout India and effectively controlled *O. dillenii* and *O. elatior* (Rao *et al.*, 1971).

Lantana camara was introduced as an ornamental plant to the Calcutta Botanic Garden around 1809 and later on to other places. It escaped cultivation and became a serious weed in different ecological regimes throughout India. Ramachandra Rao (1920) surveyed India and Burma and reported 148 local species of insects recruited by *L. camara*. The lantana bug, *Orthezia insignis* Browne was observed on lantana in the Nilgiri hills in 1915. Since it is a polyphagous insect, efforts to eradicate it in the early stages of its establishment were not successful. This insect was accidentally introduced from Sri Lanka (Fletcher, 1917). In 1921, *Ophiomyia lantanae* (Froggatt) was received from Hawaii and released in Bangalore. Twelve years after release, in 1933, the flies were recovered in the field. Since then it has spread throughout India.

The lantana lace bug, *Teleonomia scrupulosa* Stal was imported from Australia in 1941. Host specificity tests conducted at Dehra Dun showed this insect to feed on flowers of teak and hence the laboratory culture was destroyed in 1943. A few individuals escaped and established in the field (Roonwal, 1953). Field observations proved that this insect was not feeding on teak and hence it was distributed to central and eastern parts of India. In 1969 and 1970 the then Commonwealth Institute of Biological Control (CIBC) substation in Bangalore imported and distributed it throughout South India. In 1971, the leaf mining beetles *Octotoma scabripennis* Guerin and *Uroplata girardi* Pic. were imported from Australia and established in northern India (Sen-Sarma and Thapa, 1981). Muniappan and Viraktamath (1986) found *Epinotia lantana* (Bursck) to be fortuitously established in India.

Ageratina adenophora is found mostly in the hills above 1000 m in India. In 1963, the CIBC substation in Bangalore received shipments of *Procecidochares utilis* Stone from New Zealand and released them in the Nilgiris and West Bengal (Rao *et al.*, 1971). Now it has established throughout the hill stations in India. The maggots of this fly cause galls in the stems but the local parasitoids have reduced their effectiveness.

Chromolaena odorata was introduced as an ornamental plant to Calcutta Botanical Gardens in 1845 (Hooker, 1882). It escaped and established throughout the humid tropical regions of India. Attempts made in 1971 and 1978 to establish the natural enemy, *Pareuchaetes pseudoinsulata* Rego Barros were not successful. In 1984, a culture of this insect brought in from Sri Lanka was released and established in Kerala and Karnataka (Jayanth and Ganga Visalakshy, 1998). In 1995, an eriophyid mite, *Acalitlus adoratus* has fortuitously established.

Parthenium hysterophorus was accidentally introduced in 1956 and it has invaded all the arid and semiarid regions in India. A chrysomelid beetle, *Zygogramma bicolorata* Pallister was imported from Mexico in 1983 and field released in 1984. It was first considered as a threat to sunflower crop, however, further observations proved it to be of no concern. This insect has proven effective in suppressing the weed in certain regions (Jayanth and Ganga Visalakshy, 1994).

Salvinia molesta, a neotropical fern, was accidentally introduced. It has spread and clogged channels, rivers and waterways in Kerala. In 1974, an acridid grasshopper, *Paulinia acuminata* De Geer from Trinidad was introduced but it was not effective. A weevil, *Cyrtobagous salviniae* Calder and Sands was imported from Australia in 1983. In 1984 it was field released and in three years it has effectively suppressed the fern (Julien and Griffiths, 1998).

Eichhornia crassipes is a native of Brazil and it was introduced as an ornamental plant to Bengal around 1896. It is one of the major weeds clogging the water tanks. In 1982, the weevils, *Neochetina eichhorniae* Warner and *Neochetina bruchi* Hustache and the mite, *Orthogalumna terebrantis* Wallwork were imported from Florida and field released in 1983. The weevils were more effective in suppressing the weed population (Jayanth, 2000).

Biological control has been utilized for suppression of a few invasive weeds in India. A number of exotic weeds have invaded India and some of them are good targets for this programme. *Mikania micrantha* H.B.K. in the southwestern and northeastern India, *Mimosa diplotricha* (C. Wright ex Suavalle) in Kerala, *Ulex europaeus* L. and *Cytisus scoparius* (L.) Link in the Nilgiris and *Alternanthera philoxeroides* (Martius) Grisebach in the water ponds are some of the invasive weeds for which biological control programmes exist in other countries.

References

Drake, J. A., Mooney, H. A., Di Castri, F., Groves, R. H., Kruger, F. J., Rejmanek, M. & Williamson, M. *Biological Invasions: A Global Perspective* (John Wiley & Sons, New York, 1989), pp. 525.
Fletcher, T. B. Lantana (*Lantana aculeata*). Proc. 2nd Entomol. Meet., Pusa, 38–40 (1917).
Hooker, J. D. *Flora of British India*. Vol. III (Shottiswoode, London, 1882).

Jayanth, K. P. Biological control of water hyacinth: the Indian experience. *Biocontrol News and Information* **21,** 58N (2000).

Jayanth, K. P. & Ganga Visalakshy, P. N. Dispersal of parthenium beetle *Zygogramma bicolorata* (Chrysomelidae) in India. *Biocontrol Sci. Technol.* **4,** 363–365 (1994).

Jayanth, K. P. & Ganga Visalakshy, P. N. Current status of biological control trials against *Chromolaena odorata* in India. Proc. Fourth Intern. Workshop on Biological Control and Management of *Chromolaena odorata*, pp 93–96 (1998).

Julian, M. H. & Griffiths, M. H. Biological control of weeds. *A World Catalogue of Agents and their Target Weeds* (CABI Publishing, 1998), 223 pp.

Muniappan, R. & Viraktamath, C. A. Status of biological control of the weed, *Lantana camara* in India. *Tropical Pest Manage.* **32,** 40–42 (1986).

Muniappan, R. & Viraktamath, C. A. Invasive alien weeds in the western Ghats. *Curr. Sci.* **64,** 555–557 (1993).

Ramachandra Rao, Y. *Lantana* insects in India. Report on an inquiry into the efficiency of indigenous insect pests as check on the spread of *Lantana* in India. *Mem. Dep. Agric. India, Entomol. Ser.* **5,** 239–314 (1920).

Ramakrishnan, P. S. *Ecology of Biological Invasion in the Tropics* (International Scientific Publ., New Delhi, 1991), 195 pp.

Rao, V. P., Ghani, M. A., Sankaran, T. & Mathur, K. C. A review of the biological control of insects and other pests in South-East Asia and the Pacific Region. *Commonwealth Agric. Bureau Tech. Comm.* **6,** 149 (1971).

Roonwal, M. L. Further remarks on the distribution of the lantana bug, *Teleonomia scrupulosa* Stal (Hemiptera, Tingidae) in India, since its introduction in 1941 from Australia. *Indian For.* **79,** 628–629 (1953).

Sen-Sarma, P. K. & Thapa. R. S. Recent advances in forest entomology in India. in *Recent Advances in Entomology in India* (ed. Ananthakrishnan, T. N.) 21–36 (Entomol. Res. Inst., Madras, 1981).

Tadulingam, C. & Venkatanarayana, G. *A Handbook of Some South Indian Weeds* (Government Press, Madras, 1932) 356 pp.

Biodiversity and Ecosystem Functioning

Tropical Ecosystems: Structure, Diversity and Human Welfare.
Proceedings of the International Conference on Tropical Ecosystems
K. N. Ganeshaiah, R. Uma Shaanker and K. S. Bawa (eds)
Published by Oxford–IBH, New Delhi. 2001. pp. 465–468.

Diversity and ecosystem functioning in managed tropical communities

A. J. Hiremath[*†] and J. J. Ewel*[**]

*Ashoka Trust for Research in Ecology and the Environment,
Bangalore 560 024, India
**USDA Forest Service, Institute of Pacific Islands Forestry, Honolulu,
Hawaii, USA
[†]e-mail: hiremath@atree.org

The high productivity, nutrient retention, and stability (resistance and resilience in response to pests, pathogens, and invasive weeds) observed in natural systems are frequently attributed to their high diversity (Tilman, 2000). High productivity, nutrient retention, and stability are also associated with ecosystem sustainability. In much of the temperate world – as also in parts of the tropics – these aspects of ecosystem functioning have been achieved in highly simplified human-managed systems through subsidies in the form of fertilizers and pesticides. Over much of the tropical world, however, such fossil-energy-based subsidies continue to be an economically unviable option. Understanding the ecological underpinnings of the diversity–functioning relationship, therefore, is crucial to the design of sustainable human-managed tropical systems.

There is some evidence for increased productivity associated with high diversity (the so called 'intercropping advantage'; Willey, 1985). The mechanism proposed to explain this phenomenon is akin to the idea of niche partitioning in animal communities (Vandermeer, 1981), that a diverse

---

Keywords: Diversity, lifeforms, nutrient retention, productivity, tropics.

species mixture can more completely capture and use available resources than can a single species on its own. There is both empirical and theoretical support for resource partitioning by diverse plant communities as demonstrated by: a) the spatial partitioning of resources by leaves (Trenbath, 1986) and roots (Jackson *et al.*, 1995), (b) the temporal partitioning of resources by species of varying phenologies (Felker, 1978) or lifespans (Rao, 1986), (c) the use of different forms of the same resource, as in mixtures of legumes and non-legumes (Binkley *et al.*, 1992), and d) the use of resources in different proportions (as demonstrated theoretically by Trenbath (1976) and Tilman (1988).

Just as there is evidence in support of the proposed diversity–functioning relationship in human-managed systems, there are also exceptions. An example comes from an experiment (by Ewel and co-workers) in which monocultures were compared with successional vegetation and with high-diversity mimics of successional vegetation. The monocultures, it turned out, were as productive (Ewel, 1999), and as effective at resource capture (Berish and Ewel, 1988) as the more diverse systems. The diverse systems, on the other hand, demonstrated greater responsiveness to herbivory damage (Brown and Ewel, 1987), and greater nutrient retention (Ewel *et al.*, 1991). These findings suggest that greater diversity may be more crucial for risk aversion and ecosystem resilience (in the face of unforeseen perturbations, or nutrient losses) than for augmenting productivity (Ewel, 1986).

The idea of resource partitioning by plant communities also has certain conceptual and practical limitations. First, all plants rely on the same basic suite of resources, so greater diversity need not inevitably lead to resource partitioning unless there are inherent differences among species in their architecture, habit, or physiology of resource use. Thus, apparent spatial separation of resource acquisition by species occurring together may actually be an effect of competition, rather than a cause of resource partitioning. Second, augmentation of species numbers in experimental systems is often confounded with increased density of planting. Thus, increased resource capture or use by an increased number of species may in fact only be an artefact of incomplete resource use in the original community.

To assess the possible role of species diversity in ecosystem functioning, monocultures dominated by three different tree species were compared with polycultures in which each of those tree species was co-planted with two large, perennial monocots, which are important components of tropical forests. One monocot was a palm (thus had an apical meristem, therefore indeterminate height growth) and the other a heliconia (having a basal meristem, thus biomechanically constrained height growth). The idea of combining different lifeforms was that if there are differences in species' modes of accessing and using resources, these differences are most likely to be manifest in species that differ in their allocation to resource capture and use. Ecosystem nutrient use efficiency – a measure of ecological functioning that integrates productivity and nutrient retention – was one of many response variables assessed in an effort to investigate the relationships between diversity and ecosystem functioning.

The presence of the additional lifeforms increased nutrient use efficiency in two of the three experimental systems, in two of four years. These results indicate that it is not a greater diversity of lifeforms, per se, but the mix of species and lifeforms that determines efficiency at the ecosystem level. Furthermore, the impact of lifeform diversity on ecosystem functioning is not a static phenomenon; instead, it varies with the growth (thus nutrient uptake) of the community's components, rising when growth is vigorous and declining when growth slows. In addition, although productivity and nutrient uptake varied over time, total nutrient accrual (a measure of ecosystem nutrient retention) remained high in the more diverse systems in all four years, despite the large-scale die-back of one of the lifeforms early in the experiment, with nutrient accrual by the other lifeforms compensating. Although lifeform diversity significantly affected ecosystem functioning across all systems and years, ecosystem nutrient use efficiency was better related to soil nutrient supply than to species' resource-use characteristics.

Just as there are notable exceptions to the diversity–functioning relationship in natural systems (as attested to by the existence of some very successful low-diversity natural systems: the monodominant *Gilbertiodendron* forests of Africa, mangroves on tropical coasts, and the *Shorea* swamp forests of SE Asia), so also there are exceptions in managed systems. The role of diversity may be most critical in ensuring ecosystem resilience in the long term, rather than high productivity in the short term. Ecosystem functioning results from the interaction of both bottom-up and top-down factors, the former related to traits of individual species as well as the mix of species or lifeforms composing the community, and the latter related to external factors such as soil nutrient supply.

Acknowledgements

The ideas developed in this paper came, in part, from research supported by the US National Science Foundation, the Organization for Tropical Studies, the Tropical Conservation and Development Program at the University of Florida, and the Andrew W. Mellon Foundation.

References

Berish, C. W. & Ewel, J. J. Root development in simple and complex tropical successional ecosystems. *Plant Soil* **106**, 73–84 (1988).

Binkley, D., Dunkin, K. A., DeBell, D. & Ryan, M. G. Production and nutrient cycling in mixed plantations of *Eucalptus* and *Albizzia* in Hawaii. *For. Sci.* **38**, 393–408 (1992).

Brown, B. J. & Ewel, J. J. Herbivory in complex and simple successional tropical ecosystems. *Ecology* **68**, 108–116 (1987).

Ewel, J. J. Natural systems as models for the design of sustainable systems of land use. *Agroforestry Systems* **45**, 1–21 (1999).

Ewel, J. J. Designing agricultural ecosystems for the humid tropics. *Annu. Rev. Ecol. Syst.* **17**, 245–271 (1986).

Ewel, J. J., Mazzarino, M. J. & Berish, C. W. Tropical soil fertility changes under monocultures and successional communities of different structure. *Ecol. Appl.* **1**, 289–302 (1991).

Felker, P. State of the art: *Acacia albida* as a complementary permanent intercrop with annual crops. USAID, Washington, DC, USA (1978).

Jackson, P. C., Cavelier, J., Goldstein, G., Meinzer, F. C. & Holbrook, N. M. Partitioning of water resource use among plants of a lowland tropical forest. *Oecologia* **101**, 197–203 (1995).

Rao, M. R. Cereals in multiple cropping. in *Multiple Cropping Systems* (eds Francis, C. A.) 96–132 (MacMillan, New York, USA, 1986).

Trenbath, B. R. Plant interactions in mixed crop communities. in *Multiple Cropping* (eds Papendick, R. I., Sanchez, P. A. & Triplett, G. B.) 129–170 (American Society of Agronomy Special Publication Number 27, 1976).

Trenbath, B. R. Resource use by intercrops. in *Multiple Cropping Systems* (eds Francis, C. A.) 57–87 (MacMillan, New York, USA, 1986).

Tilman, D. Causes, consequences, and ethics of biodiversity. *Nature* **405**, 208–211 (2000).

Tilman, D. *Plant Strategies and the Dynamics and Functioning of Plant Communities* (Princeton University Press. Princeton, New Jersey, USA, 1988).

Vandermeer, J. The interference production principle: An ecological theory for agriculture. *BioScience* **31**, 361–364 (1981).

Willey, R. W. Evaluation and presentation of intercropping advantages. *Exper. Agric.* **21**, 119–133 (1985).

Tropical Ecosystems: Structure, Diversity and Human Welfare.
Proceedings of the International Conference on Tropical Ecosystems
K. N. Ganeshaiah, R. Uma Shaanker and K. S. Bawa (eds)
Published by Oxford–IBH, New Delhi. 2001. pp. 469–472.

Plant pathogens and disease development in a Mexican tropical rain forest

Graciela García-Guzmán* and Rodolfo Dirzo

Departmento de Ecología Evolutiva, Instituto de Ecología, UNAM, A.P. 70-275, CU,
México, D.F. 04510, Mexico
*e-mail: mggarcia@miranda.ecologia.unam.mx

Most studies regarding pathogens and plants have been in agricultural systems, where pathogens play an important role in the population dynamics of a wide variety of crop species (Agrios, 1997). In contrast, the study of plant–pathogen interactions under natural conditions has been very much neglected, particularly in the tropics (Coley and Barone, 1996; Gilbert and Hubbell, 1996). Reports suggest that environmental conditions in tropical areas are more favourable for development of a wide range of diseases of crop plants (Waller, 1976). But evidence does not exist to assess if such observations in agronomic systems are applicable to natural ecosystems. Studies in a Mexican rainforest indicate that ca. 65% of the understorey plant species are affected by fungal pathogens (García-Guzmán and Dirzo, in press), but no quantitative assessments on diseases development exist. Therefore, in this paper we present data regarding disease incidence, levels of leaf damage by fungal pathogens and diseases development rates in ten of the most abundant plant species in the Los Tuxtlas tropical rainforest (Mexico), according to García-Guzmán (1990). Especially, we tried to explore if disease incidence and development were affected by the rainfall seasonality of this forest.

Keywords: Disease development, fungal pathogens, tropical rain forest, Los Tuxtlas.

Four sites of closed mature forest were selected on relatively flat areas at the Los Tuxtlas Research Station to survey disease incidence in the ten selected plant species (Table 1). In each site we established sampling plots during the rainy (November) and the dry (April) seasons. We collected all leaves from the surveyed plants and visually estimated the percentage of area lost to pathogens according to Dirzo and Dominguez (1995). Thereafter, undamaged leaves from the ten plant species were marked in the forest, collected 50 days later, and checked for any disease symptom. Leaf area damaged by pathogens as well as total leaf area were measured with a leaf area meter. Disease development rates (DDR) were defined as: DDR = (%DLAi − %DLAf)/d; where DLAi = initial leaf area damaged, DLAf = final leaf area damaged and d = number of days between the two measurements.

Percentage of diseased leaves per plant species was very similar in both seasons (cf. Table 1). *Astrocaryum mexicanum* was the plant species with the highest percentage of damaged leaves (~ 80%) (Table 1). *Philodendron guttiferum* was disease-free during the dry season, while all leaves of *Salacia megistophylla* were healthy in the rainy season. Levels of leaf area damaged per plant in all species were similarly low in both seasons (< 6%). However, *A. mexicanum* showed ≤ 20% of leaf area damaged per plant during the rainy season (cf. Table 1).

Table 1. Number of analyzed leaves (N), percentage of diseased leaves (%DL) and percentage of leaf area damaged per plant (%LAD) during the dry and rainy seasons.

Species	Dry season			Wet season		
	N	%DL	%LAD	N	%DL	%LAD
Aphelandra aurantiaca (Scheidw.) Lundell (Acanthaceae)	261	1.53	1.11	187	2.14	2.70
Astrocaryum mexicanum Liebm. ex Mart. (Arecaceae)	39	84.62	5.28	20	80.00	19.98
Diplazium lonchophyllum Kunze (Polypodiaseae)	1660	61.93	2.49	2101	60.73	3.75
Monstera acuminata G. Koch. (Araceae)	152	42.11	2.40	179	43.58	1.38
Nectandra ambigens (Blake) Allen (Lauraceae)	179	71.51	3.42	73	63.01	3.24
Omphalea oleifera Hemsl. (Euphorbiaceae)	33	30.30	2.10	1	100	3
Philodendron guttiferum Kunth in H.B.K. (Araceae)	3	0	0	24	16.67	0.81
Rhodospatha aff. *wendlandii* Schott (Araceae)	219	21.92	1.74	259	26.25	1.62
Salacia megistophylla Standley (Hippocrateaceae)	25	20.00	3.51	8	0	0
Syngonium podophyllum Schott (Araceae)	97	67.01	4.05	167	52.10	3.72

Disease development rates in all plant species were remarkably low (< 1% of leaf area affected per day) (Figure 1). On an average, the daily rate for the dry season was 0.065% and 0.032% for the rainy season, but was not significant between-seasons (ANOVA; $F = 1.466$, $P > 0.227$). *A. mexicanum, Diplazium lonchophyllum, Monstera acuminata, Nectandra ambigens* and *Salacia megistophylla* showed higher DDR during the dry season (Mann-Whitney's U; $P < 0.05$), whereas *Rhodospatha* aff. *wendlandii, Aphelandra aurantiaca* and *P. guttiferum* had higher rates during the rainy season (Mann–Whitney's U; $P < 0.05$) (Figure 1).

The results show that most of the surveyed species were affected by disease (90% in both seasons). However, the severity of disease was relatively low (< 6%), and only *A. mexicanum* had more than 19% of leaf area damaged per plant during the rainy season. High damage levels in some species, and the lack of damage in others may be explained by stochastic reasons (probability of a pathogen of finding a suitable host as a function of its abundance), as well as due to variation in defensive mechanisms (Agrios, 1997). The seasonality present at Los Tuxtlas did not affect disease incidence in the studied species. A previous study suggests that environmental conditions in the understorey of Los Tuxtlas are favourable for development of fungal pathogens throughout the year (García-Guzmán and Dirzo, in press). Nonetheless, we detected seasonal variation in disease development rates in most of the studied species. While 50% of the species had higher rates during the dry season, 30% had higher rates in the rainy season. Some studies (e.g. Thrall and Jarosz, 1994) have shown that plant–pathogen interactions have a very strong environmental component. Changes in specific physical factors

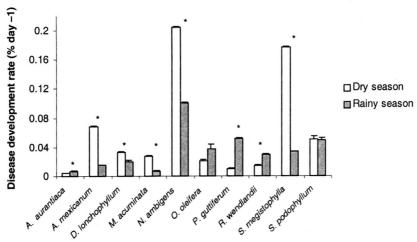

Figure 1. Mean daily disease development rates (± 1 SE) in ten plant species during the dry and rainy seasons. Mann–Whitney U tests, * = significant ($P < 0.05$) difference between seasons.

may affect the host, the pathogen, or their interaction, by altering the germination of spores and growth of pathogens, as well as the germination, growth or susceptibility of host plants, disease expression, the survival of infected plants (Colhoun, 1973), or behaviour of vectors (García–Guzmán and Dirzo, in press). Disease development rates in all species were low in both seasons of the year, and very similar to the ones documented in tropical crop systems. Van der Plank (1963) reported rates of infection of ~ 0.001 (per unit of host tissue per day) for banana infected by *Fusarium oxysporum*. However, these studies have been performed under controlled conditions (artificially infecting plants), or in agricultural systems characterized by their ecological and genetic simplicity, increasing the probabilities for a plant of becoming successfully infected. In contrast, plant pathogens in a tropical ecosystem have to deal with a variety of plant species, each having different patterns of defence mechanisms or resistance against pathogens. Undoubtedly the role of biotic and abiotic factors in explaining interspecific variation in pathogen damage and disease development warrants further study.

References

Agrios, G. N. *Plant Pathology* (Academic Press, New York, 1997).
Coley, P. D. & Barone, J. A. Herbivory and plant defences in tropical forests. *Annu. Rev. Ecol. Syst.* **27**, 305–335 (1996).
Colhoun, J. Effects of environmental factors on plant disease. *Anun. Rev. Phytopathol.* **11**, 343–364 (1973).
Dirzo, R. & Domínguez, C. A. Plant–herbivore interactions in Mesoamerican tropical dry forests. in *Seasonal Dry Tropical Forests* (eds Bullock, S. H., Mooney, H. A. & Medina, E.) 304–325 (Cambridge University Press, Cambridge, UK, 1995).
García-Guzmán, G. Estudio sobre ecología de patógenos en el follaje de plantas en la selva de Los Tuxtlas. Msc. Thesis. Universidad Nacional Autónoma de México, Mexico, Mexico (1990).
García-Guzmán, G. & Dirzo, R. Patterns of leaf-pathogen infection in the understorey of a Mexican rain forest: incidence, spatio temporal variation, and mechanisms of infection. *Am. J. Bot.* in press.
Gilbert, G. S. & Hubbell, S. P. Plant diseases and the conservation of tropical forests. *BioScience* **46**, 98–106 (1996).
Thrall, P. H. & Jarosz, A. M. Host–pathogen dynamics in experimental populations of *Silene alba* and *Ustilago violacea*. I. Ecological and genetic determinants of disease spread. *J. Ecol.* **82**, 549–559 (1994).
Van der Plank, J. E. *Plant Diseases: Epidemic and Control* (Academic Press, New York, 1963).
Waller, J. M. The influence of climate on the incidence and severity of some diseases of tropical crops. *Rev. Plant Pathol.* **55**, 185–194 (1976).

Tropical Ecosystems: Structure, Diversity and Human Welfare.
Proceedings of the International Conference on Tropical Ecosystems
K. N. Ganeshaiah, R. Uma Shaanker and K. S. Bawa (eds)
Published by Oxford–IBH, New Delhi. 2001. pp. 473–476.

Is the trichome-rich leaf surface of the aggressive weed *Broussonetia papyrifera* (paper mulberry) a successful means of impeding herbivores?

Megha Shenoy and Renee M. Borges

Centre for Ecological Sciences, Indian Institute of Science, Bangalore 560 012, India
e-mail: cloudu2@usa.net

The various defense mechanisms adopted by plants reflect the kind of interactions they have with other organisms. Most plants have physical and chemical defenses to prevent exploitation by insects. However, insects in turn have developed strategies to circumvent this problem. Hence, the kinds of plant–insect interactions observed are the result of a fine balance between the strategies used by plants and insects for their survival. One successful and ubiquitous defense tactic employed by plants is the utilization of trichomes on the leaf surface. Glandular trichomes actively secrete toxins whereas non-glandular trichomes may either entangle insects or in some cases pierce through the cuticle of the infesting insect (Jeffree *et al.*, 1986).

Broussonetia papyrifera (Moraceae) is a native of the Himalayas. It is found in large numbers in and around the Indian Institute of Science campus in Bangalore, where it has achieved weed status. It is a large, gregarious,

---

Keywords: *Broussonetia*, insect leg modifications, trichomes.

evergreen tree. This plant is of particular interest since it has virtually no insects on its surface. Hand-made paper has been made from its bark in North Thailand for over 700 years. A large number of chemicals toxic to many organisms have been isolated from this plant, e.g. antifungal compound against *Aspergillus niger* – broussonin A (Iida *et al.*, 1999), a new isoprenylated aurone, broussoaurone A (Fang *et al.*, 1995), a novel isoprenylated flavan, broussoflavan A, as well as six known compounds, butyrospermol acetate, erythrinasinate, kazinols A and B, broussochalcones A and B were isolated and characterized from the cortex of Formosan *B. papyrifera* (Fang *et al.*, 1994). *Broussonetia* also has leaf and stem surfaces rich in non-glandular trichomes. Hence the belligerent physical and chemical defense mechanisms employed by *B. papyrifera* may be responsible for its aggressive abundance in this region.

Although aphids, homopteran-tending ants and many other herbivorous insects inhabit plants that surround paper mulberry, the leaf surface of this plant has virtually no insects. In this study, we investigated whether the trichome-rich leaf surface could contribute to the absence of insects on the leaves of *B. papyrifera*. In order to test this, we measured the rate of locomotion of some insect species on both the dorsal and ventral leaf surfaces.

We measured the trichome density on the dorsal and ventral surface of each test leaf by counting the number of trichomes in 1 cm^2 under 25x using a stereozoom microscope. We then collected individuals of species with complex leg modifications (the psyllid *Heteropsylla cubana* and the white fly *Aleurodicus dispersus*) and those with simple legs (the ant *Monomorium* sp. and the coccinellid beetle *Curinus coeruleus*). We measured the locomotion rates (m/s) of 10 randomly chosen individuals of each species on the dorsal and ventral surface of the test leaves. This was done by allowing the insects to walk on the leaf surface covered by a glass petri dish. As the insects walked, their path was tracked on the glass surface with a marker. The time taken for the walk was noted, and a rate was then calculated. Since the insects did not walk continuously, the rate measurement was repeated for 10 trials for each insect. The mean of all 10 trials was taken as the locomotion rate of that individual insect.

The mean trichome density (trichomes/cm^2) on the dorsal surface of the leaf of paper mulberry was 14.93 ± 4.81 (sd) ($n = 9$) and that on the ventral surface was 53.56 ± 19.20 (sd) ($n = 9$). The trichome density on the dorsal surface was significantly lower than the trichome density of the ventral surface (*t*-test, *t* value $= -6.31$, $n = 9$, $P < 0.05$) (Figure 1).

For insects with complex legs, we found that the locomotion rates on the dorsal surface were significantly higher than on the ventral surface only in the adult psyllid *H. cubana* (one tailed *t*-test, *t* value $= 6.35$, $n = 10$, $P < 0.001$), while there was no significant difference between the locomotion rates on the two surfaces in the immature psyllid (one tailed *t*-test, *t* value = 6.77, *n* = 3, $P > 0.001$) and the adult white fly *A. dispersus* (one tailed *t*-test, *t* value = 1.59, *n* = 10, $P > 0.05$). For insects with simple legs, we found that the locomotion rates on the dorsal surface were significantly higher than on the ventral surface only in the coccinellid beetle *C. coeruleus* (one tailed *t*-test, *t* value = 4.85, *n* = 10, $P < 0.001$). The rate of locomotion of the immature psyllid on the ventral surface was zero, since the insects got tangled and impeded in the trichomes. The rate of locomotion of *Monomorium* could not be measured on the dorsal surface since the individuals moved too fast (Figure 2). We were, therefore, only able to compare the locomotion rates of insects in the two groups on the ventral surface. We found that the

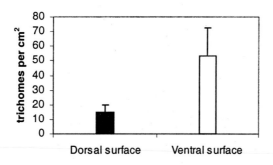

Figure 1. Average trichome density on the dorsal and ventral leaf surface of *Broussonetia papyrifera* (dorsal *n* = 9, ventral *n* = 9).

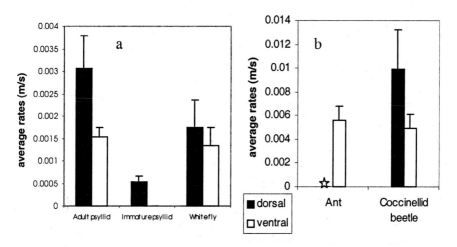

Figure 2. Rate of locomotion of insects with complex legs (2a) and simple legs (2b) on the dorsal and ventral leaf surface of *Broussonetia papyrifera* (n values in text). ★ Indicates rate not measurable.

rate of locomotion on the ventral leaf surface of insects with complex legs was significantly lower than the rate of locomotion of insects with simple legs (two tailed t-test, t value $= -4.36$, $n = 43$, $P < 0.05$).

We have therefore shown that only those insects with complex legs, i.e. with spines, claws or dactyls, were markedly hindered by the presence of trichomes, while those with simple legs were not impeded in their locomotion. Furthermore, locomotion on the dorsal surface was usually faster than that on the ventral surface due to the lower densities of trichomes on the dorsal surface. Immature psyllids were entangled on the ventral surface and completely unable to move. The higher densities of trichomes on the ventral surface could be related to the need for more protection of leaf veins that are more accessible on the ventral surface. This investigation gives an insight into plant defense mechanisms that are specialized towards certain insect types.

References

Jeffree, C. E. The cuticle, epicuticular waxes and trichomes of plants, with reference to their structure, functions and evolution. in *Insects and the Plant Surface* (eds Juniper, B. & Southwood, R.) 23–55 (Edward Arnold Publishers, Suffolk, Great Britain, 1986).

Fang, S. C., Shieh, B. J. & Lin, C. N. Phenolic constituents of Formosan *Broussonetia papyrifera*. *Phytochemistry* **37**, 851–853 (1994).

Fang, S. C., Shieh, B. J., Wu, R. R. & Lin, C. N. Isoprenylated flavonols of Formosan *Broussonetia papyrifera*. *Phytochemistry* **38**, 535–537 (1995).

Iida, Y., Yonemura, H., Oh, K. B., Saito, M. & Matsuoka, H. Sensitive screening of antifungal compounds from acetone extracts of medicinal plants with a bio-cell tracer. *Yakugaku-Zasshi* **119**, 964–971 (1999).

Tropical Ecosystems: Structure, Diversity and Human Welfare.
Proceedings of the International Conference on Tropical Ecosystems
K. N. Ganeshaiah, R. Uma Shaanker and K. S. Bawa (eds)
Published by Oxford–IBH, New Delhi. 2001. pp. 477–479.

Soil macrofauna and litter dynamics in three tropical tree plantations in Puerto Rico

Matthew W. Warren* and Xiaoming Zou

University of Puerto Rico, Institute for Tropical Ecosystem Studies, PO Box 363682,
San Juan, PR 00936-3682
*e-mail: mmww@coqui.net

Tree plantations are becoming an increasingly common land cover in tropical regions. Despite the recognition that soil fauna are the most important factor contributing to the decomposition of organic matter in the humid tropics (Lavelle *et al.*, 1993; Henegan *et al.*, 1999, Gonzalez and Seastedt, in press), and that the fertility of tropical soils largely depends on the maintenance of biological systems of regulation of decomposition, the impact of soil organisms on soil fertility in tropical tree plantations remains largely unexplored. Individual plantation species may influence populations of soil macrofauna, affecting the mineralization and humification of organic matter, which largely determines the nutrient status of the soil. The objectives of this study were to quantify the effects of three contrasting plantation species (*Eucalyptus robusta*, *Casuarina equisetifolia*, and *Leucaena leucocephala*) on soil macrofauna on a degraded site in Puerto Rico. Furthermore, to elucidate correlations between the presence of soil macrofauna and nutrient dynamics in decomposing surface litter, we consider the following hypotheses: (1) Plantation species will vary in the quality and quantity of fine litter on the forest floor, which will affect the biomass and abundance of soil

Keywords: Plantations, litter, soil fauna, nutrients.

macrofauna and (2) Nutrient standing stocks in the forest floor pool will differ between plantation species and will relate to the presence or absence of soil macrofauna.

Table 1. Abundance and biomass of soil fauna in plantation treatments mean (SE); Uncommon letters following values indicate significance at $P = 0.05$.

Abundance (individuals m^{-2})	Casuarina	Leucaena	Eucalyptus
Millipedes	19.33 (5.97)^b	56.00 (14.05)^a	15.33 (7.04)^b
Earthworms	32.67 (22.26)^a	56.00 (36.15)^a	8.67 (4.55)^a
Other	10.67 (5.43)^a	15.67 (6.66)^a	8.00 (2.31)^a
Total	62.67 (27.72)^a	127.67 (60.49)^a	32.00 (12.00)^a
Biomass (g m^2)			
Millipedes	5.58 (2.59)^b	15.60 (1.49)^a	6.41 (3.61)^b
Earthworms	7.97 (7.54)^a	16.03 (16.03)^a	4.70 (2.84)^a
Other	0.32 (0.12)^a	0.66 (0.24)^a	0.22 (0.08)^a
Total	13.87 (4.94)^b	32.34 (15.61)^a	11.33 (3.27)^b

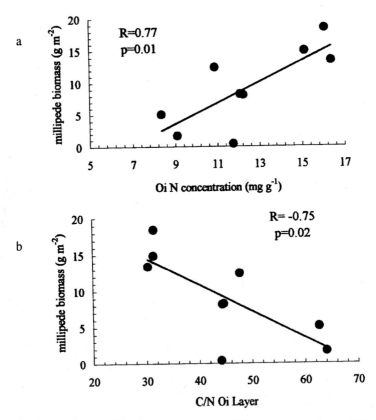

Figure 1. Correlation of millipede biomass versus (a) N concentration in the Oi litter layer and (b) C/N ratio in the Oi layer.

The abundance and biomass of soil macrofauna were measured in treatment plots of *Leucaena*, *Casuarina*, and *Eucalyptus*. Litter was separated into three categories representing organic horizons in three stages of decomposition: identifiable intact slightly decomposed (Oi), identifiable fragmented moderately decomposed (Oe), and unidentifiable highly decomposed (Oa). Horizons were weighed and analyzed for N, P, K, Ca, Mg, Mn, Fe, and Al concentration.

Abundance and biomass of the millipedes *Trigoniulus lubricinus* and *Leptogoniulus nearsii*, and total fauna biomass, were significantly higher in *Leucaena* plots than *Casuarina* and *Eucalyptus* (Table 1). Earthworm biomass also tended to be higher in *Leucaena* plots, however samples were highly variable. *Leucaena* litter had significantly higher N and P concentrations, and significantly lower C/N and C/P ratios than *Casuarina* and *Eucalyptus*; millipede biomass was highly correlated to N concentration and C/N ratio in Oi horizon (Figure 1). Higher N and P concentrations were not indicative of greater standing stocks of these nutrients in the Oi horizon, where the mass of *Leucaena* litter was lower than *Casuarina* and *Eucalyptus*. However, standing stocks of N, P, and K in the Oe horizon were greater in *Leucaena* plots. Total forest floor mass did not vary among species. These results suggest that the abundance and biomass of soil macrofauna are associated with the litter quality of the plantation species, which may lead to more rapid mineralization of organic matter.

References

Gonzalez, G. G. & Seastedt, T. R. Soil fauna and plant litter decomposition in tropical and subalpine forests. (in press) (2000).

Henegan, L., Coleman, D. C., Zou, X., Crossley Jr D. A. & Haines, B. L. Soil microarthropod contributions to decomposition dynamics: tropical–temperate comparisons of a single substrate. *Ecology* **80**, 1873–1882 (1999).

Lavelle, P., Blanchart, E., Martin, A. & Martin, S. A hierarchical model for decomposition in terrestrial ecosystems: application to soils of the humid tropics. *Biotropica* **25**, 130–150 (1993).

Tropical Ecosystems: Structure, Diversity and Human Welfare.
Proceedings of the International Conference on Tropical Ecosystems
K. N. Ganeshaiah, R. Uma Shaanker and K. S. Bawa (eds)
Published by Oxford–IBH, New Delhi. 2001. pp. 480–482.

Increase in earthworm diversity and recovery of native earthworms during secondary succession in old tropical pastures

Yaniria Sanchez*[†], Xiaoming Zou* and Sonia Borges[‡]

*Institute for Tropical Ecosystem Studies and Department of Biology, University of Puerto Rico, P. O. Box 363682, San Juan, PR 00936, USA
[‡]Department of Biology, University of Puerto Rico, Mayaguez, PR, USA
[†]e-mail: yaniria@yahoo.com

Regeneration of secondary forests is recognized to be an important means for the recovery of native species in human-disturbed tropical lands (Brown and Lugo, 1990). Native earthworms are often replaced with exotic species in tropical forests and the recovery of native earthworm may indicate the restoration of forest functions (Gonzalez *et al.*, 1996; Zou and Gonzalez, 1997). We studied changes in earthworm species composition, density and fresh weight along a chronosequence of abandoned pastures in the Cayey Mountains of Puerto Rico. This chronosequence was represented by active pastures, young secondary forests of 25–40 years old and mature secondary forests of greater than 77 years old. The exotic soil-feeding earthworm *Pontoscolex corethrurus* dominated in the pastures and in the young secondary forests (Figure 1). However, five native earthworm species were found in the

---

Keywords: Exotic species, native earthworms, Puerto Rico, restoration, tropical forests.

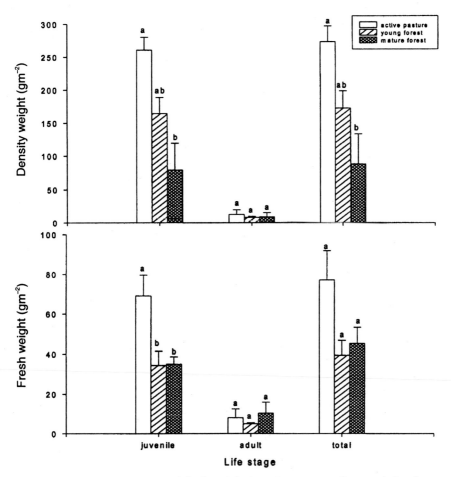

Figure 1. Earthworm density and fresh weight in active pastures (0 year of abandonment), young secondary forest (25–40 year of abandonment) and mature secondary forest (> 77 year of abandonment) in the Cayey Mountains, Puerto Rico. Bars represent means and error bars represent standard error (SE) ($n = 3$). Common letters within a life stage indicate no significant differences among sites.

mature forests together with the exotic earthworm. These native earthworms were litter feeding *Borgesia sedecimsetae*, *Estherella* sp., *Onychochaeta borincana*, *Neotrigaster rufa* and *Trigaster longisimuss* (Table 1). Earthworm density was the highest in the active pastures (273 individuals m^{-2}) and the lowest in the mature forests (88 ind. m^{-2}). Our results suggest that native earthworms can be recovered through the regeneration of mature secondary forests in old tropical pastures.

Table 1. Earthworm species density (ind. m^{-2}) and fresh weight (g m^{-2}) and standard deviation (in parenthesis) along depth and active pastures (pasture), young secondary forests (young) and mature secondary forests (mature) in the Cayey Mountains, Puerto Rico.

Species (family)	Soil depth (cm)	Density (ind. m^{-2})			Fresh weight (g m^{-2})		
		Pasture	Young	Mature	Pasture	Young	Mature
P. corethrurus	0–10	245 (29)	151 (37)	53 (69)	66 (18)	33 (14)	16 (21)
(Glossoscolecidae)	10–25	28 (14)	21 (12)	14 (22)	11 (7)	6 (2)	4 (7)
A. gracilis	0–10	1 (2)					0.6 (1.0)
(Megascolecidae)	10–25						
B. sedecimsetae	0–10			14 (25)			7 (12)
(Megascolecidae)	10–25			2 (2)			1 (1)
Estherella sp.	0–10			2 (3)			13 (22)
(Glossoscolecidae)	10–25						
O. borincana	0–10			2 (3)			2 (3)
(Glossoscolecidae)	10–25						
N. rufa	0–10			1 (2)			0.6 (1)
(Megascolecidae)	10–25						
T. longissimus	0–10						
(Megascolecidae)	10–25			1 (2)			3 (4)

Acknowledgements

This research was supported by the NASA-IRA program and the University of Puerto Rico. We thank Francisco Torregrosa for allowing us to work on his lands. We thank Mitchell T. Aide, Elvira Cuevas and the students from the graduate course Scientific Writing for commenting on an earlier version of this manuscript. We thank Javier Lugo, Josue Casillas, and José Vega for their help in the field.

References

Brown, S. & Lugo, A. E. Tropical secondary forest. *J. Trop. Ecol.* **6**, 1–32 (1990).

González, G., Zou, X. & Borges, S. Earthworm abundance and species composition in abandoned tropical croplands: Comparisons of tree plantations and secondary forest. *Pedobiologia* **40**, 385–391 (1996).

Zou, X. & González, G. Changes in earthworm density and community structure during secondary succession in abandoned tropical pastures. *Soil Biol. Biochem.* **29**, 627–629 (1997).

Mammal–Plant Interaction/Role of Mammals in Ecosystem Processes

Tropical Ecosystems: Structure, Diversity and Human Welfare.
Proceedings of the International Conference on Tropical Ecosystems
K. N. Ganeshaiah, R. Uma Shaanker and K. S. Bawa (eds)
Published by Oxford–IBH, New Delhi. 2001. pp. 485–488.

Hotspots in nutrient landscapes and herbivory by giant squirrels in an Indian seasonal cloud forest

*Renee M. Borges**[*,†] *and Subhash Mali*[‡]

*Centre for Ecological Sciences, Indian Institute of Science, Bangalore 560 012,
India
‡Foundation for the Revitalisation of Local Health Traditions, Anandnagar,
Bangalore 560 032, India
[†]e-mail: renee@ces.iisc.ernet.in

Tropical forests usually have a high diversity of tree species which constitute a highly variable food resource landscape for arboreal herbivores. Here, nutrient variability could be viewed at the level of individual tree species. Much research on food selection by arboreal herbivores has focused on species selection and has determined various phytochemical parameters responsible for food choice, e.g. McKey *et al.* (1981), Davies *et al.* (1988), Borges (1992), Waterman and Kool (1994). Only a few studies have documented choice of individual trees within these selected species, e.g. Glander (1978), Ernest (1994). Very few studies (Lawler *et al.*, 1998; Lawler *et al.*, 2000) have examined factors responsible for hyper-feeding or low-feeding on these individual trees. In this study we examine physical and chemical factors responsible for such inter-tree variation in feeding by the Malabar giant squirrel *Ratufa indica* in a low diversity tropical seasonal cloud forest in India. Here the low diversity enabled us to focus on two dominant tree species within this species-poor forest.

Keywords: Intraspecific variation, phytochemistry, *Ratufa indica.*

The giant squirrel is a territorial, strictly arboreal herbivore. We documented feeding behaviour using focal animal sampling for 11 giant squirrels over a two-year period, each squirrel being continuously observed for two consecutive days each month, in the Bhimashankar Wildlife Sanctuary of Maharashtra, India (Borges, 1992). We focused on two constantly available resources of the two most dominant species in the forest, i.e. the leaf centres of *Memecylon umbellatum* (Melastomataceae) and the inner bark of *Mangifera indica* (Anacardiaceae). We selected these resources as they constitute a major component of the squirrel daily diet, and preliminary observations had indicated that there was inter-tree variation in resource consumption within each squirrel territory. We observed squirrels in a gridded area where we mapped trees of all species > 30 cm GBH and measured their girths ($n = 5680$). We used girth as a surrogate for tree height, and considered tree height as an important factor as giant squirrels are vulnerable to both aerial avian predation as well as predation from leopards. Foraging at safe distances away from the canopy and the ground would, therefore, only be possible in the tallest trees. We chemically analysed samples of *Memecylon* leaf centres from 16 consumed and 10 non-consumed trees and of *Mangifera* inner bark from 9 consumed and 10 non-consumed trees. We measured energy, ash, non-structural carbohydrates, nitrogen, fat and fibre contents, as described in Borges (1992). The minerals Ca, K, P and Na, total phenolic content, condensed tannins, and tannin astringency were analysed as described in Mali (1999).

For both resources, the utilisation of individual trees in each squirrel territory was highly skewed, with most trees not being consumed at all; even within the set of consumed trees, there was a significantly non-normal distribution of consumption between trees (Kolmogorov-Smirnov tests, $p \leq 0.05$).

The utilised *Memecylon* and *Mangifera* trees had significantly larger girths than the non-utilised trees, indicating that squirrels prefer to feed on taller trees (Table 1). In *Memecylon*, there was no significant difference in nutrient and secondary compound content between utilised and non-utilised trees (Mann–Whitney U tests; $n_{consumed} = 16$; $n_{non\text{-}consumed} = 10$; $p > 0.05$). Within the subset of consumed trees ($n = 16$), there was a significant negative relationship only between consumption and astringency (Kendall's $\tau = -0.4048$; $p \leq 0.05$) as well as between consumption and condensed tannin content (Kendall's $\tau = -0.4004$; $p \leq 0.05$) (consumption pooled across squirrels and across years). For *Mangifera* inner bark, there was no significant difference in nutrient and secondary compound content between utilised and non-utilised trees with the exception of fibre components (Mann–Whitney U tests; $n_{consumed} = 9$; $n_{non\text{-}consumed} = 10$). The consumed trees had significantly lower fibre levels in their inner bark (for neutral detergent fibre NDF, acid detergent fibre ADF and acid detergent lignin ADL; $U = 17.5$, 21.0, 19.0 respectively; $p \leq 0.05$) than the non-consumed trees. Within the subset of consumed trees ($n = 9$), mean consumption levels correlated positively only

Table 1. Comparison of girth between consumed and non-consumed conspecifics of *Memecylon umbellatum* and *Mangifera indica* by *Ratufa indica*.

Resource	Year 1 N	Year 1 C	Mann–Whitney U value Year 1	Year 2 N	Year 2 C	Mann–Whitney U value Year 2	Level of significance Year 1	Level of significance Year 2
Memecylon leaf centres	916	165*	52082.0	1324	260*	103931.5	0.00001	0.00001
Mangifera inner bark	903	83*	27601.0	1427	91*	45558.0	0.0001	0.00001

N = number of non-consumed trees; C = number of consumed trees; *indicates higher mean values of the pair.

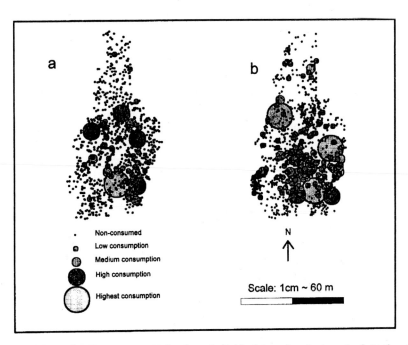

Figure 1. Maps of resource consumption from individual trees by giant squirrels in the study area. A, *Mangifera indica* inner bark in year 1. b, *Memecylon umbellatum* leaf centres in year 2. Low, medium, high and highest consumption levels indicate four equal divisions of percentage consumption from > 0 to 100 (pooled across squirrels).

with nitrogen values (Kendall's $\tau = 0.5508$; $p \leq 0.05$), with the ratio of nitrogen to fibre (N/ADF) (Kendall's $\tau = 0.5353$; $p \leq 0.05$) and with nitrogen to fibre + condensed tannin content (N/(ADF + CT)) (Kendall's $\tau = 0.5353$; $p \leq 0.05$) (consumption pooled across squirrels and across years).

Since (i) these data result from 1290 and 1593 hours of total observation in the two years, which corresponds to an average of approximately 107 and 133 observation hours on each squirrel in the two years, (ii) *Mangifera* bark and *Memecylon* mature leaves are consumed daily, and all chemically investigated trees were chosen randomly, these findings are unlikely to be due to under-sampling and are very likely to be indicative of a real phenomenon.

Our study has, therefore, demonstrated that individual squirrels use individual trees within species in their own territories based on both physical characteristics such as height and chemical characteristics such as nitrogen, fibre, astringency and condensed tannin contents. The results with astringency are especially important, as astringency measures the biological activity of phenolic compounds irrespective of their chemistry.

Lawler *et al.* (2000) pointed out that determining the nutrient patchiness of a forest using chemical method is time-consuming and expensive; they recommended faster screening methods such as infra-red spectroscopy. Yet, they also found that the best measure of the nutritional quality of *Eucalyptus* foliage in species-poor *Eucalyptus* forests was the voluntary dry matter intake of their folivorous marsupial experimental subjects in no-choice feeding trials. We have mapped the consumption by squirrels of all individual *Memecylon* and *Mangifera* trees in the gridded observation area (for only the two plant parts investigated) (Figure 1). These maps indicate hyper-feeding on only a few individual trees. Behavioural maps such as these can serve to map nutrient hotspots in the forest. Since squirrel foraging is also dictated by predation risk (Schmidt 2000), these maps also indicate relatively risk-free feeding locations.

References

Borges, R. M. A nutritional analysis of foraging in the Malabar giant squirrel (*Ratufa indica*). *Biol. J. Linn. Soc.* **47,** 1–21 (1992).

Davies, A. G., Bennett, E. L. & Waterman. P. G. Food selection by the South-east Asian colobine monkeys (*Presbytis rubicunda* and *Presbytis melalophos*) in relation to plant chemistry. *Biol. J. Linn. Soc.* **34,** 33–56 (1988).

Ernest, K. Resistance of creosotebush to mammalian herbivory: temporal consistency and browsing-induced changes. *Ecology* **75,** 1684–1692 (1994).

Glander, K. E. Howling monkey feeding behavior and plant secondary compounds: a study of strategies. in *The Ecology of Arboreal Folivores* (ed. Montgomery, G. G.) 561–574 (Smithsonian Institution, Washington DC, USA, 1978).

Lawler, I. R., Foley, W. J., Eschler, B. M., Pass, D. M. & Handasyde, K. Intraspecific variation in *Eucalyptus* secondary metabolites determines food intake by folivorous marsupials. *Oecologia* **116,** 160–169 (1998).

Lawler, I. R., Foley, W. J. & Eschler, B. M. Foliar concentration of a single toxin creates habitat patchiness for a marsupial folivore. *Ecology* **81,** 1327–1338 (2000).

Mali, S. Plant chemical profiles and their influence on food selection in the Malabar giant squirrel *Ratufa indica*. Ph.D. diss., Univ. of Bombay (1999).

McKey, D. B., Gartlan, J. S., Waterman, P. G. & Choo, G. M. Food selection by black colobus monkeys (*Colobus satanas*) in relation to food chemistry. *Biol. J. Linn. Soc.* **16,** 115–146 (1981).

Schmidt, K. A. Interactions between food chemistry and predation risk in fox squirrels. *Ecology* **81,** 2077–2085 (2000).

Waterman, P. G. & Kool, K. M. Colobine food selection and food chemistry. in *Colobine Monkeys. Their Ecology, Behaviour and Evolution* (eds Davies, A. G. & Oates, J. F.) 51–284 (Cambridge Univ. Press, Cambridge, England, 1994).

Tropical Ecosystems: Structure, Diversity and Human Welfare.
Proceedings of the International Conference on Tropical Ecosystems
K. N. Ganeshaiah, R. Uma Shaanker and K. S. Bawa (eds)
Published by Oxford–IBH, New Delhi. 2001. pp. 489–492.

As good as it gets: Large herbivore densities in south Asian forests

K. Ullas Karanth[†] and N. Samba Kumar[‡]*

*Wildlife Conservation Society (India Program), 403, SEEBO Apartments,
26-2, Aga Abbas Ali Road, Bangalore 560 042, India
[‡]Centre for Wildlife Studies, 177, Pavillion Road, Jayanagar First Block East,
Bangalore 560 011, India
[†]e-mail: karanth@blr.vsnl.net.in

South Asian forests are recognized as harbouring a diverse and rich fauna, of which large herbivorous mammals (ungulates and primates) are important components. Large herbivores play a critical role in the dynamics of these tropical forests by modifying their structures, influencing succession and seed dispersal patterns (Crawley, 1983). Furthermore, among different taxonomic groups of animals, large herbivores are among the most critically threatened due to over-hunting and habitat destruction (Karanth and Sunquist, 1992; Madhusudan and Karanth, 2000). Most published studies of tropical forest mammal densities have come from Africa and Latin America. They indicate low densities of large herbivores, particularly ungulates (Eisenberg, 1980). Data from the tropics in southern Asia are scarce.

This study aimed at estimating the densities and biomass of large herbivores (> 5 kg body weight) – ungulates and primates – in a variety of south Asian tropical forest vegetation types in India. The field surveys were carried out at 10 different sites: Pench and Kanha National Parks (NP), Madhya Pradesh;

Keywords: Tropical forests, ungulates, primates, distance sampling, densities.

Kaziranga NP, Assam; Ranthambore NP, Rajasthan and in five sites in Karnataka (Kalkere in Bandipur, NP, Lakkavalli and Muthodi in Bhadra Wildlife Sanctuary, Arkeri, Sunkadakatte and Nalkeri in Nagarahole NP). These sites represented tropical forests of the following types (Meher-Homji, 1990): moist deciduous forests of the *Tectona grandis–Terminalia alata* (Pench), *Shorea robusta–Terminalia alata–Adina cordifolia* (Kanha), *Tectona grandis–Dillenia pentagyna–Lagerstroemia microcarpa* (in Nagarahole, Bhadra and Bandipur); dry deciduous forests of the *Terminalia alata–Anogeissus latifolia–Tectona grandis* (in Nagarahole and Bandipur); semi-arid dry forests of the *Anogeissus pendula–Acacia catechu* series (Ranthambore) and Alluvial flood-plain grasslands (Kaziranga). We also tried to sample and represent the observed variations in terms of anthropogenic interference such as logging, plantation activities and differential hunting pressures in Bhadra and Nagarahole sites.

A major methodological problem in estimating the abundances of tropical forest mammals is incomplete and/or varied detectability of target species, caused by dense vegetation, inter-observer differences, clustering behaviour of species and other factors (Karanth and Sunquist, 1992). We dealt with this problem using a field survey approach that employed powerful Distance Sampling methods (Buckland *et al.*, 1993, Thomas *et al.*, 2001). Transect lines that representatively sampled each areas were walked on foot (and on elephant back at Kaziranga) by research teams who put in substantial sampling efforts. For each detection, the species observed, number of animals in the cluster, the observer-to-cluster radial distance, and, the sighting angle were recorded (Buckland *et al.*, 1993). The line transect survey data were analyzed using program Distance in a three-step process: data exploration and truncation; model fitting/selection and parameter estimation (Thomas *et al.*, 2001). We estimated the means, standard errors, and the 95% confidence intervals of the following parameters: effective strip width sampled, cluster size, cluster density and animal density.

These distance sampling methods using line transects worked reasonably well for most species in most forest types, except where low sample sizes (resulting from low densities) precluded reliable estimation. In our line transect surveys we detected two species of primates and 13 species of ungulates (including elephants). The results of the line transect surveys (Ahrestani, 1999; Kumar, 2000; Karanth and Nichols, 2000) are summarized below in terms of the spatial distribution and densities of each species. The densities are reported as estimates and their standard errors $\hat{D}\,(SE\,[\hat{D}\,])$ expressed in number of animals/km^2.

Among the two primates, Hanuman Langurs (*Presbytis entellus*) were present in all the sites except alluvial grasslands. Their estimated densities ranged from 16.4 [3.00]/km^2 to 81.8 [6.71]/km^2 in well-protected moist deciduous forests. Langur densities were also lower in semi-arid dry forests at 8.7

[1.57]/km^2. Bonnet macaques (*Macaca radiata*) occurred at six sites, being absent in the four northern Indian sites. Their densities were too low for estimation at Bandipur. In other five sites, macaque densities ranged from 2.0 [0.61]/km^2 to 4.6 [1.36]/km^2.

Among the five deer species, Muntjac (*Muntiacus muntjac*) did not occur in the semi-arid forests and were too low for estimation in alluvial grasslands. Among the other 8 sites, muntjac densities ranged from 0.6 [0.14]/km^2 to 7.7 [1.22]. Sambar (*Cervus unicolor*) occurred in all 10 sites, although their densities were too low to be estimated in alluvial grasslands. At other sites Sambar densities ranged from 1.5 [0.35]/km^2 to 10.7 [1.20]/km^2. Chital (*Axis axis*) were also widely distributed, occurring in all the sites except alluvial grasslands. Their densities varied widely, ranging from 1.00(0.34)/km^2 to 51.3 [3.02]/km^2. The distribution of Barasingha (*Cervus duvaceli*) was restricted to two sites: anthropogenic grasslands in Kanha (crude density of 3.0 [0.43])/km^2) and alluvial grasslands in Kaziranga (density 14.2 [5.42]/km^2). The Hog Deer (*Axis porcinus*) occurred only in alluvial grasslands at a density of 38.6 [3.46]/km^2.

Among the three antelope species, Nilgai (*Boselaphus tragocamelus*) occurred only at Pench (0.7 [0.17]/km^2) and Ranthambore (6.6 [0.65]/km^2). Chinkara (*Gazella gazella*) was present only in the semi-arid dry forests of Ranthambore, at a low density of 1.3 [0.25]/km^2.

Two species of wild cattle were recorded during the surveys. The Asiatic wild buffalo (*Bubalis bubalis*) was found only in the alluvial grasslands of Kaziranga at a density of 2.7 [0.89]/km^2. Gaur (*Bos gaurus*) were not found in semi-arid forests of Ranthambore, and their encounter rates were too low for density estimation in Kanha and Kaziranga. At other sites, Gaur densities ranged from 0.19 [0.07]/km^2 to 18.0 [4.66]/km^2.

Wild pig (*Sus scrofa*) occurred in all sites, generally at low densities of 0.4 [0.15]/km^2 to 3.60 [0.68]/km^2. Rhinoceros (*Rhinoceros unicornis*) were restricted to the alluvial grasslands of Kaziranga where they occur at a high density of 7.3 [0.66]/km^2.

Elephants (*Elephas maximus*) did not occur in Ranthambore, Kanha and Pench. Their densities could be estimated at six sites, and ranged from 0.44 [0.19]/km^2 to 5.0 [0.87]/km^2.

The highest biomass density of large herbivores was supported by the forests in Nagarahole and the most species-rich herbivore assemblage was in Kaziranga. In this paper, we examine the ecological and anthropogenic factors that may shape the distributional patterns and density gradients observed during the survey. We also explore the implications of this study for the conservation of the rich assemblages of large herbivores in southern Asia.

Acknowledgements

We thank the Wildlife Conservation Society and US Fish and Wildlife Service (Division of International Conservation) for funding support; the Ministry of Environment and Forests, Government of India and the State Forest Departments of Karnataka, Madhya Pradesh, Assam and Rajasthan for facilitating this study. Analytical assistance rendered by Farshid Ahrestani, Siva Sundaresan and V. Srinivas as well as field assistance by many volunteer assistants are acknowledged.

References

Ahrestani, F. Population density estimates for mammalian herbivores in Bhadra Wildlife Sanctuary, South India. M.S. Thesis, State University of New York, College of Environmental Science and Forestry, Syracuse, NY, USA (1999).

Buckland, S. T., Anderson, D. R., Burnham, K. P. & Laake, J. L. *Distance Sampling: Estimating Abundance of Biological Populations* (Chapman and Hall, New York, 1993).

Crawley, M. J. *Herbivory: The Dynamics of Animal–Plant Interactions* pp. 437 (University of California Press, Berkeley, 1983).

Eisenberg, J. F. The density and biomass of tropical mammals. in *Conservation Biology: An Evolutionary – Ecological Perspective* (eds Soule, M. E. & Wilcox B. A.) 34–55 (Sinauer Associates, Sunderland, 1980).

Karanth, K. U. & Sunquist, M. E. Population structure, density and biomass of large herbivores in the tropical forests of Nagarahole, India. *J. Trop. Ecol.* **8**, 21–35 (1992).

Karanth, K. U. & Nichols, J. D. Ecological status and conservation of tigers in India. Final technical report to the Division of international conservation, US Fish and Wildlife Service, Washington DC and Wildlife Conservation Society, New York. Centre for Wildlife Studies, Bangalore, India (2000).

Kumar, N. S. Ungulate density and biomass in the tropical semi-arid forest of Ranthambore, India. M.S. thesis, Pondicherry University, Salim Ali School of Ecology and Environmental Sciences, Pondicherry, India, pp 1–72 (2000).

Madhusudan, M. D. & Karanth, K. U. Hunting for an answer: Is local hunting compatible with large mammal conservation in India? in *Hunting for Sustainability in Tropical Forests* (eds Robinson, J. G. & Bennett, E. L.) 339–354 (Columbia University Press, 2000).

Meher-Homji V. M. Vegetation types of India in relation to environmental conditions. in *Conservation in Developing Countries: Problems and Prospects.* Proceedings of the centenary seminar of the Bombay Natural History Society (eds Daniel, J. C. & Serrao, J. S.) 95–110 (Oxford University Press, Bombay, India, 1990).

Thomas, L., Laake, J. L., Strindberg, S., Marques, F., Borchers, D.L., Buckland, S.T., Anderson, D. R., Burnham, K. P., Hedley, S. L. & Pollard, J. H. Distance 4.0. Beta 1. Research Unit for Wildlife Population Assessment, University of St. Andrews, UK. http://www.ruwpa.st-and.ac.uk/distance/ (2001).

Tropical Ecosystems: Structure, Diversity and Human Welfare.
Proceedings of the International Conference on Tropical Ecosystems
K. N. Ganeshaiah, R. Uma Shaanker and K. S. Bawa (eds)
Published by Oxford–IBH, New Delhi. 2001. pp. 493–494.

Ecological and evolutionary interactions between nectar-feeding bats and columnar cacti in the American tropics and subtropics

Theodore H. Fleming

Department of Biology, University of Miami, Coral Gables, FL 33124, USA
e-mail: tfleming@fig.cox.miami.edu

Columnar cacti of tribes Pachycereeae, Cereeae, and Trichocereeae are dominant floristic elements in arid regions ranging from the southwestern United States southward throughout much of Mexico, Central America, northern and western South America, and the Caribbean. Many cactus species of at least eight genera produce nocturnal flowers that conform to a 'bat pollination' syndrome. The relative importance of bats as pollinators of these species varies geographically. Bats are nearly the exclusive pollinators of tropical species but are co-pollinators along with birds and insects of extra-tropical species. Two groups of phyllostomid bats have independently evolved cactophilic habits: one (including four species) in the northern neotropics and the other (including one species) in the Peruvian Andes. Results of detailed studies conducted on bat–cactus interactions in Arizona, Mexico, Curacao, Venezuela, and Peru will be summarized to address the question: Why have 'bat-pollinated' columnar cacti evolved more genera-lized pollination systems outside the tropics? These results are presented in Petit (1995), Fleming *et al.* (1996), Sahley (1996), and Nassar *et al.* (1997). Other

Keywords: Nectar-feeding bats, columnar cacti, pollination biology, America.

aspects of bat–cactus evolution, including the evolution of non-hermaphroditic breeding systems and the genetic consequences of bat pollination, will also be summarized based on results in Fleming *et al.* (1998) and Hamrick *et al.* (2002). Our results show that although they are highly effective pollinators, cactus-visiting bats vary strongly in abundance in geographically predictable patterns. This variation strongly affects the evolution of specialized *vs* generalized pollination systems in columnar cacti, as well as the occurrence of trioecious or gynodioecious breeding systems in the Sonoran Desert cactus *Pachycereus pringlei*. Finally, bat pollination results in higher levels of between-population gene flow and lower levels of genetic subdivision compared with cacti that are pollinated by moths or bees.

References

Fleming, T. H., Tuttle, M. D. & Horner, M. A. Pollination biology and the relative importance of nocturnal and diurnal pollinators in three species of Sonoran Desert columnar cacti. *Southwest. Nat.* **41**, 257–269 (1996).

Fleming, T. H., Maurice, S. & Hamrick, J. L. Geographic variation in the breeding system and the evolutionary stability of trioecy in *Pachycereus pringlei* (Cactaceae). *Evolution. Ecol.* **12**, 279–289 (1998).

Hamrick, J. L., Nason, J. D., Fleming, T. H. & Nassar, J. Genetic diversity in the Cactaceae. in *Columnar Cacti and their Mutualists: Evolution, Ecology, and Conservation* (eds Fleming, T. H. & Valiente-Banuet, A.) (University of Arizona Press, Arizona, 2002).

Nassar, J., Ramirez, N. & Linares, O. Comparative pollination biology of Venezuelan columnar cacti and the role of nectar-feeding bats in their sexual reproduction. *Am. J. Bot.* **84**, 918–927 (1997).

Petit, S. The pollinators of two species of columnar cacti on Curaçao, Netherlands Antilles. *Biotropica* **27**, 538–541 (1995).

Sahley, C. T. Bat and hummingbird pollination of an autotetraploid columnar cactus, *Weberbauerocereus weberbaureri* (Cactaceae). *Am. J. Bot.* **83**, 1329–1336 (1996).

Tropical Ecosystems: Structure, Diversity and Human Welfare.
Proceedings of the International Conference on Tropical Ecosystems
K. N. Ganeshaiah, R. Uma Shaanker and K. S. Bawa (eds)
Published by Oxford–IBH, New Delhi. 2001. pp. 495–497.

Diversity and function in ecosystems

S. J. McNaughton

Syracuse University, Syracuse NY 13244-1220 USA
e-mail: sjmcnaug@syr.edu

One of the most active, and contentious, areas of research and commentary in ecology at present is the relationship, if any, between biodiversity and ecosystem functional properties. My interest in this, stimulated by the writing of Ramon Margalef (1961, 1963), preceded 1965–66, when I did research on the problem at Stanford University's Jasper Ridge Experimental Area (McNaughton, 1968).

There is considerable, sometimes heated, discussion about whether it is individual species or 'functional groups' that contribute to any mechanisms connecting diversity and ecosystem function. However, what is a species, or even an ecotype, if it is not a functional group? It has been now documented for over a century that almost all widespread species, whether plant or animal, consist of a series of subgroups that are genetically and functionally distinct, call them races or ecotypes or what you will. By 'widespread' I mean occupying quantifiably distinct habitats, be they the sweep of climate from subarctic to equatorial or localized, closely juxtaposed soils (McNaughton, 1966). Similarly, there can be no doubt that species composition, biodiversity, and the environment are inextricably coupled, so it is fruitless to argue about whether it is the environment or species identification or biodiversity that link to ecosystem process; of course they do (Sankaran and McNaughton, 1999).

Keywords: Biodiversity, ecosystem function, grassland, grazing.

Some have studied natural communities (Hurd *et al.*, 1971; Mellinger and McNaughton, 1975), others have examined artificial 'communities' in uniform environments ranging from experimental field plots (Tilman and Downing, 1994; Tilman *et al.*, 1996; Hooper and Vitousek, 1997) to laboratory microcosms (Naeem *et al.*, 1994; McGrady-Stead *et al.*, 1977). The latter very artful guides allow precise control over both species composition and the environment so they must be given thoughtful respect. However, nearly a century has been devoted diligently to restoring the Curtis Prairie at the University of Wisconsin and it still lacks many uncommon species. And, rare species can have strong and indirect effects through the food web or the environment (Berlow, 1999). Therefore, it is prudent to recall that humans have nowhere, at any time, restored a diverse community, whether of animals or plants, to its precise original composition and diversity. Ecosytems we can destroy; ecosystems we cannot, so far, recreate. Nature is both wiser, and more technically adept, than are we.

Therefore, I have let Nature shape the community, taking what it gives me, and searched for links between structure and function. I have done this at Jasper Ridge, at old fields in Central New York State, USA (Mellinger and McNaughton, 1975), and in the tropical grazing ecosystem of the Serengeti, Tanzania (McNaughton, 1985). Studies in the Serengeti document:

- Increased stability of green biomass associated with greater biodiversity inside exclosures accompanying variable rainfall.
- Increased stability of green biomass outside exclosures due to variable rainfall and grazing.
- Resistance to dry season grazing by a single ungulate species positively related to diversity.
- Resilience of grassland biomass in previously grazed plots at the onset of the rainy season positively related to diversity.
- Resistance to wet season grazing by a single ungulate species positively related to diversity.
- Grazing resistance in both wet and dry seasons related to foraging selectivity by species.
- Green biomass was not resistant to grazing by multiple grazing species in either the wet or dry seasons.
- Resistance to grazing by a single ungulate species was greater in more mature swards than in the rapidly growing swards maintained by grazers.

In addition,

- Although the net aboveground primary productivity of ungrazed grassland increased with rainfall, there was no evidence that this key ecosystem variable was influenced by biodiversity.

- Complete removal of a single grass species in plots varying in biodiversity indicated that extinction was not buffered by diversity.

Resolution of complex scientific problems, whether identification of the fundamental particles of matter or the connection between biodiversity and ecosystem function, will be solved by the slow, but sure, progress of science, not by a single experiment or the force of a particular argument. We must take the variety of evidence as it builds up until someone, somewhere, synthesizes it. The day-by-day testing of hypotheses in which we engage does not constitute a theory; the theory will be a harmonious synthesis of the results of well-designed and executed hypothesis-testing.

References

Berlow, E. L. Strong effects of weak interactions in ecological communities. *Nature* **398**, 330–334 (1999).

Hooper, D. U. & Vitousek, P. M. The effects of plant composition and diversity on ecosystem processes. *Science* **277**, 1302–1305 (1997).

Hurd, L. E., Mellinger, M. V., Wolf, L. L. & McNaughton, S. J. Stability and diversity at three trophic levels in terrestrial ecosystems. *Science* **173**, 1134–1136 (1971).

Margalef, R. Communication of structure in planktonic populations. *Limnol. Oceanogr.* **6**, 124–128 (1961).

Margalef, R. *Perspectives in Ecological Theory.* University of Chicago Press, Chicago (1963).

McGrady-Stead, J., Harris, P. M. & Morin, P. Biodiversity regulates ecosystem predictability. *Nature* **390**, 162–165 (1997).

McNaughton. Ecotype function in the *Typha* community-type. *Ecol. Monogr.* **36**, 297–325 (1966).

McNaughton, S. J. Structure and function in California grasslands. *Ecology* **49**, 962–972 (1968).

McNaughton, S. J. Ecology of a grazing ecosystem: the Serengeti. *Ecol. Monogr* **55**, 259–294 (1985).

Mellinger, M. V. & McNaughton, S. J. Structure and function of successional vascular plant communities in central New York. *Ecol. Monogr.* **45**, 161–182 (1975).

Naeem, S., Thompson, L. J., Lawlor, S. P., Lawton, J. H. & Woodfin, R. M. Declining biodiversity can alter the performance of ecosystems. *Nature* **368**, 734–737 (1994).

Sankaran, M. & McNaughton, S. J. *Nature* **401**, 691–693 (1999).

Tilman, D. & Downing, J. A. Biodiversity and stability in grasslands. *Nature* **367**, 363–365 (1994).

Tilman, D., Wedin, D. & Knops, J. Productivity and sustainability influenced by biodiversity in grassland ecosystems. *Nature* **379**, 718–720 (1996).

Tropical Ecosystems: Structure, Diversity and Human Welfare.
Proceedings of the International Conference on Tropical Ecosystems
K. N. Ganeshaiah, R. Uma Shaanker and K. S. Bawa (eds)
Published by Oxford–IBH, New Delhi. 2001. pp. 498–500.

Neotropical mammal home ranges and cross scale seed dispersal and survivorship

J. M. V. Fragoso*, K. M. Silvius and J. A. Correa

Biological Sciences, Florida Atlantic University, 777 Glades Road,
Boca Raton, Florida, USA
*e-mail: jfragoso@fau.edu

By radio-tracking small and large-bodied mammals in the tropical forests of the Amazon we found that collared peccaries, white-lipped peccaries and agouti rodents moved to different spatial scales (Fragoso 1998a, 1999; Silvius, 1999). White-lipped peccary herds (individual weight 40 kg; group weight approximately 4,000 kg) had home ranges to 10,000 ha, collared peccaries (individual weight 20 kg; group weight 200 kg) to 1,000 ha and agouti rodents (3 kg, solitary to pair groupings) to 5 ha. These differences in range size suggested that each species was responding to different spatial levels of vegetation organization (Fragoso, 1999; 1998b). I correlated scale of movement with vegetation pattern and found that white-lipped peccaries were responding to landscape level (100,000 ha) habitat mosaics, and collared peccaries to lower plant communities embedded within the mosaic level (Fragoso, 1999). Agoutis responded to the individual tree level, a sub-unit of the other two vegetation categories (Silvius, 1999). It appears that collared peccaries and agoutis delineate their habitats at finer levels of resolution than white-lipped peccaries. Over flights of the study region followed by

---

Keywords: Seed dispersal, frugivores, spatial scale, tapirs, tropical vegetation pattern.

vegetation mapping confirmed that vegetation organization occurred at a number of spatial scales.

To understand the processes responsible for generating scale-dependent vegetation pattern we hypothesized that plant–animal interactions played a key role in pattern generation (Fragoso, 1997). We tested the hypothesis by examining relationships between tropical frugivores and granivores, seed dispersal patterns, seed survivorship rates, recruitment to seedling and saplings and the spatial dispersion pattern generated by the different consumers. Seeds dispersed by small-bodied organism remained within 30 m of parent trees, where they experienced high rates of mortality through consumption by Bruchid beetles and to a lesser degree by other granivores (Fragoso, 1997). Tapirs (250 kg) were the only frugivores to swallow large seeds in large numbers and to defecate seeds in clumps at latrines thousands of meters away from parent plants and parent aggregations. Tapir seed dispersal resulted in higher seed survivorship, higher germination rates and higher recruitment into subsequent age classes. Tapirs also generated a unique sapling-dispersion pattern which appears to result in the formation of conspecific plant aggregations. Rodent dispersal alone, at best appeared to be responsible for maintaining and/or increasing the size of parent tree aggregations (Fragoso, 1997). However, rodents also secondarily dispersed seeds first defecated by tapirs at latrines, thus extending the seed shadows generated by tapirs.

We simulated the landscape level seed-dispersal dynamics generated by the lowland tapir (*Tapirus terrestris*) and supported the hypothesis of a qualitative difference between survival dynamics at local and landscape levels. Mature seeds of *Attalea maripa* without bruchid beetle eggs, larvae or adults were collected from whole-ripe fruit. Around 13 adult *A. maripa*, we placed in a paired treatment: (1) seeds in tapir feces and (2) bare seeds on the ground, both located within 5 m of a conspecific fruit-dropping tree located in conspecific aggregations. We repeated the experiment by placing seeds in the two treatments within 5 m of 13 emergent forest trees (simulated tapir latrines) all located from 4 to 7 km from conspecific tree aggregations and reproductive age conspecifics. All treatments were covered with a protective cage and left in the forest for 7 months. Significant differences in survivorship were observed between seeds in feces around forest emergents (92% survirorship) and seeds in feces around conspecific adults (61%); and between bare seeds around emergent trees (75% survivorship) relative to those around conspecific adults (9% survivorship). The differences in survivorship between the site types were statistically significant, as were those between seeds in feces and in bare piles. The magnitude of difference between seeds in feces versus those left on the ground was less for the treatments located farther from conspecific aggregations, but great for the treatments within aggregations (61% versus 9%). Landscape level dispersal moved seeds away from conspecific aggregations and into new within-forest colonization zones dominated by non-conspecific plant communities. Seeds were also more likely to survive within *Attalea* aggregations if they passed through tapir guts and thus were protected by burial in feces.

At our site seed–seedling–animal interactions mitigated by tapirs is a process that leads to higher plant survivorship and the generation of vegetation pattern at large scales (Fragoso, 1997). Rodents and other smaller bodied seed dispersers generated pattern at smaller scales. Although we examined interactions with only the palm *Attalea maripa*, we found that tapirs dispersed large numbers of viable seeds from at least 38 plant species, many of which recruited into seedlings and saplings at tapir latrines (Fragoso and Huffman, 2000). At this Neotropical site plant–animal interactions appear to be responsible for vegetation pattern evident at different spatial scales. Movement patterns of seed consumers, such as peccaries appear to be based to some degree on the pattern generated by interactions occurring at larger scales. The tapir dominated series of interactions appear to create vegetation pattern at a meso-scale. Plant–animal interactions of the type described here may also be responsible for the meso-scale vegetation pattern identified by other researchers at other Neotropical forest sites.

References

Fragoso, J. M. V. Home range and movement patterns of white-lipped peccary (*Tayassu pecari*) herds in the northern Brazilian Amazon. *Biotropica* **30**, 458–469 (1998a).

Fragoso, J. M. V. White-lipped peccaries and palms on the Ilha de Maracá. in *Maracá: The Biodiversity and Environment of an Amazonian Rainforest* (eds Milliken, W. & Ratter, J. A.) (John Wiley & Sons Ltd., England, 1998b).

Fragoso, J. M. V. Scale perception and resource partitioning by peccaries: behavioral causes and ecological implications. *J. Mammal.* **80**, 993–1003 (1999).

Fragoso, J. M. V. & Huffman, J. Seed-dispersal and seedling recruitment patterns by the last Neotropical megafaunal element in Amazonia, the tapir. *J. Trop. Ecol.* **16**, 369–385 (2000).

Fragoso, J. M. V. Tapir-generated seed shadows: scale-dependent patchiness in the Amazon rain forest. *J. Ecol.* **85**, 519–529 (1997).

Silvius, K. M. Interactions among *Attalea* Palms, Bruchid Beetles and Neotropical Terrestrial Fruit Eating Mammals: Implications for the Evolution of Frugivory. Ph D Thesis, University of Florida, Florida, USA (1999).

Tropical Ecosystems: Structure, Diversity and Human Welfare.
Proceedings of the International Conference on Tropical Ecosystems
K. N. Ganeshaiah, R. Uma Shaanker and K. S. Bawa (eds)
Published by Oxford–IBH, New Delhi. 2001. pp. 501–504.

Impact of forest fragmentation on frugivory and primary seed dispersal in French Guiana

Sandra Ratiarison and Pierre-Michel Forget

Muséum National Histoire Naturelle, Laboratoire d'Ecologie Générale,
CNRS-MNHN UMR 8571, 4 Avenue du Petit Château, F-91800 Brunoy, France
e-mail: forget@mnhn.fr

Forest fragmentation, especially in the tropics, severely threatens the long-term survival and regeneration success of trees. Apart from direct effects attributable to the deterioration of fragmented habitats, trees dispersed by animal vectors could be worst affected due to disruption of the disperser activity.

In this study, we are particularly interested in primary seed dispersal by arboreal and flying frugivores in French Guiana. Studying fruit phenology of two *Tetragastris* spp. (Burseraceae) species, the composition of their fruit consumers assemblage, and seed dispersal and predation rates, we compared trees standing at a forest fragmented (islands and control) by the flooding of a dam storage, and at a continuous forest area. Temporal fluctuations were considered to account for food resources availability at a given time. Indeed, the irregular resources availability can generate a competition between trees for disperser attraction, and between animals for food resources partitioning. The purpose of this study was to compare the efficiency of seed dispersal depending on the level of habitat perturbation from nil (protected continuous

Keywords: Primates, birds, frugivores, *Tetragastris*, French Guiana.

forest, ca. > 10,000 ha) to medium (control in a fragmented landscape, ca. > 1000 ha) and high (islands ranging from 0.4 to 22 ha). We observed the fruiting phenology of *Tetragastris panamensis*, the composition of the assemblage of consumers and seed dispersal and predation rates in the Nouragues Reserve in 2000. We then compared these results to those obtained in 1999 at Saint Eugène for a related species *T. altissima* on forested islands isolated by the flooding of a reservoir, on the one hand, and, in a continuous forest having undergone the indirect effects of fragmentation on the other.

The two canopy tree species fruit yearly during the rainy season (March-April); during this period there is a peak in the number of fruiting trees. Fruits are capsules with seeds possessing arils rich in sugars and water. Ripe fruits have reddish-purple outer husks (one husk per seed) that dehisce to expose pure white arillate seeds against a bright red core. Fruits of *Tetragastris* spp. are part of the diet of primates and birds which function as the primary dispersal agents (Howe, 1980; Guillotin *et al.*, 1994; Simmen & Sabatier, 1996). We sampled 9 fruiting *Tetragastris altissima*, 4 located on 4 different islands and 5 on mainland control area at Saint Eugène, and 10 *T. panamensis* at Nouragues. For each tree, fruits and all fruits debris were collected weekly with traps installed after fruit production had started until the end of the study period. Traps contents were categorized into husks and seeds. We distinguished entire and empty husks, husks with seeds and husks pulled to bits by animals (monkeys, parrots). We also distinguished entire seeds, empty seeds (parasited), seeds with husks and seeds destroyed by predators. The presence of pulp was noted but it does not constitute a reliable information in-so-far as the pulp is abundantly eaten by insects in traps and deteriorate rapidly with rain. The number of seeds removed was assessed by computing the difference between the number of husks and the number of seeds collected in traps. The number of fruits/seed fallen under the parent-tree crown was recorded and referred to as 'wasted' seeds. Causes of seeds wasted can be attributed to destruction by predation, fall of entire fruits, parasitism and fall of seeds partly or wholly surrounded by pulp, due to spontaneous falls or to handlings by animals. Estimates of seed production, dispersal and waste were calculated by dividing the sample in question, i.e. respectively the number of husks, seeds removed and seeds collected, by the proportion of the crown sampled.

We also estimated the diurnal visitation patterns of the dispersal agents. Census were made each day, in the morning (7:00–9:00), in midday (11:30–13:30) and in the late afternoon (16:00–18:00). For a tree, census were continuously recorded from the day of trap installation until the end of its fruiting period. A census consisted of observation from the ground of the tree crown with binoculars, from a point at which the entire crown was visible, for 10 (Nouragues) to 15 (Saint Eugène) minutes per tree. To mitigate the differences in census duration, all observations were ascribed to an index of observations per 5 minutes of scanning.

Our results indicate that the relative contribution of birds and primate consumer species to the seed dispersal of two tree species differ between protected and unprotected sites, but not significantly within the fragmented landscape (Table 1). At the unprotected fragmented areas, seed predation rate, and birds and non-disperser animals visits, increased with fruit production at a given period. On the contrary, seed dispersal rate did not vary significantly with fruit quantity. Trends were opposite at the protected forest, with a significant increase in the level of seed dispersal, of visits by seed dispersers and a decrease in rate of seed predation and visits by non-dispersers. Comparative analysis showed that there was no

Table 1. Overall results on seed crop, rates of seed dispersal and predation, and animal sightings during regular tree crown scanning in two *Tetragastris* species at an unprotected fragmented landscape (Saint Eugène) and a protected forest (Nouragues) in 1999 and 2000, respectively.

Study site (N trees)	Unprotected fragmented (T. altissima)		Protected (T. panamensis)
	Islands (4)	Control (5)	Continuous forest (10)
Crop (seeds)	3597 ± 3652	16297 ± 11497	8149 ± 7 997
Fruit peak	End of March–April		end April–early May
Study area (ha)	≈ 30	≈ 45	≈ 50
Dispersal (%)	21.0 (6.6)	5.5 (5.1)	36.9 (14.5)
Predation (%)	25.8 (14.4)	10.2 (7.9)	12.6 (14)
Number of scans (duration)	102 (15′)	175 (15′)	493 (10′)
Number of sightings	64	100	181
Birds (%)	98.4	64	74
Dispersers (%)	26.6	45	43.1
Predators (%)	15.6	20	26.5
Non-dispersers (%)	57.8	35	30.4
Primates	*Alouatta seniculus* (1)	*A. seniculus* (7) *A. paniscus* (1 casual) *Saguinus midas* (11) *Cebus apella* (14)	*A. seniculus* (13) *A. paniscus* (34)
Rodents		*Sciurus aestuans* (3)	*S. aestuans* (1 casual)
Birds	*Cotingidae* (6) *Icteridae* (17) *Psittacidae* (19) *Ramphastidae* (5) *Trogonidae* (5) *Thraupidae* (10)	*Cotingidae* (1) *Psittacidae* (18) *Ramphastidae* (2) *Trogonidae* (1) *Penelope marail* (9) *Columbidae* (10)	*Cotingidae* (6) *Psittacidae* (44) *Trogonidae* (7) *Ramphastidae* (19) *Cracidae* (4) *Columbidae* (2) *Pandionidae* (2)

significant effect of forest fragmentation on seed dispersal and predation rates nor on assemblage composition. Mammals were rarely recorded in continuous as well as in fragments at unprotected Saint-Eugène area in 1999 compared to the protected Nouragues area in 2000. This suggests that the continous area within the fragmented landscape that was used as a control could also have been disturbed, especially by hunting activity in the time between flooding and the time of study. The lack of primates at Saint-Eugène vs Nouragues, especially the spider monkey *Ateles paniscus* (Atelidae) in fruiting trees in particular, may explain the fluctuations of seed dispersal rate in spite of temporal variations in the food resource availability. Besides this temporal heterogeneity, strong spatial variations at tree microhabitat scale could conceal the fragmentation effects. Among primates, variations between trees could be explained by competitive interactions for fruit consumption between small and large monkeys in the protected and unprotected areas. Nevertheless, frequency of visits of consumers predominantly depends on individual tree seed crop. Finally, overall fruit availability at each study site could also influence dietary behaviour of animals, thus their impact on seed fate. This study demonstrates that conservation of plant diversity in a tropical rainforest fragmented landscape requires a strong protection of fauna for sustainable seed dispersal of trees in large remnants (> 1000 ha).

Acknowledgements

This study was supported by a grant from EDF (convention EDF-MNHN 7531), UMR 8571 CNRS-MNHN, le Ministère de l'Enseignement de la recherche et de la Technologie (Programme Pluriformation Guyane) et la Fondation Jean et Marie-Louise Dufrenoy.

References

Guillotin, M., Dubost, G. & Sabatier. D. Food choice and food composition among three major primate species of French Guiana. *J. Zool. London* **233,** 551–579 (1994).
Simmen, B. & Sabatier. D. Diets of some French Guianan primates: food composition and food choice. *Int. J. Prim.* **17,** 661–693 (1996).

Tropical Ecosystems: Structure, Diversity and Human Welfare.
Proceedings of the International Conference on Tropical Ecosystems
K. N. Ganeshaiah, R. Uma Shaanker and K. S. Bawa (eds)
Published by Oxford–IBH, New Delhi. 2001. pp. 505–508.

Flying foxes and rain forest: Maintaining forest regeneration as bat populations diminish

Kim R. McConkey* and Don R. Drake

School of Biological Sciences, Victoria University, PO Box 600, Wellington,
New Zealand
*e-mail: kim.mcconkey@vuw.ac.nz

The islands of the tropical Pacific have a long history of faunal decline and extinction. Since humans first reached the region some 3000 years ago, more than 2000 bird species have gone extinct (Steadman, 1995), and, on many islands, only increasingly fragmented forests remain to support the existing bird and bat species. Situated in Western Polynesia are the islands of Tonga, where we are investigating the impact that these extinctions have had on the process of seed dispersal for animal-dispersed fruit. Tonga has lost at least 75% of its fruit-eating animal species, and two of these extinctions were of massive pigeons, larger than any surviving today (Steadman, 1993). Only one pigeon species (*Ducula pacifica*) and one bat species (*Pteropus tonganus*) now remain that can disperse the seeds from large-fruited species. We have estimated that the pigeon is only capable of swallowing seeds two-fifths the size that were swallowed by the extinct species, meaning at least 13 species (12.5% of overstorey vertebrate-dispersed plants) in the Tongan forests can now only be dispersed by flying foxes.

Flying foxes (Pteropodidae) are considered good seed-dispersers throughout Africa, Asia and the Pacific (e.g. Banack, 1998). Following the loss of the large

Keywords: *Pteropus*, Tonga, seed dispersal, bat density.

pigeons, they have been termed 'keystone' seed dispersers in the Pacific region (Rainey *et al.*, 1995). Although flying foxes feed mainly on fruit juices, spitting out all but the smallest seeds, they often carry large-seeded fruits considerable distances from the feeding tree. Flying fox populations throughout the Pacific have been declining over the last few decades. It has been predicted that these reductions will greatly reduce the number of seeds being dispersed away from the parent tree, and dispersal may not occur *at all*, at low bat densities. This is because of the foraging behaviour of the bats. Flying foxes are aggressive and defend territories in the feeding tree (Richards, 1990). When there are few bats, interactions are rare, and the bats feed within the tree and drop the fruit below the tree crown. As bat numbers increase they fight, forcing out newcomers, who may snatch a fruit to eat elsewhere, and so disperse the seeds.

We made use of the natural geographic variation in flying fox densities in Tonga to investigate the influence of this variation on the frequency of seeds being dispersed. We have collected data on primary seed dispersal by bats for 71 trees and lianas, from 13 species, in 12 sites. We predicted that we would see no dispersal beyond the crown at low bat density and there would be a threshold density at which seeds begin to be dispersed away from the crown (Figure 1).

As predicted, our data show a nonlinear relationship between the index of bat abundance and proportion of seeds being dispersed more than 5 m from the parent crown ('far', Spearman rank, $r = 0.95$, $n = 12$, $p < 0.0001$) (Figure 2). Although minimal seed dispersal did occur at low bat abundance, a threshold was apparent. Beyond this threshold the proportion of seeds being dispersed steadily increased. No relationship with bat abundance was found for the proportion of seeds dispersed by bats under the crown ('under', $r = -0.23$, $p = 0.47$), or within 5 m of the crown ('edge', $r = 0.12$, $p = 0.71$).

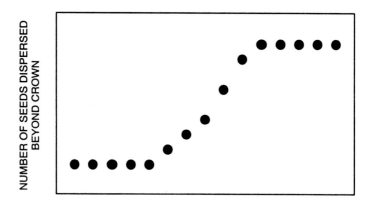

BAT INDEX

Figure 1. Hypothesised effect of bat density on number of seeds dispersed beyond the crown.

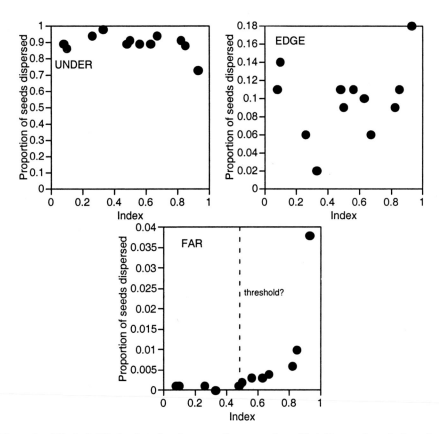

Figure 2. Effect of differing bat abundance on the proportion of bat-dispersed seeds deposited in three zones: (under crown, edge: < 5 m from crown edge, distant: > 5 m from crown edge). Each point on the graph represents the mean number of seeds dispersed for each plant at the separate sites.

The probability of seed dispersal beyond the parent tree is positively correlated to bat abundance. Although our results reveal no absolute threshold for seeds to be dispersed more than 5 m from the parent tree, such dispersal is sporadic and infrequent at low bat densities. It may occur for particularly favoured trees in which bats would be expected to congregate, creating an artificially high density at a very local scale.

Maintaining bat populations at densities sufficient for frequent seed dispersal may be crucial to maintain patterns of forest dynamics and seed-mediated gene flow of large-seeded species in the tropical Pacific. Thus, while it is unlikely that tree species would become extinct without dense bat populations, populations of large-seeded species may become very localized. There could be a shift in forest composition to smaller-seeded plants, as is found in areas of Uganda where primate populations have been reduced (Chapman and Onderdonk, 1998).

Acknowledgements

We thank Nola Parsons, Nic Gorman, Defini Tau'laupe, Leigh Bull and Hayley Meehan for their help in collecting the data. Logistical support was provided by Pat and Keith McKee, Aleiteisi Tangi, Tavake Tonga and Marlene Moa. We would like to thank the Tongan Government for permission to carry out the research. Financial support was provided by Wildlife Conservation Society, Percy Sladen Memorial Trust, Victoria University of Wellington and Polynesian Airlines.

References

Banack, S. Diet selection and resource use by flying foxes (Genus *Pteropus*). *Ecology* **79**, 1949–1967 (1998).

Chapman, C. A. & Onderdonk, D. A. Forests without primates: primate/plant codependency. *Am. J. Primatol.* **45**, 127–141 (1998).

Rainey, W. E., Pierson, E. D., Elmqvist, T. & Cox, P. A. The role of flying foxes (Pteropodidae) in oceanic island ecosystems of the Pacific. *Symp. Zoo. Soc. London* **67**, 79–96 (1995).

Richards, G. C. The spectacled flying fox, *Pteropus conspicillatus* (Chiroptera: Pteropodidae), in north Queensland. 2. Diet, seed dispersal and feeding ecology. *Austr. Mammal.* **13**, 25–31 (1990).

Steadman, D. W. Biogeography of Tongan birds before and after human impact. *Proc. Natl. Acad. Sci.* **90**, 818–822 (1993).

Steadman, D. W. Prehistoric extinctions of Pacific Isand birds: Biodiversity meets zooarchaeology. *Science* **267**, 1123–1131 (1995).

Tropical Ecosystems: Structure, Diversity and Human Welfare.
Proceedings of the International Conference on Tropical Ecosystems
K. N. Ganeshaiah, R. Uma Shaanker and K. S. Bawa (eds)
Published by Oxford–IBH, New Delhi. 2001. pp. 509–512.

Changes in the seed shadow generated by gibbons in the rain forests of central Borneo

Kim R. McConkey

Department of Anatomy, University of Cambridge, Cambridge CB2 3DY, UK
Present address: School of Biological Sciences, Victoria University, PO Box 600,
Wellington, New Zealand
e-mail: kim.mcconkey@vuw.ac.nz

Frugivorous animals that disperse the seeds of forest plants are crucial to the functioning of tropical rain forests. Seed dispersal involves a series of sequential stages that must be understood to fully evaluate the effectiveness of a particular dispersal agent (Jordano and Herrera, 1995). After the frugivore has removed the fruit from the plant and dropped or defaecated the seed, the seed must then escape the attention of predators, possibly be moved by secondary dispersers, germinate and establish, before it can have a chance of surviving to adulthood. It is the primary seed shadow (e.g. spatial deposition of seeds and characteristics of the scats) created by frugivores that will determine the fate of the dispersed seeds.

Borneo's interior forests support fewer frugivores than richer coastal areas. Large-bodied animals, such as orangutans and pig-tailed macaques, and the hornbills that congregate in large flocks are particularly scarce in the interior, because of lower fruit availability. Sources of fruit in these forests tend to be small and scattered, which is ideal for the gibbon groups that are found in

Keywords: Seed dispersal, *Hylobates,* lowland dipterocarp forest, Borneo.

substantial numbers here. Consequently gibbons were thought to be the most important disperser of fleshy fruits in the interior of Borneo (McConkey, 2000).

I followed two gibbon *(Hylobates muelleri × agilis)* groups over 12 months in the lowland dipterocarp forest of Barito Ulu, Central Kalimantan, Indonesia. The aim of the research was to describe the primary seed shadow generated by their ranging and feeding behaviour. I also monitored 403 gibbon scats, that were collected during the gibbon follows, for up to one year. The effect of four scat characteristics and two environmental variables on seed predation rates were examined using logistic regression; (1) seed density, (2) number of seed species, (3) quantity of non-seed faeces present in scats (i.e. the olfactory 'attractant'), (4) distance to nearest fruiting tree, (5) fruit availability in the area, and (6) rainfall.

The gibbons dispersed up to 81% of the species they consumed and destroyed the seeds of only 12%. Size of dispersed seeds ranged from tiny *Ficus* seeds, to seeds up to 20 mm wide. Each scat had a median of four seeds (larger than 3 mm wide) (range: 1–51 seeds) and a median of three species (range: 1–6 species).

The gibbons' main benefit to the plant was removal from frequently high mortality directly under the parent crown; this is known as the escape hypothesis (Janzen, 1970; Connell, 1971). The gibbons' ranging and behavioural patterns meant only 4% of defaecations were deposited under the parent trees after a passage rate of nearly 30 hours, while 28% were defaecated under the crown of fruiting trees of other species. Most primates drop some seeds when feeding (e.g. Estrada and Coates-Estrada, 1984) and the gibbons were no exception. For most species this was confined to only a few seeds whereas large numbers were swallowed, but they consistently dropped the seeds of nearly 3% of feeding species.

Post-dispersal processes significantly altered the primary seed shadow created by gibbons at Barito Ulu. Seed fall followed a clumped distribution, while the pattern of surviving seeds was spatially random. Only 8% of seeds in gibbon scats escaped predation from seed predators. Overall, the number of seeds successfully dispersed to the seedling stage was 13 seedlings ha^{-1} year^{-1}. Mammals killed or removed 86% of seeds and insects 2.2%, while 1.8% of germinated seeds died at the seedling stage. Rodents were the main seed predators, visiting 64% of scats. Civets, pigs, mousedeer and muntjac also frequently visited the gibbon scats and more rarely terrestrial birds and cats; combined they visited 98% of scats within one week of deposition.

Seed predation by mammals was unresponsive to seed number within scats, but increased with increasing scat density (scats ha^{-1}). Thus, the tendency for gibbons to use particular ranging routes and defaecate in the same area increased seed mortality. Insect seed predators responded positively to seed density within scats rather than scat density. Mammals ate fewer seeds from scats with a single species, and seedling survival was also higher.

Mammals were able to locate seed piles that were almost devoid of non-seed matter and they usually avoided scats with lots of non-seed matter except when food was scarce. Subsequent experiments indicated that although faeces attracted mammals to the possibility of a seed reward, where an adequate, faeces-free, food source existed the mammals preferentially consumed that. Insect predation showed the opposite response, with higher levels found in scats with more non-seed faecal matter.

The distance the seeds were deposited to a fruiting tree had a consistent effect on the chances of the seeds being located and killed by insect and mammalian predators. Insect infestation was consistently higher under fruiting trees, but this was only for certain plant species and the rates of infestation were low. In contrast, mammals showed no consistent effect with predation in this region, although rates of mammal predation were much higher than for insects. Overall, there seemed to be a benefit in actually being deposited under fruiting trees, again because mammals preferentially consumed seeds from fresh fruit.

Fruit availability was the only environmental variable that had an effect on rates of seed predation. This effect was significantly stronger than any scat characteristic examined. Highest rates of mammalian seed predation were observed when food availability was low (when animals had fewer food options), and when food availability peaked (when rodent populations probably also peaked). Insect predation was highest when availability of figs were highest, probably due to a build-up of insect populations.

Acknowledgements

I would like to thank the Ministry of Forestry and LIPI in Indonesia for their support and permission to conduct the research at the Barito Ulu site. The work would not have been possible without the logistical help of Rupert Ridgeway and D. J. Chivers and the unlimited field assistance of Bapaks Suriantata, Nurdin, Mulyadi, Kursani and Arbadi. Financial assistance was provided by the New Zealand Embassy in Jakarta, Cambridge Commonwealth Trust, Primate Conservation Inc., New Zealand Federation of University Women, Leakey Foundation, Cambridge University Board of Graduate Studies, Sophie Danforth Conservation Biology Fund of the Roger Williams Park Zoo and Rhode Island Zoological Society and Selwyn College.

References

Connell, J. H. On the role of natural enemies in preventing competitive exclusion in some marine animals and in rain forest trees. in *Dynamics of Populations* (eds Den Boer, P. J. & Gradwell, G.) 298–312 (Wageningen: PUDOC, 1971).
Estrada, A. & Coates-Estrada, R. Fruit eating and seed dispersal by howling monkeys (*Alouatta palliata*) in the tropical rain forest of Los Tuxtlas, Veracruz, Mexico. *Am. J. Primatol.* **6**, 77–91 (1984).

Janzen, D. H. Herbivores and the number of tree species in tropical forests. *Am. Nat.* **104,** 501–527 (1970).

Jordano, P. & Herrera, C. M. Shuffling the offspring: uncoupling and spatial discordance of multiple stages in vertebrate seed dispersal. *Ecoscience* **2,** 230–237 (1995).

McConkey, K. R. Primary seed shadow generated by gibbons in the rain forests of central Borneo. *Am. J. Primatol.* **52,** 13–29 (2000).

Tropical Ecosystems: Structure, Diversity and Human Welfare.
Proceedings of the International Conference on Tropical Ecosystems
K. N. Ganeshaiah, R. Uma Shaanker and K. S. Bawa (eds)
Published by Oxford–IBH, New Delhi. 2001. pp. 513–516.

Fruit removal patterns and dispersal of *Emblica officinalis* (Euphorbiaceae) at Rajaji National Park, India

Soumya Prasad, Ravi Chellam[†] and
Jagdish Krishnaswamy[‡]*

[†]Department of Endangered Species Management, [‡]Department of Habitat Ecology,
Wildlife Institute of India, Post Box # 18, Chandrabani, Dehradun 248 001, India
*e-mail: soumyap@usa.net

Fruit removal refers to the quantity of fruit taken by frugivores. Patterns of fruit removal have both ecological and evolutionary implications. The ecological effects of fruit removal can be defined in terms of the quantity of fruits or seeds removed from a parent tree, the phenological stage at which these food items are selected, methods of seed handling, and the manner in which a given species deposits seeds within its geographic range (Schupp, 1993). The evolutionary effects are seen in the way these interactions shape the fruiting season, fruit crop size, fruit type and size, nutritive value of fruit or seed and other strategies adopted by the plant to ensure effective dispersal and the possible coevolutionary interactions between fruiting trees and their animal dispersers. A fraction of the total fruit removed goes into seed dispersal. Disperser effectiveness is defined as the contribution a disperser makes to the future reproduction of a plant and disperser effectiveness has both quantitative (proportion of fruit crop dispersed by a disperser) and qualitative (quality of treatment given in the mouth and in the gut, quality of

Keywords: Fruit removal, frugivory, seed dispersal, *Emblica officinalis*, mammals, Rajaji National Park.

seed deposition determined by the probability that a deposited seed will survive and become an adult) components (Schupp, 1993). Such plant–frugivore interactions determine the abundance of tree and frugivore species and consequently define the management protocol for conservation of these systems (Howe, 1993).

Emblica officinalis has a long history of use by humans. The fruits are edible and used in pickles, hair oil and ayurvedic medicines (Wood, 1992). *E. officinalis* is an important non-timber forest produce of India's tropical deciduous forests. Previous studies on *E. officinalis* have focused on socio-economic aspects of extraction of its fruit by humans and on aspects of demography of *E. officinalis* populations (Uma Shankar *et al.*, 1996; Koliyal, 1997; Musavi, 1999). There are no studies dealing with its ecological aspects like frugivory, seed dispersal, germination or seedling establishment. To understand the impact of fruit extraction by humans on *E. officinalis* populations, we need to look at natural patterns of fruit removal, follow fate of fruit/seed dispersed and seedling/saplings and then consider this in the context of use by humans.

The main questions explored in our study are:

- What are the patterns (diurnal, nocturnal and across the fruiting seasons) of fruit removal of *Emblica officinalis*?
- Who are the vertebrate removers of *E. officinalis*? Are the vertebrate removers of *E. officinalis* fruit, its effective dispersers?

The study was conducted at Dhaulkand (78°E and 30°N), Rajaji National Park, Uttaranchal, India from November 1999 to April 2000. To answer the first question we followed fruit removal patterns (both on the tree and on the ground) for nineteen *E. officinalis* trees in the study area. To address the second question we determined the identity of the fruit removers and made observations on fruit handling using methods such as tree watches, camera trapping and track plots. Further, we monitored ungulate boluses with *E. officinalis* endocarps and followed the fate of the endocarps and seeds in them. Germination experiments were also carried out to investigate the impact of passage through ungulate rumens on germination success.

Nineteen trees with a range of fruit crops (from 17×10^3 fruit to less than 100 fruit) were chosen for the study. The total fruit crop size was estimated for each of the study trees at the start of the study. Eleven of these trees (3 with fruit crop size between 50 and 1000 fruit, 3 with fruit crop size between 1000 and 5000 fruit, 5 with fruit crop size above 5000 trees) were monitored to ensure that there was no human removal of the fruit. Eight study trees (3 with fruit crop size between 50 and 1000 fruit, 2 with fruit crop size between 1000 and 5000 fruit, 3 with fruit crop size above 5000 trees) were left unmonitored, and fruits were extracted from three of these trees (each had fruit crop size above 5000 fruit) by humans.

For each tree, all fruit on the ground in the fruit fall area and all fruit on 5–7 marked branches were counted thrice every ten days in 12 h intervals to arrive at nocturnal and diurnal removal patterns for fruit. The fruit were counted on the tree and on the ground till less than 5% of the initial number of fruit remained on the marked branches. Patterns of fruit persistence (number of days fruit persisted on the marked branches) on the trees differ considerably in the high fruit crop bearing trees between the monitored and unmonitored trees. On the ground, fruit removal was largely during the night.

A range of direct and indirect methods – tree watches, track plots and camera trapping were used to determine the fruit removers and to observe the mode of fruit handling by these removers. Langur (*Semnopithecus entellus*), ungulates such as chital (*Axis axis*) and barking deer (*Muntiacus muntjac*) were observed feeding on *E. officinalis* fruit. We also have evidence that the gerbil, *Tatera indica*, scatter hoards the fruit. Ungulates regurgitate the endocarp with the seeds inside intact. Langurs feed mostly on the pulp and sometimes damage seeds. *Tatera indica* was found to be feeding only on the pulp and so far there is no evidence for seed predation by the same.

Ungulate boluses with *E. officinalis* endocarp with seeds were monitored on two 100 m transects each with ten 1 m × 1 m plots set at 10 m intervals. The endocarps dehisced within 2–3 days and released 5 to 6 seeds each. Post-dispersal, seeds were subjected to further movement due to trampling by animals and wind. These seeds were also predated by insects and rodents.

Germination trials were conducted with 95 seeds collected from endocarps found in ungulate boluses and 95 control seeds from fruit whose pulp was removed manually. Each seed was buried in sand within a socket lined by tissue paper in a germination tray. All sockets contained the same volume of sand and all seeds were placed in the same position and buried under the same depth of sand. The trays were watered once in one or two days and each seed received the same volume of water. 54% of the control and 22% of the ungulate regurgitated seeds germinated by the end of April 2001. Seeds set up in early December 2000 started germinating towards the end of February, while seeds set up in February started germinating in late March. The time taken to germinate did not differ between the control and ungulate regurgitated seeds. For comparison, germination trials were also done in a similar fashion with *Zizyphus mauritiana* (Rhamnaceae) another fruit whose endocarp was frequently encountered in ungulate boluses along with *E. officinalis*. In the case of *Z. mauritiana*, 2% of the control and 51% of the ungulate regurgitated seeds germinated by the end of April 2001.

Though *E. officinalis* or its frugivore community might be far from extinction, plant–animal interactions like these are threatened and can never be recreated once lost.

Acknowledgements

We wish to thank the Uttaranchal Forest Department for providing permits and logistic support at Dhaulkand; Ministry of Environments and Forests through the Wildlife Institute of India and the Wildlife Conservation Society-India Programme for funds; faculty, researchers, students and staff at the Wildlife Institute of India for all the support and encouragement and Divya Mudappa and Aparajita Dutta for comments on this abstract.

References

Howe, H. F. Specialized and generalized dispersal systems: where does the 'paradigm' stand? in *Frugivory and Seed Dispersal: Ecological and Evolutionary Aspects* (eds Fleming, T. H. & Estrada, A.) 3–13 (Kluwer Academic Publishers, Boston, 1993).

Koliyal, A. Extraction of Non Timber Forest Produce from selected tree species in Betul Forest Division and its impact on the population structure of these species. M.Sc Dissertation, Saurastra University, Rajkot (1997).

Musavi, A. A Socio-economic study of Tribes and Non-Tribes in Melghat Tiger Reserve and adjoining areas. Ph.D. thesis. Department of Sociology and Social Work, Aligarh Muslim University, Aligarh (1999).

Schupp, E. W. Quantity, quality, and the effectiveness of seed dispersal by animals. in *Frugivory and Seed Dispersal: Ecological and Evolutionary Aspects* (eds Fleming, T. H. & Estrada, A.) 15–30 (Kluwer Academic Publishers, Boston, 1993).

Uma Shankar, Murali, K. S., Uma Shaanker, R., Ganeshaiah, K. N. & Bawa, K. S. Extraction of NTFP in the forests of Biligiri Rangan Hills, India. 3. Productivity, Extraction and Prospects of Sustainable Harvest of Amla *Phyllanthus emblica* (Euphorbiaceae). *Econ. Bot.* **50**, 270–279 (1996).

Wood, C. V. B. *Trees in Society in Rural Karnataka, India.* (Natural Resource Institute, Kent, UK, 1992).

Measuring and Mapping Biodiversity

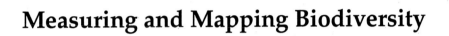

Tropical Ecosystems: Structure, Diversity and Human Welfare.
Proceedings of the International Conference on Tropical Ecosystems
K. N. Ganeshaiah, R. Uma Shaanker and K. S. Bawa (eds)
Published by Oxford–IBH, New Delhi. 2001. p. 519.

Spatial models to estimate patterns of diversity from incomplete data sets: Tiger beetle species richness in Northwestern South America (Coleoptera: Cicindelidae)

David L. Pearson* and Steven S. Carroll

Department of Biology, Arizona State University, Tempe, AZ 85287-1501, USA
*e-mail: dpearson@asu.edu

Species richness patterns of tiger beetles (Coleoptera: Cicindelidae) were analyzed using a grid of 407 squares (137.5 km or 1.2 degrees per side) across northwestern South America (Guyana, Venezuela, Colombia, Ecuador, Perú, Bolivia and western Brazil). Reliable data on species numbers were available for only 149 of the squares. Using a trend surface model (a model used to represent the mean of a spatial process by a polynomial function of spatial coordinates) as well as altitudinal range, precipitation and biogeographical influence for each square, we were able to predict the number of tiger beetle species likely to occur in intermediate squares for which no or unreliable data were available. The resultant spatial patterns of species richness were compared with similar analyses for temperate areas of North America; the differences should be useful in understanding general diversity patterns and in the environmental management of species richness as well as eventually applying other pertinent parameters of diversity such as species turnover.

Keywords: Spatial models, diversity patterns, tiger beetle, South America.

Tropical Ecosystems: Structure, Diversity and Human Welfare.
Proceedings of the International Conference on Tropical Ecosystems
K. N. Ganeshaiah, R. Uma Shaanker and K. S. Bawa (eds)
Published by Oxford–IBH, New Delhi. 2001. p. 520.

Beta diversity: The forgotten component?

Kevin J. Gaston* and Patricia Koleff

Biodiversity & Macroecology Group, Department of Animal & Plant Sciences,
University of Sheffield, Sheffield S10 2TN, UK
*e-mail: k.j.gaston@sheffield.ac.uk

Three components to species diversity are regularly distinguished, alpha (or local) diversity, gamma (or regional) diversity, and beta diversity (or spatial turnover). Of these, alpha diversity and gamma diversity have received by far the most attention. However, it is clear that understanding of the patterns in beta diversity, and of the processes that give rise to those patterns, is central to developing a fuller picture of how biodiversity changes from place to place. Resolution of major methodological issues, and consideration of the role of scale, reveals significant patterns in beta diversity, and highlights its interaction with alpha and gamma diversities. It also reveals that some perceived wisdom about beta diversity is simply wrong.

Keywords: Species diversity, beta diversity, alpha diversity, gamma diversity.

Tropical Ecosystems: Structure, Diversity and Human Welfare.
Proceedings of the International Conference on Tropical Ecosystems
K. N. Ganeshaiah, R. Uma Shaanker and K. S. Bawa (eds)
Published by Oxford–IBH, New Delhi. 2001. p. 521.

Time series data, extinction risk modeling, and the design of conservation areas

William Fagan

Department of Biology, Arizona State University, Tempe, AZ 85287-1501, USA
e-mail: bfagan@asu.edu

The study of biodiversity is rife with practical problems requiring theoretical and quantitative approaches. One of these involves the estimation of extinction risk from time series data. I present an overview of the strengths and weaknesses of extinction risk analyses, emphasizing the difficulties involved in parameter estimation for population models, and outlining some potential solutions. Two possibilities include (1) recasting the problem of extinction probabilities into categories of risk, and (2) new parameter estimation techniques that can reduce the influences of observation error. I summarize by clarifying the kinds of insights risk modeling can and cannot provide in the context of conservation planning, and outline ways in which temporal and spatial data can be combined to design nature reserve systems.

Keywords: Time series data, risk modeling, conservation planning.

Tropical Ecosystems: Structure, Diversity and Human Welfare.
Proceedings of the International Conference on Tropical Ecosystems
K. N. Ganeshaiah, R. Uma Shaanker and K. S. Bawa (eds)
Published by Oxford–IBH, New Delhi. 2001. p. 522.

Convincing decision makers: Organisation and presentation of complex biodiversity data sets

John Pilgrim

Center for Applied Biodiversity Science, Conservation International, 1919 M Street,
NW, Suite 600, Washington, DC 20036, USA
e-mail: j.pilgrim@conservation.org

When collecting biodiversity data that could be valuable for conservation, biologists often fail to consider the audiences for this information. Presentation of complex, lengthy, biogeographically-defined data sets in peer-reviewed journals can be the best medium for communicating results to other scientists, but may be unintelligible to the general public, confusing to policy-makers or boring to funders. There is a need to organise, target and present accurate, scientifically justified data in a simple, concise, interesting, and attractive fashion.

At the early stages of data collection or collation, it is important to consider the questions that will be asked by wider audiences. Data will be most useful if they are standardised and comparable across any scale, whether contrasting sites within a country or regions of the world. Data storage should be flexible enough to cope with changing needs of the audience and changes in the data accumulated over time. For presentation to non-scientists, complex information may need to be translated into popular language. In particular, maps are increasingly useful to conservationists as powerful and effective communication tools that transgress language barriers.

Keywords: Biodiversity datasets, biodiversity maps, data storage.

Tropical Ecosystems: Structure, Diversity and Human Welfare.
Proceedings of the International Conference on Tropical Ecosystems
K. N. Ganeshaiah, R. Uma Shaanker and K. S. Bawa (eds)
Published by Oxford–IBH, New Delhi. 2001. pp. 523–525.

Mapping biodiversity from the sky

Kamaljit S. Bawa*[†], *J. Rose*, K. N. Ganeshaiah*[†,‡,§],
M. C. Kiran[†], *Narayani Barve*[†] *and R. Uma Shaanker*[†,‡]

[*]Biology Department, University of Massachusetts-Boston, 100 Morrissey
Boulevard, Boston, MA 02125, USA
[†]Ashoka Trust for Research in Ecology and the Environment, No. 659, 5th A Main,
Hebbal, Bangalore 560 024, India
[‡]University of Agricultural Sciences, GKVK Campus, Bangalore 560 065, India
[§]e-mail: kng@vsnl.com

An efficient conservation planning is contingent upon assessment and mapping of biological diversity to identify the local, regional and global hotspots (Wilson, 2000). The existing methods to map biodiversity on a large spatial scale are time consuming as they require intensive ground surveys. In this paper we demonstrate the use of remote sensing imagery to assess and map the biodiversity in a tropical ecosystem and discuss the use of this technique in monitoring as well as conservation planning.

Rationale for our approach stems from two established relationships. While there exists a strong relationship between species richness and productivity (Tucker and Sellers, 1986; Adams and Woodward, 1989; Curie, 1991), it has been demonstrated that the latter can be measured through annual sum of Normalized Difference Vegetation Index (NDVI) derived from satellite imagery at global and regional scales (Holben, 1986; Eidenshink and Faundeen, 1994). Thus we expected a positive relationship between plant species richness and NDVI values.

Keywords: Mapping, biodiversity, NDVI, conservation, monitoring, species richness, species diversity.

We tested for such relation in Biligiri Rangaswamy Temple (BRT) Wildlife Sanctuary, in Karnataka, India (11°43′ to 12°08′N latitude and 77° to 77°16′E longitude). The sanctuary was divided into grids of 2×2 km and their tree species richness was enumerated from plots of 80×5 m laid in the center of each grid ($n = 134$). From the Indian Remote Sensing Satellite (IRS) 1C LISS III imagery we calculated NDVI for each grid. The association between NDVI and tree species richness (and between NDVI and Shannon diversity for the tree species) was calculated for the grids of the entire sanctuary and of the three major vegetation types.

There was a positive correlation between mean NDVI and tree species richness for the entire sanctuary and for dry deciduous forest types (Figure 1; Table 1). Similarly the Shannon diversity index was also correlated with the NDVI for the entire sanctuary and the dry deciduous forest type (Table 1). For scrub forest, the relation was not significant perhaps due to the low

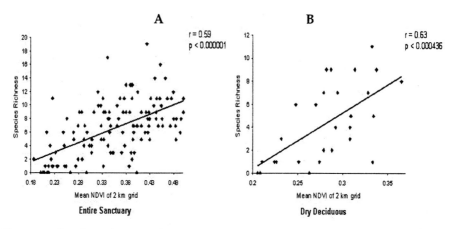

Figure 1. Correlation between the NDVI and tree species richness of the grids of the entire BRT Sanctuary (A) and of the dry deciduous forest vegetation (B).

Table 1. Association of NDVI with Shannon diversity index and species richness of trees in BRT sanctuary and its vegetation types. Evergreen vegetation was not included as the number of samples grids was very less.

	Sample size	Shannon index	Species richness
Entire sanctuary	134	0.66**	0.59**
Dry deciduous	27	0.63**	0.63**
Moist deciduous	39	0.34*	−0.11 NS
Scrub	13	0.42 NS	0.43 NS

**Significant at 1%.
*Significant at 5%.

sample size; for the moist deciduous forests the Shannon diversity index was correlated with the NDVI while species richness was not (Table 1).

Despite the well-established relationship between the productivity and species richness, remotely sensed imagery that can detect differences in measures of productivity has not been used for assessment of biodiversity. Our results demonstrate that NDVI from remotely sensed images can identify areas of high species richness and diversity. This technique if standardized for specific ecosystems can also be used for regular monitoring of the biodiversity. Thus the technique, besides contributing to the planning of conservation strategies would also help in a more frequent monitoring of the ecosystems.

References

Adams, J. M. & Woodward, F. I. Patterns in tree species richness as a test of the glacial extinction hypothesis. *Nature* **339**, 699–701 (1989).

Curie, D. J. Energy and large-scale patterns of animal and plant species richness. *Am. Nat.* **137**, 27–49 (1991).

Eidenshink, J. C. & Faundeen, J. L. The 1 km AVHRR global land data set: An up data, *IGBP Global Change Newslet.* **27**, 5–6 (1994).

Holben, B. N. Characteristics of maximum-value composite images from temporal AVHRR data. *Intl. J. Rmt Sens.* **7**, 1417–1434 (1986).

Tucker, C. J. & Sellers, P. J. Satellite remote sensing of primary production. *Intl. J. Rmt Sens.*, **7**, 1395–1416 (1986).

Wilson, E. O. A global biodiversity map. *Science* **289**, 2279 (2000).

Tropical Ecosystems: Structure, Diversity and Human Welfare.
Proceedings of the International Conference on Tropical Ecosystems
K. N. Ganeshaiah, R. Uma Shaanker and K. S. Bawa (eds)
Published by Oxford–IBH, New Delhi. 2001. pp. 526–528.

The ant fauna of a tropical rainforest: Estimating species richness three different ways

John T. Longino*[†], Jonathan Coddington[‡]
and Robert K. Colwell[§]

*The Evergreen State College, Olympia WA 98505, USA
[‡]Department of Entomology, National Museum of Natural History, Smithsonian
Institution, Washington DC 20560, USA
[§]Department of Ecology and Evolutionary Biology, University of Connecticut, U-43,
Storrs, CT 06269-3043, USA
[†]e-mail: longinoj@evergreen.edu

Species richness is an important characteristic of ecological communities, but it is difficult to quantify (Soberón and Llorente, 1993; Colwell and Coddington, 1994; Palmer, 1995). We report here a thorough inventory of a tropical rainforest ant fauna, and use it to evaluate species richness estimators. The study was carried out in approximately 1500 ha of lowland rainforest at La Selva Biological Station, Costa Rica (McDade *et al.*, 1993). Diverse methods were used, including canopy fogging, Malaise traps, Berlese samples, Winkler samples, baiting, and manual search (Longino and Colwell, 1997). Workers of 437 ant species were encountered. The abundance distribution was clearly lognormal, and the distribution emerged from a veil line with each doubling of sampling effort (Figure 1). Three richness estimates were calculated: the area under the fitted lognormal distribution,

Keywords: Ants, Costa Rica, formicidae, richness estimation, tropical rainforest.

the asymptote of the Michaelis–Menten equation fit to the species accumulation curve, and the Incidence-based Coverage Estimator (ICE). The performance of the estimators was evaluated with sample-based rarefaction

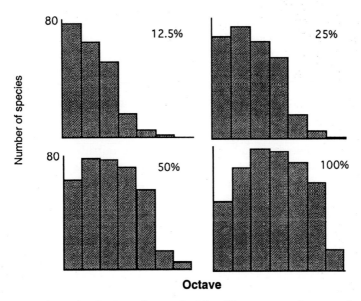

Figure 1. Abundance distributions of ants at La Selva. Histograms are for successive doublings of sampling effort, beginning with 12.5% of samples randomly drawn from the full dataset. Abundance is measured as number of species occurrences in samples. Assignment to octaves follows the method of Preston (1948).

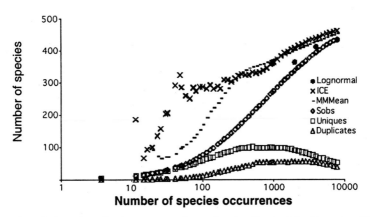

Figure 2. Sample-based rarefaction curve and associated estimators for the ants of La Selva. Lognormal: richness estimate based on the area of an estimated lognormal distribution. ICE: incidence-based coverage estimator. MMMean: asymptote of Michaelis–Menten curve, estimated from the smoothed, sample-based rarefaction curve. Sobs: observed species richness. Uniques: number of species each known from only one sample. Duplicates: number of species each known from exactly two samples. The abscissa is scaled logarithmically.

plots (Figure 2). The inventory was nearly complete because the species accumulation curve approached an asymptote, the richness estimates were very close to the observed species richness, and the uniques and duplicates curves were declining. None of the richness estimators were stable in sample-based rarefaction plots, but regions of stability of estimators occurred, and may represent subcommunities that are efficiently sampled by the combination of methods used. Fifty-one species (12% of the total) were still uniques (known from one sample) at the end of the inventory. The rarity of six of these was explained by 'methodological edge effects. They were possibly abundant at the site but difficult to sample because of their microhabitat (the subterranean fauna). The rarity of fourteen species was possibly due to geographic edge effects, because the species were known to be common in habitats or regions outside of La Selva. Most of the 51 rare species were known from additional collections outside of La Selva, both other parts of Costa Rica and in other countries. Only 6 species were global uniques, known from only one sample on Earth.

Acknowledgements

First and foremost we thank the ALAS staff who made the project possible: Danilo Brenes, Nelci Oconitrillo, Maylin Paniagua, and Ronald Vargas. Nichole Barger and Chris Thompson provided two of the methodological datasets while carrying out undergraduate independent research projects. The Directors and staff of the La Selva Biological Station have been extremely helpful. This work has been supported by National Science Foundation grants BSR-9025024, DEB-9401069, and DEB-9706976.

References

Colwell, R. K. & Coddington, J. A. Estimating terrestrial biodiversity through extrapolation. *Phil. Trans. R. Soc. London* **B345,** 101–118 (1994).

Longino, J. T. & Colwell, R. K. Biodiversity assessment using structured inventory: capturing the ant fauna of a lowland tropical rainforest. *Ecol. Appl.* **7,** 1263–1277 (1997).

McDade, L. A., Bawa, K. S., Hespenheide, H. A. & Hartshorn, G. S. (eds) *La Selva, Ecology and Natural History of a Neotropical Rainforest* (University of Chicago Press, Chicago, IL, USA, 1993).

Palmer, M. W. How should one count species? *Nat. Areas J.* **15,** 124–135 (1995).

Preston, F. W. The commonness, and rarity, of species. *Ecology* **29,** 254–283 (1948).

Soberón M. J. & Llorente, J. B. The use of species accumulation functions for the prediction of species richness. *Conserv. Biol.* **7,** 480–488 (1993).

Tropical Ecosystems: Structure, Diversity and Human Welfare.
Proceedings of the International Conference on Tropical Ecosystems
K. N. Ganeshaiah, R. Uma Shaanker and K. S. Bawa (eds)
Published by Oxford–IBH, New Delhi. 2001. pp. 529–531.

Distribution of medicinal trees in Maharashtra

Raghunandan Velankar*, Suresh Jagtap, Anagha Ranade and Supriya Dandekar

Medicinal Plants Conservation Center, 425/84, T.M.V Colony, Mukundnagar,
Pune 411 037, India
*e-mail: rcmpcc@vsnl.com

We analyse here geographic, habitat and microhabitat distribution of 200 medicinal tree species found in Medicinal Plants Conservation Areas (MPCAs) established by the Forest Department of Maharashtra State. These MPCAs have been established under the UNDP-Government of India Project on *'In-situ* Conservation of Medicinal Plants' being implemented by Foundation for Revitalization of Local Health Traditions (FRLHT), Bangalore and Rural Communes. There are 13 MPCAs, each of about 200 to 300 hectares in less disturbed forests representing different bio-climatic zones of the state (Table 1). Seasonal botanical surveys were carried out during the year 2000 to document the plant diversity in these areas. Botmast – a database on medicinal plants of India at FRLHT was referred to sort out medicinal species from the checklists of the MPCAs. We consider about 200 tree species belonging to 128 genera and 46 families found in these 13 MPCAs. These include a few sparsely distributed species such as *Canarium strictum, Knema attenuata,* and *Radermachera xylocarpa,* as well as the widely distributed species like *Terminalia crenulata, Lannea coramandelica* and *Garuga pinnata.*

Keywords: Medicinal plants, unique species, vegetation types, bioclimatic zones.

Evergreen forests in Maharashtra are confined to the Western Ghats and are constrained by prolonged dry period. Semi-evergreen forests are found scattered in the higher ridges and also on the hill slopes of Western Ghats. Together semi-evergreen and evergreen forests occupy about 2–3% of the forest area of the state. Moist deciduous forests occupy comparatively larger area than the former. Coastal plains and the slopes of Western Ghats harbour moist deciduous forests. Warmer pockets in eastern parts of Maharashtra plateau occasionally shelter patches of moist deciduous forests. Dry deciduous forests occupy the largest portion of the forest area in the state. Thorny scrubs are confined to Central Maharashtra (Dixit, 1986; Anonymous, 2000; Lakshminarashimhan and Prasanna, 2000).

We analysed the distribution of medicinal plants diversity and uniqueness among different vegetation types of Maharashtra (Figure 1). Unique species

Table 1. Distribution of MPCAs across climatic and vegetation types.

| Vegetation types | Average no. of species | Average no. of medicinal species | Biogeographic details | | |
			Zone	Average annual rainfall mm	No. of MPCAs
Evergreen	35	24	Western Ghats	3600–6000	2
Semi-evergreen	28	21	Western Ghats	2000–3600	1
Moist deciduous	46	37	Coast	1500–2000	2
			Deccan	1500–2000	1
Dry deciduous	40	33	Western Ghats	1200–1500	2
			Deccan	1200–1500	3
Thorn scrub	41	34	Deccan	500–800	2

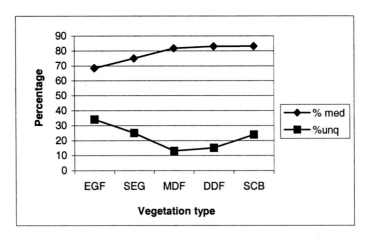

Figure 1. Medicinal plants and uniqueness of MPCAs in Maharashtra.

are those that are found in only one MPCA. Moist deciduous forests were found to have the highest percentage of medicinal species, while evergreen forests representing one extreme end along the moisture gradient were found to support lower percentage of medicinal plants. On the other hand, the evergreen and scrub forests had the highest proportion of unique species.

Our study indicates that not only diversity but also uniqueness of medicinal plants at a site need to be considered while evaluating the conservation worthiness of a site. The reasons that might explain the differences in the richness or otherwise of medicinal plants across the vegetation types are not immediately forthcoming. However the pattern could be related to the underlying species richness and or diversity of the site.

Acknowledgements

We are thankful to Darshan Shankar, Director, FRLHT, Bangalore for allowing us to make use of Botmast – the database on medicinal plants of India. Muneer Alavi, Director, Rural Communes, Satish Elkunchwar, Director, MPCC, Rajashree Joshi and MPCC staff are acknowledged for sharing ideas. Thanks are due to concerned forest officers for local hospitality and Utkarsh Ghate, FRLHT Bangalore for valuable comments on the draft.

References

Botmast – Database on Medicinal Plants of India. Foundation for Revitalisation of Local Health Traditions, Bangalore (1993).
Dixit, K. R. *Maharashtra in Maps* 43–51 (1986).
Anonymous. *Forest Survey of India*. State of Forest Report–1999 (2000).
Lakshminarashimhan, P. & Prasanna, P. V. *Flora of Maharashtra State, Dicotyledons*, Vol. I, 25–50 (2000).

THEME III

BIODIVERSITY HOT-SPOTS:
Western Ghats: Ecology, Natural History and Conservation

Tropical Ecosystems: Structure, Diversity and Human Welfare.
Proceedings of the International Conference on Tropical Ecosystems
K. N. Ganeshaiah, R. Uma Shaanker and K. S. Bawa (eds)
Published by Oxford–IBH, New Delhi. 2001. pp. 535–538.

Decentralised biodiversity assessment in the Western Ghats

G. Utkarsh*, P. Pramod, K. Kunte, K. P. Achar, G. K. Bhatta, P. Pandit and K. G. Sivaramakrishnan

Western Ghats Biodiversity Network, Centre for Ecological Sciences, Indian Institute
of Science, Bangalore 560 012, India
*RANWA, C-26/1, Ketan Heights, Kothrud, Pune 411 029, India
e-mail: ranwa@pn3.vsnl.net.in,

The enormity of magnitude of biodiversity, heterogeneity in its spatial and temporal distribution, limited expertise in its assessment and management constitute one of the most pressing challenges in assessing and managing the biodiversity in the tropics (Gadgil, 1996). To address these concerns, a feasible approach may involve sampling selected organisms in selected ecological strata with the help of parataxonomists. Toward this end, the Indian Institute of Science (IISc) has been coordinating since 1994 the Western Ghats Biodiversity Network (WGBN), involving 15 undergraduate colleges scattered all along the Western Ghats (Achar *et al.*, 2000), one of the mega-diversity hot-spots in India.

WGBN has tapped the interest and promoted expertise of two dozen teachers and over 300 students to investigate and monitor biodiversity. Balancing their desire to study remote, unexplored areas with operational efficacy, each college has been studying a nearby village landscape about 25 km^2 in size. This flexible and exploratory design rather than hypothesis-driven approach

Keywords: Western Ghats, species richness, endemism, floristic types.

has facilitated in launching a broadband, participatory monitoring programme beginning with a benchmark inventory of human impacts and covariation across taxonomic groups.

The methodology involves land use classification and mapping including through remote sensing (Nagendra and Gadgil, 1999). The sampling design in each landscape consists of nearly 20 transects each measuring 500 m in length, scattered in about 5 to 7 natural and human influenced vegetation types except grasslands and fields (Achar et al., in press). Along these transects, species richness and abundance of trees, birds, butterflies and ants have been sampled (Kunte et al., 1999). Other taxa such as aquatic insects and fish have been sampled along streams and rivers using nets (Sivaramakrishnan et al., in press).

Table 1 shows significant covariation amongst organismic groups in terms of their diversity across vegetation types and concomitant human influence (Kunte et al., 1999). For instance, human habitations harbour high bird diversity but not of trees or butterflies or ants. Scrub/savanna harbours peak diversity of ants but not of other taxa. On the whole, semi-evergreen forests harbour peak biodiversity value while monoculture plantations shelter lowest level. Thus, the prevalent conservation prioritisation algorithms, which presume that hotspots would harbour peak diversity of various organismic groups, may hold only for geographical zones but not habitat types. Thus, taxawise and composite biodiversity values may be assigned to each habitat types and management strategy may focus on maximising total biodiversity value of the mosaic of habitat types within the landscape (Pramod et al., 1996b). For instance, nurturing of keystone species like figs (Ficus spp.) is both socially appealing and biodiversity friendly (Utkarsh and Almeida, 1999).

To assess transformations in biodiversity values, habitat transformations over last few decades were also gauged through social mapping of present and past landuse (Chandran, 1996). Table 2 lists various transformations and corresponding loss of biodiversity value, as evident from Table 1. Conversion

Table 1. Covariation in biodiversity values across organismic groups and vegetation types.

Vegetation type	% Area of landscape	Trees (a)	Birds (b)	Butterflies (c)	Ants (d)	Total (a+b+c+d)
Evergreen forest	5	10	7	5	9	31
Semi-evergreen forest	10	9	10	10	8	37
Deciduous forest	10	7	5	6	5	23
Scrub/savanna	15	6	6	1	10	23
Plantations	15	1	1	2	1	5
Human habitations	30	5	8	3	2	18

The values depict mean rank of Shanon's diversity index for each organismic group per type per landscape.

of semi-evergreen forest into monoculture plantations is the most prevalent transformation, which is also associated with maximum loss of biodiversity value. However, conversion of deciduous forests into scrub/savanna does not appear to cause any such value loss, despite changes in species composition. Few transformations may even enhance biodiversity value, such as conversion of evergreen forests into semi-evergreen, under low disturbance or ecorestoration of scrub/savanna into forests due to recent reduction in human pressure.

Besides promoting scientific understanding, WGBN has also helped in capacity building at the grassroots level – colleges and villages – in assessing biodiversity. This has led college teachers to publish monographs and field guides (Bhatta, 1998) or underlining the importance that human-influenced areas such as habitations may have for sheltering highly valued biodiversity elements (Bhatta, 1997). Some collaborators have helped in standardising sampling techniques, thereby permitting efficient and decentralised monitoring even at order and family levels (Sivaramakrishnan et al., in press). Students have helped in developing ecosystem health indicators such as mean bark thickness of trees in a given habitat (Hegde et al., 1998). To translate these assessments into biodiversity-friendly management practices, corresponding social appraisal schemes have been launched with the involvement of local government offices and media (Gadgil et al., 2000). WGBN approach is now boosted by complimentary provisions for local biodiversity assessments in the proposed Indian biodiversity legislation and global efforts at Millennium Ecosystem Assessments.

Table 2. Ongoing landscape transformations and loss of ecological value, averaged for study landscapes.

From	To	Value loss (a)	% Area change (b)	Total value lost (a×b)
Evergreen forest	Semi-evergreen forest	–6	25	–150
Evergreen forest	Scrub/savanna	8	25	200
Evergreen forest	Plantation	26	10	260
Evergreen forest	Habitation	13	5	65
Semi-evergreen forest	Deciduous forest	14	5	70
Semi-evergreen forest	Scrub/savanna	14	10	140
Semi-evergreen forest	Plantation	32	40	1280
Semi-evergreen forest	Habitation	19	5	95
Deciduous forest	Semi-evergreen forest	–14	5	–70
Deciduous forest	Scrub/savanna	0	10	0
Deciduous forest	Plantation	18	25	450
Scrub/savanna	Semi-evergreen forest	–14	5	–70
Scrub/savanna	Plantation	18	25	450
Scrub/savanna	Habitation	5	5	25
			Total	2745

Acknowledgements

The studies described here owe their origin and evolution to Prof. Madhav Gadgil. Over 20 college teachers and 300 students in the WGBN and BCPP enriched this understanding. Colleagues M. D. S. Chandran, Harini Nagendra, K. A. Subramaniam and V. V. Sivan helped greatly. Prof. N. V. Joshi helped with data analysis. The activities described received financial assistance from the PEW Foundation, the World Wide Fund for Nature, India (WWF-I), and the Department of Biotechnology (DBT), and the Ministry of Environment and Forests (MoEF), Government of India. We are obliged to them all.

References

Achar, K. P., Bhatta, G. K., Bhat, D. D., Ganguly, A., Gokhale, Y., Nagendra, H., Pramod, P. & Utkarsh, G. Linking college education to environmental monitoring and management: A case study from India. in *Communicating Sustainability* (ed. Filho W. Peter) 205–225 (Lan Scientific Publishers, 2000).

Achar, K. P. *et al.*, Tree communities and human influence in the Western Ghats, India. *J. Indian Inst. Sci.* (in press).

Bhatta, G. K. Caecilian diversity of the Western Ghats: In search of the rare animals. *Curr. Sci.* **73**, 183–187 (1997).

Bhatta, G. K. A field guide to the caecilians of the Western Ghats, India. *J. Biosci.* **23**, 73–85 (1998).

Chandran, M. D. S. On the ecological history of the Western Ghats. *Curr. Sci.* **73**, 146–155 (1996).

Gadgil, M. Documenting diversity: An experiment. *Curr. Sci.* **70**, 36–44 (1996).

Gadgil, M., Rao, P. R. S., Utkarsh, G., Pramod, P. & Chhatre, A. New meanings for old knowledge: The people's biodiversity registers programme. *Ecol. Appl.* **10**, 1307–1317 (2000).

Hegde, V., Chandran, M. D. S. & Gadgil, M. Variation in bark thickness in a tropical forest community of Western Ghats in India. *Funct. Ecol.* **12**, 313–318 (1998).

Kunte, K., Joglekar, A., Utkarsh, G. & Pramod, P. Patterns of butterfly, bird and tree diversity in the Western Ghats. *Curr. Sci.* **77**, 577–586 (1999).

Nagendra, H. & Gadgil, M. Biodiversity assessment at multiple scales: Linking remotely sensed data with field information. *Proc. Nat. Acad. Sci.* **96**, 9154–9158 (1999).

Pramod, P., Joshi, N. V., Ghate, U. & Gadgil, M. On the hospitality of the Western Ghats habitats for the bird communities. *Curr. Sci.* **73**, 122–127 (1996a).

Pramod, P., Daniels, R. J. R., Joshi, N. V. & Gadgil, M. Evaluating bird communities of Western Ghats to plan for a biodiversity friendly development. *Curr. Sci.* **73**, 156–162 (1996b).

Sivaramkrishnan, K. G., Venkataraman, K. Moorthy, R. K. & Utkarsh, G. Aquatic insect diversity and ubiquity of the streams of the Western Ghats, India. *J. Indian Inst. Sci.* (in press).

Utkarsh, G. & Almeida. M. R. Figs. *Resonance* **4**, 90–100 (1999).

Tropical Ecosystems: Structure, Diversity and Human Welfare.
Proceedings of the International Conference on Tropical Ecosystems
K. N. Ganeshaiah, R. Uma Shaanker and K. S. Bawa (eds)
Published by Oxford–IBH, New Delhi. 2001. pp. 539–544.

Patterns of richness and endemism of arborescent species in the evergreen forests of the Western Ghats, India

B. R. Ramesh

French Institute, PB 33, Pondicherry 605 001, India
e-mail: ramesh.br@ifpindia.org

The Western Ghats form a more or less continuous mountain chain over a distance of about 1600 km along the western coast of India. The western slopes and summits of these reliefs experience the full intensity of the summer monsoon, which, as a result of orographic effect, brings abundant rainfall, sometimes more than 7000 mm. Such conditions have enabled the growth of dense, humid forest formations from the coast up to the summit of the Ghats. However, the climatic conditions are not uniform throughout the Ghats. Since the monsoon arrives from the south and retreats in the reverse direction, the rainy season is longer in the south than in the north. Thus, beyond about 16°N, the dry season is too long for the growth of evergreen forests except in the moist pockets on the hills. A second aspect is that the monsoon rains diminish very rapidly once they cross the Ghats ridge. From there onwards, the interior regions, which become increasingly dry eastwards are mostly covered by deciduous formations, except in the moist valleys, which harbour evergreen forests. The third climatic gradient governing the distribution of evergreen formations is the fall in temperature with altitude.

Keywords: Western Ghats, species richness, endemism, floristic types.

This diversity in climatic conditions is expressed by a large variety of plant formations and high species richness. Nearly 4000, or 27% of the total plant species in India, have been recorded from the Western Ghats (Nayar, 1996). The evergreen forests of the Western Ghats are characterised by a very high percentage of species endemic to the region. The total number of endemic plant species is estimated to be 1500 (MacKinnon & MacKinnon, 1986). Therefore, the Western Ghats are considered as one of the biodiversity hot spots of the world (Myers, 1988).

Species richness and endemism are, however, not uniform along the Ghats. Based on the distribution patterns and association of certain characteristic species, in relation to bioclimate and topographic variations, 23 floristic types have been identified in the Western Ghats (Pascal, 1982a,b, 1984; Ramesh *et al.*, 1997). Out of these, 19 types belong to evergreen forests (Table 1). The present study examines the richness and endemism of arborescent evergreen species within the evergreen floristic types.

The species, which can attain girth of more than 10 cm is considered as an arborescent. By going through the literature and our own field observations 645 arborescent evergreen species belonging to 68 families and 230 genera are listed from the evergreen forests of the Western Ghats. The distribution patterns of these species across 19 floristic types were studied.

From the ecological point of view, 78% of the total number of species are primary, which are generally found in climax forests and 22% are secondary that are predominantly found in secondary or disturbed forests. Species, which can potentially reach canopy and sub-canopy level, constitute 13 to 15%, middle strata account for 29% and the lowermost strata are very rich with 49%.

Representation of species in the higher taxonomic groups (family and genera) is highly skewed. 10% of the families are dominant with more than 25 species and 50% are represented by only 1 or 2 species. Of the 230 genera, 73% have less than 3 species. Euphorbiaceae, Lauraceae, Rubiaceae and Myrtaceae are the most common families (> 42 species in each), represented in all the floristic types, thus indicating their tolerance to wide ecological amplitude. Berberidaceae, Ericaceae and Vaciniaceae are specialised families generally confined to high altitude types (> 1800 m).

Species richness across the different wet evergreen formations (Table 1) indicates that the Dipterocarp types (DKS, DDS, DKH) of low elevation and Cullenia types (CMPG, CMP) of medium elevation are relatively rich (> 43%) compared to others. These types are confined to the windward side of the southern part of the Ghats where the terrain is rugged and length of the dry season is short (2–4 months). Among low elevation evergreen forests, DP and DDH ('Kan' forests in Soraba region) types, which are in the relatively less

540

Table 1. Relationships between forest types, bioclimate, species richness and endemics (floristic types in each group arranged in increasing latitudes).

Forest types	Annual rainfall (mm)	Temp-erature* (C°)	Dry season (months)	Species Total	%	Endemics Total	%
Wet evergreen climax forests							
Low elevation types							
Dipterocarpus indicus–Kingiodendron pinnatum–Strombosia ceylanica (DKS)	2000–5000	>23	2–3	283	44	163	45
Dipterocarpus indicus–Dipterocarpus bourdilloni–Strombosia ceylanica (DDS)	2000–5000	>20	2–3	318	49	187	52
Dipterocarpus indicus–Kingiodendron pinnatum–Humboldtia brunonis (DKH)	2000–6000	>20	4–5	278	43	160	44
Dipterocarpus indicus–Humboldtia brunonis–Poeciloneuron indicum (DHP)	5000–8000	>20	4.5 slope 5-5.5 plateau	183	28	103	28
Dipterocarpus indicus–Persea macrantha (DP)	> 2000	20–23	5–6	69	11	50	14
Dipterocarpus indicus–Diospyros candolleana–Diospyros oocarpa (DDD)	3500–7000	>20	5–6 slope 6–7 plateau	202	31	101	28
Persea macrantha–Diospyros spp.–Holigarna spp. (PDH)	2000–6000	>23	6–7	175	27	83	23
Diospyros spp.–Dysoxylum malabaricum–Persea macrantha ('Kan' forest) (DDP)	1500–1800	23–24.5	6–7	117	18	43	12
Medium elevation types							
Cullenia exarillata–Mesua ferrea–Palaquium ellipticum–Gluta travancorica (CMPG)	2000–5000	16–23	2–3	284	44	179	49
Cullenia exarillata–Mesua ferrea–Palaquium ellipticum (CMP)	2000–5000	16–23	2–4	384	60	236	65
Mesua ferrea–Palaquium ellipticum (MP)	2000–5000	17–22	4–5	205	32	132	36
Poeciloneuron indicum–Palaquium ellipticum–Hopea canarensis (PPH)	5000–7000	18–20	4–5	130	20	72	20
Memecylon umbellatum–Syzygium cumini–Actinodaphne angustifolia (MSA)	5000–6500	17–22.5	5–7	108	17	55	15
High elevation type							
Bhesa indiaca–Gomphandra tetrandra–Litsea spp. (BGL)	3000–5000	13.5–16	2–3	185	29	107	30
Litsea spp.–Syzygium spp.–Microtropis spp. (LSM)	900–6000	<13.5	0–4	202	31	116	32
Schefflera spp.–Meliosma arnottiana–Gordonia obtusa (SMG)	>2000	13.5–17	3–6	85	13	62	17

Continued

Table 1. *(Continued)*

Dry evergreen climax forests							
Diospyros foliolosa–Mitreophora heyneana–Miliusa spp.– *Kingiodendron pinnatun* (DMMK)	1200– 1500	>23	4–5	211	33	130	36
Diospyros foliolosa–Mitreophora heyneana–Miliusa spp. (DMM)	1200– 1500	>23	4–6	197	31	116	32
Diosyros ovalifolia–Memecylon lushingtonii–Olea glandulifera (DMO)	1200– 1500	16–23	4–6	93	14	32	9

*t = mean temperature of the coldest month.

rainfall zone (around 2000 mm) and with 5–7 months dry period have low richness (11–18%). Of the medium elevation types, *Memecylon* forests (MSA), which are predominantly found on the basaltic Maharashtra plateau, has only 17% richness. Among the high elevation types, LSM (> 1800 m) is found only in the Nilgiris and Palni-Anamalaias and contain 31% of the total species.

Dry evergreen forests are confined to valleys along the steep eastern slopes of the Western Ghats where the rainfall is less than 1500 mm. In spite of this constraint, DMMK and DMM have significant richness (31 to 33%). DMO is basically an Eastern Ghat type, found in the eastern slope of the Nilgiris and has very low (14%) richness.

In spite of variations in the number of species, the appearance and disappearance of species in accordance with latitude and altitude determine the floristic composition of each type. The passage from one forest type to another is not abrupt. The Jaccard similarity index indicates that the similarity of adjacent types of wet evergreen forests along latitudinal gradient varies from 51 to 65% in the low elevation types, 45 to 55% in medium elevation types and 28 to 41% in high elevation types. However, the similarity of adjacent types along an altitudinal gradient is only 10 to 14% between low and medium elevation types and 4 to 17% between medium and high elevation types. This implies that the effect of temperature is more pronounced in bringing out the dissimilarity among different floristic types than that of duration of dry season, which is more gradual. Apart from these general trends, the types in the specialised habitats as in dry zones (DMMK, DMM, and DMO) and edaphic types (DDP, MSA) show marked dissimilarity compared to other types. DMMK type in the eastern part of the Agastyamalai region is unique as a meeting point of the Western and Eastern Ghats elements. It has 10 to 14% affinities with CMP and CMPG of the west and 10% with the DMO of the east.

In the Western Ghats none of the families is endemic. Among tree species only 5 are of endemic genera (*Blepharistemma, Meteoromyrtus, Otoneohelium,*

Poeciloneuron and *Pseudoglochidion*). If only evergreen arborescent species are considered, 56% of the total tree species (645) are endemic to the Ghats. Except in DDP and DMO, in all other types 50 to 75% of their total species are endemic of the Western Ghats. The richness of endemic species across different floristic types shows more or less the same trend as in the analysis of total richness of each type (Table 1). LSM has 27 species exclusive to the type and is followed by CMPG (14), CMP (12) and DMM (11). In others, species endemic to the type are less than 5.

However, actual distribution of the endemics across the Ghats, irrespective of the type, displays strong correlation with the relief and bioclimate. A study (Ramesh and Pascal, 1997) on the distribution of 352 endemic evergreen arborescent species indicates that in the Agastyamalai region, south of Ariankavu pass, endemism is highest with nearly 70%. North of this region, up to Kodagu, it varies between 60 and 65%. Between Kodagu and Uttara Kannada, endemism declines drastically from 38 to 25%. Further north, in the Konkan and the Deccan Trap escarpment region, endemism reduces to less than 10%.

The high level of endemism in the southern Western Ghats (south of Kodagu region) is mostly attributed to the high elevation and steepness of the hills, which impose drastic changes in the bioclimate, especially in the length of the dry season, which declines from 2 to 6 months within a few kilometres from the windward to the leeward sides of the Southern Ghats ridges (very strong foehn effect during the monsoon season). Because of these rapid variations in environmental conditions in all three directions of space, the distribution of endemic species in this region is highly variable and correspond to very restricted habitats and niches that do not occur in such diversity north of the Palghat Gap. Nearly one third of the total endemics has narrow ecological amplitude and may be confined to less than one-degree square area. Several of them (like *Vateria macrocarpa*) are confined to a single valley and some show disjunct distribution (e.g., *Hopea racophloea*). Altitude-wise, endemism also decreases as a consequence of the decreasing temperature. In high-elevation forests in the Nilgiri, Anamalai and Palni Hills, endemism varies between 5 and 15%.

References

MacKinnon, J. & MacKinnon, K. *Review of the Protected Areas System in the Indo-Malayan Realm,* (IUCN, Gland, Switzerland, 1986).
Myers, N. Threatened biotas: 'hotspots' in tropical forests. *Environmentalist* **8**, 1–20 (1988).
Nayar, M. P. *Hot spots of Endemic plants of India, Nepal and Bhutan* (Tropical Botanical Garden and Research Institute, Thiruvananthapuram, 1996).
Pascal, J. P. (with the collaboration of S. Shyam Sunder and V. M. Meher-Homji). Forest Map of South India – sheet: Mercara–Mysore. Published by the Karnataka and Kerala Forest Departments and the French Institute of Pondicherry. Inst. Fr. Pondichéry, Trav. Sec. Sci. Tech. Hors série 18a (1982a).
Pascal, J. P. (with the collaboration of S. Shyam Sunder and V. M. Meher-Homji). Forest Map of

South India – sheet: Shimoga. Published by the Karnataka Forest Department and the French Institute of Pondicherry. Inst. Fr. Pondichéry, Trav. Sec. Sci. Tech. Hors série 18b (1982b).

Pascal, J. P. (with the collaboration of S. Shyam Sunder and V. M. Meher-Homji). Forest Map of South India – sheet: Belgaum–Dharwar–Panaji. Published by the Karnataka and Goa Forest Departments and the French Institute of Pondicherry. Inst. Fr. Pondichéry, Trav. Sec. Sci. Tech. Hors série 18c (1984).

Ramesh, B. R. & Pascal, J. P. Atlas of Endemics of the Western Ghats (India): Distribution of tree species in the evergreen and semi-evergreen forests. Institut Français de Pondichéry, Publications du département d'écologie vol. 38. 403 (1997).

Ramesh, B. R., De Franceschi, D. & Pascal, J. P. Forest Map of South India – sheet: Thiruvananthapuram–Tirunelveli. Published by the Kerala and Tamil Nadu Forest Departments and the French Institute of Pondicherry. Institut Français de Pondichéry, Publications du départment d'écologie. Hors série 22a (1997).

Tropical Ecosystems: Structure, Diversity and Human Welfare.
Proceedings of the International Conference on Tropical Ecosystems
K. N. Ganeshaiah, R. Uma Shaanker and K. S. Bawa (eds)
Published by Oxford–IBH, New Delhi. 2001. pp. 545–548.

Patterns of distribution and diversity of vertebrates in the Western Ghats, India

R. J. Ranjit Daniels

Care Earth, No 5, 21st Street, Thillaiganganagar, Chennai 600 061, India
e-mail: careearth@usa.net

Peninsular India was part of the Gondwanaland till about 150 million years ago, after which it split and started drifting north. Major geologic transformations took place as the drifting peninsula experienced a domal uplift 120–130 million years ago. The uplifted crust broke along the centre, pushing the western segment into the sea, giving rise to the Western Ghats during the Eocene (between 45 and 65 million years ago) (Radhakrishna, 1991).

After peninsular India became part of the mainland Asia 45 million years before present, rapid colonisation by Palearctic, Ethiopian and Oriental vertebrates took place. Topography, climate and vegetation have favoured a greater prevalence of vertebrates of the Indo-Malayan biogeographical realm in the Western Ghats. During the past 5,000–12,000 years human interference in and around the Western Ghats (Subash Chandran, 1997) has apparently influenced the patterns of distribution of vertebrates. Patchy distribution of species, except those with populations which are isolated due to specialised habitat requirements, is a clear evidence of local extinctions during recent history (Daniels, 1997). Presently, excluding the migratory birds, there are 938 species of vertebrates in the Western Ghats, 324 (35%) being endemic (Table 1).

Keywords: Western Ghats, vertebrate diversity, endemic species.

Table 1. Vertebrate diversity and endemism in the Western Ghats

Class	Total species	Endemic species	% endemism
Fishes	218	108	50
Amphibians	120	90	75
Reptiles	156	93	54
Resident birds	324	19	6
Mammals	120	14	12
Total	938	324	35

Source: Swengel (1991); Daniels (1992 & 1997); Easa (1998); Nameer (1998)· Rema Devi (pers. comm.).

Hill-stream fishes of the Western Ghats tend to show a greater Malayan affinity (Hora, 1944). The streams and rivers in the southern Western Ghats are more diverse including a larger number of endemic species of fish than those in the north. Easa and Shaji (1997), based on a study of fishes in the Nilgiri Biosphere Reserve have suggested that the east- and west-flowing rivers do not differ in species richness.

Family Ranidae has the largest number of species amounting to 42% of the amphibian fauna of the Western Ghats (Daniels, 1992). Many amphibians are highly restricted in range. Interestingly, species restricted to the south (< 13°N latitude) are more frequently patchily distributed with a preference for moist forests. Most amphibian species are found in the altitudinal range of 0–1200 m asl. The highest diversity of species is at 800–1000 m asl (Daniels, 1992). Number of amphibian species in any locality in the Western Ghats may however be low and at very local scales it has been observed that species richness is determined by the proximity to water – most species tending to aggregate closer to a source of water (Vasudevan et al., 2001).

Endemism in reptiles is highest amongst snakes (58 species), especially with the family Uropeltidae contributing 32 species. 50% of the reptilian species in a study reported by Easa (1998) are of snakes. Inger et al. (1987) suggested that more species and individual snakes were found in terrestrial compared with aquatic habitat. It is interesting to note that snakes dominate the forest floor reptilian communities at altitudes of 1200 m asl and above (Ishwar et al., 2001).

Most of the resident, endemic and typically forest birds are not found north of Goa. Endemic bird species are primarily birds of the rain forests and the higher elevation shoal–grassland complexes (Daniels, 1997). Locally, when equal areas are compared, there are more species of birds per unit area in the central parts of the Western Ghats. This is due to mixing of migrants and generalist species of birds with the resident specialists and endemics. Wet evergreen forests and montane sholas, despite providing habitat to a number of specialists and endemic birds with greater conservation value, are comparatively less diverse in bird species than secondary/disturbed evergreen and moist deciduous forests (Daniels *et al.*, 1991; 1992).

Human interference of forests has led to the disappearance of birds locally in the Western Ghats. However, when large landscapes are considered, the avifauna has remained stable during the past 100 years (Daniels *et al.*, 1990a). Bird species diversity is inversely related to woody plant species diversity (Daniels *et al.*, 1992). Monocultures may support an assemblage of birds as diverse as (or even more diverse than) evergreen forests. However, birds that inhabit the monocultures are often generalist habitat users drawn from a wide range of neighbouring habitats (Daniels *et al.*, 1990b; Pramod *et al.*, 1997).

Mammalian fauna of the Western Ghats is dominated by bats (41 species), rodents (27 species including the porcupine) and insectivores (11 species) (Nameer, 1998). Early attempts made to understand the factors governing the distribution of wild mammals in the Western Ghats have suggested that evergreen forests are particularly suited to frugivorous arboreal primates and squirrels while the deciduous forests offer the best habitat for the larger grazing herbivores like the gaur and deer. Drought-resistant ungulates, particularly antelopes are specially adapted to the open dry scrub (Prasad *et al.*, 1979). Large herbivore biomass was highest in moist deciduous forests and adjacent teak plantations whereas it was lowest in the dry deciduous forests (Karanth and Sunquist, 1992).

Generally, the southern Western Ghats, with their varied topography and shorter dry season support a greater diversity of vertebrates including the endemic species. The presence of some of the endemic birds (Malabar grey hornbill, Rufous babbler and Crimsonbacked sunbird) locally indicates greater abundance of endemic mammals such as Nilgiri langur and Lion-tailed macaque (Prasad *et al.*, 1998). Such concordance is not apparent in the distribution of amphibians and birds (Daniels, 1992).

References

Daniels, R. J. R. Geographical distribution patterns of amphibians in the Western Ghats, India. *J. Biogeogr.* **19**, 521–529 (1992).
Daniels, R. J. R. *A Field Guide to the Birds of Southwestern India* (Oxford University Press, New Delhi, 1997) pp. 217.
Daniels, R. J. R., Joshi, N. V. & Gadgil, M. Changes in the bird fauna of Uttara Kannada, India, in relation to changes in the land use over the past century. *Biol. Conserv.* **52**, 37–48 (1990a).
Daniels, R. J. R., Hegde, M. & Gadgil, M. Birds of the man-made ecosystems: the plantations. *Proc. Indian Acad. Sci. (Anim. Sci.)* **99**, 79–89 (1990b).
Daniels, R. J. R., Hegde, M., Joshi, N. V. & Gadgil, M. Assigning conservation value: a case study from India. *Conserv. Biol.* **5**, 464–475 (1991).
Daniels, R. J. R., Joshi, N. V. & Gadgil, M. On the relationship between bird and woody plant species diversity in the Uttara Kannada district of south India. *Proc. Natl. Acad. Sci. USA* **89**, 5311–5315 (1992).

Easa, P. S. Survey of reptiles and amphibians in Kerala part of Nilgiri Biosphere Reserve. KFRI Research Report no. 148, pp. 40 (1998).

Easa, P. S. & Shaji, C. P. Freshwater fish biodiversity in Kerala part of the Nilgiri Biosphere Reserve. *Curr. Sci.* **73**, 180–182 (1997).

Hora, S. L. On the Malayan affinities of the freshwater fish fauna of peninsular India, and its bearing on the probable age of the Garo-Rajmahal gap. *Proc. Nat. Inst. Sci. India* **9**, 423–439 (1944).

Inger, R. F., Shaffer, H. B., Koshy, M. & Badke, R. Ecological structure of a herpetological assemblage in south India. *Amphibia-Reptilia* **8**, 189–202 (1987).

Ishwar, N. M., Chellam, R. & Kumar, A. Distribution of forest floor reptiles in the rainforest of Kalakad-Mundanthurai Tiger Reserve, south India. *Curr. Sci.* **80**, 413–418 (2001).

Karanth, U. & Sunquist, M. E. Population structure, density and biomass of large herbivores in the tropical forests of Nagarhole, India. *J. Trop. Ecol.* **8**, 21–35 (1992).

Nameer, P. O. Checklist of Indian Mammals. Kerala Forest Department, pp. 90 (1998).

Pramod, P., Joshi, N. V., Ghate, U. & Gadgil, M. On the hospitality of Western Ghats habitats for bird communities. *Curr. Sci.* **73**, 122–127 (1997).

Prasad, N. S., Nair, P. V., Sharatchandra, H. C. & Gadgil, M. On factors governing the distribution of wild mammals in Karnataka. *J. Bombay Nat. Hist. Soc.* **75**, 718–743 (1979).

Prasad, N. S., Vijayan, L., Balachandran, S., Ramachandran, V. S. & Verghese, C. P. A. Conservation planning for the Western Ghats of Kerala. I. A GIS approach for location of biodiversity hot-spots. *Curr. Sci.* **75**, 211–219 (1998).

Radhakrishna, B. P. An excursion into the past – 'the Deccan volcanic episode'. *Curr. Sci.* **61**, 641–647 (1991).

Subash Chandran, M. D. On the ecological history of the Western Ghats. *Curr. Sci.* **73**, 146–155 (1997).

Swengel, F. B. The Nilgiri Tahr Studbook, Minnesota Zoo, USA, pp. 26 (1991).

Vasudevan, K., Kumar, A. & Chellam, R. Structure and composition of rainforest floor amphibian communities in Kalakad-Mundanthurai Tiger Reserve. *Curr. Sci.* **80**, 406–412 (2001).

Tropical Ecosystems: Structure, Diversity and Human Welfare.
Proceedings of the International Conference on Tropical Ecosystems
K. N. Ganeshaiah, R. Uma Shaanker and K. S. Bawa (eds)
Published by Oxford–IBH, New Delhi. 2001. pp. 549–551.

Land-use change and conservation priorities in the Western Ghats

S. Menon*,[#], K. S. Bawa**, K. N. Ganeshaiah[†] and R. Uma Shaanker[‡]

*Department of Biology, Grand Valley State University, Allendale, MI 49401, USA
**University of Massachusetts, Boston, MA 02125, USA
‡Department of Genetics and Plant Breeding, ‡Department of Crop Physiology,
University of Agricultural Sciences, GKVK, Bangalore 560 065, India
[#]e-mail: menons@gvsu.edu

Conservation planning has historically been unsystematic because it was not fueled by the unifying goal of maximizing the representativeness and persistence of biodiversity. For example, the establishment of new reserves is more economically and politically feasible in remote areas than in regions where development and biodiversity compete for land; thus, new reserves did not always contribute to the representation of biodiversity (Pressey *et al.*, 1996). Requirements for systematic conservation planning and priority setting for a region include identification of biodiversity indicators and conservation goals, compilation of a biodiversity database, review of existing conservation areas and patterns of land-use change, and selection of additional conservation areas. Systematic conservation planning is effective when it efficiently uses limited resources to achieve conservation goals, is defensible and flexible in the context of competing land uses, and is accountable in allowing critical review of decisions (Margules and Pressey, 2000). Both land-use change and conservation priority-setting are strongly

Keywords: Western Ghats, land use change, conservation priorities, gap analysis.

spatial phenomena that have benefited from applications of spatial information technologies (geographic information systems and remote sensing) and spatially explicit modeling. We present some examples of studies of land-use change and conservation priority-setting in the Western Ghats, India, a biodiversity hotspot with a fairly high rate of deforestation (Jha *et al.*, 2000). Specific methodologies and applications include gap analysis of the protected area network, vegetation indices applied to satellite imagery for biodiversity assessment, spatial analysis of the drivers of land-use change, spatially explicit modeling to predict future land-use change, and analyses of the patterns and implications of habitat fragmentation.

The Western Ghats, India, is one of the biologically richest areas in south Asia and is characterized by high levels of biodiversity and endemism. More than 63% of India's evergreen tree taxa are endemic to the Western Ghats (Ramesh and Pascal, 1991) and the number of endemic plant species in the Western Ghats is estimated to be 1500 (MacKinnon and MacKinnon, 1986). The forests of the Western Ghats are some of the best representatives of non-equatorial tropical evergreen forests in the world. However, forest loss, degradation, and fragmentation are severe threats to the unique biota of the Western Ghats.

Gap analysis programs and data have been used to determine biodiversity hotspots, identify gaps in existing protection, and for conservation prioritization (Jennings, 2000; Kiester, 1996). We performed a vegetation-based gap analysis of the Agastyamalai hills, the southernmost region in the Western Ghats (Ramesh *et al.*, 1997). This region has some of the least fragmented forests and high levels of plant endemism. The study area was unique in that all of it was afforded some level of protection ranging from a national park and wildlife sanctuaries to reserved forests, managed for timber resources. We used a detailed map of floristic types, developed from both intensive field work and interpretation of satellite imagery, to generate spatial data layers corresponding to floristic species richness, zones of floristic endemism, floristically unique areas, and habitat distribution of representative endemic faunal species. We derived a conservation value map from these layers and compared the correlation of conservation value with protection status. The conclusions from this study reinforced the observations that protected area designation rarely coincides with biodiversity representation and that generation of detailed biodiversity data bases requires exhaustive inventories of species and ecosystem diversity and distribution.

Unfortunately, detailed inventories in the field are constrained by severe resource and time limitations, further compounded by the accelerated rate of biodiversity loss. Urgent conservation prioritization and action in areas of extremely high vulnerability should not be obstructed by the immediate unavailability of detailed information on biodiversity distribution over extensive geographical areas. Examples of alternative interim strategies to help focus efforts in collecting detailed biodiversity data and in targeting conservation action include vegetation analyses of satellite imagery and spatial analysis of existing and future land-use change.

We conducted a study correlating the annual sum of Normalized Difference Vegetation Index (NDVI) derived from satellite imagery with biodiversity richness for the Biligiri Rangaswamy Temple Wildlife Sanctuary, in Karnataka, India (Bawa *et al.*, 2001). NDVI has been demonstrated to be a useful indicator of primary productivity on global and regional scales but has not been used in the past to characterize biodiversity richness. We demonstrated a positive relationship between plant species richness and NDVI values based on the well-known relationship between plant species richness and productivity.

Spatially explicit modeling can predict areas susceptible to future deforestation and therefore help in conservation priority-setting. In addition to habitat loss, habitat fragmentation and degradation are major contributors of biodiversity loss. Landscape ecology and patch analysis studies can provide insight into fragmentation. We examined characteristics of forest fragments within and outside protected areas in the Western Ghats. Since most tropical biodiversity losses are a direct result of habitat loss due to land-cover and land-use changes, we need to understand the process and drivers of land-use change. Drivers of land-use change are often spatially auto-correlated and we used spatial statistics techniques such as kriging to obtain insight into the extent of autocorrelation between various drivers of change. We conclude with a review of efforts towards systematic conservation planning for the Western Ghats.

References

Bawa, K. S., Rose, J., Ganeshaiah, K. N., Kiran, M. C., Barve, N. & Uma Shaanker, R. Assessing biodiversity from space (submitted) (2001).
Jennings, M. D. Gap analysis: concepts, methods, and recent results. *Landscape Ecol.* **15**, 5–20 (2000).
Jha, C. S., Dutt, S. & Bawa, K. S. Deforestation in Western Ghats, India. *Curr. Sci.* **79**, 231–238 (2000).
Kiester, A. R., Scott, J. M., Csuti, B., Noss, R. F., Butterfield, B., Shar, K. & White, D. Conservation prioritization using GAP data. *Conserv. Biol.* **10**, 1332–1342 (1996).
MacKinnon, J. & MacKinnon, K. Review of the protected areas system in the Indo-Malayan realm. IUCN, Gland, Switzerland (1986).
Margules, C. R. & Pressey, R. L. Systematic conservation planning. *Nature* **405**, 243–253 (2000).
Pressey, R. L., Ferrier, S., Hager, T. C., Woods, C. A., Tully, S. L. & Weiman, K. M. How well protected are the forests of north-eastern New South Wales? – Analyses of forest environments in relation to formal protection measures, land tenure, and vulnerability to clearing. *For. Ecol. Manage.* **85**, 311–333 (1996).
Ramesh, B. R. & Pascal. J. P. Distribution of endemic, arborescent evergreen species in the Western Ghats. Proceedings of the Symposium on Rare, Endangered, and Endemic Plants of the Western Ghats. Kerala Forest Department, 20–29 (1991).
Ramesh, B. R., Menon, S. & Bawa, K. S. A vegetation-based approach to biodiversity gap analysis in the Agastyamalai region, Western Ghats, India. *Ambio XXVI(8)*, 529–536 (1997).

Tropical Ecosystems: Structure, Diversity and Human Welfare.
Proceedings of the International Conference on Tropical Ecosystems
K. N. Ganeshaiah, R. Uma Shaanker and K. S. Bawa (eds)
Published by Oxford–IBH, New Delhi. 2001. pp. 552–556.

A regional approach for the conservation of the biological diversity of the Western Ghats

K. N. Ganeshaiah*[†], R. Uma Shaanker**, Narayani Barve[‡], M. C. Kiran[†] and K. S. Bawa[‡,§]

*Department of Genetics and Plant Breeding, **Department of Crop Physiology,
University of Agricultural Sciences, GKVK, Bangalore 560 065, India
[‡]Ashoka Trust for Research in Ecology and the Environment, Hebbal,
Bangalore 560 024, India
[§]University of Massachusetts, Boston, MA 02125, USA
[†]e-mail: kng@vsnl.com

Biological resources of a region constitute a range of layers such as intra-specific genetic diversity, diversity of species, their assemblages and interactions and, the diversity of ecosystems of the region (Figure 1). A systematic regional level conservation plan ought to incorporate steps to conserve all these components; but most of the present day conservation efforts are usually focused on specific layers rather than taking a comprehensive approach. Consequently, most of these programs are often overlapping, redundant and are cost ineffective in conserving the biological diversity of a region. Systematic conservation plans that aim at an effective conservation of as many components as possible with least overlap and minimal resources are being rigorously sought for (Ganeshaiah et al., 1999, 2001; Margulis and Pressey, 2001). In this paper we propose a program aimed

Keywords: Biological diversity, Western Ghats, regional approach, ecosystem approach, species-based approach, management plan.

at the conservation of the gamut of biological resources of the entire Western Ghats region. The proposed protocol, a modification of that proposed for the conservation of the forest genetic resources (Ganeshaiah *et al.*, 2001), also incorporates certain elements from a similar program suggested by Margulis and Pressey (2001) for the conservation of the biological diversity of a region. The suggested program has three important steps: (a) Preliminary identification of the conservation areas based on two complementary approaches, viz. TOP-DOWN or Ecosystem approach and BOTTOM-UP or species-based approach (Ganeshaiah *et al.*, 2001); (b) Incorporation of the existing protected areas into the program and identifying the complementary sites for conservation (Margulis and Pressey, 2001) and; (c) Mapping threat patterns of the selected target areas of conservation and arriving at a management plan.

Step 1: Preliminary identification of the conservation areas based on TOP DOWN and BOTTOM UP approaches. We suggest two complementary approaches to identify preliminary set of areas of conservation (Ganeshaiah *et al.*, 2001). First of these, the 'Top-Down' approach aims to identify the sites that could conserve the diversity of a wide range of forest species from diverse habitat types of the entire ecosystem. The second, 'Bottom-UP' approach targets the groups of species that are economically or taxonomically linked and aims to identify common areas for conservation of biological and genetic resources of a range of related species.

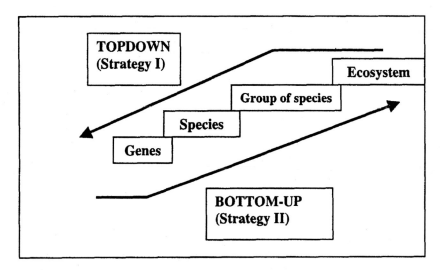

Figure 1. The pyramid of the different hierarchies of the genetic resources of a region or of an ecosystem. The present approach focuses on each species and begins at the base of the pyramid. The two new strategies suggested focus the conservation efforts at the ecosystem level and proceeds downwards (strategy I) or at a group of related species and proceeds upwards (strategy II).

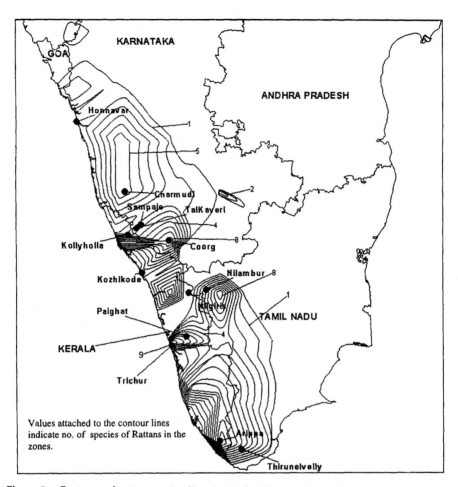

Figure 2. Contours of rattan species diversity in the Western Ghats. Note two areas enclaved by the contours depicting the high species diversity (near Coorg-Kozhikode and Trichur Nilgiris) are the candidate areas for conservation of the genetic resources of the rattans. These were further sampled for the genetic diversity of focal species.

The top-down approach relies on stratification of the area to identify distinct bio-geographic habitats that are likely to harbor diverse biological and genetic resources. Stratification can be based on a set of important climatic, physical, biological and stress factors that might together shape the structure of the genetic and biological resources. For instance, we could stratify the Western Ghats into 45 distinct bio-ecological zones based on layers such as rainfall, temperature regimes, altitude range, vegetation types, soil types, diversity of plant in general and of the threatened species in particular and other social and economic threats. From the 45 habitat types, we culled out those less than 300 km^2 as unimportant in the context of the conservation of the genetic and biological diversity of the entire region. From the remaining

16 types, we identified areas that are in the top five per cent of the size category for further working. The identified large patches shall be sampled across their geographic range for evaluating the diversity of flora and a few important faunal elements, populations sizes or densities of economically important species and their genetic diversity (Ganeshaiah et al., 2001) or any such components. Some of data sets for this step can also be derived from the existing literature as has been suggested by Margulis et al. (2001). Sites that harbor best combination of these components are identified as conservation worthy.

In 'Bottom-up' approach, we propose that conservation sites are identified keeping in focus the groups of plants species that are either economically important and or threatened and rare. The steps involve constructing contours of diversity of the species richness of the target groups based on the occurrence and abundance of the species (Ganeshaiah and Uma Shaanker, 1998, 1999; Ravikanth et al., 1999) (Figure 2). From such density contours, hot-spots of species richness are identified. From within these identified hot-spots, patches containing the viable populations and high genetic diversity of the focal species are identified.

Finally, the complementary sites suggested by both the bottom-up and top-down approach are combined to arrive at the preliminary set of areas for conservation. Together the two approaches are likely to offer a comple-mentary sets of sites that address conservation concerns from the genetic to ecosystem level (Figure 3).

Step 2: Incorporation of the existing protected areas: It is likely that most of the areas identified in the above steps might already be having sets of protected

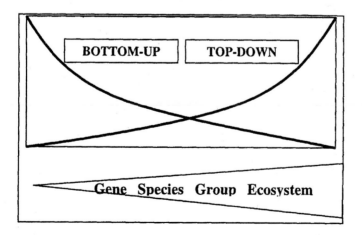

Figure 3. Efficiency of Bottom-up and Top-down approaches in addressing the conservation of the components at the levels of ecosystem and specific target group. Sites chosen via Top-down approach are very effective in conserving the major ecosystem components while those selected via Bottom-up approach are very effective in conserving the target group of species.

areas in them; conversely, some of the habitat types identified in the first step may correspond to the existing protected area networks. Thus as suggested by Margulis and Pressey (2001), these protected areas are incorporated into the program to serve as the basal set of areas for conservation. Following this it is essential to reevaluate the areas that are not represented. Accordingly fresh areas that are complementary to these protected areas be identified so as to address the range of conservation areas identified in the first step. These together with the selected protected areas constitute the regional network areas for conservation.

Step 3: Mapping threats of the selected target areas of conservation and arriving at the management plan: For a proper management of the selected areas we suggest that the threat maps are prepared for each of the target area. The threat maps incorporate information on the settlements in and around the areas, human-derived stresses, the developmental activities, patterns of extraction of the forest. Accordingly the most sensitive areas are identified and suitable management plans drawn to protect the conservation areas. We have derived such threat maps for two areas of the Western Ghats and have shown that the intensity of threat indeed affects the biological resource status of the forest. The target areas can then be continuously monitored to assess the impact of management and a feedback correction process needs to be followed such that the conservation becomes a dynamically responsive process that incorporates the lessons learnt and accommodates the changing scenarios.

References

Ganeshaiah, K. N. & Uma Shaanker, R. Mapping genetic diversity of *Phyllanthus emblica*: Forest gene banks as a new approach for *in situ* conservation of genetic resources. *Curr. Sci.* **73**, 163–168 (1997).

Ganeshaiah, K. N. & Uma Shaanker, R. Contours of conservation – A national agenda for mapping biodiversity. *Curr. Sci.* **75**, 292–298 (1998).

Ganeshaiah, K. N. & Uma Shaanker, R. Planning conservation strategies. *GIS Dev.* **3**, 67–69 (1999).

Ganeshaiah, K. N., Uma Shaanker, R., Narayani Barve, Kiran, M. C., Bawa, K. S. & Ramanatha Rao, V. *In situ* conservation of forest genetic resources at regional level: Two complementary programs using GIS approach (2001) (in press).

Margules, C. R. & Pressey, R. L. Systematic conservation planning, *Nature* **405**, 243–253 (2001).

Ravikanth, G., Chaluvaraju, Uma Shaanker, R. & Ganeshaiah, K. N. Mapping rattan species diversity in South India, IPGRI Newsletter for Asia, *The Pacific and Oceania* **28**, 10 (1999).

Tropical Ecosystems: Structure, Diversity and Human Welfare.
Proceedings of the International Conference on Tropical Ecosystems
K. N. Ganeshaiah, R. Uma Shaanker and K. S. Bawa (eds)
Published by Oxford–IBH, New Delhi. 2001. pp. 557–560.

Plant biodiversity and conservation of tropical evergreen forest in the Anamalais, Western Ghats, India

N. Parthasarathy, N. Ayyappan, S. Muthuramkumar and J. Annaselvam*

Salim Ali School of Ecology and Environmental Sciences, Pondicherry University,
Pondicherry 605 014, India
*e-mail: partha@pu.pon.nic.in

Plant diversity assessment and monitoring programs of tropical forests, especially of large-scale research plots, have been on the increase in the past two decades. Large-scale plots capture species which are represented by very low populations and also stabilize species-area curve. Understanding the variation in diversity and how an ecosystem functions requires careful study of species across spatial and temporal scales. Most ecological inventories on plant diversity have focussed on tree flora and a few on lianas too, while the forest understorey and epiphytes have largely been ignored. Investigations involving whole plot diversity are fewer (e.g. Whitmore *et al.*, 1986; Gentry and Dodson, 1987; Kelly *et al.*, 1994; Killeen *et al.*, 1998), although the scales of such inventories vary considerably. The prime objective of this research was to ascertain the plant diversity of all principal life-forms and analyze their diaspore types and dispersal modes, and to monitor the plot for biodiversity changes and forest functioning. This paper summarizes the results of initial inventories of total vascular flora and that of two recensuses.

Keywords: Biodiversity, dispersal mode, permanent research plot, plant life-forms, population density, Western Ghats.

Research was conducted in the tropical evergreen forest at Varagalaiar (elevation 600 to 660 m asl) within the Indira Gandhi National Park and Wildlife Sanctuary (Lat. 10°25'N and Long. 76°52'E) located in Coimbatore district of Tamil Nadu, south India. Mean annual rainfall is 1840 mm. The study site lies between the two perennial rivers, Kurampalliyar and Varagalaiar, although water flow is meager during summer. The vegetation is a closed canopy tropical evergreen forest.

Fieldwork was carried out in two phases, September–October 1997 – and from December 1997 to May 1998. Recensuses were carried during March – May of 1999 and 2000. A 30-ha (500 m × 600 m) permanent plot was established to inventory all trees ≥ 30 cm gbh, lianas ≥ 1 cm dbh (in the whole plot), understorey plants (in 3000 4 m × 4 m quadrats placed regularly one in each 100 m² quadrat) and herbaceous vascular epiphytes on all woody species ≥ 30 cm gbh. To capture small trees, all stems ≥ 1 cm dbh were sampled in a total of 320 5 m × 5 m quadrats (120 regularly spaced and 200 contiguous quadrats). To facilitate inventory each hectare was subdivided into 100 10 m × 10 m quadrats. Girth measurements were taken for trees and lianas at 1.3 m and those with protuberance or buttress, measurements were taken above them. Species-area curves were plotted by sequential arrangement of 30 1-ha plots by serpenting between one side to the other, and by using program Estimates (Colwell, ·1997) based on the mean accumulation of 100 times randomization of sample order. Dispersal modes of plants were ascertained based on field observation of diaspore types, supplemented with details from herbarium collections and floras.

A total of 431 species of vascular plants were recorded within the 30-ha plot, representing 338 genera and 117 families (Table 1). Among them, 153 species were trees ≥ 30 cm gbh in a total density of 13,415 stems. Trees ≥ 1 cm dbh totalled 3,386 stems (in 0.8 ha) representing 125 species and 41 families, of which 22 species were exclusively small trees (< 30 cm gbh). The plot contained 75 species of lianas in a sample of 11,200 individuals, 155 species of understorey plants with a total density of 2,18,471 individuals, and 26 species

Table 1. Summary of plant diversity inventories (trees, lianas, understorey plants and herbaceous vascular epiphytes) in a large-scale permanent plot of tropical evergreen forest at Varagalaiar, Anamalais, Western Ghats, India.

Diversity inventories	Trees		Lianas	Understo-	Epiphytes	Total
	≥ 30 cm gbh	≥ 1 cm dbh*	≥ 1 cm dbh	rey plants		
Species (range ha⁻¹)	153 (52–79)	125	75 (26–48)	155 (17–83)	26 (2–13)	431
Genera	123	102	65	132	19	338
Families	50	41	36	53	10	117
Population density	13,415	3,386	11,200	2,18, 471	–	–
Basal area (m²)	1094.1	34.88	13.5	–	–	–

*Sampled in a total of 0.8 ha.

of herbaceous vascular epiphytes in 3,392 individuals that were screened on a total sample of 13,445 trees and 348 liana individuals ≥ 30 cm gbh.

Whole plant diversity inventories at 0.05 to 1.5 ha scales recorded a range of 219 to ~ 500 species in various tropical forest sites (see Richards, 1996; Chapter 11), while a quantitative inventory in a 400-ha block of semideciduous forest of Bolivia yielded 310 species. Although wide variations in inventory methods, plot size, dimension, study protocols (whole/subsample inventories), life-forms and their size classes considered render comparisons difficult, available data reveal that our relatively undisturbed Varagalaiar forest is lower in species richness compared to many neotropical and southeast Asian forests sites, possibly due to present forest status, and past history, besides its geographical location.

The predominant species based on density/importance value include (by life-forms) *Drypetes longifolia*, *Dipterocarpus indicus*, *Poeciloneuron indicum* and *Reinwardtiodendron anamallayanum* (trees); *Olax scandens*, *Piper nigrum*, *Chilocarpus atrovirens*, *Combretum latifolium* (lianas); *Nilgirianthus barbatus*, *Pellionia heyneana*, *Ecbolium viride* (understorey plants); *Pholidota pallida*, *Oberonia iridifolia* and *Cottonia peduncularis* (epiphytes). Of the total of 153 tree species, *Drypetes subsessilis* (Euphorbiaceae) and *Prismatomeris tetrandra* subsp. *malayana* (Rubiaceae) formed new records to the flora of peninsular India and the liana *Trichosanthes anaimalaiensis* (Cucurbitaceae) was a new record to southern Western Ghats.

The observed species-area curve stabilized at 26th ha for trees, 27th for lianas and 28th ha for understorey plants and epiphytes, while the estimated curves (Chao 2 and ICE) stabilized earlier than that observed for lianas (16th ha), herbs (13th), epiphytes (18th), but only at 29th ha for trees. The cumulative family-area curves stabilized at 21st hectare for lianas and understorey plants and 23rd ha for trees, indicating the sufficiency of plot size to capture the vascular plant diversity. The population structure of trees and lianas revealed an exponential decrease in density with increasing size class.

Fruit features and dispersal modes of fruits of the 431 species revealed that 51% (218 species) are zoochorous (Figure 1). Among woody species, 70% of trees as well as lianas are zoochorous with fleshy fruits or seeds with accessory tissues (aril, caruncle, etc.), indicating the faunal dependence in forest functioning and the need for a holistic approach in biological conservation. 31% of plants are autochorous, while 14% (fruits winged/seeds comose) are anemochorous. Most understorey plants (61%) are autochorous, while epiphyte diaspores are largely anemochorous (54%) and autochorous (42%).

Recensus of tagged trees monitored over 2 years revealed that one species *Ficus beddomei* was lost. A significant change (40% reduction) in the

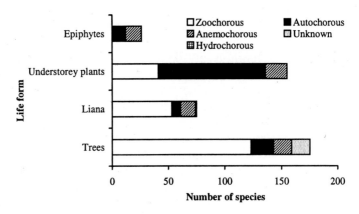

Figure 1. Dispersal modes of principal life-forms of plants of the tropical evergreen forest at Varagalaiar, Anamalais, Western Ghats, India.

population density of the once in 7-year blooming *Nilgirianthus barbatus* (Acanthaceae) is notable. Resource extraction levels in the forest by the local people are minimal.

Potential areas of further research designed in the permanent plot include: (i) monitoring for patterns of tropical tree growth, recruitment and mortality and forest regeneration, (ii) addressing the biology of selected plant species such as the 7-year living *Nilgirianthus barbatus* and its impact on forest dynamics and (iii) understanding the forest phenology and production ecology – foliar and fruit production and frugivore resource use and forest conservation.

Acknowledgements

We thank the DST, New Delhi for financing the study and Tamil Nadu Forest Department for site research permission.

References

Colwell, R. K. Estimates: statistical estimation of species richness and shared species from samples, version 5. Users guide and application published at http://viceroy.eeb.uconn.edu/estimates (1997).

Gentry, A. H. & Dodson, C. Contribution of non-trees to species-richness of a tropical forest. *Biotropica* **19**, 149–156 (1987).

Kelly, D. L., Tanner, E. V. J., Nic Lughadha, E. M. & Kapos, V. Floristics and biogeography of a rain forest in the Venezuelan Andes. *J. Biogeogr.* **21**, 421–440 (1994).

Killeen, T. J., Jardim, A., Mamani, F. & Rojas, N. Diversity, composition and structure of a tropical semideciduous forest in the Chiquitania region of Santa Cruz, Bolivia. *J. Trop. Ecol.* **14**, 803–827 (1998).

Richards, P. W. *The Tropical Rainforest: An Ecological Study* (Cambridge University Press, London, 1996).

Whitmore, T. C., Peralta, R. & Brown, K. Total species count in a Costa Rican tropical rain forest. *J. Trop. Ecol.* **1**, 375–378 (1986).

Tropical Ecosystems: Structure, Diversity and Human Welfare.
Proceedings of the International Conference on Tropical Ecosystems
K. N. Ganeshaiah, R. Uma Shaanker and K. S. Bawa (eds)
Published by Oxford–IBH, New Delhi. 2001. pp. 561–564.

Regeneration of woody flora in the sacred landscapes of Kodagu, Karnataka, South India

K. T. Boraiah*, Shonil A. Bhagwat[†], C. G. Kushalappa** and R. Vasudeva*[‡]

*Department of Forest Biology, College of Forestry, Sirsi 581 401, India
[†]Linacre College, Oxford, OX1 3JA, UK
**Department of Forest Biology and Wildlife, College of Forestry, Ponnampet, Kodagu 571 216, India
[‡]e-mail: vasukoppa@yahoo.com

Sacred groves are the patches of forests traditionally preserved without much human intervention and monitored through social institutions since several centuries. They represent a functional link between social life and ecological system of a region (Hughes and Chandran, 1995). According to several reports, these are the treasure houses of rare/endemic plants and refugia for relic flora of a region and act as centers of seed dispersal as well (Gadgil and Vartak, 1975; Chandrashekara and Sankar, 1998). However, there are very few assessments of regeneration in these landscapes in comparison to large stretches of state-owned, formally conserved forest patches (Chandrashekara and Sankar, 1998). In this communication we provide a comparative assessment of regeneration of woody flora between the sacred groves and reserve forests. This work is part of a larger study being conducted on sacred forests of Kodagu, which is in the central part of the

Keywords: Western Ghats, species richness, endemism, floristic types.

Western Ghats. With over 1200 sacred groves Kodagu is regarded as a 'global hotspot of sacred groves' (Bhagwat and Kushalappa, 2000).

The sacred groves are locally called 'Devarakadus' meaning 'forests of god' These are, however, not immune from anthropogenic pressures such as collection of minor forest produce, cattle grazing, etc. We selected two sacred groves in close proximity but which differ with respect to their disturbance levels. We assessed the regeneration of wood, flora in these and compared with that of a nearby reserve forest. Disturbance levels were assessed by recording the extent of canopy opening, quantifying human activities such as tree cutting, fire, etc. The study was replicated at two sites, viz. one near 'Heggala' village and another 20 km away near 'Tora' village. In both the sites, the sacred groves selected and reserve forests were within a radius of 10 km and had similar vegetational associations (Ramesh and Pascal, 1997). The average rainfall in these sites varies between 2500 and 3500 mm during May to November.

Permanent transects (5 m width and a maximum of 100 m length) randomly laid in an earlier study (Bhagwat and Kushalappa, 2000) were used for assessing the regeneration. In each of the sacred grove, five regeneration plots of size 2.5 m × 10 m were laid in randomly selected permanent transects. All plants, less than 30 cm GBH, were identified to the species level and classified into four height/girth classes. Their density, frequency were estimated and Shannon–Wiener diversity index calculated. The species were classified as rare/endemic based on Nayar (1996).

The density of regenerating individuals was higher in the sacred groves compared to the reserve forests of their respective localities (about two and five times more in Heggala and Tora sites respectively; Table 1). At both sites the number of species regenerating (51) were higher compared with the reserve forests (40 and 42, respectively for two sites). Interestingly, these differences were due to presence of a higher number of shrub species in the sacred groves. Further, a higher number of rare/endemic plant species was found regenerating in the sacred groves compared with the reserve forests. Some of the very important rare and endemic species found regenerating in the sacred groves include *Artocarpus hirsutus*, *Nathopodytes foetida*, *Persea macrantha*, and *Symplocos racemosa*. There was a higher species diversity among the regenerating individuals in the sacred groves compared with that in the reserved forests.

In general, our results indicate that sacred groves possessed a higher regeneration, more number of species, a higher number of endemics and a higher species diversity compared with the reserve forests. The size class distribution of seedlings pooled over both the sites suggests that there was more number of saplings (i.e. > 100 cm in height and < 10 cm girth) in sacred groves compared to the reserve forests (Figure 1).

Table 1. Density, species richness and diversity of regenerating individuals in disturbed, well conserved sacred groves and reserve forest in two localities.

	Heggala site			Tora site		
	Reserve forest	Disturbed sacred grove	Well-conserved sacred grove	Reserve forest	Disturbed sacred grove	Well-conserved sacred grove
Density (individuals per ha)						
Trees	30,560	15,280	29,840	8,800	44,400	67,840
Shrubs	2,320	36,720	10,240	3,680	9,440	12,640
Total	32,880	50,400	40,080	12,480	53,840	80,480
Species richness (number of species)						
Trees	34	31	38	30	37	38
Shrubs	8	14	13	10	14	13
Total	42	45	51	40	51	51
Number of rare/endemic species						
Trees	17	9	19	18	13	20
Shrubs	4	5	5	2	2	4
Total	21	14	24	20	15	24
Species diversity (H)						
Trees	1.89	1.10	2.11	0.79	1.41	1.18
Shrubs	0.24	0.84	0.67	0.30	0.61	0.52
Total	2.02	1.84	2.77	1.11	2.05	1.71

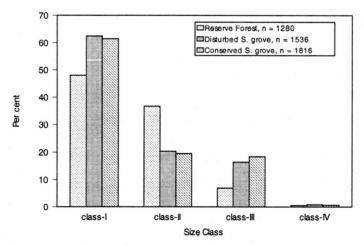

Figure 1. Distribution of regenerating individuals into size class in reserve forest, disturbed and well-conserved sacred groves in Kodagu. Data pooled over two localities. (Class I, < 40 cm height; Class II, 41–100 cm height; Class III, > 100 cm height, < 10 cm GBH; Class IV, > 100 cm height, 10–30 cm GBH).

Thus our study further reinforces the notion that sacred groves, though small in size, are important repositories of endemic flora and have a high conservation value.

References

Bhagwat, S. A. & Kushalappa, C. G. Families of plants, birds, fungi and Coorgs: Ecology of *Devarakadus* in Kodagu. Proceedings of the National Workshop on Community Strategies on the Management of Natural Resources. IGRMS, Bhopal (2000).

Chandrasekhara, U. M., & Sankar, S. Ecology and management of sacred groves in Kerala, India. *For. Ecol. Manage.* **112**, 165–177 (1998).

Gadgil, M. & Vartak, V. D. Sacred groves of India: a plea for continued conservation. *J. Bombay Natl. History Soc.* **72**, 314–320 (1975).

Hughes, J. D. & Chandran, M. D. S. Sacred groves around the earth: an overview. in *Conserving the Sacred for Biodiversity Management* (eds Ramakrishnan, P. S., Saxena, K. G. & Chandrasekhara, U. M.) 69–86 (Oxford and IBH Publishing Co., New Delhi, 1995).

Nayar, M. P. *'Hot Spots' of Endemic Plants of India, Nepal and Bhutan* (Tropical Botanical Garden and Research Institute, Thiruvananthapuram, 1996).

Ramesh, B. R. & Pascal, J. P. *Atlas of Endemics of the Western Ghats (India)* (French Institute, Pondicherry, 1997).

Tropical Ecosystems: Structure, Diversity and Human Welfare.
Proceedings of the International Conference on Tropical Ecosystems
K. N. Ganeshaiah, R. Uma Shaanker and K. S. Bawa (eds)
Published by Oxford–IBH, New Delhi. 2001. pp. 565–569.

Conservation and management of sacred groves of Kodagu, Karnataka, South India – A unique approach

C. G. Kushalappa*[†], Shonil A. Bhagwat** and K. A. Kushalappa[‡]

*University of Agricultural Sciences (Bangalore), College of Forestry, Ponnampet,
Kodagu 571 216, India
**Oxford Forestry Institute, Department of Plant Sciences, University of Oxford,
OX1 3RB, UK
[‡]No. 666, I Block, Ramakrishnanagar,
Mysore 570 023, India
[†]e-mail: kushalcg@blr.india

The practice of protecting forests, plants and animals in the names of Gods is followed in many parts of the world. The concept, which began as nature worship, underwent many transformations and these practices have either been totally lost or are in practice in a few ethnic groups in the world today. India is one such country where the concept of worship of sacred groves, plants and animals is still being practised. Though these practices are common all over the country, there are certain micro sites within India where these traditional worship practices have sustained over the years due to the support by native communities. One such area where the concept of sacred groves is still prevalent is in Kodagu district in Karnataka, South India.

Keywords: Sacred groves, community management, joint forest planning and management, conservation.

Kodagu is the second smallest district in Karnataka and lies entirely in the Western Ghats, one of the hotspots of biodiversity in the world. Nearly 69% of Kodagu's geographical area is under tree cover. This district contains 8% of the floristic diversity of India and 32% of the floristic diversity of Karnataka. The flora is not only diverse but also highly endemic. Ramesh *et al.* (1997) report Kodagu as an important microcenter of endemic plants in Western Ghats. The district has 1214 sacred groves covering an area of 2550.45 ha. The density of these groves in the landscape is very high with one grove for every 300 ha of land. Every village has one and many cases more than one grove. There are 14 villages with more than ten groves and Thakari village in Somwarpet taluk has the largest number of 17 groves. Though the district has large number of groves, most of them are today very small islands. Out of 1214 groves in Kodagu, 997 groves (80%) are less than 2 ha and there are only 123 groves which are more than 4 ha (Kushalappa and Kushalappa, 1996).

These groves have been protected in the names of 165 different deities. The most common deities are Iyappa, Bhagavathi, Bhadrakali, Mahadeva, Basaweshwara, etc. All the 18 native communities including muslims called 'Jammamapilas' are part of this unique tradition. Most of the sacred groves have a village level committee consisting of representatives of all the major communities in the village with 'Devathakka', an important person, to look after the functioning of the temple. There are others like 'Bhandara Thakka' (keeper of money and jewels) and Uruthakka (head of village) and Nadthakka (head of group of village) who assist the temple committees. All these responsibilities are traditionally given to certain families within a village. Even the responsibilities of performing music, dance and other religious functions at the festival are given to members of certain communities within the village. Hence this unique system of nature worship has evolved locally for local gods by local people.

Over the years the sacred groves have undergone considerable changes in their physical extent and worship traditions. The extent of the groves has reduced from 6277 ha in 1905 to 2550 ha in 1985. But their numbers have increased from 873 to 1214 during the same period indicating that the reduction in the total area has resulted in their fragmentation (Kalam, 1996). This reduction in area was mainly due to encroachments for cultivation and habitation. Partly such encroachments are due to the confusion regarding their legal ownership, which was presumed to be with the revenue department from 1905 till 1985 when they were transferred to forest department. But this administrative lapse where government orders were issued for their transfer from forest to revenue department in 1905 and later to forest department in 1985 resulted in a dual ownership with both these departments failing to protect the forests.

The loss in their physical area and conversion of the sacred groves to coffee estates and habitation also brought in a marked change in the social set up where the restrained resource utilization concept practised earlier gave way to rapid resource utilization. Alongside these developments there was a drastic change in the worship concept. The concept, which originated as nature worship transformed to installation of deities and later construction of structures or temples. The construction of temples and installation of deities resulted in regular worship and the process of 'Sanskritization' took over the annual or seasonal worship concept. This transformation resulted in the temple becoming more important than the grove. Hence the temple committee began mainly looking after the management of temples rather than the groves. These changes in physical and social structure of sacred groves of Kodagu have raised important concerns regarding the role of sacred groves in conserving the biological diversity of the Western Ghats.

The College of Forestry, Ponnampet of the University of Agricultural Sciences (Bangalore) has initiated studies on sacred groves with the following objectives: (a) To undertake inventory of sacred groves, (b) To undertake biodiversity inventory in sacred groves and assess their conservation value, (c) To initiate efforts for conservation of the groves by involving communities and governmental and non-governmental agencies, and (d) To develop a long-term management plan for the conservation of sacred groves.

Information was collected from the forest and revenue departmental records regarding the number and extent of sacred groves. Later field visits were undertaken in 78 large sacred groves to know their present status, growing stock and regeneration. Detailed maps and land records available were also collected. From the data collected, size, class distribution, village-wise distribution of sacred groves, important deities were prepared. Based on their present status the groves were classified as: Fully intact grove (30), Partially encroached groves (30), Fully encroached groves (9), Groves worked and enriched by forest department (9). The results indicate that a large number of groves are either fully or partially intact but some groves have been encroached and few have been worked by the forest department.

Twenty-one sacred groves were studied for their tree diversity, growing stock and regeneration. The growing stock in terms of basal area varied from 86.83 m^2/ha to 4.98 m^2/ha. This indicates that there is a large difference in the basal area and this parameter provides quantitative indication of the present status of sacred groves. Most sacred groves had basal area in the range of 25 to 40 m^2/ha (Kushalappa and Kushalappa, 1996). There was large variation among the sacred groves with respect to number of regenerations/ha (1793/ha to 96/ha).

In a collaborative biodiversity inventory undertaken by College of Forestry, Ponnampet and Oxford Forestry Institute, UK tree, bird and fungal diversity were studied in 13 villages of Virajpet taluk. The results indicate that though the sacred groves on an average occupy only 1.22% of village landscape area they contain 47.34% of woody plants, 44.91% of bird families and 21.11% of fungal families that could potentially be found in the wet evergreen forests of Kodagu (Shonil Bhagwat and Kushalappa, 2000).

The inventory and biodiversity studies undertaken indicate that a large majority of the sacred groves in the Kodagu district are small islands of less than 2 ha. But since they are distributed in the entire district they still offer a wide spectra of the biological diversity of the Western Ghats.

Encroachment of sacred groves in Kodagu is mainly dependent on their accessibility and the availability of alternative resource base in the vicinity of a devarakadu (Kalam, 2000). It is suggested that these coupled with other social issues need to be addressed to arrive at a management strategy for the conservation of the sacred groves.

It is clear from the existing situation that many diverse groups need to be involved for effective conservation of sacred groves. Forest department as legal owners, temple committees as managers, community leaders, non-governmental organisations, researchers, teachers and students and media need to join hands to evolve a conservation strategy. In this regard a festival of sacred groves was held on 15 and 16 October 2000 at Virajpet to formulate an action plan for the conservation of sacred groves of Kodagu. Members of 85 temple committees, community leaders, forest officials, researchers, teachers and students from 6 colleges, 7 non-governmental organizations and representatives from 20 local and state mass media took part in the festival. This was the first effort to bring together government departments and other agencies that are needed for an effective conservation on a common platform.

The action plan proposed that: 1. Sacred groves should be conserved as an important element of the bio-cultural landscape and should not be looked at as a means of generating revenue. 2. For effective conservation it is necessary to form a federation of temple committees to strengthen the traditional management. 3. It is essential to devise a mechanism for the Joint Forest Planning and Management of sacred groves involving all the major players like government departments, temple committees, community leaders, researchers and non-government organizations. 4. College of Forestry, Ponnampet will host a working group on the sacred groves to co-ordinate the various activities for management of sacred groves. 5. Non-government organizations and educational institutions need to participate in documenting information on sacred groves and undertake awareness generation programmes. 6. The forest department will resurvey and demarcate the sacred groves and provide the needed legal help for future protection and eviction of encroachments.

Based on these proposals the following conservation steps have been drawn up: 1. A working group on sacred groves has been established at the Forestry College. 2. A draft JFPM (Devarakadu) was prepared by the Karnataka Forest Department and legal experts and was discussed by the working group. This draft proposal was approved by the temple committee and community leaders on 17 March 2001 and will be submitted to the forest department for approval. This shall be the first such document prepared in India for the sacred groves. 3. The forest department has published a tentative list of sacred groves and Directorate of Survey has proposed a time bound survey of the sacred groves.

References

Kalam, M. A. Sacred Groves in Kodagu district of Karnataka (South India): A socio-historical study. *Pondy Papers in Social Sciences 21* (French Institute Pondicherry, 1996).
Kalam, M. A. Sacred groves and encroachments. in *Mountain Biodiversity, Land use Dynamics and Traditional Ecological Knowledge* (eds Ramakrishna, P. S. *et al.*) 98–119 (Oxford and IBH Publishing Co. Pvt. Ltd. New Delhi, 2000).
Kushalappa, C. G. & Kushalappa, K. A. Preliminary report. Impact assessment of working in the Western Ghat Forest. Unpublished, (1996).
Ramesh, B. R., Pascal, J. P. & Novguier, C. *Atlas of Endemic Evergreen Tree Species of the Western Ghats)* 403 p. (French Institute Pondicherry, 1997).
Shonil Bhagwat, A. & Kushalappa, C. G. Families of plants, birds, fungi and the coorgs. Ecology of devarkadus in Kodagu. in *Proceedings of National Workshop on Community Management of Natural Resources* (Indira Gandhi Rashtriya Manava Sangrahalaya, Bhopal, 2000).

Tropical Ecosystems: Structure, Diversity and Human Welfare.
Proceedings of the International Conference on Tropical Ecosystems
K. N. Ganeshaiah, R. Uma Shaanker and K. S. Bawa (eds)
Published by Oxford–IBH, New Delhi. 2001. pp. 570–573.

Management of *Kans* in the Western Ghats of Karnataka

Yogesh Gokhale

Centre for Ecological Sciences, Indian Institute of Science. Bangalore 560 012, India
e-mail: yogesh@ces.iisc.ernet.in

K*ans* are patches of evergreen forests in the Western Ghats of Karnataka. These forests are reported from Uttara Kannada district and Old Mysore State districts like Shimoga and Chikmagalur. I review here the historical management status of *kans* in Uttara Kannada and Shimoga districts with reference to Sorab taluk and Siddapur taluk respectively. I propose that the management system followed during the Old Mysore State could be helpful in designing present day management. I also identify the joint relationship of local people and the state forest department in the earlier management system and compare that with the present day programme of Joint Forest Management. I discuss the potential of *kans* for a range of NTFPs (non-timber forest products) based on the vegetation survey of about 29 *kans* in Siddapur (17) and Sorab (12) taluks.

The treatment of *kans* by British Government was different in the erstwhile Bombay Presidency and the Old Mysore State. The Bombay Presidency curtailed the rights of local people on the *kans* and treated these forests for timber exploitation. In the Old Mysore State local landlords enjoyed the rights over the *kans* till 1970s and the state forest department also recognized *kans* as separate forest management regime until 1990. These historical

---

Keywords: Kan forests, Western Ghats, NTFPs, joint forest management, Karnataka.

management differences have contributed to present vegetation patterns in these remnant evergreen forest patches.

I describe here status of *kans* in mainly two ranges of Sirsi forest division – Siddapur and Kyadagi administratively falling into Siddapur taluk.

Siddapur taluk shows large number of *kans* – about 113 *kans* – (84 in Siddapur range and 29 in Kyadagi range) according to the records of Village Forest Registers (VFRs). Most of the *kans* form a contiguous forest patch by merging with the *kan* of neighbouring village. Degradation of *kan* forests in Siddapur taluk has history of about 200 years. The *kans* growing near Siddapur towns like Kondli, Haladkatte, Kunaji, Ballatte were further degraded after the selective felling by the forest department and their subsequent failure to regenerate The majority of the population in the area is even today dependent on agriculture and lately on horticulture like arecanut gardens. The cultivation of arecanut requires large amounts of green fodder in the form of green leaves which ultimately is taking toll of standing woody areas.

The degradation of *kans* in Siddapur may be attributed to the following three major reasons:

Loss of rights of local people in Bombay presidency: The forest department established by British Government in the Bombay Presidency denied right of local people on their 'Sacred *kan* lands' (Chandran and Gadgil, 1993; Buchanan, 1870). This decision had very adverse impact on the local management of not only *kans* but also overall natural resources in Malnaad area.

Plantations of Acacia auriculiformis: During the period 1966 to 1985, selective felling was done in the evergreen *kan* forests in Siddapur taluk. The plantations of *Acacia auriculiformis* were done in place of clearfelled areas. People depend on these *kans* for several needs like dry leaves, several non-timber forest products like pepper, wild nutmegs, etc. apart from the daily requirements like firewood. The plantations were unable to meet the needs of the people. Obviously the earlier untouched *kans* became the only available resources to meet the daily requirements of the growing population. The village Kadkeri lost its *kans* to plantations of *Acacia auriculiformis*; hence, people of the village turned to the *kan* land of the neighbouring villages to source their requirements.

Arecanut cultivation: The last decade of 20th century witnessed a huge rise in market prices for the arecanuts. It prompted even the marginal farmer to convert the paddy land or encroach the *kans* in the valleys for the cultivation of arecanut. The vegetation in these encroached *kans* could be easily distinguished from undisturbed *kans* due to the selective protection to *Hopea ponga*. Also in many *kans* selective protection has been given to *Garcinia gummigutta, G. morella* and *G. indica* due to the economic value; and especially to *G. gummigutta* in recent years.

In contrast to *kans* in Siddapur taluk, *kans* in Sorab taluk exhibit a better potential. There were 116 *kans* in the taluk but according to the forest department the present number of *kans* is 65. However, the total number of *kans* in Sorab taluk could be more than 65 as many earlier *kans* are now under the status of Minor Forest or District Forest. The following reasons could be attributed to the present day condition of *kan* forests in Sorab taluk.

The Shimoga circle of the Karnataka State Forest Department had *Kans* as a separate management regime till 1990, i.e. until the last reorganization of the forests in the circle. There were official prescriptions followed for the maintenance of the *kans* since the time of the Old Mysore State under the management of British Government. The management of *kans* and sharing of benefits were vested with the local landlords like Gowda of the village. There was a system of tax (*shisht*) to be paid by the local Gowdas in whose name the *kans* were leased out. The state forest department continued the system till the local landlords lost their rights on *kans* mainly due to the land tenancy act. Unlike Siddapur taluk the *kans* in Sorab taluk do not favour the arecanut cultivation and thus face lesser impact.

The forest department auctions the non-timber forest produces like pepper, *Cinnamomum*, resins, etc. which are basically collected from the *kans*. Local Idiga community get supplementary income from the collection of NTFPs. But the contract system for purchasing these NTFPs is eroding the resources like *Cinnamomum* where the bark is recklessly removed or the pepper vine is uprooted instead of carefully plucking the berries by the collector.

Sorab taluk even today has tremendous social disparity in terms of agricultural land holdings in the villages. The dominant Idiga community owns lesser proportion of land in comparison to Lingayat/Gowda who were the local landlord and unfortunately retain the status even today due to the disparity in the land holdings.

The Joint Forest Management scheme of the state forest department has totally neglected the dependence of local people on the *kans* in both the districts. The major emphasis of the scheme is more on converting degraded lands under plantations. For the Malnaad region the emphasis could have been to jointly manage the existing stands of forest as the region is bestowed with evergreen forest tracts known for the endemic flora and fauna. Local people historically enjoyed limited rights over these *kans*. The government also earned reasonable revenue from the *kans*. Even today *kans* have the potential of yielding good revenue due to the potential of *Piper nigrum*, *Cinnamomum malabathrum* as well as NTFPs like *Artocarpus gomezianus*, *Zanthoxylum rhetsa*, *Z. ovalifolium*, etc. To conserve these species in the rich evergreen forest pockets in the central Western Ghats there is an urgent need to extend hands with local people for joint *kan* management.

The current JFM programme fails to identify and address the local needs. The state forest department could manage the *kan*s, i.e. the vestiges of evergreen forest patches in the central Western Ghats by improving the collection, regeneration and marketing of NTFPs in collaboration with local people instead of having the timber-based approach for JFM. The *kan* forests could be of immense use in managing the forests because of their potential to yield the highly required immediate economic benefits to local people.

Acknowledgements

I am thankful to Winrock-Ford Small Grant Program for supporting this work. I am also thankful to Mr. M. R. Almeida, Prof. Madhav Gadgil, and Dr. M. D. S. Chandran for the valuable guidance. I am thankful to Bombay Natural History Society, Mumbai for hosting the project. Thanks are also due to Sirsi and 'Sagar forest divisions especially the range offices of Siddapur, Kyadagi and Sorab.

References

Buchanan, F. *Journey Through the Northern Parts of Kanara (1801–2)*, vol. 2 (Higginbothams, Madras, 1870).
Chandran, M. D. S., & Gadgil, M. *Kans-safety forests of Uttara Kannada* (ed. Brandl, M.) Proceedings of the IUFRO Forest History Group Meeting on Peasant Foresty 2–5 September 1991, No. 40. Forstliche Versuchs-und Forschungsanstalt, Freiburg (1993), pp. 49–57.

South and South–East Asian Dipterocarp Forests

Tropical Ecosystems: Structure, Diversity and Human Welfare.
Proceedings of the International Conference on Tropical Ecosystems
K. N. Ganeshaiah, R. Uma Shaanker and K. S. Bawa (eds)
Published by Oxford–IBH, New Delhi. 2001. pp. 577–580.

Edaphic specialisation of *Shorea* section *Doona* across a topographic catena in Sri Lanka: Responses to nutrient availability

D. F. R. P. Burslem, C. V. S. Gunatilleke** and
T. R. H. Pearson**

*Department of Plant and Soil Science, University of Aberdeen, Cruickshank
Building, St. Machar Drive, Aberdeen AB24 3UU, Scotland, UK
**Department of Botany, University of Peradeniya, Peradeniya, Sri Lanka
e-mail: bot160@abdn.ac.uk

The mechanisms that maintain species richness in plant communities remains one of the most controversial issues in plant community ecology. Some ecologists argue that persistence is facilitated by regeneration niche partitioning and competitive interactions between species, while others highlight the importance of the stochastic processes of chance and history. Understanding these mechanisms is important for improving our theoretical understanding of the functioning of ecosystems, and has implications for attempts to manipulate the diversity of natural ecosystems for biodiversity conservation.

Tropical rain forests are some of the most species-rich terrestrial communities, with immense importance for conservation, and represent the ultimate challenge for protagonists in this debate. Empirical research on

Keywords: Sri Lanka, edaphic specialization, nutrient availability, *Shorea, Doona*.

regeneration niche partitioning among tropical trees has focussed on the importance of competition for light in canopy gaps, whilst ignoring the importance of competition for other potentially limiting resources such as water and nutrients. Below-ground resources can vary over very small scales along topographic gradients in tropical rain forest, and this variation might be equivalent to the patchiness in resource availability imposed by classical gap phase regeneration. Based at Sinharaja Forest Reserve, Sri Lanka, we have tested the hypotheses that (a) seven species of *Shorea* section *Doona* are differentiated along a topographic gradient of soil nutrient availability, (b) soil and foliar nutrient concentrations show correlated variation across the catena, and (c) seedling growth is most limited by the availability of P, as predicted on theoretical grounds and by research in other tropical forests.

Sampling and experimentation were conducted in Sinharaja Forest Reserve (6°21–26′N, 80°21–34′E) in the wet zone of Sri Lanka, a forest that is protected as a Man and the Biosphere Reserve and a World Heritage site and has been described in detail elsewhere. In August 1997, foliar samples of seven species of *Shorea* section *Doona* (*S. affinis*, *S. congestiflora*, *S. cordifolia*, *S. disticha*, *S. megistophylla*, *S. trapezifolia*, *S. worthingtonii*) were collected along a transect from ridge to valley in natural lowland tropical rain forest at Sinharaja. At the same time, soil samples from 0 to 10 and 10 to 20 cm depth were collected from ridge, mid-slope and valley positions along three transects in the same part of the forest. Foliar and soil samples were air-dried and transferred to Aberdeen, UK, for chemical analysis. The sampling period coincided with a major fruiting of some of the *Shorea* species being studied, and this event was exploited for establishment of a pot bioassay experiment of nutrient limitation for *Shorea worthingtonii*, a species characteristic of ridge-top locations at Sinharaja, and the valley-specialist species *S. megistophylla* in soil taken from ridge-top and valley-bottom sites. For each species growing in each soil type there were six nutrient amendment treatments: an unfertilized control, the same with additions of N as NH_4NO_3, P as Na_2HPO_4, K as KCl, Ca as $CaCl_2$ and Mg as $MgSO_4$. Newly germinated seedlings of the two test species were collected from cohorts of seedlings recently established in the forest understorey, and were transplanted into pots containing soil taken from 0 to 30 cm depth from either a ridge or a valley site. Nutrient additions were made to the transplanted seedlings and they were then transferred to a shadehouse and arranged in a completely randomised design with respect to species, soil type and nutrient addition treatment. Non-destructive measurements of seedling growth were made monthly, and a final harvest was conducted to obtain dry mass yield and biomass allocation ratios, relative growth rates, leaf area and tissue nutrient concentrations. A three-way analysis of variance was performed to determine the significance of differences among species, soil type treatments and nutrient addition treatments, and of interactions among these factors.

Analysis of variance showed significant differences between the two species, soil types and nutrient addition treatments on leaf production; significant differences were also obtained between the two species in their pattern of response to the nutrients. *S. worthingtonii* produced more leaves than *S. megistophylla* and produced significantly more leaves when growing in the valley soil than the ridge soil (Figure 1). *S. megistophylla* did not respond differently to the two soil types.

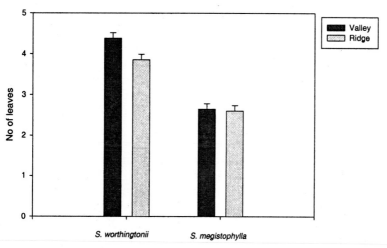

Figure 1. Mean (and standard error) leaf number for seedlings of *Shorea worthingtonii* and *Shorea megistophylla* grown in ridge and valley soils from Sinharaja forest reserve.

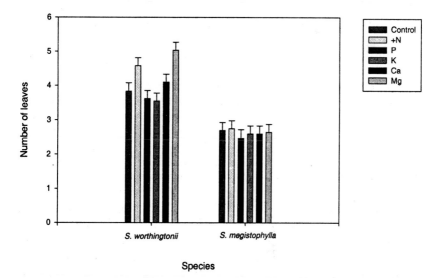

Figure 2. Mean (and standard error) leaf number for seedlings of *Shorea worthingtonii* and *Shorea megistophylla* in ridge and valley amended with N, P, K, Ca or Mg or without amendment (C, Control) in Sinharaja forest reserve.

When compared to the unfertilized control, seedlings of *S. worthingtonii* showed a significantly greater production of leaves following fertilization with a solution containing either N or Mg, but not when fertilized with solutions containing P, K or Ca. *S. megistophylla* seedlings did not respond to the addition of any nutrient solution (Figure 2).

Seedlings of *S. worthingtonii* showed a greater leaf production and growth when grown in soil taken from a valley than soils from a ridge-top site. This response reflects the higher pH and availability of mineral nutrients in the valley soils than the ridge soils and was a predicted outcome of this experiment. The lack of response of *S. megistophylla* seedlings to treatments may reflect its greater physiological capacity for efficient uptake of nutrients, or a greater morphological plasticity in response to environmental conditions. Biomass allocation data from the final harvest will help to distinguish between these possibilities. In either case it is likely that these characteristics represent adaptations to growth in relatively nutrient-rich valley soils.

The response of *S. worthingtonii* seedlings to N and Mg suggests that these elements were the key limiting nutrients in both soil types. Limitation of plant growth by N supply is a common feature of many ecosystems, although it has been argued that N limitation is less likely in the aseasonal tropics than in temperate ecosystems because of the higher rates of decomposition and nutrient cycling in the tropics. This experiment provides the first unequivocal evidence of limitation by N for tropical tree seedlings growing in a natural forest soil. Limitation by the availability of Mg has not been detected previously in any undisturbed forest ecosystem, and is unexpected because plants require only small amounts of this element to sustain their growth and metabolism, and concentrations in most forest soils are adequate. However, this key finding builds on related recent research that suggests an important role for Mg in determining the growth and distribution of dipterocarp trees. Future research will focus on the implications of these findings for competitive interactions between seedlings of coexisting *Shorea* species at Sinharaja and elsewhere.

Acknowledgements

We are grateful to the Nuffield Foundation for funding this research, to the Sri Lankan Forestry Department for permission to sample plants and soils in Sinharaja and to R. M. Ratnayake for assistance with data entry and maintenance of the experiment.

Tropical Ecosystems: Structure, Diversity and Human Welfare.
Proceedings of the International Conference on Tropical Ecosystems
K. N. Ganeshaiah, R. Uma Shaanker and K. S. Bawa (eds)
Published by Oxford–IBH, New Delhi. 2001. pp. 581–584.

Crossing barriers within rainforest tree species can arise over small geographic scales

E. A. Stacy*[#], I. A. U. N. Gunatilleke**,
C. V. S. Gunatilleke**, S. Dayanandan[†], C. J. Schneider*
and P. D. Khasa[‡]

*Department of Biology, Boston University, 5 Cummington Street, Boston,
Massachusetts 02215, USA
**Department of Botany, Faculty of Science, University of Peradeniya, Peradeniya,
Sri Lanka
[†]Department of Biology, Concordia University, 1455 de Maisonneuve Blvd.-West,
Montreal, Quebec, H3G1M8, Canada
[‡]Centre de recherché en biologie forestiere, Universite Laval, Faculte de foresterie et
de geomatique, Quebec G1K7P4, Canada
[#]e-mail: estacy@bio.bu.edu

Assessing variation in cross-fertility, and its underlying causes, across a plant species' range allows estimation of the spatial scales and strengths of inbreeding and outbreeding depression, within the species. Both inbreeding and outbreeding depression, have important implications for the evolution and conservation of tropical forest trees. For example, knowing the spatial scale at which outbreeding depression becomes significant within a species' range allows insight into the geographic scales at which conditions for speciation may arise. On the other hand, knowledge of the spatial scale of

Keywords: Cross-fertility, genetic structure, microsatellites, Sri Lanka, tropical trees.

inbreeding depression may be more relevant to conservation efforts in fragmented forest landscapes, where the distribution of outcrossing events may be biased in favor of shorter-distance crosses. To assess the spatial scales and causes of variation in cross-fertility within tropical rainforest tree species, we examined the relationships among cross-fertility, genetic similarity, and geographic distance between trees (and populations) occurring in Sri Lanka's rainforests. Sri Lanka's rainforests occur in the wet zone of the island's southwest corner, a small region ($\sim$ 15,000 km^2) with striking topographic heterogeneity (Gunatilleke and Ashton, 1987). The highly fragmented forests of the wet zone are characterized by modest tree species diversity and exceptional tree species endemism (Gunatilleke and Gunalilleke, 1980).

The study species [*Syzygium rubicundum* (Myrtaceae) and *Shorea cordifolia* (Dipterocarpaceae)] were selected to represent two of Sri Lanka's major canopy-dominant, economically important genera. For each species, three maternal trees were each hand-crossed with five pollen donors ranging from the maternal tree itself to donors occurring in separate forests 12 km (*S. rubicundum*) and 35 km (*Sh. cordifolia*) away (Stacy, 2001). Cross-fertility was estimated as rates of fruit set, seed germination, and seedling survivorship and height at 1 yr. Mean values of most cross-fertility measures increased steadily with outcrossing distance, peaking at the distant, within-forest crosses (1–2 km for *S. rubicundum* and 1–10 km for *Sh. cordifolia*) and then declining in the between-forest crosses. The relationship between outcrossing distance and cumulative cross-fertility (comprising all four fitness measures) was roughly quadratic for both species. The exceptions were seed germination and seedling height for *Sh. cordifolia*, both of which suggested hybrid vigor in the between-forest crosses. The fitness cost of mating with nearest neighbors relative to crossing with more distant neighbors was strong, though variable, for *S. rubicundum* (mean $\pm$ SE; 45 $\pm$ 20%), but ambiguous for *Sh. cordifolia* (0 $\pm$ 28%). Persistence of the near-neighbor mating effect beyond the stage of seed set strongly suggested the influence of inbreeding depression over genetic incompatibility between near neighbors. In contrast, the mean fitness cost of between-forest crosses was substantial for both species (52% and 70% for *S. rubicundum* and *Sh. cordifolia*, respectively). The dramatic drop in cross-fertility (i.e., outbreeding depression) observed for between-forest crosses relative to within-forest crosses may be due to genetic divergence of separate forest populations through drift and/or the influence of a selectively heterogeneous environment.

To address both the variability in biparental inbreeding depression in *Sh. cordifolia* and the relative roles of genetic drift and natural selection underlying the observed outbreeding depression, we developed seven polymorphic microsatellite loci for this species (Stacy et al., 2001). A total of 65 alleles at five loci (per locus mean: 13 alleles, range: 4–19 alleles) were scored in 190 individuals sampled from seven populations selected to cover the full geographic range of the species. With populations pooled, gene diversity estimates (H_E) ranged from 0.27 to 0.91 across loci (mean: 0.69). Genetic differentiation among populations was low ($F_{ST} = 0.031$), but significant, and did not fit a strict isolation by distance model (i.e., genetic distance was slightly, but not significantly, correlated with the geographic distance between populations). Genetic differentiation between populations occurring in separate forests did not exceed that between populations that co-occurred within forests.

To address the problem of variable biparental inbreeding depression, we examined patterns of genetic relatedness within populations (occurring in ≥ 25 yr-old logged forest). Although an inverse relationship between intertree distance and genetic relatedness was apparent, near neighbors were not significantly more related than pairs of trees selected at random. The variability in near-neighbor relatedness was consistent with the variability observed in biparental inbreeding depression for this species. To assess the relative roles of drift and selection underlying the observed outbreeding depression in between-forest crosses, we compared the degree of change in genetic differentiation at microsatellite loci (assumed to be selectively neutral) and cross-fertility (a measure of biologically meaningful differentiation) as functions of the geographic distance between populations. We used migration rate (Nm) to represent genetic differentiation, such that two populations that are highly differentiated have a low pairwise Nm value, and vice versa. Regressing Nm onto geographic distance revealed an 83% decrease in Nm over a 40 km distance, coincident with an 86% decrease in cross-fertility. Moreover, if cross-fertility is measured as fruit set alone (i.e., ignoring seed germination and seedling growth, where hybrid vigor was observed in the long-distance crosses), the drop in cross-fertility increased to 92% over the 40 km. Variation in cross-fertility between trees, particularly in fruit set, appears to exceed variation at neutral loci across the species' range. This observation, coupled with that of weak isolation by distance, suggests that the genetic divergence underlying the partial reproductive isolation between distant populations in this species is likely attributable to the combined effects of drift and selection. Thus the major conclusions of the study are:

- Biparental inbreeding depression occurs in tropical tree populations, and its intensity in near-neighbour crosses can vary substantially between species as a function of near-neighbour relatedness.
- For tropical rainforest trees, fitness may be optimized at outcrossing distances of 1 to 10 km. The significant and near significant fitness differences between the 'optimal' and shorter-distance crosses for the study species suggest that these longer-distance gene flow events may be critical for long-term population fitness.
- Conditions favorable for speciation in tropical trees may arise over a scale of only several to tens of kilometers in a heterogeneous environment.

- Some degree of reproductive isolation exists among the separate forests of Sri Lanka's wet zone, due to both natural selection and genetic drift. This result is unexpected given the large stature of the study species and the small size of the wet zone, and it may be a consequence of the unusual fine-scale geographical heterogeneity of southwest Sri Lanka. From a conservation perspective, observation of partial reproductive isolation between forest reserves suggests that each forest represents a distinct genetic resource worthy of preservation.
- Differentiation at microsatellite loci may not be predictive of biologically meaningful differences between populations. Therefore, caution is advised when using microsatellites for the identification of evolutionarily significant units or related entities for conservation planning. In many cases, quantitative traits are likely to be a better measure of biologically meaningful differences between populations.

Acknowledgements

We thank the Sri Lanka Forest Department for permission to conduct field work in and around the Sinharaja World Heritage Site. We are also grateful to the many people who provided excellent field assistance, especially S. Harischandran, M. Gunadasa, and H. Gamage. B. Dancik, and S. Nadeem greatly facilitated the development of microsatellite loci. This work has resulted from stimulating discussions with P. Ashton, K. Bawa, J. Hamrick, and L. Kaufman.

References

Gunatilleke, C. V. S. & Ashton, P. S. New light on the plant geography of Ceylon. II. The ecological biogeography of the lowland endemic tree flora. *J. Biog.* **14,** 295–327 (1987).

Gunatilleke, C. V. S. & Gunatilleke, I. A. U. N. The floristic composition of Sinharaja – a rain forest in Sri Lanka with special reference to endemics. *Sri Lan. For.* **14,** 171–180 (1980).

Stacy, E. A. Cross-fertility in two tropical tree species: evidence of inbreeding depression within populations and genetic divergence among populations. *Am. J. Bot.* **88** (2001).

Stacy, E. A., Dayanandan, S., Dancik, B. & Khasa, P. D. Novel microsatellite DNA markers for the Sri Lankan tree species, *Shorea cordifolia* (Dipterocarpaceae), and cross-species amplification in *S. megistophylla. Mol. Ecol. Notes* **1,** A20 (2001).

Tropical Ecosystems: Structure, Diversity and Human Welfare.
Proceedings of the International Conference on Tropical Ecosystems
K. N. Ganeshaiah, R. Uma Shaanker and K. S. Bawa (eds)
Published by Oxford–IBH, New Delhi. 2001. pp. 585–587.

Age structure and disturbance history of a seasonal tropical forest

Patrick J. Baker

Silviculture Laboratory, College of Forest Resources, Box 352100, University of
Washington, Seattle, WA 98195, USA
e-mail: pjbaker@silvae.cfr.washington.edu

Forest ecologists, temperate and tropical, have increasingly recognized the central role of disturbance in determining the composition and structure of forest stands. Consequently, the study of forest stand dynamics has focused heavily on characterizing disturbance patterns and determining their influence on forest development patterns. The study of forest stand dynamics has rested on two distinct, but complementary, methodologies: (1) permanent study plots and (2) historical (stand) reconstructions using annual tree rings. Studies in tropical forests have relied exclusively on permanent study plots due to the lack of annual growth rings in many tropical tree species; this has led to an inadvertent bias towards small-scale, low-intensity disturbances such as individual tree fall gaps in interpreting the role of disturbance in tropical forests.

The lack of reliable annual growth rings in tropical forest tree species is almost axiomatic in the tropical ecology literature. Nevertheless, wood anatomists have described the presence of annual rings in tropical tree species since the late 1800s (e.g., Gamble, 1902 for south Asia). To date few studies have used tree ring data to study the temporal dynamics of tropical

Keywords: Annual tree rings, stand dynamics, establishment, growth release, intermediate disturbance.

forests. Dendroecological studies of tropical forests could provide empirical data on the timing of establishment and growth release of individual trees, long-term growth patterns, the prevalence of suppression, the influence of climate and climatic change on individual tree's growth behaviour, and the role of disturbance on forest stand dynamics.

To evaluate the role of disturbance in a tropical forest, I conducted dendroecological analyses of a seasonal dry evergreen forest in western Thailand. Sampling was conducted in three study plots (FHP, SHP, YH) that were approximately evenly spaced along a 3 km transect. Two to four cores were taken from ~ 120 canopy trees of 6 species (*Afzelia xylocarpa, Chukrassia tabularis, Melia azederach, Neolitsea obtusifolia, Toona ciliata,* and *Vitex peduncularis*) and were prepared and analyzed following standard dendrochronological procedures.

Overall establishment patterns varied among study plots. In plot FHP tree establishment occurred in two discrete pulses (1900–1920 and 1940–1960) during the past century. Establishment patterns in plot SHP were similar to those in plot FHP although a third pulse of establishment occurred in the 1960s. In contrast, at plot YH there is one period of establishment (1940–1960) that includes 75% of the sampled trees. With the exception of *Afzelia xylocarpa*, no species had individuals > 120 years. The age of several individuals of *Afzelia xylocarpa* exceeded 200 years, although in one plot (YH) the majority of *Afzelia xylocarpa* were < 55 years.

Establishment patterns varied within and among species. For example, 60% of the *Toona ciliata* established during 1940–1950 in plot SHP; whereas in plot FHP establishment of *Toona ciliata* was uniformly distributed over the period 1900–1960. *Neolitsea obtusifolia* showed the opposite pattern. *Melia azederach,* an extreme pioneer species, established during the 1960s and 1970s in plots FHP and SHP, and did not occur in plot YH. It was not possible to identify the establishment dates of *Afzelia xylocarpa* at plot FHP because of the large size of the trees, although every individual was > 135 years. In contrast, at plot YH with one exception all of the *Afzelia xylocarpa* were < 55 years. *Chukrassia tabularis* had two peaks of establishment (1900–1910 and 1950–1960) at plot FHP, one peak (1930–1940) in plot SHP, and no peaks at plot YH. *Vitex peduncularis,* which was only sampled at plot SHP, had a single peak of establishment in 1900–1910.

Growth release was defined as a standardized ratio of the mean growth rate of the decade prior to a given year and the decade after a given year. Major or extreme releases (> 100% growth increase in two or more cores from the same tree) occurred in every decade after 1930 with a minor peak in 1940–1950 (10% of all releases) and a major peak in 1960–1980 (49% of all releases). The 1940–1950 peak was particularly pronounced in plot FHP, whereas the

1960–1980 peak was most prominent in plot SHP. All species showed releases in one or more decades.

It is often assumed that the probability of a gap occurring in the forest canopy is uniform in time and space. The results of this study suggest that such an assumption of stationarity is not always appropriate. In the seasonal dry evergreen forest examined here, the occurrence of canopy disturbances appears to be non-random in time and space. Over the past century gap-creating disturbances have been concentrated in one or two short periods but have been widespread across the landscape. Therefore it appears that disturbances intermediate in intensity between random individual treefall gaps and catastrophic stand-replacing disturbances have played an important role in determining the structure and composition of the current seasonal dry evergreen forest formation.

Acknowledgements

I would like to thank Chad Oliver, Peter Ashton, Sarayudh Bunyavejchewin and Jeremy Wilson for constructive comments throughout the development of this project. The staff of the Huai Kha Khaeng Wildlife Sanctuary and the scientists and staff of the Royal Thai Forest Department and Kasetsart University, Faculty of Forestry were gracious and supportive throughout the duration of the project. The research was supported by the National Research Council of Thailand, the Landscape Management Project (Chad Oliver, PI), a National Science Foundation grant to Peter Ashton, and a Grant in Aid of Research from Sigma Xi.

Reference

Gamble, J. S. *A Manual of Indian Timbers* (Sampson, Low, Marston & Company, London, UK, 1902) pp. 856.

Tropical Ecosystems: Structure, Diversity and Human Welfare.
Proceedings of the International Conference on Tropical Ecosystems
K. N. Ganeshaiah, R. Uma Shaanker and K. S. Bawa (eds)
Published by Oxford–IBH, New Delhi. 2001. pp. 588–590.

Species spatial clusters – Moving in space?

Matthew D. Potts*,†, Joshua B. Plotkin** and William H. Bossert*

*Division of Engineering and Applied Sciences, Harvard University, Cambridge, MA 02138, USA
**Institute for Advanced Study & Princeton University, Princeton, NJ 08540, USA
†e-mail: potts@deas.harvard.edu

Hubbell's community drift model (1979, 1997) predicts that species in tropical tree communities are randomly walking through space and time. In this null model, all individuals are competitively equal, that is, an individual's chances of mortality and recruitment are not affected by its species identity. Thus, over time species should be observed to randomly drift across the landscape.

Until now, lack of long term spatially explicit data sets and inadequate statistical methodology have precluded investigating changes in species spatial patterning over time. Spurred by our recent investigations (Plotkin *et al.*, 2000; 2001) of static species spatial patterning and clustering along with access to the first four censuses from the 50 ha Forest Dynamics Plot (FDP) on Barro Island Colorado (BCI), we quantify changes in species spatial patterning over the time. In particular, we first identify and then quantify changes in size and location of species spatial clusters. Our investigation is guided by the following questions:

Keywords: Diversity, null model, tropical rainforest, percolation theory.

- Do species spatial clusters move over time? If so, do some species move more than others?

- Does species-specific dispersal syndrome correlate with cluster movement?

- Is there spatial variation in the degree of cluster movement within the BCI FDP plot?

- Are species cluster size distributions in equilibrium?

Previous studies (Condit *et al.*, 2000; Plotkin *et al.*, 2000) have demonstrated that almost all species in tropical forests are highly aggregated or clumped in space. In many species, this aggregation is realized in a varying number of distinct and highly compact clusters of individuals. It is aggregated species of this type that are the focus of our study. Changes in spatial patterning are most easily visualized and quantified in these extremely clumped species (Figure 1).

The continuum percolation method (Plotkin *et al.*, 2001) was used to identify species spatial clusters. The strength of the continuum percolation method lies in its ability that with no *a priori* assumptions to identify species spatial clusters it closely matches our complex visual ability to identify clusters.

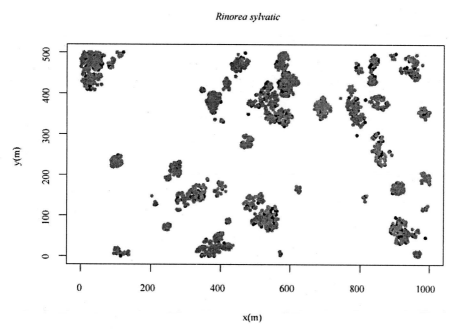

Figure 1. An example of a highly aggregated species. Black dots denote individuals alive in 1982 census, while gray dots indicated individuals alive in the 1995 census.

For each study species, spatial clusters in the 1982 and 1995 census were identified. For every pair of corresponding clusters in the 1982 and 1995 census a number of measures of spatial movement were calculated. Species movement was compared against a null model of random mortality and localized dispersal.

We found that at least for some species, spatial clusters did move. In addition, the amount of cluster movement was related to dispersal syndrome with dispersal limited species moving less than highly dispersed species. Finally, we found that not all species cluster size distributions were in equilibrium.

These results help to shed light on the degree to which species spatial patterning are determined by stochastic and deterministic forces.

Acknowledgements

The Forest Dynamics Plot of Barro Colorado Island has been made possible through the generous support of the U.S. National Science Foundation, The John D. and Catherine T. MacArthur Foundation, and the Smithsonian Tropical Research Institute. Peter Ashton provided comments on early drafts of this paper. Matthew Potts was supported in part by an EPA STAR Graduate Fellowship.

References

Condit, R., Ashton, P. S., Baker, P., Bunyavejchewin, S., Gunatilleke, S., Gunatilleke, N., Hubbell, S. P., Foster, R. B., Itoh, A., Lafrankie, J. V., Lee, H. S., Losos, E., Manokaran, N., Sukumar, R. & Yamakura, T. Spatial patterns in the distribution of tropical tree species. *Science* **288,** 1414–1418 (2000).

Hubbell, S. P. Tree dispersion, abundance, and diversity in a tropical dry forest. *Science* **203,** 1299–1309 (1979).

Hubbell, S. P. A unified theory of biogeography and relative species abundance and its application to tropical rain forests and coral reefs. *Coral Reefs* **16,** S9–S21 (1997).

Plotkin, J. B., Chave, J. & Ashton, P. S. Spatial patterns of plant species in Peninsular Malaysia: beyond spatial statistics. Submitted to *American Naturalist* (2001).

Plotkin, J. B., Potts, M. D., Leslie, N., Manokaran, N., Lafrankie, J. & Ashton, P. S. Species-area curves, spatial aggregation, and habitat specialization in tropical forests. *J. Theor. Biol.* **207,** 81–99 (2000).

Tropical Ecosystems: Structure, Diversity and Human Welfare.
Proceedings of the International Conference on Tropical Ecosystems
K. N. Ganeshaiah, R. Uma Shaanker and K. S. Bawa (eds)
Published by Oxford–IBH, New Delhi. 2001. pp. 591–594.

The role of dipterocarps, their population structures and spatial distributions in the forest dynamics plot at Sinharaja, Sri Lanka

C. V. S. Gunatilleke, N. Weerasekera, I. A. U. N. Gunatilleke and H. S. Kathriarachchi

Department of Botany, Faculty of Science, University of Peradeniya, Sri Lanka
e-mail: head@botony.pdn.ac.lk

In 1991 a 25 ha plot across an elevational gradient of 424–575 m, in an undisturbed part of the Sinharaja Mixed Dipterocarp forest, was demarcated to monitor the long-term dynamics of species within it, as part of the Center for Tropical Forest Science Network Program of the Smithsonian Institution. In this plot all individuals > 1 cm dbh were mapped, their dbh recorded and each identified to its respective species. A detailed elevation map of the area was also prepared.

The present study attempts to understand the role of the Dipterocarps in this plot. Specifically, we examine the following: (i) The collective and individual contribution of Dipterocarp taxa to the density and basal area of this forest stand; (ii) the density distribution of taxa in five different elevation classes across the elevational range and iii) their spatial distribution.

Keywords: Sri Lanka, Sinharaja, dipterocarps, spatial heterogeneity, population structure.

The 13 Dipterocarp species identified in this plot comprised 7 *Shorea* spp. of section *Doona*, *S. stipularis* of section Anthoshorea, 2 each of *Dipterocarpus* and *Hopea* species and *Stemonoporus canaliculatus*. While all of them represent canopy species, *Shorea cordifolia* is a subcanopy speceis and the last is an understorey tree. All of them are endemic to the rain forests of Sri Lanka. These taxa represented 15% of the 215 species, 14% of the 206,666 individuals and 21% of the 1572.5 m^2 basal area in this plot. *Shorea* species in section Doona, except one, had relatively high densities, ranging between 2000 and 7500 individuals in the plot, compared to the remaining species. Although individuals > 30 cm dbh were relatively few in each Dipterocarp species, they contributed over 60% of the family basal area.

The population structure of each study species was examined, by assigning their individuals to one of six stem diameter classes (1–1.9 cm, 2–4.9 cm, 5–9.9 cm, 10–29.9 cm and > 30 cm). The two lowest dbh classes constituted 61–89% of individuals in each species compared to 11–39% in the highest size classes, except in the *Dipterocarpus* taxa and *Shorea stipularis*.

The spatial distributions of the Dipterocarp taxa in the plot were studied by examining their relative abundance in five different elevation classes, based on its microtopography. These classes comprise the valley ranging between 424 and < 430 m, lower, middle and upper slopes ranging between 430 and 459 m, 460 and 489 m, 490 and 520 m, respectively and the ridge between 520 and 575 m elevation. The 20 m × 20 m subplots sampled were assigned to their respective elevation classes, taking the elevation at the center of the subplot as its mean.

The abundance of each *Shorea* species in each 20 m × 20 m subplot within an elevation class was also compiled. Using species and elevation classes as two treatments, these data were subjected to a two-way analysis of variance. The analysis indicated significant differences in abundance among species, and among elevation classes. Their interaction was also significant. Analyses were also conducted separately by species and by elevation. Means were compared using Tukey's least significant difference test. *Shorea megistophylla* in the valley, *S. megistophylla* and *S. trapezifolia* on the lower slope, *S. disticha* on the middle slope, *S. disticha* and *S. worthingtonii* on the upper slope and *S. disticha*, *S. worthingtonii* and *S. affinis* on the ridge were significantly more abundant compared to the remaining Dipterocarp species at each elevation (Figure 1 and Table 1).

Earlier studies have shown, using seedlings of potted plants, the differential performance of selected *Shorea* species to light and moisture (Ashton, 1995a,), nutrients (Gunatilleke *et al.*, 1997), elevation (Gunatilleke *et al.*, 1998) and soil (Gunatilleke *et al.*, 1995). The performance of transplanted *Shorea* seedlings in gaps across the topography has also been reported (Ashton, 1995b). This study demonstrates, for the first time, how this group of closely related

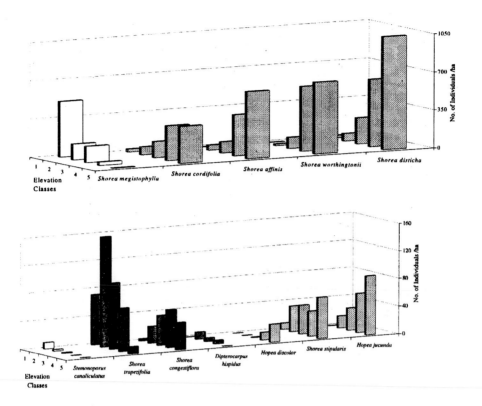

Figure 1. Density distribution of the 13 Dipterocarp taxa, across 5 elevational classes (1 = < 430, 2 = 430–459, 3 = 460–489, 4 = 490–519 and 5 = > 520 m) in the Sinharaja Forest Dynamics Plot in Sri Lanka. Species performing best at higher, middle and lower elevations shown by dotted, grey and white bars respectively.

sympatric species partition the environment within an elevational range, as small as 150 m. Variations in soil depth, moisture and fertility, and light environment in canopy gaps are also closely associated with the topography. Therefore, although the present study relates the variation in the abundance of these species to the different elevation classes, the species in fact are also responding to the interaction of all the environmental factors (soil, light and moisture) within a given elevation class.

The study also demonstrates the competitive ability of each individual species. When growing in the most suited environment, each species has a competitive edge over the others, so that each of the elevation is dominated by one or a few of the species. The wide ecological amplitude of *Shorea distica*, compared to all other species is also evident in this study. Further

Table 1. Statistical differences in abundance across elevational classes (1–5 = 424 – < 430 m, 430–459, 460–489, 490–519, 520–575 respectively) for (i) *Shorea* species collectively given in upper case letters within round brackets and (ii) for species individually row-wise in lower case letters on the left part of Table; (iii) differences among species in the plot as a whole are given in upper case letters within square brackets along-side the name of each species and (iv) that within each elevation class column-wise in lower case letters on the right part of the table. In each comparison, means sharing the same letter are not significantly different ($P > 0.05$, Tukey's least significance difference test). Degree of significance indicated by ***$P < 0.0001$, ** and $P < 0.001/$ 0.002.

Shorea species	Elevation classes					Elevation classes				
	1 (D)	2 (D)	3 (C)	4 (B)	5 (A)***	1	2	3	4	5
S. megistophylla [CD]	a***	b	bc	bc	c	a***	a***	a***	c	d
S. trapezifolia [D]	ab	a***	ab	ab	b	*b*	*a*	*bc*	*c*	*d*
S. disticha [A]***	d	cd	c	b	a***	*b*	*b*	*ab*	*a***	*a***
S. cordifolia [BCD]	c	c	b	a	a***	*b*	*b*	*ab*	*b*	*c*
S. worthingonii [B]	b	b	b	a	a***	*b*	*b*	*bc*	*a*	*b*
S. affinis [BC]	d	cd	c	b	a***	*b*	*b*	*bc*	*b*	*b*
S. congestiflora [E]	b	ab	a	a	ab**	*b*	*b*	*c*	*c*	*d*
S. stipularis [E]	b	ab	a	ab	a**	*b*	*b*	*c*	*c*	*d*

investigations are required to understand how *S. disticha*, *S. worthingtonii* and *S. affinis* co-occur together on the higher slopes and ridges, without out-competing each other.

Acknowledgements

Financial assistance provided by John D. and Catherine T. MacArthur Foundation and the Smithsonian Tropical Research Institute, USA is gratefully acknowledged.

References

Ashton, M. S. Seedling and growth of co-occurring *Shorea* species in the simulated light environments of a rain forest. *For. Ecol. Manage.* **72**, 1–12 (1995a).

Ashton, P. M. S., Gunatilleke, C. V. S. & Gunatilleke, I. A. U. N. Seedling survival and growth of four *Shorea* species in a Sri Lankan rain forest. *J. Trop. Ecol.* **11**, 263–279 (1995b).

Gunatilleke, C. V. S., Perera, G. A. D., Ashton, P. M. S., Ashton, P. S. & Gunatilleke, I. A. U. N. Seedling growth of *Shorea* section *Doona* (Dipterocarpaceae) in soils from topographically different sites of Sinharaja rain forest in Sri Lanka. in *Man and the Biosphere Series* (ed. Swaine, M. D.) **17**, 245–263 (UNESCO, Paris. Parthenon Publishing, Carnforth, UK, 1995).

Gunatilleke, C. V. S., Gunatilleke, I. A. U. N., Perera, G. A. D., Burslem, D. F. R. P., Ashton, P. M. S. & Ashton, P. S. Responses to nutrient addition among seedling of eight closely related species of *Shorea* in Sri Lanka. *J. Ecol.* **85**, 301–311 (1997).

Gunatilleke, C. V. S., Gunatilleke, I. A. U. N., Ashton, P. M. S. & Ashton, P. S. Seedling growth of *Shorea* (Dipterocarpaceae) across an elevational range in Southwest Sri Lanka. *J. Trop. Ecol.* **14**, 231–245 (1998).

Tropical Ecosystems: Structure, Diversity and Human Welfare.
Proceedings of the International Conference on Tropical Ecosystems
K. N. Ganeshaiah, R. Uma Shaanker and K. S. Bawa (eds)
Published by Oxford–IBH, New Delhi. 2001. pp. 595–598.

Ecophysiology of two *Shorea* species in a *Pinus caribaea* enrichment trial in the buffer zone of Sinharaja, Sri Lanka

H. S. Kathriarachchi[*][†]*, K. U. Tennakoon*[*]*,
C. V. S. Gunatilleke*[*] *and P. M. S. Ashton*[**]

*Department of Botany, University of Peradeniya, Peradeniya, Sri Lanka
**School of Forestry and Environmental Studies, Yale University, New Haven, USA
[†]e-mail: hashendra@hotmail.com

S*horea megistophylla* (Thw.) Ashton and *Shorea disticha* (Thw.) Ashton belong to section *Doona* in the family Dipterocarpaceae. These two endemic and late successional canopy dominants co-exist in the lowland hill rain forests in the southwest of Sri Lanka. They are partially sympatric species differentiated from each other by their growth characteristics, leaf anatomy, morphology and physiological features. The species has been reported to co-exist within the same forest landscape in relation to different environmental conditions (Gunatilleke *et al.*, 1996). Among members of section – *Doona*, these two species belong to the Beraliya group, members of which are locally recognized as multiple use species. Their fruits have edible cotyledons and they provide an additional source of carbohydrate to villagers in the vicinity of the forest. They also provide medium hardwood (Gunatilleke and Gunatilleke, 1993).

These two *Shorea* species have the potential to respond to a range of ecologically different habitats, natural as well as manipulated systems,

Keywords: *Shorea megistophylla, Shorea disticha,* Canopy removal treatments, *Pinus* enrichment.

within the Sinharaja forest. An understanding of the biological, physiological, ecological and silvicultural features of these species in different ecological habitats is important. This information will facilitate their management in restoration trials of the degraded lowland rain forests in Sri Lanka. This study attempts to examine the growth performance and some physiological attributes of these two *Shorea* species in different size canopy openings of a *Pinus* enrichment trial at Sinharaja MAB reserve in order to understand the species adaptability to different light environments.

Growth performance and physiological attributes of the study species were examined using plants established in 1991 under four different light regimes created by canopy removal in a *Pinus caribaea* plantation in the buffer zone of Sinharaja forest. The canopy removal treatments and the daily photosynthetic photon flux (DPPF) received initially were as follows: 3 pine rows removed (22 mol/m^2/day), 1 pine row removed (10 mol/m^2/day), 3 pine rows under planting (5 mol/ m^2/ day) and the closed canopy (3 mol/m^2/day). This trial was set up as a split plot design with three replicates, twenty individuals per replicate per treatment as reported by Ashton *et al.* (1997). The allometric measurements related to growth of individuals were recorded for 8 years annually. The physiological measurements (photosynthetic rate, stomatal conductance, transpiration rate and water use efficiency) of the study species were recorded in February–March 2000 from 9.00 am to 3.00 pm on sunny days using a LiCor 6400 portable photosynthesis system under ambient conditions.

The results showed that root collar diameter (RCD), height and diameter at breast height (DBH) after 8 years and the annual increments of root collar diameter and height were significantly higher among the canopy removal treatments compared to the closed canopy control for both species (Table 1). In both *Shorea* species greatest DBH, RCD and its increment were in the three pine rows removal treatment and least in the closed canopy under planting. No significant difference in height was observed among the three pine rows and one pine row removal treatments and the three pine rows under planting treatment. Both study species raised under the pine removal treatments showed more or less similar trends where higher growth rates were associated with increasing light levels (Table 1).

In the physiological studies, *Shorea megistophylla* showed significant differences in the transpiration rate, stomatal conductance and water use efficiency when grown under different light regimes. *Shorea disticha* on the other hand did not show any significant difference in these measurements among the canopy removal treatments (Table 1). Overall, with one exception, *S. megistophylla* always showed higher water use efficiency than that of *S. disticha*. If this trend continues up to the adult stage, the former would be a better species for *Pinus* enrichment programs.

Table 1. Growth and physiological measurements of *S. megistophylla* and *S. disticha* grown under different canopy removal treatments in an enrichment *P. caribaea* plantation. Letters qualitatively indicate significant differences ($a > b > c$) among the treatments for each species according to Duncan's Multip Range Test ($P < 0.05$)

	Canopy removal treatments			
	3 pine rows removed	1 pine row removed	3 pine rows under planting	Closed understorey (control)
Growth measurements after 8 years				
Mean height (m)				
Shorea megistophylla	8.1 ± 0.37ᵃ	7.8 ± 0.45ᵃ	7.2 ± 0.27ᵃ	4.3 ± 0.27ᵇ
Shorea disticha	7.3 ± 0.39ᵃ	7.6 ± 0.31ᵃ	6.7 ± 0.34ᵃ	5.0 ± 0.20ᵇ
Mean RCD (cm)				
Shorea megistophylla	6.75 ± 0.23ᵃ	6.09 ± 0.32ᵃᵇ	5.69 ± 0.27ᵇ	3.68 ± 0.17ᶜ
Shorea disticha	6.16 ± 0.37ᵃ	5.61 ± 0.23ᵃᵇ	5.11 ± 0.26ᵇ	3.19 ± 0.14ᶜ
Mean DBH (cm)				
Shorea megistophylla	6.05 ± 0.26ᵃ	5.68 ± 0.34ᵃᵇ	5.13 ± 0.26ᵇ	2.62 ± 0.19ᶜ
Shorea disticha	5.88 ± 0.29ᵃ	5.47 ± 0.23ᵃᵇ	4.60 ± 0.28ᵇ	2.92 ± 0.13ᶜ
Annual Growth Increments*				
Mean height increment (cm)/year				
Shorea megistophylla	92 ± 4ᵃ	87 ± 5ᵃ	74 ± 4ᵇ	38 ± 3ᶜ
Shorea disticha	78 ± 4ᵃ	80 ± 4ᵃ	70 ± 4ᵃ	44 ± 3ᵇ
Mean RCD increment (cm)/year				
Shorea megistophylla	0.72 ± 0.05ᵃ	0.68 ± 0.03ᵃᵇ	0.61 ± 0.03ᵇ	0.33 ± 0.02ᶜ
Shorea disticha	0.72 ± 0.07ᵃ	0.66 ± 0.05ᵃᵇ	0.53 ± 0.04ᵇ	0.26 ± 0.02ᶜ
Mean DBH increment (cm)/year				
Shorea megistophylla	0.67 ± 0.05ᵃ	0.75 ± 0.04ᵃ	0.66 ± 0.04ᵃ	0.57 ± 0.06ᵃ
Shorea disticha	0.65 ± 0.04ᵃ	0.66 ± 0.04ᵃ	0.59 ± 0.04ᵃ	0.58 ± 0.05ᵃ
Physiological measurements†				
Net photosynthetic rate ($\mu mol\ CO_2\ m^{-2}s^{-1}$)				
Shorea megistophylla	1.67 ± 0.19ᵃ	1.43 ± 0.17ᵃ	1.34 ± 0.14ᵃ	1.81 ± 0.19ᵃ
Shorea disticha	1.17 ± 0.14ᵃ	1.12 ± 0.09ᵃ	1.14 ± 0.12ᵃ	1.39 ± 0.16ᵃ
Stomatal conductance (mol $H_2O\ m^{-2}s^{-1}$)				
Shorea megistophylla	0.050 ± 0.011ᵃ	0.022 ± 0.002ᵇ	0.035 ± 0.003ᵃᵇ	0.031 ± 0.003ᵇ
Shorea disticha	0.021 ± 0.002ᵃ	0.026 ± 0.002ᵃ	0.026 ± 0.003ᵃ	0.028 ± 0.002ᵃ
Transpiration rate (mol $H_2O\ m^{-2}\ s^{-1}$)				
Shorea megistophylla	0.80 ± 0.05ᵃ	0.49 ± 0.04ᵇ	0.65 ± 0.05ᵃ	0.67 ± 0.05ᵃ
Shorea disticha	0.61 ± 0.04ᵃ	0.56 ± 0.04ᵃ	0.53 ± 0.05ᵃ	0.61 ± 0.05ᵃ
Water use efficiency ($\mu mol\ CO_2/mol\ H_2O$)				
Shorea megistophylla	2.33 ± 0.21ᵇ	3.30 ± 0.32ᵃ	2.48 ± 0.27ᵇ	3.01 ± 0.29ᵃᵇ
Shorea disticha	1.97 ± 0.19ᵃ	2.37 ± 0.22ᵃ	2.80 ± 0.37ᵃ	2.70 ± 0.29ᵃ

*Measurements were taken from all surviving individuals (ranging between 12 and 20 individuals per replicate per treatment) annually during the 8-year period.
†Measurements were taken from 3 individuals per replicate per treatment from 9.00 am to 3.00 pm at an ambient CO_2 concentration of approximately 340 $\mu mol\ mol^{-1}$, relative humidity 50–55% and when photon flux density was between 350 and 1200 $\mu mol\ m^{-2}\ s^{-1}$.

The spatial and temporal light variation of the understorey is a major factor that affects the physiological processes of the two species investigated. It also demonstrates that changes between the initial light intensities at the start of the experiment and that of present (8 years) due to the growth of the introduced plants can also affect the physiological responses of these plants. Studies are in progress to determine the effects of the variations in light intensities at different height levels (top and middle of the crown and at ground level) on the growth of introduced species.

Previous studies demonstrate that seedlings of late successional canopy species can be established on formerly cleared forests by planting beneath the canopy of a *P. caribaea* plantation (Ashton *et al.*, 1997; Gamage, 1997). After 8 years both *S. megistophylla* and *S. disticha* showed a higher growth rate in canopy removal treatments compared to those in the *Pinus* underplanting (closed understorey). For these *Shorea* species the three pine rows removed treatment was the best, but in most instances was not significantly different from the other removal treatments.

Furthermore, the morphological plasticity and the ecophysiological responses of the two *Shorea* species to different canopy removal treatments revealed in this study will be beneficial to identify the silvicultural practices required to promote the introduction of these two endemic species to *P. caribaea* monoculture plantations in the lowland wet zone of Sri Lanka.

Acknowledgements

Thanks are due to Anura Tennakoon for field assistance. Financial assistance provided by the NAGAO Natural Environment Foundation, Japan, MacArthur Foundation, USA and the National Science Foundation, Sri Lanka are gratefully acknowledged.

References

Ashton, P. M. S., Gamage, S., Gunatilleke, I. A. U. N. & Gunatilleke, C. V. S. Restoration of a Sri Lankan rainforest using Caribbean pine *Pinus caribaea* as a nurse for establishing late successional tree species. *J. Appl. Ecol.* **34,** 915–925 (1997).

Gamage, S. Feasibility studies on underplanting multiple use species in buffer zone pine plantations of the Sinharaja MAB reserve. M. Phil. thesis (University of Peradeniya, Sri Lanka, 1997).

Gunatilleke, I. A. U. N. & Gunatilleke, C. V. S. Underutilized food plants of Sinharaja rain forest, Sri Lanka. in *Food and Nutrition in the Tropical Rain Forest, Biocultural Interactions. Man and Biosphere Series* (eds Hladik, C. M., Hladik, A., Pagazy, H., Linares, O. F. & Hadley, M.) Vol. 15, pp 183–198 (UNESCO, Paris and Parthenon Publishing, Camforth, UK, 1993).

Gunatilleke, I. A. U. N., Ashton, P. M. S., Gunatilleke, C. V. S. & Ashton, P. S. An overview of seed and seedling ecology of *Shorea* (section *Doona*) Dipterocarpaceae. in *Biodiversity and the Dynamics of Ecology, DIWPA Series 1* (eds Turner, I. M., Dlong, C. H., Lim, S. S. L. & Ng, P. K. L.) 81–102 (1996).

Tropical Ecosystems: Structure, Diversity and Human Welfare.
Proceedings of the International Conference on Tropical Ecosystems
K. N. Ganeshaiah, R. Uma Shaanker and K. S. Bawa (eds)
Published by Oxford–IBH, New Delhi. 2001. pp. 599–602.

Dipterocarps in a sacred grove at Nandikoor, Udupi district of Karnataka, India

V. K. Vasanth Raj, P. V. Shivaprasad and K. R. Chandrashekar*

Department of Applied Botany, Mangalore University,
Mangalagangotri, Mangalore 574 199, India
*e-mail: bkvasanthraj@yahoo.com

Sacred groves are the most important areas of conservation of indigenous plants which are owned both by private land owners and the forest department. They are usually considered as protected areas or reserve forests. These groves are reported to have a great diversity and density of different trees than adjacent comparable reserve forests. The biomass of these forests per unit area are likely to be higher than the neighbouring reserve forests.

We undertook a detailed vegetation analysis of a sacred grove in Nandikoor village, Udupi District, Karnataka to study the species richness, diversity and other attributes of the sacred grove. The grove extends to about 5 ha and is located at about 110 msl. The vegetation is semi-evergreen with dominance of members of Dipterocarpaceae, viz. *Hopea ponga* (Dennst.) Mabberely and *H. parviflora* Bedd. Due to the construction of a coastal Konkan railway track parts of the grove are threatened.

Keywords: Dipterocarps, sacred grove, density, endemism.

One transect covering an area of 1000 m^2 was laid and divided into 25 quadrats of 10 × 4 m. The GBH and the height of the individuals above 10 cm GBH were measured and identified with the help of floras and field keys (Gamble, 1923; Pascal and Ramesh, 1987). Seedlings in the area were identified and their frequency recorded.

The individuals were categorized into three representative class of past, present and future growth based on their architecture (Halle *et al.*, 1978). The frequency, density, dominance, Importance Value Index (IVI) for each species were calculated (Pascal, 1988). The Family Importance Value (FIV) index was calculated by adding the IVI of different species belonging to the same family. The dominance was measured by the Simpson's index and the diversity by Shannon-Wiener's index.

A total of 22 species belonging to 18 genera and 14 families were identified. Only two species of Dipterocarps, viz. *H. parviflora* and *H. ponga* were recorded in this grove. A total of 335 individuals were recorded from the 1000 m^2 study area (Table 1). *H. ponga* was represented by 163 individuals followed by *H. parviflora* by 39 individuals. A total of 179 (about 53%) individuals belong to the set of future growth. Of this, about 114 and 7 individuals were of *H. ponga* and *H. parviflora* respectively. The remaining 58 individuals were represented by 20 other species. Most of the *H. parviflora* (82%) belongs to the set of present while, *H. ponga* was represented by only 30% of individuals belonging to that of the present. Only about 5% of the individuals belong to the set of past.

The total stand density of the grove was 37.59 m^2/ha (3350 individuals). Of these, members of Dipterocarps constitute nearly 3/4th (about 28.9 m^2/ha with 2020 individuals) which is low when compared to that of Pilarkan reserve forest, a secondary semi-evergreen forest in the same district (Shivaprasad *et al.*, 2001). *H. parviflora* alone represents a stand density of 21.61 m^2/ha with only 390 individuals and *H. ponga* about 7.29 m^2/ha with about 1630 individuals.

The IVI of *Hopea parviflora* was highest in this grove (84.22) followed by *H. ponga* (80.56). Only eight species exhibited IVI more than 10 and 15 species less than 10. The FIV of Dipterocarpaceae was highest being 164.78. This was followed by Myrtaceae (25.13), Melastomataceae (25.54), Euphorbiaceae (19.07), Anacardiaceae (13.99), Ebenaceae (11.36), Opiliaceae (10.23), Cluciaceae (5.82), Rubiaceae (4.55), Moraceae (4.42), Sapotaceae (4.25), Lauraceae, Santalaceae and Verbenaceae (1.17 each).

About 80% of the individuals were within 0–8 m height range which indicate that there has been good regeneration of the forest in the recent past. Only 2% of the individuals come under the height class above 20 m and most of them belong to Dipterocarpaceae. About 84% of the individuals belong to the

Table 1. The frequency (f), density (D), dominance (d), relative frequency (rf), relative density (rD), relative dominance (rd) and importance value index (IVI) of a sacred grove of Nandikoor

Species	f	D	d	rf	rD	rd	IVI
Hopea parviflora Bedd.*	72.0	39	2.161	15.12	11.62	57.48	84.22
Hopea ponga (Dennst.) Mabb.*	60.0	163	0.729	12.60	48.57	19.39	80.56
Memecylon terminale Dalz.*	56.0	28	0.167	11.76	8.34	4.44	24.54
Syzygium caryophyllatum Gaertn.	40.0	18	0.160	8.40	5.36	4.26	18.02
Aporosa lindleyana (Wight) Baillon	44.0	16	0.058	9.24	4.77	1.64	15.55
Holigarna ferruginea Marchand *	24.0	7	0.168	5.04	2.09	4.47	11.60
Diospyros paniculata Dalz.*	32.0	11	0.051	6.72	3.28	1.36	11.36
Cansjera rheedii *Gmel.*	28.0	13	0.018	5.88	3.87	0.48	10.23
Syzygium sp.	20.0	7	0.031	4.20	2.09	0.82	7.11
Garcinia xanthochymus Hook. ex Anders.	4.0	2	0.117	0.84	0.60	3.11	4.55
Ixora brachiata Roxb.*	12.0	6	0.009	2.52	1.79	0.24	4.55
Mimusops elengi L.	12.0	5	0.009	2.52	1.49	0.24	4.25
Sepium insigne Benth. And Hook.	12.0	3	0.004	2.52	0.89	0.11	3.52
Mangifera indica L.	8.0	2	0.004	1.68	0.60	0.11	2.39
Ficus sp.	4.0	2	0.021	0.84	0.60	0.56	2.00
Calophyllum apetalum Willd. *	4.0	1	0.005	0.84	0.30	0.13	1.27
Artocarpus hirsutus L.*	4.0	1	0.004	0.84	0.30	0.11	1.25
Leea indica L.	4.0	1	0.001	0.84	0.30	0.03	1.17
Ficus benghalensis L.	4.0	1	0.001	0.84	0.30	0.03	1.17
Santalum album L.	4.0	1	0.001	0.84	0.30	0.03	1.17
Lantana camara L.	4.0	1	0.001	0.84	0.30	0.03	1.17
Unidentified	12.0	4	0.006	2.52	1.19	0.16	3.87
Dead	12.0	3	0.033	2.52	0.89	0.88	4.29

*Endemic species.

GBH class of 10–40 cm and nearly 12% come under the class 40–80 cm. This also suggests that the forest is regenerating. The indices of diversity are given in Table 2. The Simpson's index is 0.73 and low compared to those of Pilarkan reserve forest and Jadkal forest (Shivaprasad *et al.*, 2001; Vasanth Raj *et al.*, 2001). *N/S* ratio, Shannon-Wiener's index, and the equitability ratio were all low, indicating a moderate diversity in this area.

About 20% of the seedlings are represented by *H. ponga* followed by *Ixora coccinea* L. (13%), *Holigarna ferruginea* Marchand (8%), *Pothos scandens* L. (6.6%), *Psychotria dalzelli* Hook. f. (6%), etc. *H. parviflora* is represented by only 2% and has low regeneration similar to that in the Pilarkan reserve forest (Shivaprasad *et al.*, 2001).

The high density of Dipterocarps species (about 60%) in this grove indicates the dominance of this species. This is again proved by its high stand density. Although the density of *H. parviflora* is very low, the stand density is comparatively higher than any other species because of the presence of stout boles occupying larger area. However, the regeneration of this species is less as seen in Pilarkan reserve forest (Muralikrishna and Chandrashekar, 1997;

Table 2. Diversity indices of a sacred grove of Nandikoor.

Area (m²)	Number of individuals = N (G ≥ 10 cm)	Number of species = S (G ≥ 10 cm)	N/S	Shannon-Wiener's Index			
				Simpson's index	H′	H_{max}	E = H′ H_{max}
1000	335	23	14.57	0.73	3.02	4.52	0.71

Shivaprasad *et al.*, 2001). The less number of individuals belonging to the set of the past indicates less disturbance in the forest. The presence of more number of individuals of *H. ponga* belonging to the set of future indicates a very good regeneration in the recent past.

The sacred groves are regarded as shelters for biodiversity (Meher-Homji, 1997). Our results indicate that the diversity in this forest is moderately high as observed for most of the sacred groves. About 8 species (34%) which are endemic to the Western Ghats are present in this grove, indicating that the endemism is fairly high. Out of 335 individuals, 256 (about 76%) belong to eight endemic species. In other words, the sacred grove merits high conservation value, given that it stands amidst large scale perturbation in the adjoining landscape.

Acknowledgements

The financial support given by University Grants Commission in the form of a major research project is gratefully acknowledged. The help provided by our colleague Ramakrishna Marati, is also gratefully acknowledged.

References

Gamble, J. S. *Flora of the Presidency of Madras*. III Vols. 1027 (1921–1935).

Halle, F., Oldeman, R. A. A. & Tomlinson, P. B. *Tropical Trees & Forests: An Architectural Analysis* (Springer-Verlag, New York, 1978).

Meher-Homji, V. M. Biodiversity conservation priorities. *Indian J For.* **20**, 1–7 (1997).

Muralikrishna, M. & Chandrashekar, K. R. Regeneration of *Hopea ponga*: Influence of wing loading and viability of seeds. *J. Tropical Forest Sci.* **10**, 58–65 (1997).

Pascal, J. .P. *Wet Evergreen Forests of the Western Ghats of India – Ecology, Structure, Floristic Composition and Succession*, p. 345 (French Institute, Pondicherry, 1988).

Pascal, J. P. & Ramesh, B. R. A Field Key to the Trees and Lianas of the Evergreen Forests of the Western Ghats (India) p 236 (French Institute, Pondicherry, 1987).

Shivaprasad, P. V., Vasanthraj, B. K. & Chandrashekar, K. R. Studies on the structure of Pilarkan reserve forest, Udupi District of Karnataka. *J. Tropical Forest Sci.* (2001).

Tropical Ecosystems: Structure, Diversity and Human Welfare.
Proceedings of the International Conference on Tropical Ecosystems
K. N. Ganeshaiah, R. Uma Shaanker and K. S. Bawa (eds)
Published by Oxford–IBH, New Delhi. 2001. pp. 603–607.

Sapling allometry related to shade tolerance and crown exposure in two dipterocarp species in Sabah, Malaysia

Martin Barker*†, Michelle Pinard** and Reuben Nilus‡

*School of Forest Resources and Conservation, University of Florida, PO Box
110410, Gainesville, FL 32611-0410, USA
**Department of Agriculture and Forestry, University of Aberdeen,
Aberdeen AB24 5UA, UK
‡Forest Research Centre, Sepilok, Sabah, Malaysia
†e-mail: mbarker@aris.sfrc.ufl.edu

Tree allometry influences forest structure and function, and in turn is strongly affected by forest structure. Biomass allocation patterns within trees are determined by both environmental and developmental constraints. For example, differences in allometry among coexisting saplings may be explained by spatial and temporal variations in the light environment, differences in potential size of adult trees within a given species, regeneration niche, and tree lifespan (Kohyama, 1991; King, 1996).

In this study, our aim was to compare sapling allometry in two, late-successional species in the Dipterocarpaceae which differ in shade tolerance. Specifically, our objective was to examine relationships between allometric variables and crown exposure, to explore how the two species respond to different light environments through changes in allometry. Our basic precept was that species that are more shade-tolerant are likely to allocate more

Keywords: Tree allometry, Dipterocarpaceae, *Shorea macroptera*, crown exposure, Malaysia.

resources to crown development than the less shade-tolerant species, which probably devote more resources to height growth.

The study was conducted in dipterocarp forest on sandstone ridges in Kabili-Sepilok Forest Reserve in eastern Sabah. The climate is aseasonal and tropical with mean annual rainfall of 3150 mm/year. Soils on the ridges and spurs are red yellow podzols derived from sandstones. *Shorea beccariana*, *Shorea multiflora* and *Dipterocarpus acutangulus* dominate the canopy of the forest on the sandstone ridges in the reserve, with emergent trees up to 40 m in height.

Two focal species were selected from the range (c. 20) of dipterocarp species occurring in the plots. We used, respectively, the highest and lowest mean wood densities as a surrogate of growth rate and shade tolerance. The selected species were *Shorea macroptera* Dyer (air dry density 0.48–0.55 g cm^{-3}) and *Vatica micrantha* van Slooten (0.83–0.93 g cm^{-3}). *S. macroptera* grows to be a large tree (3 m dbh, 60 m tall) and is considered to be relatively fast growing, whereas *V. micrantha* is a small tree (up to 27 m tall). The species were also chosen on the basis of similar leaf sizes. Both species were plagiotropic in architecture, with relatively small leaves held in generally flat (planar) arrays by mainly horizontally-oriented branches.

Ten saplings (5.0–9.9 cm dbh) each of *S. macroptera* and *V. micrantha* were selected in a restricted random fashion within existing permanent plots, with a minimum distance of 20 m between conspecifics. Only saplings with intact, unobscured crowns were used. For each sapling we measured dbh, total height, height to lowest branch, height to bottom of crown, crown width, and leaf number. We also determined diameter at 10% of stem height.

For measurements within tree crowns, access was by a climbing ladder; six saplings were measured per species. Counts were made of the relative numbers of primary (i.e. proximal, or lowest order) branches nearest the trunk, and secondary and tertiary (i.e. distal, higher-order) branches, moving away from the trunk. Primary and secondary branch counts were used to calculate bifurcation ratios (sensu Veres and Pickett, 1982) for each tree. Total numbers of leaves within each crown were counted, and six 'representative' leaves were collected for leaf area measurements. Total crown leaf area was estimated and then used to calculate leaf area index (LAI) of each tree.

The more shade-tolerant species, *V. micrantha*, had deeper crowns than *S. macroptera*, but crown widths (and crown areas) and height to lowest branch were similar (Table 1). For *V. micrantha*, the crown occupied, on average, 44% of total height, whereas for *S. macroptera*, the crown occupied only 23% of total height. Mean crown volume for *V. micrantha* was more than twice that for *S. macroptera*. Mean numbers of primary, secondary and tertiary branches were higher in *V. micrantha* saplings than in *S. macroptera*. LAI (and mean number of leaves) was higher in *V. micrantha* than in *S. macroptera* saplings.

Bifurcation ratio and LAI were marginally different for the two species; however, small sample size limits the power of this analysis to reject the null hypothesis of no difference.

For saplings of *V. micrantha*, as the level of crown exposure increases, crown width and crown depth to tree height ratios decrease. The associations between allometric variables and crown exposure for *S. macroptera* were not significant.

The allometric differences observed between *V. micrantha* and *S. macroptera* were primarily related to crown depth and leaf packing. The differences may be functional, and related to maximizing carbon gain, but may also relate to differences in developmental stage. Although similar sized saplings were compared in the study, *V. micranthra* saplings were approaching sizes typical of mature trees, whereas *S. macroptera* saplings were still small relative to mature trees.

The similarity in lowest branches between *V. micrantha* and *S. macroptera* in this study may at least partly explain similarities in crown width and area

Table 1.   Attributes of trees sampled for two species. Mean values (SD, N) are presented, except for crown illumination index where median (min, max, N) is presented. T-tests were used to compare species for most response variable, however, non-parametric Mann–Whitney U tests were used for comparisons involving bifurcation ratio and leaf area index due to skewed data.

	Shorea macroptera	*Vatica micrantha*	Level of significance
dbh (cm)	7.4 (1.2, 10)	7.0 (1.3, 10)	NS
Diameter at 10% stem height (cm)	7.4 (1.0, 10)	7.3 (1.2, 9)	NS
Total height (m)	13.7 (5.4, 10)	13.8 (3.8, 10)	NS
Crown depth (m)	2.9 (0.9, 10)	5.8 (1.2, 10)	***
Height to first branch (m)	10.8 (5.3, 10)	7.9 (3.5, 10)	NS
Crown illumination index	2.5 (1.0-4.0, 10)	2.0 (1.0-3.0, 10)	NS
Crown depth/crown height (m m^{-1})	0.23 (0.09, 10)	0.44 (0.13, 10)	***
Crown area (m^2)	14.0 (5.1, 10)	15.1 (3.5, 10)	NS
Crown volume (m^3)	44.0 (27.3, 10)	90.2 (32.1, 10)	***
Mean leaf area (cm^2)	157 (33, 6)	85 (14, 6)	***
Number of leaves	533.2 (248.1, 6)	2174.1 (1172.1, 6)	***
Leaf area index[a]	0.6 (0.3, 6)	1.2 (0.7, 6)	*
Number of primary branches	18 (6.8, 6)	47 (8.6, 6)	***
Number of secondary branches	73 (35, 6)	426 (160, 6)	***
Number of tertiary branches	45 (21, 6)	238 (175, 6)	**
Bifurcation ratio[b]	1.16 (0.10, 6)	1.07 (0.05, 6)	0.06

[a]Leaf area index was calculated as total leaf area divided by crown area. Total leaves were counted and mean leaf area was estimated from a sample of 10 leaves per tree.
[b]Bifurcation ratio (R_b) was calculated as N_u/N_{u+1}, where N_u is the number of branches in a given branch order, N_{u+1} is the number of branches of the next branch order in the hierarchy; lower R_b values indicate more branching.
***P-value < 0.001; **P-value between 0.01 and 0.001; *P-value between 0.05 and 0.01, NS, P-value > 0.08.

between the species. As King (1998) points out, the height to the lowest (i.e. 'first') branch is an important attribute because it marks the point at which prolonged lateral growth is possible. Instead, the more shade tolerant, slower growing species achieved a higher crown volume by developing, or at least maintaining, a deeper crown than *S. macroptera*. It is possible that the primarily vertical differences in crown structure between the two species reflect an intense competition among saplings (Kohyama and Hotta, 1990). There may be adaptive advantages to *V. micrantha* saplings developing a narrow rather than broad crown shape. For example, a narrower crown might confer greater protection from fallen branches, especially if the sapling has a prolonged presence in the understorey (see King, 1990).

The deeper crowns of *V. micrantha* saplings also incorporate more branches and leaves and, despite having generally smaller leaves, *V. micrantha* achieves greater LAI than *S. macroptera*. The slightly lower bifurcation ratio (R_b) values in *V. micranta* saplings confirm more branching in that species. If more branching allowed lateral expansion (King, 1998), greater crown widths might have been expected in *V. micrantha* compared with *S. macroptera* though we did not observe this, possibly because *V. micrantha* sapling instead utilize a strategy of vertical expansion.

A study of saplings in Costa Rica (Oberbauer *et al.*, 1993) showed that mean leaf area tends to be larger in saplings of more shade-tolerant species. We observed the opposite relationship, though our results are consistent with Corner's rule, that number of apices, or degree of branching (see above), is inversely related to leaf size (White, 1983). The number of leaves was greater in the more shade-tolerant species (*V. micrantha*). The greater packing of leaves in *V. micrantha* explains the higher LAI in that species, given the similarity of crown area between the two species.

Higher LAI implies a greater proportion of overlapping and, hence, self-shading in the more light-tolerant species (*V. micrantha*), as expected from modeling (Kohyama, 1991). Our results show that, at a given time (and for similar-sized saplings) *V. micrantha* allocates more resources to leaf production than the less shade-tolerant *S. macroptera*.

The species appear to differ in their response to understorey light environments, with *V. micrantha* saplings showing greater plasticity in the development of crown traits than the less shade-tolerant *S. macroptera* saplings. But, given that it may take several years for changes in crown allometry related to a change in light environment to develop (Sterck *et al.*, 1999), and that we were unable to find saplings growing under a wide range of light conditions, our results must be taken with caution.

References

King, D. A. Allometry of saplings and understorey trees of a Panamanian forest. *Funct. Ecol.* **4,** 27–32 (1990).

King, D. A. Allometry and life history of tropical trees. *J. Trop. Ecol.* **12,** 25–44 (1996).

King, D. A. Influence of leaf size on tree architecture: first branch height and crown dimensions in tropical rain forest trees. *Trees* **12,** 438–445 (1998).

Kohyama, T. A functional model describing sapling growth under a tropical forest canopy. *Funct. Ecol.* **5,** 83–90 (1991).

Kohyama, T. & Hotta, M. Significance of allometry in tropical saplings. *Funct. Ecol.* **4,** 515–521 (1990).

Oberbauer, S. F., Clark, D. B., Clark, D. A., Rich, P. M. & Vega, G. Light environment, gas exchange, and annual growth of saplings of three species of rain forest trees in Costa Rica. *J. Trop. Ecol.* **9,** 511–523 (1993).

Sterck, F. J., Clark, D. B., Clark, D. A. & Bongers, F. Light fluctuations, crown traits, and response delays for tree saplings in a Costa Rican lowland rain forest. *J. Trop. Ecol.* **5,** 83–95 (1999).

White, P. S. Corner's rules in eastern deciduous trees: allometry and its implications for the adaptive architecture of trees. *Bull. Torrey Bot. Club* **110,** 203–212 (1983).

Eastern Himalayan Biodiversity: Distribution, Human Impacts and Conservation Strategies

Tropical Ecosystems: Structure, Diversity and Human Welfare.
Proceedings of the International Conference on Tropical Ecosystems
K. N. Ganeshaiah, R. Uma Shaanker and K. S. Bawa (eds)
Published by Oxford–IBH, New Delhi. 2001. pp. 611–614.

Biodiversity status and conservation initiatives in Kangchenjunga conservation area in Nepal

Ghana Shyam Gurung* and Janita Gurung

WWF Nepal Program, P.O. Box 7660, Kathmandu, Nepal
*e-mail: ghana@wwfnepal.org.np

The Kangchenjunga Conservation Area (KCA) is situated in the northern part of Taplejung district in northeastern Nepal. With an area of 2,035 km^2, the KCA covers approximately 60% of the total land area of Taplejung district and includes the four Village Development Committees (VDCs) of Lelep, Tapethok, Walangchung Gola and Yamphudin. The population of the KCA is 4874 (of which females comprise 48%) distributed among 884 households (Dhakal, 1996). The average literacy rate among KCA residents is 38%, but female literacy rate (23%) is less than half the male literacy rate (48%). Farming is the major occupation of KCA residents, but more than half the population face food-scarcity annually.

The Kangchenjunga Conservation Area is situated in a strategic location in the Eastern Himalaya Ecoregion; it is bordered by Sikkim-India in the east and the Tibet Autonomous Region (TAR) of China in the north. Three ecoregions are represented in the KCA: (1) Eastern Himalayan Broadleaf forests, (2) Eastern Himalayan Subalpine Conifer forest, and (3) Eastern Himalayan Alpine Shrub/Meadows (Wikramanayake *et al.*, 1998).

Keywords: Kangchenjunga, community-based conservation, sustainable development.

A wealth of natural and cultural resources exists in the KCA. While the physiography of the area is dominated by rocks and ice (65%), forests, shrubs and grasses constitute 14%, 10% and 9% of the area, respectively (Amatya *et al.*, 1995). Altitude in the KCA varies from a low of 1,200 m to 8,586 m – the height of Mt. Kangchenjunga, the third highest peak in the world. The area hosts the only extensive pure stands of Himalayan larch (*Larix griffithiana*) in Nepal, and harbors 15 of the country's 28 endemic flowering plants as well as 24 of Nepal's 30 known rhododendron species (Shrestha, 1994). Among the numerous faunal species, the area is home to the endangered snow leopard (*Uncia uncia*), red panda (*Ailrus fulgens*) and musk deer (*Moschus chrysogaster*) (Amatya *et al.*, 1995). Numerous ethnic groups, including the Limbu, Bhotia/Sherpa/Lama, Gurung, Rai and occupational castes such as Kami, Damai and Sarki, along with their traditions and ways of life, add to the cultural wealth of the KCA (Uprety, 1994). In recognition of the rich natural and cultural resources, and in support of WWF's Living Planet Campaign, His Majesty's Government of Nepal (HMG/Nepal) declared the Kangchenjunga area a 'Gift to the Earth' on 29 April 1997. The area was conferred protected area status on 21 July 1997.

The natural diversity of the KCA faces numerous threats from the local populace (Amatya *et al.*, 1995). Land encroachment for residential/tea house construction, farming and grazing purposes threaten floral and faunal habitat. Residents are also highly dependent on forests for fuelwood, timber, fodder and forest litter. Poaching of animals and plants, as well as unsustainable harvesting of non-timber forest products, particularly of lokta (*Daphne papyracea* and *D. bholua*) and chiraito (*Swertia chiraita*), also threatens the KCA's natural resources.

Some key causes of biodiversity threats in the KCA are poverty, lack of conservation awareness, limited infrastructure, difficulty in enforcing regulations and spontaneous tourism development. With more than half the population under poverty level, and the relative paucity of alternative income-generation and energy sources, there is added pressure on the area's natural resources. Furthermore, the remoteness of the KCA has resulted in minimal infrastructure development. Collectively, these factors threaten the viability of the very resource base that resulted in the KCA's recognition as a 'Gift to the Earth'.

To address the environmental and socio-economic issues in the KCA, the Kangchenjunga Conservation Area Project (KCAP) was launched on 22 March 1998. The Project is a joint undertaking of HMG/Nepal's Department of National Parks and Wildlife Conservation (DNPWC) and WWF Nepal Program. The primary goal of the project, which implements community-based conservation and development programs, is to conserve the biodiversity of the KCA by integrating natural resource conservation with sustainable development. This is achieved by strengthening the capacity of local communities, community-based organizations, and government personnel to manage the area's natural resources, and by enabling the local populace to improve their socio-economic conditions. Since its inception, the project has focused on program activities including nature and culture conservation, ecotourism, community services, and capacity-building of local women and men. Among these, ecotourism will be significant in achieving poverty reduction in the KCA.

The achievements of the project to date include establishing a Project headquarter office at Lelep and sector offices at Ghunsa, Walangchung Gola, and Yamphudin, all equipped with wireless and Motorola communication systems; extension of the Conservation Area boundary in 14 September 1998, from 1,650 km^2 to 2,035 km^2; and gazetting of the Conservation Area Regulations 2057 by HMG/Nepal. Formation and institutionalization of local user groups and institutions include seven Conservation Area User Committees, 30 Mother Groups, and three School Eco Clubs. The Project has also exposed 86 KCA residents to other protected areas in Nepal, provided kitchen gardening and other skill development training to 354 residents, distributed stipends to 18 girl students, and provided basic literacy to 185 residents (WWF Nepal Program 2001).

The Project recognizes that poverty reduction and trans-boundary initiatives are crucial to achieving the goal of biodiversity conservation. Poverty reduction is proposed through three major activities – production and harvesting of non-timber forest products, agro-forestry promotion, and integrated ecotourism development. The recently produced Tourism Plan for the KCA (Schellhorn and Simmons, 2000) recommends 'staged' tourism development in the area for maximizing 'low volume, high yield' quality ecotourism ideals. This is achieved by development of a unique product mix, supply quality products and services, impact management, regional liaison and cooperation development, and strengthening management capacity of community-based organizations. This tourism development approach is intended to address the needs of the communities for improving their quality of life, while concurrently achieving the Project's objectives of natural resource conservation – the very product on which tourism is based. Along with District Development Committee Taplejung, the Asian Development Bank has shown interest for investment under the Ecotourism Project to materialize the plan. In fact, Nepal has demonstrated that it is possible to establish a symbiotic relationship between tourism and nature for long-term biodiversity conservation and sustainable development.

Furthermore, regional trans-boundary initiatives with Sikkim-India and TAR-China are crucial for maintaining continuous biodiversity habitat landscapes. A biodiversity assessment and gap analysis of the Himalayas (Wikramanayake *et al.*, 1998) found that the Kangchenjunga complex deserves priority for the protection of Himalayan broadleaf forests and populations of *Ovis ammon* (argali), *Equus kiang* (kiang) and *Procapra picticaudata* (Tibetan gazelle). Participatory biodiversity conservation among the three countries is required to achieve this goal of trans-boundary conservation in the Kangchenjunga complex.

References

Amatya, D. B., Brown, T., Sherpa, L. N., Shrestha, K. K. & Uprety, L. P. Feasibility Study for the proposed Kanchenjunga Conservation Area. WWF Nepal Program Report Series #21 (1995).

Dhakal, N. H. Socio-economic Survey and Integrated Conservation and Development Plan of the Proposed Kanchenjunga Conservation Area. WWF Nepal Program Report Series #24 (1996).

Schellhorn, M. P. & Simmons, D. R. (eds). Kangchenjunga Conservation Area Tourism Plan 2001–2006. Prepared for His Majesty's Government of Nepal, Ministry of Forests and Soil Conservation, and Department of National Parks and Wildlife Conservation by Lincoln International (1995) Ltd., New Zealand (2000).

Shrestha, K. K. Floristic Diversity, Vegetation and Ethnobotany of the Proposed Kanchenjunga Conservation Area. WWF Nepal program Report Series #6 (1994).

Wikramanayake, E. D., Dinerstein, E., Allnutt, T., Loucks, C. & Wettengel. W. A Biodiversity Assessment and Gap Analysis of the Himalayas. Conservation Science program, World Wildlife Fund-US (1998).

WWF Nepal Program. Kangchenjunga Conservation Area Project: Half-Yearly Technical Progress Report July 01-December 31, 2000 (2001).

Tropical Ecosystems: Structure, Diversity and Human Welfare.
Proceedings of the International Conference on Tropical Ecosystems
K. N. Ganeshaiah, R. Uma Shaanker and K. S. Bawa (eds)
Published by Oxford–IBH, New Delhi. 2001. pp. 615–618.

Eastern Himalayan biodiversity: Importance, threats and conservation initiatives

*Eklabya Sharma**[*†]* and R. C. Sundriyal***

*G. B. Pant Institute of Himalayan Environment and Development,
Sikkim Unit, P.O. Tadong, Gangtok 737 102, India
**G. B. Pant Institute of Himalayan Environment and Development,
NE-Unit, Vivek Vihar, Itanagar 791 113, India
[†]e-mail: gbp.sk@sikkim.org

Indian subcontinent is one of the 12 mega-biodiversity centers of the world. It has two biodiversity hot-spots namely the Eastern Himalayas and the Western Ghats. The Eastern Himalaya constitutes parts of eastern Nepal, India, Bhutan and China. Northeast India comprising of the Brahmaputra valley and Assam hills of India, Chittagong hills of Bangladesh and northern parts of Myanmar are also included in the Eastern Himalayan Complex. This complex is influenced by elements from Palaeartic, Mediterranean, Sino-Japanese, Indo-Malayan and Peninsular Indian biogeographical realms that have given rise to a very rich biodiversity (MacKinnon and MacKinnon, 1986). The complex has the largest number of endemics and endangered species (Khoshoo, 1992). Three major features such as longitude (east to west), latitude (south to north) and altitude (low to high) contribute to a great variety of microclimates in the complex.

Keywords: Biosphere reserves, ecoregion, endangered species, endemism, traditional practices.

The Eastern Himalayan Complex is more extensively spread in India having 2,43,001 km² area and of which 38% is covered by dense forests (> 40% crown cover). There are five vegetation types, viz. tropical evergreen forests, subtropical forests, temperate forests, alpine vegetation and cold deserts. Natural swamps and grasslands in the Brahmaputra valley provide habitats for large herbivores typical of alluvial grasslands such as rhinoceros, buffalo, swamp deer, hog deer, pygmy hog and hespid hare (Rodgers and Panwar, 1988). In the region, 136 native plant species and 14 mammals are considered threatened as reported in the IUCN Red Data Book. Some of the critically endangered plant taxa are *Aconitum ferox*, *Aphyllorhis parviflora*, *Bulleya yunnaensis*, *Calanthe alpina*, etc. (Maiti and Chauhan, 2000). *Coelogyne treutleri* is reported to be extinct. Some of the endangered mammals are snow leopard, red panda, musk deer, marbled cat, Himalayan yellow-throated marten, Himalayan thar, great Tibetan sheep and clouded leopard, and among the birds are forest eagle owl, Himalayan golden eagle, monal pheasant, lammergeier, sparrow hawk and Tibetan snowcock (Chettri, 2000).

This complex is very rich in biodiversity and harbours the largest number of endemics and Schedule I species than anywhere else in India (MacKinnon and Mackinnon, 1986). Micro-climatic variations of the complex and high species richness suggest high degree of evolutionary activity resulting in pockets of high degree of endemism (Khoshoo, 1992). Broadly, endemics are either relict as last remnants of old taxa whose distribution has shrunk or those of recent origin which have not extended their range; in the Eastern Himalayan Complex, the latter type of endemism is more prevalent. The complex is ecologically and phytogeographically highly diversified. It is the ground for high evolutionary activity which is clear from the cytogeographic studies on rhododendrons (JanakiAmmal, 1950). In the rhododendron, diploids have a very wide range followed by progressively smaller geographic ranges of tetraploids, hexaploids, octoploids and do-decaploids (Figure 1; Source: JanakiAmmal, 1950). It is clear from this study that higher ploidy level species have narrower geographic range and they are essentially endemics confined to narrow region.

Primary direct threats to the unique biodiversity of the Eastern Himalayan Complex are land use transformation, habitat degradation, forest-fires and landscape fragmentation. Activities like grazing, unregulated tourism, development projects, land use conversion, unsustainable harvest of biodiversity products and poaching in some instances are the root causes of degradation and loss of biodiversity. Breaking down of some excellent traditional practices has also aggravated the situation. Deferred grazing in Jhumsa system in the alpine areas of Sikkim and Jhum cultivation in Northeast India are some good examples of traditional practices that have helped biodiversity conservation. Jhumsas are slowly overtaken by newer systems of administration and in the case of Jhum cultivation the cycles have shortened, rendering them unsustainable. The Eastern Himalayan Complex

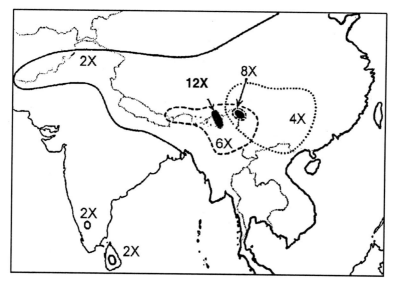

Figure 1. Distribution of polyploids of rhododendrons in the Eastern Himalayan complex.

comprises many countries where large tracts of habitats are continuous across the boundaries. Transboundary issues of high conservation concern prevail in the region and this has been one of the factors attributable to biodiversity threats and loss.

Because of the extremely high biodiversity richness of the area, the Eastern Himalayan Complex has been identified as a priority area of biodiversity conservation in the National Policy of India. National and regional institutions and NGOs are involved in various biodiversity programmes in the region. State Governments are encouraged to bring more areas under protected area network in the form of national parks and wildlife sanctuaries. At the initiatives of UNESCO's Man and Biosphere Programme, the biosphere reserve concept was initiated by the Ministry of Environment and Forests, Government of India in which Nokrek in Meghalaya, Manas and Dibru-Saikhowa in Assam, Dehang-Debang in Arunachal Pradesh and Khangchendzonga in Sikkim are managed as biosphere reserves. Biodiversity Institute for addressing issues on conservation is being established in Arunachal Pradesh by the Ministry of Environment and Forests, Government of India. Regional consultation on conservation of the Khangchendzonga Mountain Ecosystem has been suggested (Rastogi *et al.*, 1997). Several programmes are being conceived on regional collaboration for biodiversity conservation. Participatory biodiversity conservation involving joint forest management institutions is slowly gaining ground in the region.

References

Chettri, N. Impact of habitat disturbances on bird and butterfly communities along Yuksam-Dzongri Trekking Trail in Khangchendzonga Biosphere Reserve. Ph.D. thesis, North Bengal University, India (2000).

JanakiAmmal, E. K. Polyploidy in the genus *Rhododendron*. *Rhododendron Year Book*, pp. 92–96 (1950).

Khoshoo, T. N. Plant diversity in the Himalaya: Conservation and utilization. Pandit Gobind Ballabh Pant Memorial Lecture II. G.B. Pant Institute of Himalayan Environment and Development, Kosi–Katarmal, Almora, India, pp. 129 (1992).

MacKinnon, J. & MacKinnon, K. Review of the Protected Areas System in the Indo-Malayan Realm. IUCN, Gland (1986).

Maiti, A. & Chauhan, A. S. Threatened plants in the Sikkim Himalaya. *Himalayan Paryavaran* 113–120 (2000).

Rastogi, A., Shengji, P. & Amatya, D. Regional Consultation on Conservation of the Kanchanjunga Mountain Ecosystem. International Centre for Integrated Mountain Development, Kathmandu, Nepal (1997).

Tropical Ecosystems: Structure, Diversity and Human Welfare.
Proceedings of the International Conference on Tropical Ecosystems
K. N. Ganeshaiah, R. Uma Shaanker and K. S. Bawa (eds)
Published by Oxford–IBH, New Delhi. 2001. pp. 619–621.

Agrobiodiversity: Need for conservation in Northeast India

D. K. Hore

National Bureau of Plant Genetic Resources, Regional Station,
Barapani 793 103, India
e-mail: navinkverma@123india.com

The Northeastern region of India occupies 5.6% of total geographical area of India. Politically, the region is located in a strategic position and is comprised of eight states, which are bounded by China and Tibet in the North, Myanmar in the East and Bangladesh in the South. Demographically, more than 120 diverse ethnic groups, who are maintaining their own cultural heritage, inhabit this region. Physiographically and climatically the region encompasses wide variation in altitude, climate, soil and rainfall. The cropping system, cropping pattern, cultivation and storage practice differ from state to state. Nearly 70% of land area is cultivated at one time or the other each year.

Floristically, the region accounts for 43% of Indian plant species with about 48% of them being endemic. The region has been recognized as primary/secondary center of origin/diversity or center for regional diversity for many cultivated crops. Within the Indian gene center, the region represents:

- A primary center of diversity for rice, minor millets, jobs tear, tree cotton, banana, cucurbits and canes.

Keywords: Northeast India, agrobiodiversity, novelty species, conservation.

- A secondary center of diversity for crops of new world such as maize, chillies, tea, arecanut.
- Regional (Asiatic) diversity for crops like buckwheat, foxtail millet, finger millet, rice bean, *Cucumis, Benincasa, Solanum, Citrus, Sachharum, Colocasia, Curcuma, Piper, Dioscorea* and Bamboo.

The region represents at least 138 herbaceous and 59 tree species of medicinal and aromatic importance, which are used to cure various diseases.

The NBPGR Regional Station, Shillong, since its establishment in 1978 has initiated activities to assess the crop genetic resources in the region. The station has undertaken about 70 exploration trips, in 58 districts of the region, in order to collect the genetic resource. Table 1 indicates the richness of various crop diversities.

A number of unique and important plant species occur in this region. A few of them are: *Gossypium arboreum, Rynchostylis retusa, Schumannianthus dichotomus, Nepenthes khasiana, Citrus indica, Atylosia elongata, Moghania vestita, Ensete superba, Mangifera sylvatica, Parkia roxburghii I, Heliconia dasyantha, Zizania latifolia, Mangifera* sp. (Dwarf mango), *Lilium mackliniae, Coptis teeta, Musa velutina, Psidium guinensis* and *Antherium* species.

Among the cultivars, cereals (rice, maize), pseudocereals (buckwheat) and millets (finger millet, foxtail millet, pearl millet) are given top priority for cultivation. Daily staple food and drinks are prepared from these crops. The conscious selections of rice landraces are preferred for brewing, glutinous characteristics, colored rice (i.e. black kerneled rice of Manipur), soft cooking quality, aroma, etc. Buckwheat is used as staple food in high hills (Tawang District) of Arunachal Pradesh, where because of low temperature there is very poor seed setting in rice. Popping quality of maize, soft-shelled jobs tears and 'Raisan' (*Digitaria cruciata* var. *esculenta*) are also important and preferred crops, which cater the need of local tribes. The region is also known

Table 1. Estimated diversities of major crops in NE region of India.

Crop(s)	Estimated diversities	Diversities collected till year 2000
Rice	9650 +	4300
Maize	15 races and 3 subraces – 1200 +	760
Taros	300	272
Yams	230	200
Citrus	17 spp. + 52 vars	80
Banana	16 taxa	120
Orchids	700 taxa	15 + NRC Orchids
Sugarcane	19 taxa	*SBI, Coimbatore
Bamboo	78 taxa	*ICAR Complex, Basar

for its horticultural crop diversities, particularly banana, *Citrus*, cucurbits, brinjal, tea, sugarcane, bamboo and orchids.

Under the *in situ* conservation program, about 25,400 km^2 of forest has been declared as reserve forest area and includes Biosphere reserve, National Park, Sanctuary and protected areas.

With regard to *in situ* conservation of plant genetic resources, there is hardly any attempt in this region except the case of *Citrus* gene sanctuary of Garo Hills and *Nepenthes khasiana* habitat of Jerrain, Jaintia hills. 'On farm conservation' is being practised in remote areas of this region. 'Khamatis' and 'Apatanis' of Arunachal Pradesh are quite conversant with the preservation of their local rice landraces. Similarly, Mizos of Mizoram are fully aware to cultivate the ecosystem-based and adapted local rice cultivars. Suitable genetic stock of scented rice landraces are cultivated in few pockets in Assam. It is argued that emphasis is to be given on 'cultivation of wild forms' (domestication) or 'as is where is' basis for such landraces where the specific agrohabitats and microclimates or niche are already existing.

Ex-situ conservation is gradually gaining momentum in the region. Various extension programmes, NGOs activities, initiatives of government organizations have encouraged the public at all levels for tree planting, creation of parks, botanic gardens, etc. Plant genetic resources conservation activities have been initiated through the establishment of two gene banks in the region. The Botanical Survey of India, the NBPGR, Shillong and the Forest Department of the respective states of the region have established their respective field gene banks for various plant species, cultivars and cultigens.

Tropical Ecosystems: Structure, Diversity and Human Welfare.
Proceedings of the International Conference on Tropical Ecosystems
K. N. Ganeshaiah, R. Uma Shaanker and K. S. Bawa (eds)
Published by Oxford–IBH, New Delhi. 2001. pp. 622–626.

Extended biodiversity 'hotspot' analysis: A case of eastern Himalayan region, India

Ajay Rastogi and Nakul Chettri

Ashoka Trust for Research in Ecology and the Environment (ATREE), Eastern
Himalaya Programme, Bungalow #2, Bhujiapani,P.O.: Bagdogra,
Darjeeling 734 422, India
e-mail: atree@dte.vsnl.net.in.

Various approaches for biodiversity assessment, conservation planning and designing protected area network (PAN) are being developed, e.g. rapid biodiversity assessment approach (Oliver and Beattie, 1996); an indicator and surrogate species approach (Kremen, 1994); rarity and complementary set approach (Scott and Csutti, 1996); key ecoregion approach (Olson and Dinerstein, 1998) and richness, endemic and threat 'hotspot' approach (Myers, 1990; Bibby, 1992; Myers et al., 2000). Eastern Himalaya figures in most of these assessments. Applying largely the same principles, it is important to identify high priority zones within a region for an effective followup.

Eastern Himalaya forms a part of the Indo-Burma 'hotspot' region (Myers et. al., 2000). Within Indian borders, the region stretches along extremely rugged mountains between 26°30' to 28° north latitude and 87° to 97°30' east longitude covering the Singalila ridge and the Khanchendzonga massif in the west to the Patkai hills in the south-east. Darjeeling and Sikkim Himalayas are contiguous, separated from Arunachal Pradesh by Bhutan. The area of

Keywords: Biodiversity hotspot, Eastern Himalaya, protected areas, threatened and endemic species, India.

Darjeeling district is 3264 km² with 47% forest cover and 11% of geographical area under PAN. The same for the state of Sikkim is 7096 km²; 43% total forest cover and 39% under PAN; and that for Arunachal is 83,743 km²; 62% forest cover and 12% geographical area under PAN.

Significant amount of forest cover and substantive area under coverage of PAN gives a false sense of security. The delineation of protected areas in the past had less scientific rigour. Therefore, an examination of the effectiveness of the existing PAN in conserving a range of species was undertaken in different altitude zones. The distribution of endemic and rare, threatened and endangered species amongst mammals, birds, and certain plant groups (rhododendrons, orchids and other angiosperms) is better documented in the region. An extensive literature search was carried out to compile data on status, location and distribution of species and protected area coverage in various altitude zones (Rodgers and Panwar, 1988; Ali, 1989; BSI, 1990; BSI, 1996; WCMC, 1997; Khaling et al., 2000; Myint et al., 2000; Pradhan and Bhujel, 2000).

The species distribution pattern belonging to rare, threatened, endangered and endemic fauna and flora in each of the three units is provided in Table 1.

Table 1. Distribution of threatened plants and animals in different ecological zones of eastern Himalaya.

| Region | Places | Threatened animals | | Threatened and endemic plants | | | | |
| | | | | | Orchids | | Other plants | |
		Mammals	Birds	Rhodo-dendron	En	T	En	T
Tropical (<900 m)	Darjeeling	38	19	–	9	2	16	9
	Sikkim	44	21	–	35	11	19	18
	Arunachal	44	56	–	11	7	29	39
Warm temperate (900–1800 m)	Darjeeling	23	10	–	2	1	11	18
	Sikkim	29	21	–	22	13	21	20
	Arunachal	39	20	–	11	8	18	17
Cool temperate (1800–2700 m)	Darjeeling	32	12	2	3	–	13	8
	Sikkim	36	17	3	31	8	34	28
	Arunachal	26	25	1	2	2	14	24
Sub-alpine (2700–3500 m)	Darjeeling	13	3	4	8	-	6	8
	Sikkim	25	16	5	7	4	9	7
	Arunachal	13	10	6	2	1	8	7
Alpine (>3500 m)	Darjeeling	7	3	6	2	–	–	–
	Sikkim	19	18	8	2	4	4	3
	Arunachal	8	4	4	-	–	–	1

En, Endemic to eastern Himalayan region; T, Threatened, rare and endangered.

It is observed that the species distribution in warm temperate and cool temperate region (mid hills) is close to 50% for Sikkim and Darjeeling. While precise landuse data is not available, extensive travel in the field gives the impression that the mid hills region has suffered maximum habitat loss and conversion. The fact that only small fragments of forest is left in the mid hills zone is also reflected in the PAN coverage (Table 2). In Sikkim only two PAs of sizes 52 km^2 (Fabong Lho) and 35 km^2 (Maenam) exist. There are others (Khanchedzonga and Barsey), which begin from cool temperate zone stretching to sub-alpine and alpine areas, with negligible coverage of warm temperate zone. In Darjeeling, again only two PAs of sizes 129 km^2 (Mahanada) and 88 km^2 (Neora) comprise the warm temperate and cool temperate zone. Close to 60% of the area of Mahananda consists of old plantations. In case of Arunachal, there is relatively less proportion of

Table 2. Existing Protected Area Network and Proposals (bold and italized) in Eastern Himalaya.

Ecological zones	Elevation (m)	Sikkim	Darjeeling (area in sq. km.	Arunachal
Tropical	< 900	*Kitam WS (13)*	Mahananda WS (129), *Mahananda-Neora Elephant corridor (200)*	D'Ering (190), Koronu grassland (15), *Namdapha NP (1985)*, *Pakhui WS (861)*, Kamlang (783), Mehao WS (281.50)
Warm Temperate	900–1800	*Khecheopalri (12)*, *Chuzachen-Asam-Lungze-Bushuk RFs*	Neora Valley NP (88), Jorepokhari (0.04) *Neora Extension (32)*	
Cool temperate	1800–2700	Khanchendzonga BR (2619.92)	Senchel WS (39)	Pakhui , Namdapha, Kamlang, Eagle's Nest WS (217), Tale valley WS (337), Mouling NP (483), Mehao, Dibang-Dihang BR (4149)
Warm-Cool temperate	900–2700	Fabong Lho WS (51.76), Maenam WS (35.34)	*Neora Extension*	
Cool temperate-Sub-alpine	1800–3500	Singba RS (43), Barsey RS (104), Pangolakha NP (108), Khanchendzonga BR	Singhalila NP (109)	Yordi Subse WS (491.6), Mehao, Dibang-Dihang, Namdapha, Mehao, Kamlang
Sub-alpine-Alpine	>2700	Kyongnosla WS (31) Khanchendzonga BR		
Trans-Himalayan		*Cold Desert National Park*		*Tawang National Park (100)*

NP, National Park, WS, Wildlife Sanctuary, RS, Rhododendron Sanctuary, BR, Biosphere reserve.

conservation-dependent species in mid hills zone. Better conservation status is duly reflected in the level of coverage with as many as six fairly large PAs providing protection in this zone. In Arunachal, 40% of all species recorded belong to tropical zone. The protected area coverage in this zone needs to be enhanced. Tropical zone is extremely limited in case of Sikkim and thus fewer number of species (26%) are recorded. In Darjeeling a relatively higher proportion of species (32%) are spread over a larger area under tropical zone. Based on the analysis and field experience in the region, the following recommendations are made to enhance long-term conservation measures in the identified important zones for each of the units.

The newly declared Pangolakha NP comprises mainly subalpine to the cool temperate zones. To the north it connects with Chuzachen Reserve Forest (RF). Also contiguous are the Asam–Lungze and Bushuk RFs to the west and north-west. This is an excellent stretch of warm temperate and cool temperate forests fed by Rora Chhu, Lungze Chhu and Rangpo Chhu rivers and criss crossed by many streams. Anthropogenic pressure is low and potential exists to bring nearly 100 km^2 (approx. 50% of current extent) under PAN. This unit would potentially connect to Torsa Strict Nature reserve of Bhutan. In West Sikkim, Khecheopalri Lake ecosystem (1700–1800 m) extending to about 12 km^2 is another important area in this zone. For the protection of tropical area the Government of Sikkim has already proposed Kitam WLS. The Government has also proposed to declare a Cold Desert National Park in the Chho Lhamo Plateau area to bring currently uncovered Trans-Himalayan zone in PAN.

The inadequate coverage of warm and cool temperate zones can be enhanced by an eastern extension of Neora. The extent of good primary forest cover in parts of Paren, Chichu, Khumani and Ruka blocks is more than fifty percent (32 km^2) and the rest is old plantations. The area of Neora can be brought to the range of 120 km^2 making it much more effective. The entire tropical belt of Darjeeling district is an important elephant corridor. The contiguity of forest cover exists from Neora to Mahananda (Sakam–Dalingkot–Ambiok–Fagu–Noam–Lethi–Ramthi–Churonthi–Lish–Mongpong forest blocks). The direct distance in-between the two PAs is 28 km and the total forest cover in these blocks is close to 190 km^2. These forests have relatively higher natural forest cover (54%) than the Mahananda WLS and besides facilitating elephant migration would contribute to overall biodiversity conservation in the tropical zone.

Tropical Semi-Evergreen and Evergreen Forests, and the Floodplain Grassland ecosystem abound only in Arunachal Pradesh in the entire eastern Himalayan region. The floodplain grasslands of D'Ering and Koronu recorded seven globally threatened species of birds. Therefore, Koronu area needs to be brought under PAN. In addition, greater area of South Bank Tropical Evergreen forests needs to be protected. There is no Trans-

Himalayan protected area in Arunachal and therefore earlier proposal of Tawang National Park (Rodgers and Panwar, 1988) also needs to be implemented.

An attempt has been made to recommend gaps in the existing PAN in the three geographic units comprising eastern Himalayan region in India, based on analysis of available information in various published, unpublished reports and field experiences. Detailed feasibility of each of the proposals would need to be undertaken for further follow-up.

Acknowledgements

The authors acknowledge the support of Director and other colleagues at ATREE, Eastern Himalaya Programme.

References

Ali, S. *The Birds of Sikkim. Second impression* (Oxford University Press, 1989).
Bibby, C. J. *Putting Biodiversity in Map: Priority Areas for Global Conservation* (International Council for Bird Preservation, Cambridge, UK, 1992).
BSI. *Flora of Sikkim* (Botanical Survey of India, Calcutta, 1996).
BSI. *Red data book of Indian plants* Vol. I and II (Botanical Survey of India, Calcutta, 1990).
Khaling, S., Ganguli-Lachungpa, U., Lucksom, S. & Lachungpa, C. Biodiversity conservation in the Sikkim Himalayas. in *Biodiversity Assessment and Conservation Planning: Kanchenjunga Mountain Complex* (WWF Nepal Program, Kathmandu (2000).
Kremen, C. Biological inventory using target taxa: A case study of butterflies of Madagascar. *Eco. Appls.* **4**, 407–422 (1994).
Myers, N. The biodiversity challenge: Expanded hotspots analysis. *The Environment.* **10**, 243–256 (1990).
Myers, N., Mittermiler, R. A., Mittermiler, C. G., Gustava, A. B., Da Foseca & Kent, J. Biodiversity hotspots for conservation priorities. *Nature* **403**, 853–858 (2000).
Myint, M., Rastogi, A. & Joshi, G. Biodiversity assessment and conservation planning: Eastern Arunachal Pradesh, India. WWF Nepal Program, Kathmandu (2000).
Oliver, I. & Beattie, A. J. Designing a cost-effective invertebrate survey: A test for methods for rapid assessment of biodiversity. *Eco. Appls.* **6**, 594–607 (1996).
Olson, D. M. & Dinerstein, E. The Global 2000: A representation approach to conserving the earth's most biologically valuable ecoregions. *Conserv. Biol.* **12**, 502–515 (1998).
Pradhan, S. & Bhujel, R. Biodiversity conservation in the Darjeeling Himalayas. in *Biodiversity Assessment and Conservation Planning: Kanchenjunga Mountain Complex.* (WWF Nepal Program, Kathmandu, 2000).
Rodgers, W. A. & Panwar, H. S. *Planning a Wildlife Protected Area Network in India.* Vol. I and II (Wildlife Institute of India, Dehradun, 1988).
Scott, J. M. & Csutti, B. Gap analysis for biodiversity survey and maintainance. in *Biodiversity II: Understanding and Protecting our Biological Resources* (eds Reaka-Kudla, M. L., Wilson, D. E. & Wilson, E. O.) 320–340 (John Henry press, Washington DC, 1996).
WCMC. Status report: Sikkim, Arunachal Pradesh, Myanmar. Unpublished Report. WCMC (1997).

TROPICAL FORESTS: STRUCTURE, DIVERSITY AND FUNCTION – PART B
Insects–Plant Interaction

Tropical Ecosystems: Structure, Diversity and Human Welfare.
Proceedings of the International Conference on Tropical Ecosystems
K. N. Ganeshaiah, R. Uma Shaanker and K. S. Bawa (eds)
Published by Oxford–IBH, New Delhi. 2001. pp. 629–632.

Ant effects on seedling recruitment in *Guapira opposita* (Nyctaginaceae) in a Brazilian rainforest

Luciana Passos*[†] *and Paulo S. Oliveira***

*Department de Botânica and **Department de Zoologia, Universidade Estadual de
Campinas, C.P. 6109, 13083-970 Campinas SP, Brazil
[†]e-mail: llpassos@yahoo.com, pso@unicamp.br

In tropical forests nearly 90% of the trees and shrubs bear fleshy fruits and rely on vertebrate frugivores for seed dispersal (Jordano, 1993). However, recent studies have shown that interactions involving ants and fruits/seeds are common on the floor of tropical forests. These ant–seed interactions can modify the fate of seeds, and affect patterns of recruitment in primarily vertebrate-dispersed species that lack adaptations for ant-dispersal (e.g., Levey and Byrne, 1993; Pizo and Oliveira, 1998; Böhning-Gaese *et al.*, 1999).

Fleshy fruits of tropical forests present a plethora of sizes, shapes, colors, and chemical composition of the edible portion. Consequently, ants in tropical forests interact with a broad range of fruits/seeds differing in morphology and nutrient content (Pizo and Oliveira, 2001a). It has recently been suggested that the outcome of the interaction between ants and fruits/seeds in tropical forests can be largely determined by the size and lipid content of the latter (Pizo and Oliveira, 2001b).

Proteins are an essential food source for social insects such as ants. Colonies must get adequate protein intake to meet the dietary requirements of larvae

Keywords: Seed dispersal, ants, recruitment, *Guapira*.

and functional queens (Hölldobler and Wilson, 1990). In general, fleshy fruits are extremely poor in protein in comparison with leaves and insects, and protein-rich fruits are not common (Jordano, 1993). However, protein-rich fruits are attractive to ants, especially those in the predominantly carnivorous subfamily Ponerinae (Hölldobler and Wilson, 1990). Ponerines feed on arthropod prey and use aril or pulp of fruits as a secondary food source (Pizo and Oliveira, 1998); protein-rich fruits would presumably complement the protein intake of the colonies. Ant activity can affect recruitment patterns in species with lipid-rich fruits (e.g., Böhning-Gaese et al., 1999), but there is no information for species with protein-rich fruits. Guapira opposita is a primarily bird-dispersed tree producing protein-rich fruits. We investigated ant activity at fallen fruits of G. opposita, and the influence of ants on seed fate and seedling recruitment in this species.

Fieldwork was carried out in the rainforest of Cardoso Island, SE Brazil, during April 1998–September 2000. Interactions between ants and fruits were surveyed through systematic sampling ($N = 90$), and fruit removal by ants was experimentally assessed on the forest floor ($N = 30$). Germination tests evaluated whether pulp removal by ants had any effect on seed germination of Guapira ($N = 40$ fruits per treatment). The effect of ponerines on seedling distribution was determined by censusing seedlings and juveniles of Guapira growing in nests of Odontomachus chelifer, and in control plots without nests ($N = 40$ nests). The soil composition and penetrability of experimental plots was also assessed. To evaluate if seedlings/juveniles growing in Odontomachus nests could potentially benefit from some ant-derived protection against herbivores, we performed an experiment using live dipteran larvae ($N = 60$ larvae per treatment).

Eleven ant species were attracted to fruits of Guapira. The most frequent species were the large ponerines Odontomachus chelifer and Pachycondyla striata, that together accounted for 56% of the ant–fruit interactions. Ponerines were the main seed vectors on the forest floor, while small ants recruited nestmates to fruits and removed the pulp on the spot, without displacing seeds. Chemicals mediate the behavior of ants toward potential food items, and lipids are regarded as the major attractant factor in the interaction between ants and fruits/seeds. We suggest that protein content is also an important factor in the selection of fruits for a variety of ants, especially ponerines. Pulp removal in Guapira increases germination success (χ^2 test; $P < 0.001$). Seedlings and juveniles of Guapira are more abundant in the vicinity of nests of Odontomachus than in random plots without nests (Mann–Whitney U-test; $P < 0.0001$; see Figure 1). Although Guapira is devoid of any morphological adaptations for ant-dispersal, the distribution patterns of seedlings and juveniles suggest that dispersal by ants has a strong impact on recruitment of this species.

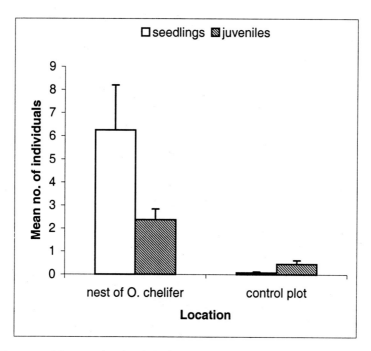

Figure 1. Mean number (+ 1 SE) of seedlings and juveniles of *Guapira opposita* in nests of *Odontomachus chelifer*, and in adjacent control plots.

Seed dispersal, germination, and early seedling growth/survival are the most critical stages in determining where plants recruit within a landscape. Ant nests are known to have specific moisture, texture, and nutrient characteristics that may be important conditions for seed germination and seedling establishment. Nests of *Odontomachus* are richer in P and Ca, and the ants also increase soil penetrability, which might improve performance of *Guapira* seedlings (see Levey and Byrne, 1993). Moreover, the association of *Guapira* seedlings with *Odontomachus* nests potentially renders for the plant some protection against herbivores, as expressed by the attack rates by ants toward dipteran larvae (χ^2 test; $P < 0.001$).

In conclusion, the nests of *Odontomachus* seem to be suitable sites for seedlings/juveniles of *Guapira*, but previous results suggest that the quality of such microsites may vary with the requirements of the plant species involved. This study indicates that protein content of fruits may be relevant for the attraction of ponerine ants on the forest floor, and illustrate the complex nature of the dispersal ecology of tropical tree species.

Acknowledgements

This study was supported by the Brazilian Research Council (CNPq), and by the Research Foundation of São Paulo (FAPESP).

References

Böhning-Gaese, K., Gaese, B. H. & Rabemanantsoa, S. B. Importance of primary and secondary seed dispersal in the Malagasy tree *Commiphora guillaumini*. *Ecology* **80**, 821–832 (1999).

Hölldobler, B. & Wilson, E. O. *The Ants* (Belknap Press, Massachusetts, USA, 1990).

Jordano, P. Fruits and frugivory, in *Seeds: the Ecology of Regeneration in Plant Communities* (ed. Fenner, M.) 105-156 (CAB International, Wallingford, 1993).

Levey, D. J. & Byrne, M. M. Complex ant–plant interactions: rain forest ants as secondary dispersers and post-dispersal seed predators. *Ecology* **74**, 1802–1812 (1993).

Pizo, M. A. & Oliveira, P. S. Interactions between ants and seeds of a nonmyrmecochorous neotropical tree, *Cabralea canjerana* (Meliaceae), in the Atlantic forest of southeast Brazil. *Am. J. Bot.* **85**, 669–674 (1998).

Pizo, M. A. & Oliveira, P. S. The use of fruits and seeds by ants in the Atlantic forest of southeast Brazil. *Biotropica* **32** (in press) (2001a).

Pizo, M. A. & Oliveira, P. S. Size and lipid content of nonmyrmecochorous diaspores: effects on the interaction with litter-foraging ants in the Atlantic rain forest of Brazil. *Plant Ecol.* (in press) (2001b).

Tropical Ecosystems: Structure, Diversity and Human Welfare.
Proceedings of the International Conference on Tropical Ecosystems
K. N. Ganeshaiah, R. Uma Shaanker and K. S. Bawa (eds)
Published by Oxford–IBH, New Delhi. 2001. p. 633.

Plant resistance against gall induction: From cells to biodiversity?

G. W. Fernandes*, T. G. Cornelissen, D. P. Negreiros and C. Saraiva

University Federal de Minas Gerais, Lab. Ecologia Evolutiva de Herbívoros
Tropicais, ICB, CP 486, 30161-970, Belo Horizonte, MG, Brazil
*e-mail: gwilson@icb.ufmg.br

Hypersensitive reaction (HR) is an important, yet neglected, type of induced defense of plants against insect herbivores. Previously thought to be an exclusive type of defense against pathogens, HR is a widespread mechanism whereby plants resist the attack by herbivores, particularly galling insects. In this long-term study we evaluated the importance of HR against a leaf-galling insect (*Contarinia* sp.) (Diptera: Cecidomyiidae) on *Bauhinia brevipes* (Leguminosae). More than 15,000 shoots and 17,000 galls were sampled during six consecutive years of study. Throughout the study period, this host-driven mortality factor killed more than 90.0% of the galls. The plant response to galling was not related to the density of attack or even to the vigor (growth) of the attacked host organ. Studies on other eight tropical savanna species corroborated the occurrence and importance of HR on insect herbivore population dynamics. HR is a common phenomenon in the plant kingdom, irrespective of plant phylogeny or growth form. This cellular host plant defense mechanism might involve cell suicide, or apoptosis, which is now under study. Because this host reaction is inhibited under high temperature, therefore under habitat and plant stress, it may be an important driving force influencing galling insect biodiversity at a global scale.

Keywords: Plant hypersensitivity, plant defenses, induced responses, *Contarinia* sp., galls.

Tropical Ecosystems: Structure, Diversity and Human Welfare.
Proceedings of the International Conference on Tropical Ecosystems
K. N. Ganeshaiah, R. Uma Shaanker and K. S. Bawa (eds)
Published by Oxford–IBH, New Delhi. 2001. pp. 634–636.

Variation among tropical forest sites in leafing phenology, seasonal abundance of insect herbivore populations, leaf herbivory and defense

R. J. Marquis,†, H. C. Morais** and I. R. Diniz‡*

*Department of Biology, University of Missouri-St. Louis, 8001 Natural Bridge Road, St. Louis, Missouri, USA
**Departamento de Ecologia, Instituto de Ciências Biológicas, Universidade de Brasilia, 70910-900, Brasilia, D.F., Brazil
‡Departamento de Zoologia, Instituto de Ciências Biológicas, Universidade de Brasilia, 70910-900, Brasilia, D.F., Brazil
†e-mail: robert_marquis@umsl.edu

Various suggestions are found in the literature that relate intersite differences in folivory levels in tropical forests to variation in environmental factors. However, there has been no attempt to bring these together into a comprehensive model. We first describe patterns of leafing phenology, seasonal abundance of insect herbivore populations, leaf herbivory, and defense for our Brazilian cerrado study site, Fazenda Agua Limpa (FAL), Brazília, Brazil. We then present a descriptive model relating these factors to among site variation in environmental factors, based on published reports, to answer the question, 'How different or similar are plant–insect herbivore relationships in cerrado?'

Keywords: Leaf herbivory, defense, cerrado, soil nutrients, tritrophic interactions.

Marquis *et al.* (2001a) found for the 25 species studied (10 trees, 10 shrubs, and 5 herbs) that damage by herbivorous insects at the end of the leaf life (one year old at the end of the dry season) ranged between 0.5% and 14.3% per plant species (mean = 6.8%). Pathogen attack was much higher, ranging from 2.0% to 52.8% (mean = 17.3%). Damage by insects peaked at two points in the life of the leaves, during the leaf expansion period at the beginning of the wet season, and again sometime in the second half of the wet season and early dry season. These two peaks in abundance coincided with known peaks in abundance of herbivorous insects in this system: the annual peak in abundance of leaf-chewing Coleoptera during the study year occurred during the leaf expansion period, while peaks in abundance of leaf-chewing Lepidoptera larvae in the years 1991–1993 occurred at the beginning of the dry season (Morais *et al.*, 1999). The seasonal pattern of attack by pathogens was quite different, in that almost no damage occurred during the leaf expansion period, but fully expanded leaves continued to accrue damage throughout their lives. Although cerrado leaves are very tough compared to those of more mesic sites, toughness was not related to interspecific differences in damage levels. Instead, protein-binding capacity was significantly negatively related to the amount of insect damage. Protein availability (nitrogen content/protein-binding capacity) and plant height were significantly positively correlated with pathogen attack. It may be that taller plant species are more susceptible to pathogen colonization because they are on average more likely to intercept windborne pathogen spores. Rate of leaf expansion was negatively correlated with pathogen attack.

As suggested previously (Coley and Barone, 1996), a literature survey demonstrates that damage by insect herbivores shows a unimodal relationship with rainfall in Neotropical sites, with damage highest at intermediate levels of rainfall (Marquis *et al.*, 2001b). However, regardless of rainfall, folivory is lowest on low soil nutrient sites (see Janzen, 1974). Our cerrado site falls into the category of low nutrient soils, resulting in much lower folivory levels than predicted by rainfall for more nutrient-rich sites. Leaf damage appears to be further lowered by the fact that many plant species produce new leaves before the rainy season begins (Morais *et al.*, 1995). In so doing, leaves are hardened, and thus pass through their most vulnerable period of attack, before herbivorous insect populations have increased with the onset of the rains. We discuss what is known about the relative importance of the third trophic level in influencing the observed patterns.

Acknowledgements

We thank the many students of the project 'Herbivores and Herbivory in Cerrado', who assisted in the collection and rearing of the caterpillars. Research was supported by FAPDF, FINATEC, CNPq, which also provided support for scientific initiation (PIBIC-CNPq-UnB) to HM and ID, and a Chancellor's Visiting Professorship to RJM. We thank Katerina Aldás, Karina Boege, Phyllis Coley, Rebecca Forkner, Nels Holmberg, Damond Kyllo, John Lill, and Eric Olson for valuable comments.

References

Coley, P. D. & Barone, J. A. Herbivory and plant defenses in tropical forests. *Annu. Rev. Ecol. Syst.* **27**, 305–335 (1996).

Janzen, D. H. Tropical blackwater rivers, animal and mast fruiting by the Dipterocarpaceae. *Biotropica* **6**, 69–103 (1974).

Marquis, R. J., Diniz, I. R. & Morais, H. C. Patterns and correlates of interspecific variation in foliar insect herbivory and pathogen attack in Brazilian cerrado. *J. Trop. Ecol.* **17**, 1–23 (2001a).

Marquis, R. J., Morais, H. C. & Diniz, I. R. Interactions among cerrado plants and their herbivores: unique or typical? in *Ecology and Natural History of a Neotropical Savanna: The Cerrados of Brazil* (eds Oliveira, P. & Marquis, R. J.) (Columbia Univ. Press, 2001b) (in press).

Morais, H. C., Diniz, I. R. & Baumgarten, L. C. Padrões de produção de folhas e sua utilização por larvas de Lepidoptera em um cerrado de Brasilia, DF. *Rev. Bras. Bot.* **13**, 351–356 (1995).

Morais, H. C., Diniz, I. R. & Silva, D. M. S. Caterpillar seasonality in a central Brazilian cerrado. *Rev. Biol. Trop.* **47**, 1025–1033 (1999).

Tropical Ecosystems: Structure, Diversity and Human Welfare.
Proceedings of the International Conference on Tropical Ecosystems
K. N. Ganeshaiah, R. Uma Shaanker and K. S. Bawa (eds)
Published by Oxford–IBH, New Delhi. 2001. pp. 637–640.

Toona ciliata and the shoot borer, *Hypsipyla robusta*: Effect of tree height and growth rate on frequency of insect damage

*S. A. Cunningham**[†], *R. B. Floyd*, M. Griffiths** and F. R. Wylie***

*CSIRO Entomology, GPO Box 1700, Canberra ACT, 2601, Australia
**Queensland Forestry Research Institute, PO Box 631, Indooroopilly, QLD, 4068, Australia
[†]e-mail: saul.cunningham@ento.csiro.au

*H*ypsipyla spp. (Pyralidae: Lepidoptera) feed almost exclusively on trees in the Swietenioideae subfamily of the Meliaceae. This subfamily includes some of the world's most highly prized tropical forest timbers, including American Mahogany (*Swietenia* spp.), African Mahogany (*Khaya* spp.), Australian Red Cedar (*Toona ciliata*), and West Indian Cedar (*Cedrela odorata*). Larvae feed on fruits and growing tips; this latter form of feeding causing great economic damage in forestry plantations. Attempts to grow these tree species in plantations have been largely unsuccessful because of damage from *Hypsipyla* spp. There is, therefore, great interest in understanding the causes of variation in insect feeding on trees.

We have established trials in Australia, Thailand, and the Philippines that include trees from more than 69 seedlots of *T. ciliata*, where a seedlot is seed

Keywords: Herbivore, forestry, Meliaceae, Pyralidae, lepidoptera.

collected from a common mother tree. Each seedlot is represented in the trials by 6 replicated 5 tree plots. These trials are all within the region of endemism of *Hypsipyla robusta, sensu stricta* (M. Horak, pers. commun.). Trials were regularly assessed for growth of trees and feeding damage by *H. robusta*. Shoot boring destroys the growing tip and causes initiation of new lateral branches, one of which then becomes the dominant growing shoot. By examining these deformations to the bole one can count the number of feeding points on the main stem, accumulated in the course of the tree's life.

The capacity of this insect to find and feed on trees was impressive. *Hypsipyla robusta* damage to trees first occurred after 6 months in the Philippines, and after 2 months in Australia and Thailand. The frequency of *H. robusta* damage in each trial then built up to high levels, such that after 12 months the percentage of *T. ciliata* trees damaged was 71% in Thailand, 60% in the Philippines, and 60% in Australia.

There was a strong positive relationship between tree height and the number of *H. robusta* attacks to the main stem. Considering trees 12 months after planting in the Australian trial, each height increment from less than 20 cm to approximately 140 cm corresponds with an increase in the mean number of damage points on the main stem (Figure 1). The same phenomenon is apparent in our trials in Thailand and the Philippines. Although *H. robusta* can feed on the shoots of trees shorter than 20 cm, they apparently prefer to feed on taller trees. This 'height effect' is consistent with observations from other sites of a correlation between tree height and the number of *H. robusta* eggs laid (Griffiths, 1997).

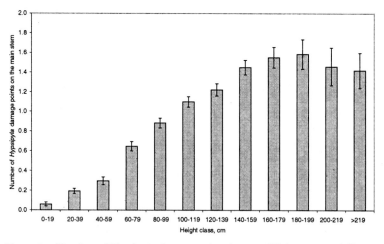

Figure 1. Number of *H. robusta* damage points (mean ±SE) for trees in different height classes, at 12 months since planting in the Australian trial. The sample size in each class ranges from 31 (> 220 cm) to 324 (80–99 cm).

There are numerous reports in the forestry literature of an influence of height on attractiveness to insect herbivores, but the cause of the pattern is not clear. One possibility is that taller trees are more apparent to moths that are seeking sites for oviposition. In this site, however, nearly 3000 potential host trees are planted over 4 ha of land with complex topography. In these circumstances it seems unlikely that 20 cm differences in height are important in determining the apparency of individual trees. That the height effect plateaus above 140 cm is also contrary to the idea that height directly determines apparency. Therefore we examined the possibility that tree height and attractiveness to insects were mutually correlated with a third variable, the growth rate of trees.

Hypsipyla robusta larvae bore into actively growing shoots, which would be expected to have a negative impact on tree growth. To examine this effect we considered trees that exceed 139 cm after 12 months, and were therefore above the height at which the 'height effect' is diminished. We found the expected negative relationship (Figure 2), such that trees with more *H. robusta* damage points to the main stem had lower relative height increases, for the period 6 to 12 months, during which most insect damage occurred. This effect runs counter to the overall positive relationship between height and insect damage (Figure 1).

Studies of plant traits indicate that plants with higher potential growth rate tend to have high nitrogen concentrations and lower density tissue (Cunningham *et al.*, 1999; Westoby, 1998). Studies of insect feeding behaviour indicate that most insects will prefer high nitrogen, low density tissue (Coley *et al.*, 1985; Mattson, 1980). This may explain the preference some insects show for attacking rapidly growing plant parts (e.g. Craig *et al.*, 1989). We

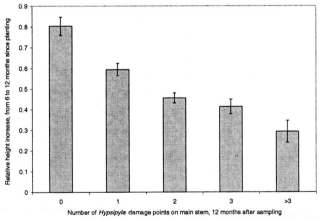

Figure 2. Relative height increase of trees in different damage classes (mean ± SE), in the Australian trial. Only considering trees > 139 cm. Sample size in each class ranges from 19 (> 3) to 166 (1).

predict a positive relationship between a plant's potential relative growth rate and its attractiveness to insect attack. We know that realized growth in the 6 to 12 month period was negatively affected by insect feeding (Figure 2), but height increase in the 2 to 6 month period (before most of the insect damage occurred) offers an estimate of the potential relative growth rate when there is little insect attack. We tested for a relationship between potential growth rate and insect damage in a conservative manner, by again excluding from analysis those trees shorter than 139 cm at 12 months, for which we had already established a strong positive association between height and attack (Figure 1). We then compared the relative height increase between 2 and 6 months of trees in the low damage class (0–1 damage points at 12 months) to that of trees in the high damage class (2–6 damage points at 12 months). Trees in the high damage class had a significantly higher early relative height increase (mean = 3.2, $N = 205$) than trees in the low damage class (mean = 2.9, $N = 246$) (t-test on log transformed data, $P = 0.035$). This result provides support for the predicted positive association between potential relative growth rate and insect attack.

Future work will examine variation in plant chemistry and tissue density to determine which particular plant traits affect *H. robusta* feeding and oviposition behaviour. As trees in the trials age we will be able to determine if the plant traits that moderate the height effect in the first 12 months continue to be important in older trees. By examining seedlot variation in growth and insect attack we can examine genetic variation for important plant attributes. Better understanding of the ecology of this relationship and the importance of plant traits will help develop these trees as a plantation forestry resource for tropical countries.

Acknowledgements

We gratefully acknowledge funding for this research from the Australian Centre for International Agricultural Research. We thank these collaborating institutions for support: the Queensland Forestry Research Institute, the Ecosystems Research and Development Bureau of the Philippines, and the Royal Forest Department of Thailand.

References

Coley, P. D., Bryant, J. P. & Chapin, F. S. Resource availability and plant antiherbivore defense. *Science* **230**, 895–899 (1985).

Craig, T. P., Itami, J. K. & Price, P. W. A strong relationship between oviposition preference and larval performance in a shoot-galling sawfly. *Ecology* **70**, 1691–1699 (1989).

Cunningham, S. A., Summerhayes, B. & Westoby, M. Evolutionary divergences in leaf structure and chemistry, comparing rainfall and soil nutrient gradients. *Ecol. Monogr.* **69**, 569–588 (1999).

Griffiths, M. The biology and host relations of the cedar tip moth, *Hypsipyla robusta* Moore (Lepidoptera, Pyralidae) in Australia. PhD Thesis, University of Queensland (1997).

Mattson, W. J. Herbivory in relation to plant nitrogen content. *Annu. Rev. Ecol. Syst.* **11**, 119–161 (1980).

Westoby, M. A leaf-height-seed (LHS) plant ecology strategy scheme. *Plant Soil* **199**, 213–227 (1998).

Tropical Ecosystems: Structure, Diversity and Human Welfare.
Proceedings of the International Conference on Tropical Ecosystems
K. N. Ganeshaiah, R. Uma Shaanker and K. S. Bawa (eds)
Published by Oxford–IBH, New Delhi. 2001. pp. 641–644.

Alarm communication in a neotropical harvestman (Arachnida: Opiliones)

Paulo S. Oliveira*, Glauco Machado and Vinícius Bonato

Departamento de Zoologia, Universidade Estadual de Campinas, C.P. 6109, 13083-970 Campinas SP, Brazil
*e-mail: pso@unicamp.br

Many arthropod species use chemical deterrents for defence. Most of these products are secreted by exocrine glands and are often complex mixtures that serve multiple roles (Blum, 1981). In many cases, compounds that were once believed to be primarily defensive have been subsequently found to function parsimoniously as sexual, aggregation, alarm, trail, and territorial pheromones. A pheromone is a substance that is secreted to the outside of an individual and causes specific reaction in another individual of the same species (Karlson and Lüscher, 1959).

Most harvestmen are nocturnal, nonacoustical, and nonvisual arthropods. They have a pair of exocrine glands on the cephalothorax that produce defensive volatile secretions (see Machado *et al.*, 2000). The combination of these features suggests that pheromones or chemicals signals can be important for intraspecific communication.

Many harvestmen species show gregarious habits and form dense diurnal aggregations consisting of nymphs and adults of both sexes. It is possible that the defensive secretion could also be used as an alarm pheromone upon disturbance of the group. Indeed group-living is a prerequisite for the

Keywords: Harvestmen, alarm, chemical communication, *Goniosoma*.

evolution of alarm pheromones, and these substances have been identified in many gregarious species of insects. With the exception of the Acari, alarm pheromones have not been discovered in any other major group in the class Arachnida.

Harvestmen species in the Neotropical genus *Goniosoma* are highly gregarious, and normally take shelter inside caves, rock crevices and tree trunks (Machado and Oliveira, 1998; Machado *et al.*, 2000). The secretions produced by *Goniosoma* are mainly quinones, a widespread predator deterrent among arthropods (Blum, 1981). We investigated the possible alarm effect of these secretions in the gregarious harvestman *Goniosoma* aff. *proximum*.

Fieldwork was carried out in the rainforest of Cardoso Island, SE Brazil, during January–July 2000. The effect of the defensive secretion as alarm pheromone was evaluated through a field experiment in which harvestmen aggregations were experimentally exposed to a cotton swab (20 cm long) soaked with the species' gland exudate (treatment, $N = 30$) or water (control, $N = 30$). After each trial the cotton swab was changed in either experimental group. The time for individuals to respond was then recorded. We counted the number of individuals of all tested aggregations both before and after each trial.

The results of the experiment revealed that the scent gland secretion unequivocally works as an alarm pheromone in *Goniosoma*. In 73% of the aggregations the individuals quickly dispersed after stimulation with the exudate. On the other hand, only 3.3% of the aggregations responded to the water-soaked cotton swab (χ^2 test, $P < 0.0001$). The individuals perceiving the odour flee promptly, and frequently bump other individuals. The alarm reaction is probably mechanically spread through the aggregation, resulting in general dispersion.

The experiments also showed that respondent aggregations were larger than non-respondent ones (20.5 ± 11.1 vs 10.5 ± 8.5 individuals; Mann–Whitney U-test; $P < 0.02$). Moreover, within respondent groups, the time to react to the scent gland secretion was inversely related with group size (Figure 1).

Harvestmen secretions can provide an effective defence against a variety of invertebrate and vertebrate predators. We have demonstrated for the first time that such secretions also elicit alarm behaviour, causing aggregated individuals to disperse from a chemically marked area. We also show for the first time among arachnids that the reaction response to the alarm signal varies with the size of the group. Larger aggregations may react faster to the chemical signal as a result of the increased number of sensorial legs used for surveillance. Such a positive relationship between number of individuals and promptness of reaction was also previously reported among visually-oriented invertebrates (Vulinec, 1990).

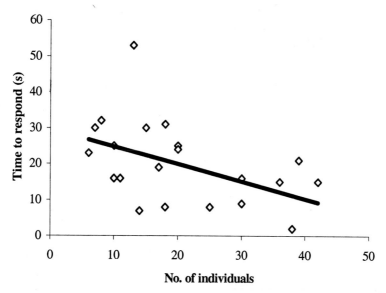

Figure 1. Group size vs time to respond in aggregations of the harvestman *Goniosoma* aff. *proximum*, when stimulated with a cotton swab soaked with the species' own secretion (Pearson's $r = -0.467$, $P < 0.03$).

Apart from the deterring function of such secretions, one might still ponder if there is any additional benefit for the sender by signalling to neighbouring conspecifics upon a predator attack. As also shown for many insects (Blum, 1985), our data suggest that scent gland secretion in *Goniosoma* functions as a defensive allomone; its pheromonal role being probably secondarily derived. The alarm effect has probably evolved as a by-product of a primarily defensive response upon predator attack. An individual's advantage of signalling, if there is any, would depend on a confusion effect on the predator due to the general fleeing, and/or on the genetic relatedness among aggregated harvestmen.

Acknowledgements

This study was supported by the Brazilian Research Council (CNPq), and by the Research Foundation of São Paulo (FAPESP).

References

Blum, M. S. *Chemical Defenses of Arthropods* (Academic Press, NY, 1981).
Blum, M. S. Exocrine systems. in *Fundamentals of Insect Physiology* (ed. Blum, M. S.) 535–579 (John Wiley & Sons, New York, 1985).
Karlson, P. & Lüscher, M. 'Pheromones' a new term for a class of biologically active substances. *Nature* **183**, 155–156 (1959).

Machado, G. & Oliveira, P. S. Reproductive biology of the neotropical harvestman *Goniosoma longipes* (Arachnida, Opiliones, Gonyleptidae): mating and oviposition behaviour, brood mortality, and parental care. *J. Zool.* **246,** 359–367 (1998).

Machado, G., Raimundo, R. L. G. & Oliveira, P. S. Daily activity schedule, gregariousness, and defensive behaviour in the Neotropical harvestman *Goniosoma longipes* (Arachnida: Opiliones: Gonyleptidae). *J. Nat. Hist.* **34,** 587–596 (2000).

Vulinec, K. Collective security: aggregation by insects as a defense. in *Insect Defenses: Adaptive Mechanisms and Strategies of Prey and Predators* (eds Evans, D. L. & Schmidt, J. O.) 251–288 (State University of New York, Albany, 1990).

Tropical Ecosystems: Structure, Diversity and Human Welfare.
Proceedings of the International Conference on Tropical Ecosystems
K. N. Ganeshaiah, R. Uma Shaanker and K. S. Bawa (eds)
Published by Oxford–IBH, New Delhi. 2001. pp. 645–648.

How effective are ants in protecting the myrmecophyte *Humboldtia brunonis* (Fabaceae) against herbivory?

Merry Zacharias[†], Laurence Gaume* and Renee M. Borges

Centre for Ecological Sciences, Indian Institute of Science, Bangalore 560 012, India
*Institut des Sciences de l'Evolution de Montpellier, Université Montpellier II, Place
E. Bataillon, BP 062.34095 Montpellier Cedex 5, France
[†]e-mail: merry@ces.iisc.ernet.in

Ant–plant mutualisms may be very specific and symbiotic or may be quite casual with many ant species interacting facultatively with many plant species (Boucher *et al.*, 1982). Myrmecophytes range from plants with complex adaptations to house and feed specialized ants, to the relatively casual occupation of partially hollow twigs and stems by various ant species. Specialised ant–plants in the genera *Acacia*, *Cecropia* and *Leonardoxa* produce foliar nectar, food bodies and domatia, which attract ants. In return for food and shelter, the ants provide protection against herbivores, e.g. Janzen (1972), Fiala (1989), Gaume and McKey (1998).

Humboldtia brunonis (Fabaceae) is an unspecialised understorey myrmecophyte, in which some individuals possess swollen, hollow internodes called domatia. Domatia, when present, attract a diversity of invertebrate associates such as ants as well as a variety of bees, wasps, centipedes and even an arboreal earthworm. Extrafloral nectaries are present on the young leaves as well as on the sepals of floral buds. The flowers are arranged in axillary

Keywords: Ant–plant mutualism, *Technomyrmex albipes*, *Crematogaster dorhni*.

inflorescences containing 25 to 30 buds. The study was conducted in the evergreen, lowland forest of Makut, Coorg, situated to the south-west of Karnataka State, India (12°5'N and 75°45'E), in the months of March and April 1999.

Of the 19 different ant species associated with *Humboldtia brunonis*, the dominant ant species were *Technomyrmex albipes* (Dolichoderinae) and *Crematogaster dorhni* (Myrmicinae). *Crematogaster dohrni* is an aggressive ant possessing a sting whereas the sting in *T. albipes* is absent or vestigial.

We did a pairwise ant exclusion experiment to look at the proportion of herbivory on flower buds in the presence and absence of ants. We located trees with at least one intact pair of inflorescence buds of similar initial sizes. We excluded ants from one inflorescence bud by the use of Tanglefoot glue and used the other bud as the control. We conducted this experiment over a 14-day period, at the end of which we noted the percentage herbivory on the buds (percentage of buds in the inflorescence with herbivory damage). We later noted number of fruits initiated on these buds. Unpatrolled floral buds from which ants had been excluded accumulated significantly more damage than patrolled buds (Wilcoxon matched pairs test, $P \leq 0.05$, $n = 13$, Figure 1a), for plants occupied by *Crematogaster dorhni*. In contrast, for plants occupied by *T. albipes*, there was no significant difference in damage to patrolled and unpatrolled floral buds (Wilcoxon matched pairs test, $P = 0.31$, $n = 16$, Figure 1b). However, fruit initiation was found to be significantly greater in buds patrolled by *T. albipes* (Wilcoxon matched pairs test, $P \leq 0.05$, $n = 16$) while there was no significant difference between control and experimental inflorescences with respect to fruit initiation in plants occupied by *C. dorhni*. Hence both species of ants provide protection against herbivores of buds, with floral buds patrolled by *T. albipes* initiating more fruits. Although *T. albipes* seems rather defenceless, yet with a mass recruiting system, it is able to attack phytophagous insects effectively.

Since fruit formation is important for the reproductive success of the plant and *T. albipes* appeared to be more successful in this regard, we did a larval manipulation experiment on floral buds patrolled by *T. albipes*. We used lepidopteran larvae of different sizes (0.3–2.0 cm) for the experiment. In each trial, we placed a single larva on a bud and we recorded the time until discovery of the larva by the first worker ant, and the fate of the larva once it was discovered. The larva was either (a) consumed entirely, (b) successfully thrown off the leaf, (c) thrown off and hung on silk, or (d) escaped by itself. Workers of *T. albipes* located the lepidopteran larvae on floral buds rapidly and eliminated them efficiently. A Mann–Whitney U test showed that there was a significant difference between the sizes of the larvae that were 'killed and eaten' and those that 'escaped' ($U = 4.0$, $n_{\text{killed and eaten}} = 7$, $n_{\text{escaped}} = 5$, $P \leq 0.05$, Figure 2) with the larvae that escaped being larger in size. There was neither any significant difference between the sizes of the larvae that were 'killed and

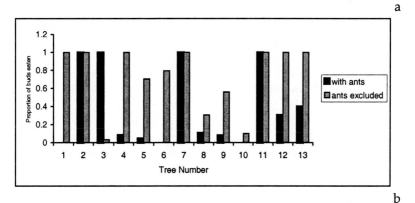

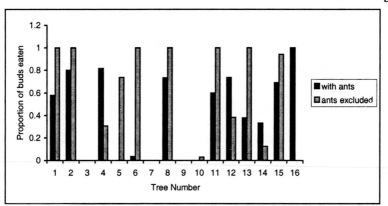

Figure 1. Effect of ant exclusion on herbivory caused by phytophagous insects on buds of *Humboldtia brunonis*: (a) Plants occupied by *Crematogaster dorhni*, (b) plants occupied by *Technomyrmex albipes*.

Figure 2. Fate of lepidopteran larvae placed on buds of *Humboldtia brunonis*, patrolled by workers of *Technomyrmex albipes*, as a function of the size of the larva.

eaten' and 'evicted' nor between the sizes of the larvae that were 'evicted' and those which 'escaped' (Mann–Whitney U tests, $P > 0.05$, Figure 2). Ants only succeeded in killing larvae on the spot if they were small, < 1.0 cm long. Most larvae > 1 cm in length were thrown off the buds. Hence, the efficiency of protection decreased with increasing size of the larva. These findings support the hypothesis that worker size in relation to size of the plant's enemies is an important component in the functioning of protective ant–plant mutualisms (Gaume et al., 1997; Meunier et al., 1999).

Our study has, therefore, demonstrated that *Technomyrmex albipes* and *Crematogaster dorhni* do indeed protect floral buds of their host plant from insect herbivores. The ants protect the tree from herbivores, by patrolling and finding phytophagous insects on its young leaves and floral buds, and by systematically deploying aggressive behaviour and mass recruitment against insects they encounter.

References

Boucher, D. H., James, S. & Keeler, K. H. The ecology of mutualism. *Annu. Rev. Ecol. Syst.* **13**, 315–347 (1982).
Fiala, B., Maschwitz, U., Tho, Y. P. & Helbig, A. J. Studies of a south-east Asian ant–plant association: protection of *Macaranga* trees by *Crematogaster dorhni borneensis*. *Oecologia* **79**, 436–470 (1989).
Gaume, L., McKey, D. B. & Anstett, M. C. Benefits conferred by 'timid' ants: active anti-herbivore protection of the rainforest tree *Leonardoxa africana* by the minute ant *Petalomyrmex phylax*. *Oecologia* **112**, 209–216 (1997).
Gaume, L. & McKey, D. B. Protection against herbivores of the myrmecophyte *Leonardoxa africana* (Baill.) Aubrev. T3 by its principal ant inhabitant *Aphomomyrmex afer* Emery. *C. R. Acad. Sci. Paris* **321**, 593–601 (1998).
Janzen, D. H. Protection of *Barteria* (Passifloraceae) by *Pachysima* ants (Pseudomyrmecinae) in a Nigerian rain forest. *Ecology* **53**, 885–892 (1972).
Meunier, L., Dalecky, A., Berticat, C., Gaume, L. & McKey, D. B. Worker size variation and the evolution of an ant–plant mutualism: Comparative morphometrics of workers of two closely related plant-ants, *Petalomyrmex phylax* and *Aphomomyrmex afer* (Formicinae). *Insect. Soc.* **46**, 171–178 (1999).

Tropical Ecosystems: Structure, Diversity and Human Welfare.
Proceedings of the International Conference on Tropical Ecosystems
K. N. Ganeshaiah, R. Uma Shaanker and K. S. Bawa (eds)
Published by Oxford–IBH, New Delhi. 2001. pp. 649–653.

Understanding ant-avoidance behaviours in the myrmecomorphic spider *Myrmarachne plataleoides*, mimic of the weaver ant *Oecophylla smaragdina*

Malvika Talwar, Merry Zacharias, Vinita Gowda,
Vena Kapoor, Sarasij Raychaudhuri*, Hari Sridhar,
Zahabia Lokhandwala and Renee M. Borges

Centre for Ecological Sciences, Indian Institute of Science, Bangalore 560 012, India
*Raman Research Institute, Bangalore 560 080, India
e-mail: tchutki@yahoo.com

The system that we are exploring involves a myrmecomorphic relationship between the weaver ant *Oecophylla smaragdina* and the jumping spider *Myrmarachne plataleoides*.

The weaver ant *O. smaragdina* is one of the two extant species of the Old World genus *Oecophylla* and ranges from India across almost all of tropical forested Asia to the Solomon Islands, Greenland and Australia. The jumping spider *M. plataleoides* belongs to the ant-mimicking genus *Myrmarachne*. This genus is one of the largest and most widespread of the salticid (jumping spiders) genera distributed primarily in tropical habitats (Jackson and Polland, 1996); the species distribution of *M. plataleoides* extends throughout

Keywords: Ant-mimicking spider, *Oecophylla*, *Myrmarachne*, myrmecomorphy.

south and south-east Asia. The spiders of this genus show behavioural and morphological resemblance to the ants they are associated with (Jackson, 1982).

Our study is based on the phenomenon of myrmecomorphy – a subset of ant mimicry. It involves the morphological and behavioural resemblance to ants of mimicking species. Ants have numerous features that make them effective models in mimicry systems. They are amongst the most common and conspicuous of insects (Hölldobler and Wilson, 1978), especially in tropical habitats making the phenomenon of myrmecomorphy a widespread one in arthropods (McIver and Stonedahl, 1993). In general, mimicry 'involves an organism (the mimic) which simulates signal properties of another organism (the model) so that the two are confused by a third organism (the operator) and the mimic gains protection, food or a mating advantage as a consequence'.

In this system, the model ant *O. smaragdina* has additional features as an effective model. These ants are arboreal, exceptionally abundant, aggressive and territorial. Earlier work with the closely related *Oecophylla longinoda* shows that *Oecophylla* major workers are strongly predaceous and highly aggressive towards intruders, including nearly all kinds of other ant species and non-nest mates of the same species. An interesting aspect of these characteristics with respect to the mimic is that in addition to strategies of mimicry, it must also be able to avoid detection by its model. In our study we explore this aspect of the system.

The spider *M. plataleoides* shows close resemblance to *Oecophylla*, however there is sexual dimorphism in its appearance. The male has enlarged chelicerae which are also positioned parallel to the plane of the body unlike that of the female. It also seems to show behavioural resemblance to *Oecophylla*. *Myrmarachne plataleoides* is found closely associated with the ant on tree trunk trails and on ground foraging trails. Hence we looked at behavioural strategies that it uses in addition to those involved in mimicking the ant's behaviour. We specifically looked at strategies that help it avoid detection/or prevent its capture by the ant.

To understand the components of behaviour involved, we designed a series of manipulative experiments. These experiments were carried out at the campus of the Indian Institute of Science, Bangalore. The timing of the experiments was kept constant. All experiments were begun at 1400 h. They were conducted at the base of a *Ficus benjamina* tree bearing *O. smaragdina* nests and the ant trunk trail used for baiting the ants to the experimental apparatus was kept constant. The spiders used in the trials were maintained in the laboratory under diurnal light conditions and fed a diet of fruit flies, egg yolk and sucrose.

Two types of manipulative experiments were carried out in the field. In one the experimental apparatus consisted of the use of a 2 m long opaque plastic tube with a diameter of 2 cm. The tube was kept in contact with the ant tree trail and positioned horizontally using stands. We used the tube as an analogue to conditions on a trunk trail and hence to gain insight into avoidance behaviours (if any). A test and a control experiment were carried out. The control experiment consisted of introducing the spider on the tube, at its centre and noting and timing of the behaviours that occurred in the period that the spider stayed on the tube. In the test experiment, an ant trail was allowed to form on the top of the tube, using egg-yolk as bait. The baiting was carried out on the upper surface and hence trails were formed on the upper surface only. Observations were carried out as above.

To investigate the reaction of the spider when there is an inevitable contact with the ant and an absence of at least one of the avoidance behaviours under examination, we carried out an experiment using a string of the same length as the tube. The diameter of the string was such that contact between spider and ants during simultaneous passage (of ants and the spider) was inevitable. The set-up remained the same as the tube experiment except for the use of the string. Experimental and control trials were again performed and behaviours timed and noted. In both the control experiments, care was taken to remove ant odour cues before trials. Nine trials were conducted in all cases, i.e nine different individual spiders were used. The following behaviours were scored for: lateral movement, reversal, dropping using silk line.

Relative proportions of time allocated by spiders to surfaces in the control and experimental trials were found to be significantly different with a greater proportion of time spent on the upper surface in the control (without ants) and also on the lower side in the experimental trial (with ants) (Table 1a). From Figure 1 it can be observed that a greater proportion of mean time was spent on the upper surface even in the experimental trial. The value for the

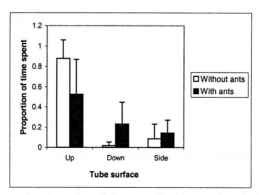

Figure 1. Proportion of time spent by *Myrmarachne plataleiodes* on different tube surfaces.

Table 1. Comparison of frequencies and durations of behaviour exhibited by *Myrmarachne platalejodes* in different experimental situations (Mann–Whitney *U*-test, *n* = 9).

a) Relative time spent on different surfaces in experiment and control trials

Surface	*U*	*p*-level
Upper	17	0.041001
Lower	12	0.012856
Sideways	26	n.s

b) Comparing behaviours in experimental and control trials

Behaviour	Experiment	*U*	*p*-level
Lateral movement	Tube	1.5	0.00574
Reversal	Tube	34	n.s
Dropped	Tube	27	n.s
Reversal	String	7.5	n.s
Dropped	String	14.5	0.038569

c) Comparing behaviours with and without ant encounters

Behaviour	Experiment	*U*	*p*-level
Reversal	Tube	34	ns
Lateral movement	Tube	13.5	0.017124
Dropped	String	16	n.s
Reversal	String	29.5	n.s

side allocation did not differ between control and experiment. Behaviours when compared between control and experiment indicated a higher frequency of lateral movements in the presence of the ant trail (Table 1*b*). Other behaviours did not show a significant difference. In the case of behaviours with and without ant encounters a greater number of lateral movements occurred when there were a greater number of ant encounters. This was also seen in the case of dropping from the string (Table 1*c*). However no such significant differences were found for reversal in the string or the tube.

Our study has indicated that lateral movement and dropping in the given experimental conditions (i.e. tube and string respectively) can be considered as avoidance behaviours while reversal does not appear to be an avoidance behaviour. The proportions of time spent on the various surfaces on the tube also are indicative of some changes in behaviour due to the presence of the ant trail. These results indicate that *M. plataleoides* does in fact show avoidance to its model when found in close proximity.

References

McIver J. D. & Stonedahl, G. Myrmecomorphy: morphological and behavioural mimicry of ants. *Annu. Rev. Entomol.* **38,** 351–379 (1993).

Hölldobler, B. & Wilson, E. O. The multiple recruitment systems of the African weaver ant *Oecophylla longinoda* (Latrielle) (Hymenoptera: Formicidae). *Behav. Ecol. Sociobiol.* **3,** 19–60 (1978).

Jackson R. R. The biology of ant-like jumping spiders: intraspecific interactions of *Myrmarachne lupata* (Araneae, Salticidae). *Zool. J. Linn. Soc.* **76,** 293–319 (1982).

Jackson R. R & Pollard, S. D. Predatory behaviour of jumping spiders. *Annu. Rev. Entomol.* **41,** 287–308 (1996).

Tropical Ecosystems: Structure, Diversity and Human Welfare.
Proceedings of the International Conference on Tropical Ecosystems
K. N. Ganeshaiah, R. Uma Shaanker and K. S. Bawa (eds)
Published by Oxford–IBH, New Delhi. 2001. pp. 654–658.

Ecophysiological significance of endothermy in *Rhizanthes lowii* and *Rafflesia tuan-mudae* (Rafflesiaceae)

Sandra Patiño*,[†], Alice A. Edwards**, Tuula Aalto[‡] and John Grace*

*Institute of Ecology and Resource Management, The University of Edinburgh,
Darwin Building, Mayfield Road, Edinburgh EH9 3JU, UK
**Chemistry Department, Universiti Brunei Darussalam, Jln. Tungku Link, Bandar
Seri Begawan BE 1410, Brunei Darussalam
‡Department of Physics, University of Helsinki, P.O.Box 9 (Siltavuorenpenger 20D),
FIN-00014 University of Helsinki, Finland
†e-mail: spatino@srv0.bio.ed.ac.uk

*R*hizanthes lowii and *Rafflesia tuan-mudae* (*Rafflesiaceae*) are rare parasitic plants on the roots and near-ground stems of a few but different species of the vine *Tetrastigma* (Vitaceae). They are sympatric in many areas of the tropical rain forest in South East Asia. In the understorey of the forest, where they are adapted to live, the ambient conditions are nearly constant (high relative humidity, low incident radiation and relatively constant air and soil temperature). These plants lack leaves, stems, or photosynthetic tissue, and produce large ephemeral flowers that are unisexual. They are characterised by gaseous emissions that attract the natural pollinators, carrion flies. Endothermy as part of the mimicry of the flower to attract the pollinating flies was hypothesised, as suggested by its appearance, mode of life, and pollination biology.

Keywords: Endothermy, Rafflesiaceae, mimicry, pollination, carbon dioxide.

Rhizanthes lowii was studied in the Batu Apoi Forest Reserve at the University of Brunei Darussalam Kuala Belalong Field Studies Centre (KBFSC) (115°8′E 4°32′N), Brunei, Borneo, S.E. Asia. *Rafflesia tuan-mudae* was studied in Taman Negara Gunung Gading, Lundu, Sarawak, Malaysia, S.E. Asia (1°40′N 109°52′E). The climate in both sites is aseasonal, with mean monthly rainfall > 100 mm for all months. The internal and surface temperatures of buds and floral structures were continuously monitored with fine thermocouples whilst radiation fluxes and microclimatic variables were recorded. In the case of *Rhizanthes lowii* there was evidence of both thermogenesis and thermo-regulation. Endothermy was detected in young and mature buds as well as in blooming flowers and even in decaying tissues three or more days after blooming. Tissue temperatures were maintained at 7–9 K above air temperature (Patiño *et al.*, 2000). In *Rafflesia tuan-mudae* it was found that the internal parts of the flower were maintained a few degrees (1–6 K) above air temperature and the maximum heating was in the evening.

To estimate the metabolic heat supply required to rise the tissue temperature above the ambient, heat transfer was modelled as a combination of forced and free convection. According to the model the heat supply necessary to produce an excess temperature 6–7 K inside a female mature bud of *Rhizanthes lowii* (7 days prior to anthesis) with a dry surface would be about 100 W m^{-2}. When the flower opens the heat supply to keep an excess temperature of 6–9 K in the centre would need to be about 180 W m^{-2}. In *Rafflesia* the heat supply necessary to produce an excess temperature of 3–6 K at a air flow of 2.4×10^{-4} m s^{-1} inside the diaphragm would be about 50–60 W m^{-2}. As the net radiation was usually less than 20 W m^{-2}, it was concluded that metabolic heat must have been an important part of the heat supply. Evidence that the energy required for the flowers to increase the temperature of the interior is endogenous is provided by the comparison of the measured net radiation in the understorey and the calculated heat supply. The net radiation in the understorey never exceeded 30 W m^{-2} while the calculated heat produced by the flower gives 180 W m^{-2}. The heat supply increased substantially when the flower was freshly open and the rate of heat was maintained for the two days of fresh flower (Figure 1).

The machinery for endothermy is present at the early stages of floral development. As they are parasitic, they have the advantage over most other species as the respiratory substrate for endothermy is derived from the host plant. It has been demonstrated that many flowers increase their temperature above the ambient as a consequence of endothermy, the heat being produced by the cyanide-insensitive respiration (Meeuse and Raskin, 1988; Skubatz *et al.*, 1990). A computational fluid dynamic model was employed for estimation of the CO_2 concentration inside the diaphragm and in the reproductive cavity of the *Rafflesia* flower, based on the geometry and the measured wind velocity near the surface of the flower. The calculations suggest that the concentration of CO_2 in the reproductive cavity is about 60 times higher than the concentration of CO_2 in the understorey of the forest.

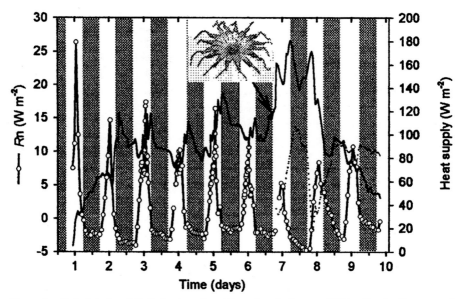

Figure 1. Calculated metabolic heat produced by a female flower of *Rhizanthes lowii* compared to the net radiation in the understorey for a period of ten days. This is the heat required to produce the observed elevation in temperature of the flower. Open symbols indicate the measured net radiation over the flower, the continuous line indicates the stalk of the column and the dotted line indicates the heat produced in the ovary. The arrow indicates the anthesis.

There is evidence that blowfly *Lucilia cuprina* has CO_2-specific sensory receptors (Stange, 1975). It has been also demonstrated that CO_2 has an anaesthetic effect in blowflies and that they are more sensitive to the CO_2 stimulus than to other anaesthetics (Diesendorf, 1975). Thus, it may be possible that the CO_2 produced by the *Rafflesia* and *Rhizanthes* flowers plays a role on the pollination by the blowflies.

The signals emitted by *Rafflesia tuan-mudae* and *Rhizanthes lowii* may be different. Despite *Rhizanthes* and *Rafflesia* attract blowflies of the same genera (*Lucilia, Chrysomya*, and *Hypopygiopsis* (Bänziger, 1991, 1996; Beaman *et al.*, 1988; Hidayati *et al.*, 2000)), it has been observed that *Rhizanthes lowii* stimulated oviposition in the flies, suggesting that *Rhizanthes lowii* is releasing specific volatiles that trick female flies. Oviposition by the blowflies on *Rafflesia tuan-mudae* was not observed in this study and has not been observed on other *Rafflesia* species (Bänziger, 1996; Beaman *et al.*, 1988).

Preliminary results from an ongoing study on the identification of the emitted volatiles by *Rafflesia tuan-mudae* and *Rhizanthes lowii* revealed that *Rhizanthes lowii* emitted 3-hydroxy-2-butanone, 2-ethyl-1-hexanol and *N,N*-diethyl-3-methyl-benzamide compounds found in other flowers (Knudsen *et al.*, 1993); and *Rafflesia tuan-mudae* emitted dimethyl disulphide and dimethyl trisulphide. These compounds have been found in various aroids (Stransky and Valterova, 1999) and in bat-pollinated flowers (Knudsen and Tollsten, 1995). Dimethyl disulphide and dimethyl trisulphide are compounds commonly related to bacterial growth on meat (Senter *et al.*, 2000) and dimethyl disulphide alone is the compound that gives the taste to the Camembert cheese (Demarigny *et al.*, 2000). It may be possible that dimethyl disulphide and dimethyl trisulphide and perhaps other compounds (not identified), mixed with CO_2 have a synergistic effect in the attraction of the pollinating blowflies. It has been found that the antennae of the female blowfly *Lucilia cuprina* have a specific receptor neuron tuned to dimethyl disulphide (Park and Cork, 1999) and it has been suggested that dimethyl trisulphide may be one of the major cues for calliphorid host finding (Nilssen *et al.*, 1996). The location of the flower by the flies may follow the classical downwind model (Stange, 1996) and once the flies have landed on the flower, mechanical and contact chemical inputs may guide the flies into the diaphragm and to the gynoecium. Although *Rafflesia* is not a 'trap' flower as defined (Dafni, 1984), it has been observed that in *Rafflesia tuan-mudae* that few flies remain inside the diaphragm for many hours (personal observations), perhaps due to the CO_2 anaesthetic effect as suggested previously (Dafni, 1984).

There may be some common characteristics among endothermic plants (Seymour and Schultze-Motel, 1997), but *Rhizanthes* and *Rafflesia* seem not to comply (Patiño *et al.*, 2000). Thus, it seems that endothermy in flowers is a homoplastic character that has evolved independently several times in different phylogenetic groups with different morphological organization and in each case it is associated with a particular pollination syndrome.

Acknowledgements

S. Patiño was funded from Colombia by a COLCIENCIAS scholarship and supported by the Instituto de Investigación de Recursos Biológicos 'Alexander von Humboldt'. We acknowledge financial support from the Davies Expedition Fund and Development Trust of the University of Edinburgh.

References

Bänziger, H. Stench and fragrance: unique pollination lure of Thailand's largest flower, *Rafflesia Kerrii* Meijer. *Nat. His. Bull. Siam. Soc.* **39**, 19–52 (1991).
Bänziger, H. Pollination of a flowering oddity: *Rhizanthes zippelii* (Blume) Spach (Rafflesiaceae). *Nat. His. Bull. Siam. Soc.* **44**, 113–142 (1996).
Beaman, R. S., Decker, P. J. & Beaman, J. H. Pollination of *Rafflesia* (Rafflesiaceae). *Am. J. Bot.* **75**, 1148–1162 (1988).
Dafni, A. Mimicry and deception in pollination. *Annu. Rev. Ecol. Syst.* **15**, 259–278 (1984).
Demarigny, Y., Berger, C., Desmasures, N., Gueguen, M. & Spinnler, H. E. Flavour sulphides are produced from methionine by two different pathways by *Geotrichum candidum*. *J. Dairy Res.* **67**, 371–380 (2000).
Diesendorf, M. *General Anaesthetic Excitation and Inhibition of Insect CO₂-Receptors: An Interpretation* (eds Denton, D. A. & Coghlan, J. P.) 195–198 (Academic Press Olfaction and Taste V. Proceedings of the fifth International Symposium. London, 1975).

Hidayati, S. N., Meijer, W., Baskin, J. M. & Walck, J. L. A contribution to the life history of the rare Indonesian holoparasite *Rafflesia patma* (Rafflesiaceae). *BTROA* **32,** 408–414 (2000).

Knudsen, J. T. & Tollsten, L. Floral scent in bat-pollinated plants: a case of convergent evolution. *Bot. J. Linn. Soc. (London)* **119,** 45–57 (1995).

Knudsen, J. T., Tollsten, L. & Bergstrom, L. G. Floral scents – a checklist of volatile compounds isolated by head-space techniques. *Phytochemistry* **33,** 253–280 (1993).

Meeuse, B. J. D. & Raskin, I. Sexual reproduction in the arum lily family, with emphasis on thermogenicity. *Sex Plant Reprod.* **1,** 3–15 (1988).

Nilssen, A. C., Tommeras, B. A., Schmid, R. & Evensen, S. B. Dimethyl trisulphide is a strong attractant for some calliphorids and a muscid but not for the reindeer oestrids *Hypoderma tarandi* and *Cephenemyia trompe*. *Entomol. Exp. Appl.* **79,** 211–218 (1996).

Park, K. C. & Cork, A. Electrophysiological responses of antennal receptor neurons in female Australian sheep blowflies, *Lucilia cuprina*, to host odours. *J. Insect Physiol.* **45,** 85–91 (1999).

Patiño, S., Grace, J. & Bänziger, H. Endothermy by flowers of *Rhizanthes lowii* (Rafflesiaceae). *Oecologia* **124,** 149–155 (2000).

Senter, S. D., Arnold, J. W. & Chew, V. APC values and volatile compounds formed in commercially processed, raw chicken parts during storage at 4 and 13 degrees C and under simulated temperature abuse conditions. *J. Sci. Food Agric.* **80,** 1559–1564 (2000).

Seymour, R. S. & Schultze-Motel, P. Heat-producing flowers. *Endeavour (Cambridge)* **21,** 125–129 (1997).

Skubatz, H., Williamson, P. S., Schneider, E. L. & Meeuse, B. J. D. Cyanide-insensitive respiration in thermogenic flowers of *Victoria* and *Nelumbo*. *J. Exp. Bot.* **41,** 1335–1339 (1990).

Stange, G. *Linear Relation Between Stimulus Concentration and Primary Transduction Process in Insects* (eds Denton, D. A. & Coghlan, J. P.) 207–210 (Academic Press Olfaction and Taste V. Proceedings of the fifth International Symposium. London., 1975).

Stange, G. Sensory and behavioural responses of terrestrial invertebrates to biogenic carbon dioxide gradients. in *Advances in Bioclimatology-4* (ed. Stanhill, G.) 223–253 (Springer, London, 1996).

Stransky, K. & Valterova, I. Release of volatiles during the flowering period of *Hydrosme rivieri* (Araceae). *Phytochemistry* **52,** 1387–1390 (1999).

Restoration Ecology

Tropical Ecosystems: Structure, Diversity and Human Welfare.
Proceedings of the International Conference on Tropical Ecosystems
K. N. Ganeshaiah, R. Uma Shaanker and K. S. Bawa (eds)
Published by Oxford–IBH, New Delhi. 2001. pp. 661–664.

Accelerating plant colonization on landslides in Puerto Rico by additions of bird perches and organic matter

A. B. Shiels* and L. R. Walker

Department of Biological Sciences, University of Nevada, Las Vegas, USA
*e-mail: shiels@nevada.edu

Landslides are among the most severe natural disturbances in Caribbean rainforests. Triggered by heavy rains, landslides result in the loss of at least the topsoil layer of the soil profile as well as the existing vegetation. The resulting disturbance represents a significant gap that alters ecological processes within the local terrestrial and aquatic ecosystems.

Conceptual models suggest that in order for plant colonization to take place on landslides, the following major factors must be present: soil stability, availability of propagules, suitable sites for germination, and the presence of organic matter and associated nutrients (Walker *et al.*, 1996). This study experimentally tests the importance of facilitating dispersal of propagules to landslides by adding artificial bird perches. Additionally, various types of organic matter have been added to landslide soil to experimentally determine the importance of various nutrient additions in the process of plant colonization.

Wind-dispersed species are often associated with early colonizing plant communities (Margalef, 1968). This is apparent on the majority of landslides

Keywords: Landslides, organic matter, artificial bird perches, restoration.

in Puerto Rico, where grasses and ferns are present. These pioneer species, especially thickets of climbing fern (Gleichenaceae), may develop monospecific populations that dominate for decades (Walker, 1994). Perch availability has been proposed as the primary factor influencing the number of bird-dispersed seedlings on an open site (Campbell *et al.*, 1990). In this study, artificial bird perches were added to six landslides with varying vegetation (grasses, ferns, bare) in Luquillo Experimental Forest (LEF), Puerto Rico, to test whether they would increase forest seed inputs. Although there have been several studies measuring seed rain dynamics in relation to perch additions in pastures (McDonnell and Stiles, 1983; McDonnell, 1986; McClanahan and Wolfe, 1993; Holl, 1998), to our knowledge there have been no similar studies on landslides.

Because of the loss of the topsoil layer and existing vegetation following landslides, a nutrient-poor substrate results. Additions of various types of organic matter will help uncover the important inputs that potentially accelerate plant colonization of landslides. We experimentally determined the impact of leaf litter from *Cyathea arborea* (tree fern) and *Cecropia scheberiana*, as well as mature forest soil, commercial fertilizer, and a control treatment on seed germination and establishment of two landslide species. Five different landslides, each with little or no vegetation present, were used for this experiment in LEF.

Although wind-dispersed seeds (primarily grasses) made up the greatest concentration of seed inputs to landslides, the input of seeds of mature forest species was much greater under perches ($n = 49$) than in controls without perches ($n = 1$). Artificial bird perches increased the number of bird-dispersed seeds (e.g., *Guarea*, *Matayba*) into landslides, showing that birds utilize artificial bird perches and defecate seeds beneath them. In addition to perch availability, our data show that there must be some vegetation (e.g., grasses, ferns) present on the landslide in order for seed deposition to take place (Figure 1), perhaps because bare areas are a deterrent to birds, or because underlying vegetation offers food or cover. These results present a problem for restoration purposes because seed germination and establishment of forest seeds may be inhibited on vegetated landslides due to competition with the early colonizing species (Walker, 1994). Additionally, seeds of mature forest species may be inhibited from germinating on non-vegetated landslides as a result of low organic matter and severe microclimates. Germination studies on the seeds collected from our artificial perch experiments are now underway to elucidate this problem.

Seed germination for both *Paspalum* and *Phytolacca* was highly variable within and between each landslide comparison. This germination variability was consistent with the more controlled bench experiment at El Verde Field Station, suggesting that the organic matter treatments do not alter the total number of germinating *Paspalum* or *Phytolacca* seeds. However, the bench

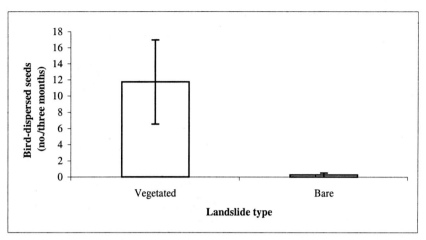

Figure 1. Mean number of bird-dispersed seeds collected over a three-month period beneath artificial bird perches on two different landslide types. Grass and climbing ferns dominate the vegetated landslides ($n = 4$). Bare landslides ($n = 2$) had recently slid and were relatively free of vegetation. Only one bird-dispersed seed was found in control plots without perches.

Table 1. Average biomass (mg) per individual *Phytolacca rivinoides* and *Paspalum millegrana* seedling harvested from organic matter treatment trays after a three-month experiment at El Verde Field Station.

	Control	Cecropia	Cyathea	Forest soil	Fertilizer
Phytolacca	2.93 ± 0.21	2.25 ± 0.13	2.36 ± 0.26	7.00 ± 0.21	47.42 ± 3.99
Paspalum	5.21 ± 0.52	4.55 ± 0.41	3.69 ± 0.70	18.18 ± 2.51	282.45 ± 53.28

experiment shows large differences in seedling biomass after a 3-month period (Table 1). This was consistent with our original hypothesis that seedling establishment would respond most to commercial fertilizer, followed by forest soil treatments, due to nutrient availability. However, with increased time, we expect that nutrients will become relatively more available in *Cecropia* and *Cyathea* leaf litter, and in mature forest soil treatments, as decomposition continues and the fertilizer is leached.

This study suggests that bird perches will accelerate the inputs of seeds of forest species to landslides. Our current and future studies on the germination success of forest seed species will help determine the importance of artificial bird perches in future vegetation restoration on landslides in the tropics. Accelerating plant colonization on landslides is possible through the inputs of commercial fertilizer and mature forest soil. Additions of leaf litter to nutrient-poor landslide soil did not have any short-term benefits for restoration of landslides.

Acknowledgements

This research is supported by the Luquillo LTER; funded by the National Science Foundation, University of Puerto Rico, and International Institute of Tropical Forestry. We would like to thank the following people for their critical assistance and/or guidance throughout this project: Laura Weiss, Fred Landau, Maria Aponte, Jakob Falk-Sorensen, Jill Thompson, Paul Klawinski, Diane Wagner, Dan Thompson, Joe Wunderle, and Tim Bragg. Additionally, we would like to thank Xiaming Zou and Honghua Ruan for the use of their laboratory facilities at El Verde Field Station.

References

Campbell, B. M., Lynman, T. & Hatton, J. C. Small-scale patterning in the recruitment of forest species during succession in tropical dry forest, Mozambique. *Vegetation* **87**, 51–57 (1990).

Holl, K. D. Do perching structures elevate seed rain and seedling establishment in abandoned tropical pasture? *Restorat. Ecol.* **6**, 253–261 (1998).

Margalef, R. Ecological succession and exploitation by man. in *Perspectives in Ecological Theory* (eds Beadle, G. W., Bogorad, L., Williams, R. C., Moscona, A. A. & Wool, I. G.) 26–38 (University of Chicago Press, London, 1968).

McClanahan, T. R. & Wolfe, R. W. Accelerating forest succession in fragmented landscape: The role of birds and perches. *Conserv. Biol.* **7**, 279–288 (1993).

McDonnell, M. J. Old field vegetation height and the dispersal pattern of bird disseminated woody plants. *Bull. Torr. Bot. Club* **113**, 6–11 (1986).

McDonnell, M. J. & Stiles. E. W. The structural complexity of old field vegetation and the recruitment of bird-dispersed plant species. *Oecologia* **56**, 109–116 (1983).

Walker, L. R. Effects of fern thickets on woodland development on landslides in Puerto Rico. *J. Vegetat. Sci.* **5**, 525–532 (1994).

Walker, L. R., Zarin, D. J., Fetcher, N., Myster, R. W. & Johnson, A. H. Ecosystem development and plant succession on landslides in the Caribbean. *Biotropica* **28**, 566–576 (1996).

Tropical Ecosystems: Structure, Diversity and Human Welfare.
Proceedings of the International Conference on Tropical Ecosystems
K. N. Ganeshaiah, R. Uma Shaanker and K. S. Bawa (eds)
Published by Oxford–IBH, New Delhi. 2001. pp. 665–669.

Short dry spells in the wet season increase mortality of tropical pioneer seedlings

B. M. J. Engelbrecht*[,†,‡], J. W. Dalling**, T. R. H. Pearson[§], R. L. Wolf[#], D. A. Galvez[†], T. Koehler[†], M. C. Ruiz[†] and T. A. Kursar*[,†]

*University of Utah, Department of Biology, Salt Lake City, USA
[†]Smithsonian Tropical Research Institute, Balboa, Panama, USA
**University of Illinois, Department of Plant Biology, Champaign-Urbana, USA
[§]University of Aberdeen, Department of Plant and Soil Science, Aberdeen, UK
[#]Yale University, School of Forestry and Environmental Studies, New Haven, USA
[‡]e-mail: engelbrb@bci.si.edu

Even in the wet tropics short-term rainfall patterns show considerable variation. While many areas have a pronounced dry season that can result in serious water stress for plants (e.g. Walsh and Newbery, 1999; Tobin et *al.*, 1999), even in aseasonal climates dry spells of several days or even weeks do occur (e.g. Burslem and Grubb, 1996). These are significant because rainfall and/or soil water availability have been shown to influence diversity and structure of tropical forests (e.g. Gentry, 1988), and germination, growth, mortality and habitat association of tropical plant species (e.g. Garwood, 1983; Mulkey and Wright, 1996; Webb and Peart, 2000).

Keywords: Pioneer seedlings, drought, gap, survival, allocation.

Forest gaps play a major role in influencing recruitment and growth of many species, and the role of gap partitioning for forest dynamics and diversity has received much attention (e.g. Brown and Whitmore, 1992; Dalling *et al.*, 1998). Pioneer species are restricted to gaps for their regeneration and have characteristically small seeds and seedlings. Small seedlings are considered to be especially prone to drought stress because of shallow/small root systems that provide only limited access to soil water. Additionally, gaps are exposed to high irradiance and consequently high temperatures. The upper soil layers may therefore dry out fast, and at the same time, evaporative demand on the seedlings will be very high. We hypothesized that even short dry spells of a few days duration increases mortality of the early seedling stages of pioneer species, due to severe drought stress. Shallow roots and/or low rooting depth relative to the transpiring leaf surface should accentuate drought stress (e.g. Sperry *et al.*, 1998) and increase the effect of dry spells on seedling mortality.

The study was conducted in Panama at the Barro Colorado Nature Monument (BCNM). The area has an annual rainfall of c. 2600 mm with a pronounced 4-month dry season. Our interest is in short dry periods that occur during the 8-month wet season. We analysed the frequency of dry spells in the wet season (days with rainfall < evaporation) from data from the Smithsonian Environmental Science Program.

We selected six abundant pioneer species for the study: *Apeiba membranaceae* Spruce ex Benth., *Ochroma pyramidale* (Cav. ex Lam.) Urban and *Luehea seemanii* Tr. & Planch. (all three Malvaceae), *Miconia argentea* (Sw.) D. C. (Melastomataceae), *Cecropia insignis* Liebm. (Cecropiaceae) and *Piper marginatum* Jacq. (Piperaceae). All of these species are characterized by gap requirement for regeneration, high maximum growth rates and small seeds (Dalling *et al.*, 1998). *Miconia, Piper* and *Cecropia* have significantly smaller seed weights (0.06–0.48 mg) than *Luehea, Ochroma* and *Apeiba* (1.66–14.02 mg; ANOVA: $p < 0.05$).

Seedling plots were established in four artificially created (20 m × 20 m) forest gaps on the Buena Vista Peninsula, and subjected to two treatments: dry and irrigated. Four plots (1.30 m × 0.80 m) of each treatment were established in each gap, and all litter was removed. Seeds were sown one month after the beginning of the wet season (late May, 2000).

The treatments started with the beginning of a natural dry spell on 07/09/2000, when 6201 seedlings were present. All plots were covered with rain-proof shelters – roofs of transparent plastic. The dry treatments were kept dry for 11 days, the wet treatments were irrigated daily (equivalent to approx. 540 mm monthly rainfall). Mortality of all seedlings was censused daily in the mornings. Exemplary soil water potentials were measured daily at noon at a random location in each plot (0.5–1 cm depth), and depth profile of soil water potentials (0–10 cm) was measured (thermocouple psychro-meters, Merrill 76-IVC). At the end of the experiment, a subsample of 7–13 seedlings per species was harvested from the wet treatment plots for the determination of biomass, leaf area and root length.

Analysis of long-term rainfall data indicates that dry spells of up to five days occur frequently in the wet season on BCNM, with dry periods of up to 30 days also recorded. During dry spells, water potentials in the upper soil layers rapidly decrease in gaps. Water potentials down to – 4.5 MPa were observed after only four days of dry treatment. Soil water potential in the upper 10 cm of soil showed a steep, exponential decrease towards the soil surface. Thus, while overall soil moisture is higher in gaps than in the understorey (e.g. Veenendaal *et al.*, 1995), steep gradients at the surface can lead to critical water potentials for tiny pioneer seedlings in gaps. Even small differences in rooting depth may lead to considerable differences in access to water, and therefore influence seedling survival.

Mortality of seedlings of six pioneer species in gaps was much higher in an experimental dry treatment, that was started with a natural dry spell, compared to an irrigation treatment (Figure 1). This trend could be observed for all species. In *Ochroma*, *Miconia*, and *Cecropia* it became significant after 6 and 7 days. For *Piper* a significant effect was only observed after 11 days, and *Luehea* showed no significant effect over the course of the experiment. *Apeiba* could not be formally analyzed, however, visual inspection of the data suggests that there was no effect. Thus, drought had an effect on seedling survival of *Ochroma*, *Miconia* and *Cecropia* within the length of dry spells that frequently occur in the wet season on BCI. This indicates that dry spells may

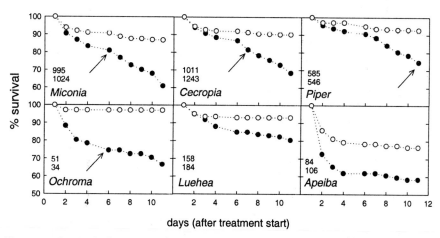

Figure 1. Survival of seedlings of six pioneer species in wet and dry treatment plots. Data are pooled for all plots and gaps. Closed symbols: dry treatment; open symbols: wet treatments. The arrows indicate, when a significant treatment effect was first observed (ANOVA: $p < 0.05$). The number of seedlings at the start of the wet and dry treatment, respectively, is also given.

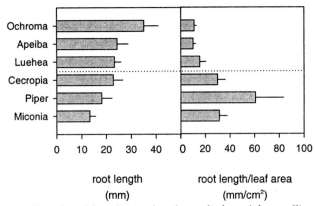

Figure 2. 'Root length', and 'root length per leaf area' for seedlings of six pioneer species at the end of the experiment. The seedlings were between 17 and 44 days old. 'Root length' was significantly different between species (ANOVA: $p < 0.0001$), and 'root length per leaf area' was different between small and tiny seeded species (t-test: $p < 0.05$). All data were log-transformed prior to analysis. The data in the graph are averages ± SE.

have a significant effect on population dynamics of these species. In contrast, *Piper*, *Luehea* and *Apeiba* did not show an effect of drought on seedling survival within dry spell lengths that are frequent, and therefore ecologically meaningful at the BCNM. Effects of drought on survival of the early seedling stages of these species are unlikely, except in very exceptional years.

'Root length', and the ratio of 'root length to leaf area' differed at the end of the experiment (Figure 2): The species with the smallest seeds had the lowest absolute root length, but one of them, *Piper*, also had the highest 'root length per leaf area'. This was the only species with very small seedlings that did not show decreased survival in normal wet season dry spells at the BCNM. In contrast, *Ochroma*, the species with the longest roots and thus the best access to soil water, nevertheless showed strongly increased mortality in the dry spell, perhaps associated with it having the lowest 'root length per leaf area'. These data strongly suggest that the combination of absolute rooting depth, determining the access to soil water, and the ratio of rooting depth to leaf area, determining the plant water balance and thus plant water potentials, is critical for the survival of these pioneer species.

This is the first time that increased mortality of pioneer seedlings due to drought has been conclusively demonstrated. The results suggest that pioneer seedling susceptibility to drought offers another niche axis for gap partitioning. They also suggest that even relatively subtle differences in rainfall patterns may govern pioneer distribution along rainfall gradients. Expected shifts of rainfall patterns in the tropics associated with future global climate change (e.g. Hulme and Viner, 1998), make our findings especially relevant.

References

Brown, N. D. & Whitmore, T. C. Do dipterocarp seedlings really partition tropical rain forest gaps? *Phil. Trans. R. Soc. London* **B335**, 369–378 (1992).

Burslem, D. F. R. P. & Grubb, P. J. Responses to simulated drought and elevated nutrient supply among shade-tolerant tree seedlings of lowland tropical forest in Singapore. *Biotropica* **28**, 636–648 (1996).

Dalling, J. W., Hubbell, S. P. & Silvera, K. Seed dispersal, seedling establishment and gap partitioning among tropical pioneer trees. *J. Ecol.* **86**, 674–689 (1998).

Garwood, N. C. Seed germination in a seasonal tropical forest in Panama: a community study. *Ecol. Mono.* **53**, 159–181 (1983).

Gentry, A. H. Changes in plant community diversity and floristic composition on environmental and geographical gradients. *Ann. Miss. Bot. Gard.* **75**, 1–34 (1988).

Hulme, M. & Viner, D. A climate change scenario for the tropics. *Climatic Change* **39**, 145–176 (1998).

Mulkey, S. S. & Wright, S. J. Influence of seasonal drought on the carbon balance of tropical forest plants. in *Tropical Plant Ecophysiology* (eds Mulkey, S. S., Chazdon, R. L. & Smith, A. P.) (Chapman & Hall, New York, 1996).

Sperry, J. S., Adler, F. R., Campbell, G. S. & Comstock, J. P. Limitation of plant water use by rhizosphere and xylem conductance: results from a model. *Plant Cell Environ.* **21**, 347–359 (1998).

Tobin, M. F., Lopez, O. R. & Kursar, T. A. Responses of tropical understorey plants to a severe drought: tolerance and avoidance of water stress. *Biotropica* **31**, 570–578 (1999).

Veenendaal, E. M., Swaine, M. D., Agyeman, V. K., Blay, D., Abebrese, I. K. & Mullins, C. E. Differences in plant and soil water relations in and around a forest gap in West Africa during the dry season may influence seedling establishment and survival. *J. Ecol.* **83**, 83–90 (1995).

Walsh, R. P. D. & Newbery, D. M. The ecoclimatology of Danum, Sabah, in the context of the world's rainforest regions, with particular reference to dry periods and their impact. *Phil. Trans. R. Soc. London* **B354**, 1391–1405 (1999).

Webb, C. O. & Peart, D. R. Habitat associations of trees and seedlings in a Bornean rain forest. *J. Ecol.* **88**, 464–478 (2000).

Tropical Ecosystems: Structure, Diversity and Human Welfare.
Proceedings of the International Conference on Tropical Ecosystems
K. N. Ganeshaiah, R. Uma Shaanker and K. S. Bawa (eds)
Published by Oxford–IBH, New Delhi. 2001. pp. 670–673.

An approach to conservation of threatened plant species through species recovery

Subhash Mali*, D. K. Ved and T. S. Srinivasmurthy

Foundation for Revitalisation of Local Health Traditions, # 50 MSH Layout,
3rd Main, 2nd Cross, Anandnagar, Bangalore 560 032, India
*e-mail: subhash.mali@frlht-india.org

Globally around 35,000 vascular plants are estimated to be threatened (IUCN Red List, 1997) and many are feared to go extinct in the next 50 years (Raven, 1987). Habitat loss and degradation, and population decline caused by rapid loss of tropical forests have been identified as major threats to species survival. The threats are especially disturbing, considering that a large number of species offer valuable resources to mankind. In this paper we are interested in the threats to medicinal plant species. We have initiated attempts to investigate the reasons for decline in medicinal plant species and to develop a conservation plan for the recovery. We present here a summary of efforts initiated by us towards bridging the existing gap in our knowledge of the medicinal plant species and the information needed for their recovery.

As a first step, we prioritized the medicinal plants in south India based on their threat assessment. One hundred and ten species have been accordingly identified. At least 80 of these species occur in the Medicinal Plant Conservation Areas (MPCA) established by the Foundation for Revitalisation of Local Health Traditions (FRLHT) in the three states of southern India.

Keywords: Tropical plant species, species extinction, species recovery, priority species, *in situ* conservation.

In second step we have attempted to seek information on the biological status of the species with an emphasis on the species distribution and demography. Information at this level helps identify species survival bottlenecks. Basically the survival bottlenecks can either be due to intrinsic or extrinsic factors, or both. Some of the extrinsic factors such as human exploitation, habitat destruction and destructive methods of harvesting can be easily controlled using simple management approaches. But for intrinsic factors related to constraints in the reproductive behaviour of the species, the solutions are much more complex and needs careful study before any intervetion can be planned.

Based on the identification of such bottlenecks, either intrinsic or extrinsic or both the final step in the process involves developing approaches that could help in the recovery of the species.

Over the last one-year we have identified various populations of priority species, and marked them for long-term research and monitoring. The systematic studies on population dynamics and phenological aspects of the species have also been commenced. Data gathered thus far are insufficient to predict explicitly any trends, however, our preliminary studies do reflect the probable problems affecting the species. We present here two case studies and disucss the possible strategies for their recovery.

Myristica dactyloides (Myristicaceae) and *Coscinium fenestratum* (Menispermaceae) are the two examples of the species targeted for the recovery exercise. The threat status of the former is vulnerable, while the latter is critically endangered.

M. dactyloides is sparsely distributed mostly in the evergreen forests of the Western Ghats. The species bears a few large fruits and is dispersed mostly through primates and large frugivores such as hornbills. The fruit aril is exploited for its medicinal use and over exploitation of the entire fruit crop for aril might have resulted in its extremely poor regeneration in the wild. Observation from the network of eight MPCA sites substantiated this finding, no saplings of *Myristica* were recovered (Figure 1). The regeneration

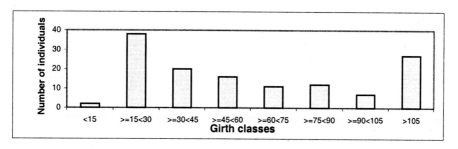

Figure 1. *Myristica dactyloides*: Distribution across girth class at Silent Valley MPCA.

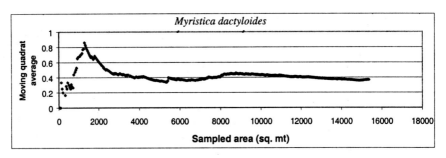

Figure 2. *Myristica dactyloides* density (mean no. of individuals/40 sq m = 0.3664) at Silent Valley MPCA.

failure may be attributed to the lack of effective seed dispersal. Given a fairly long history of over exploitation of the fruit crops, it will not be surprising that the depleted fruit resources may not be attractive enough to the seed dispersers such as primates and large frugivore birds. The relatively high density of the individuals at Silent Valley MPCA site is though encouraging (Figure 2); only detailed studies initiated in this direction would identify the cause for regeneration failure at the other sites.

Coscinium fenestratum, a dioecious liana, is reported only from two MPCA sites at Agasthymalai and Kulamavu in the Western Ghats and at both the sites there were no reproducing individuals observed. Extensive cutting of its stem for medicinal use might have depleted the plants. Frequent cutting of the stem before it bears flowers and fruits would have curtailed the reproduction and regeneration of the species. Being dioecious, the reproductive success of the species is highly influenced by sex ratio, spatial distribution of male and female plants and pollinator visitation. We premise that, the long history of stem-harvesting practices might have altered even the pollinator visitation to the lianas.

The recovery of *Myristica dactyloides* and *Coscinium fenestratum* are both challenging. In the former, not only do we need to augment the individuals in the population but also ensure their dispersal. In the latter, besides protecting the lianas from premature harvest, we need to ensure a critical reproductive populations of male and female plants, their appropriate spatial distribution and effective pollination.

It is evident that, the issue of conservation of a tropical species is quite complex and systematic conservation efforts are needed to direct us in designing appropriate species recovery program. The success of such programs would reinforce the various conservation programs on endangered species.

Acknowledgements

This is a collaborative program between FRLHT, ATREE, IFTGB and TBGRI and we sincerely acknowledge their involvement.

References

Raven, P. H. The scope of the plant conservation problem world-wide. in *Botanic Garden and World Conservation Strategy* (eds Bramwell, D., Hamman, O., Heywood, V. & Synge, H.) 19–20 (Academic Press, London, 1987).
IUCN. *Red Data Book*; 19–20 (IUCN, 1997).

Tropical Ecosystems: Structure, Diversity and Human Welfare.
Proceedings of the International Conference on Tropical Ecosystems
K. N. Ganeshaiah, R. Uma Shaanker and K. S. Bawa (eds)
Published by Oxford–IBH, New Delhi. 2001. pp. 674–676.

Dominance by shrubs and medium-sized trees in early regrowth after a hurricane in tropical forest

Nicholas Brokaw*[†], Bruce L. Haines**, D. Jean Lodge[‡], Lawrence R. Walker[§] and Jill Thompson*

*Institute for Tropical Ecosystem Studies, University of Puerto Rico, P.O. Box 363682, San Juan, Puerto Rico 00936, USA
**Department of Botany, University of Georgia, Athens, Georgia 30602, USA
[‡]USDA Forest Service, Center for Forest Mycology Research, Forest Products Laboratory, P.O. Box 1377, Luquillo PR 00773, USA
[§]Department of Biological Sciences, University of Nevada, 4505 Maryland Parkway, Las Vegas, Nevada 89154, USA
[†]e-mail: nbrokaw@lternet.edu

Forests usually contain various plant life forms, such as herbs, shrubs, and trees of various mature sizes. Forests also contain plants representing different life histories, such as pioneer and shade-tolerant species. This variety of form and history contributes to the structural and biological diversity of the forest and also plays a role in ecosystem function. After a severe hurricane in Puerto Rico in 1989 we compared regeneration among tree species that differ in mature height, and among species that were independently assessed as 'pioneer' and 'climax' tree species.

The study was carried out in 58 4-m^2 plots in a subtropical wet forest in El Verde Research Area in the Luquillo Experimental Forest (LEF) of

Keywords: Hurricane, life form, regeneration, shrub, tree.

northeastern Puerto Rico. In each plot self-supporting woody stems ≥ 1 and < 4.0 m tall ('saplings') were tallied during the first 62 months after the hurricane.

We recorded 49 species of woody plants as saplings in the 58 quadrats. Of these, 10 were large trees (mature height), 20 medium-sized trees, 7 small trees, and 12 were shrubs. Compared with the numbers of species of each of these life forms over the whole LEF (Little and Woodbury, 1976), medium-sized trees were over-represented and small trees markedly under-represented (Table 1). Small trees are the most speciose group in the LEF but the least speciose in this study.

Relative abundances among the 48 species were uneven. *Piper glabrescens* (shrub), *Cecropia schreberiana* (medium-sized, pioneer tree), *Psychotria berteriana* (shrub), and *Palicourea riparia* (shrub) reached much higher densities (counts of individuals, not stems) than other species and accounted for 352 (61%) of the 575 total sapling individuals recorded during the study. Related to this dominance medium-sized trees and shrubs heavily dominated the sapling size class in this study (Table 2).

Of the tree species in this study that were assigned to a life history class, 11 were pioneers and 9 were climax species. Pioneers dominated in terms of numbers of stems (among classified species) throughout the study, but their numbers declined sharply at the end of the study.

Table 1. Number and percentage of species among different life forms in this study and in the Luquillo Experimental Forest (LEF), Puerto Rico, as a whole.

Life form	Species (this study)	Species LEF	% (this study)	% LEF
Large tree	10	21	20.4	12.6
Medium tree	20	46	40.8	27.5
Small tree	7	60	14.3	36.0
Shrub	12	40	24.5	24.0

Table 2. Number of individuals of different life forms, including the maximum number and the final number, in plots in the Luquillo Experimental Forest, Puerto Rico, after a hurricane.

Life form	Maximum number	End number
Large tree	16	11
Medium tree	151	69
Small tree	14	12
Shrub	250	139

We make the following conclusions from this study of saplings in the regeneration following a hurricane: (1) among life forms, medium-sized trees contribute a disproportionately large share of species richness, while medium-sized trees and shrubs contribute a disproportionately large share of individuals; (2) small (mature height) trees are markedly species-poor among regeneration, compared to the overall pool of such species in the area, and this is related to the fact that there were few individuals of these species as well. Data currently being analyzed on seedlings (stems > 0.1 m and < 1.0 m tall) in this study seem to support these conclusions for saplings. We will discuss the implications of these conclusions for community structure, forest architecture, and ecosystem recovery following major disturbance.

Acknowledgements

This research was performed under grant BSR-8811902 from the National Science Foundation to the Institute for Tropical Ecosystem Studies, University of Puerto Rico, and the International Institute of Tropical Forestry, as part of the Long-Term Ecological Research Program in the Luquillo Experimental Forest. The Forest Service (U.S. Dept. of Agriculture) and the University of Puerto Rico provided additional support.

Reference

Little, E. L. Jr. & Woodbury, R. O. Trees of the Caribbean National Forest, Puerto Rico. Forest Service Research Paper, ITF-20, Institute of Tropical Forestry, Río Piedras, Puerto Rico (1976).

Tropical Ecosystems: Structure, Diversity and Human Welfare.
Proceedings of the International Conference on Tropical Ecosystems
K. N. Ganeshaiah, R. Uma Shaanker and K. S. Bawa (eds)
Published by Oxford–IBH, New Delhi. 2001. pp. 677–680.

Ecology and conservation of *Semecarpus kathalekanensis*: A critically endangered freshwater swamp tree species of the Western Ghats, India

H. B. Raghu*, R. Vasudeva*,[#], Dasappa**,
R. Uma Shaanker[†] and K. N. Ganeshaiah[‡]

*Department of Forest Biology and Tree Improvement, College of Forestry,
Sirsi 581 401, India
**Department of Forestry, [†]Department of Crop Physiology, [‡]Department of
Plant Genetics and Breeding, University of Agricultural Sciences, GKVK,
Bangalore 560 065, India
[#]e-mail: vasukoppa@yahoo.com

Freshwater swamps are one of the most threatened ecosystems in India due to the mounting anthropogenic pressures on them. Once forming an extensive network along the streams of Western Ghats, the swamps are now reduced to highly fragmented pockets. The swamps of Western Ghats, known as *Myristica* swamps, are an abode for several rare-relic plant species belonging to the family Myristicaceae. Recently a new tree species has been described from one of the *Myristica* swamps of Uttara Kannada. It has been named as *Semecarpus kathalekanensis (Species De Novo)*, because of its

Keywords: *Semecarpus kathalekanensis*, freshwater swamps, critically endangered tree, Western Ghats, India.

predominant occurrence in *Kathlekan* meaning 'Dark-Forest' swamp (Dasappa and Swaminath, 2000). Unfortunately, a large number of these swamps are being converted into arecanut gardens and as a consequence, many rare species are becoming endangered (Chandran *et al.*, 1999). The fate of *S. kathalekanensis* in the Western Ghats could be symptomatic of a large number of rare plant species distributed in small fragmented patches. We address here a few aspects of its ecology and conservation, and discuss the implications for conserving this and other rare species in the Western Ghats.

Only four populations of *S. kathalekanensis* have been recorded in the Siddapur range of Uttara Kannada; adult individuals are found in only three populations. *Kathlekan*-II possessed the highest (19 out of 38) number of flowering individuals. The *Kathlekan*-I population had 17 flowering individuals out of a total of 40. The least of 12 individuals were found in *Mundge teggu* population among which only 8 were flowering. Individuals of *S. kathalekanensis* generally bear only male or female flowers. However, in the *Kathlekan*-II population, three trees had inflorescence bearing male as well as female flowers. These individual trees ultimately set only about 4–20 fruits per tree hence they could be regarded as largely male in function.

The sex ratio (defined as the number of male individuals per female) was female biased in *Kathlekan*-I and in *Mundge teggu* populations ($M/F = 0.41$ and 0.6, respectively), whereas in the *Kathlekan*-II, the sex ratio was 1.38 hence male biased. However, if monoecious individuals are considered as functionally males, the sex ratio for all populations pooled was 0.88 (i.e. nearly 1:1).

The populations of *S. kathlekanensis* are not even-aged. Among the flowering individuals of *Kathlekan*-I, majority of them (61.91%) belonged to GBH class 51–100 cm. In *Kathlekan*-II population, there were 24 flowering individuals, among these 54.17 per cent belonged to 101–150 cm GBH range. The eight individuals of *Mundge teggu* population distributed on a wider GBH classes. Tree with largest GBH (262 cm) was also found in this population. The restricted age classes of the population were mainly due to the predation of the oil-rich cotyledons seedlings by porcupines and wild boars.

Not all individuals in a population participated in reproduction in every season. In *Kathlekan*-I population, of the five male individuals that flowered during 1999, only 4 individuals bloomed in the next season. Eleven female individuals flowered during both the years while only three individuals flowered during 2000. Irregularity of flowering is a major constraint in tropical forest trees for successful regeneration (Richards, 1998). Such variations may potentially alter the genetic structure of the cohorts arising in different seasons.

The fate of two cohorts of seedlings (1998 and 1999) was followed for different periods to study their survivorship. In *Kathlekan*-I population, 55.31 per cent of germinating seedlings were predated during rainy season, compared with 64.19 per cent in *Kathlekan*-II. The mortality rate after six months, respectively for the two populations, was 66.95 and 62.42 per cent. Only 2.58 per cent of the previous stage seedlings in *Kathlekan*-I and 12.84 per cent in *Kathlekan*-II population survived through the summer. The most important feature of the survivorship curve is its linearity when plotted on a semi-log scatter. This is a typical 'type-two survivorship curve' which suggests a constant mortality rate at various ages (Deevey, 1947). Such type-two survivorship curves have been shown in many rare plants such as *Ranuncularis acris* and *R. auricomus* (Sarukhan and Harper, 1974).

Three distinct biological features complicate the conservation of *S. kathalekanensis*. First, its high habitat specificity (only to a few freshwater swamps), second, consistently small population size (less than fifty in any population) and thirdly its dioecious nature. The human-induced habitat-destruction has further rendered the species critically endangered. Our study has raised concerns on the conservation of the rare species and initial attempts to ecologically restore the species have been made. The Karnataka Forest Department has taken an active role in the *in situ* conservation by fencing one of the populations in *Kathlekan*. There is still a great need to effectively check all human interference and of the invasive weeds to these populations.

A further possible step in the recovery of the species is to replenish existing populations and/or to introduce material into sites where the species is likely to survive (Uma Shaanker and Ganeshaiah, 1997). Such recovery plans for endangered plants often call for the creation of new, self-sustaining populations within the historic range and characteristic habitat (Griffith *et al.*, 1989; Whitten, 1990). In the present study, an attempt was made to restore the plants in their native habitats. Of the 40 plants reintroduced into two new localities, the survival was quite remarkable. At one site, all the reintroduced plants survived the hot summer and put on significant collar diameter at the end of six months. This site was just about 600 meters away from the *Kathlekan*-I population. At the second site, only two individuals could not survive. This clearly indicates that *S. kathalekanensis* can be successfully re-introduced into the new localities.

References

Chandran, M. D. S., Divakar, K., Mesta & Manjunath, B. Naik, Myristica swamps of Uttara Kannada district. *Myforest.* **35**, 207–222 (1999).

Dasappa & Swaminath, M. H. A new species of *Semecarpus* (Anacardiaceae) from the *Myristica* swamps of Western Ghats of North Kanara, Karnataka, India. *Indian For.* **126**, 78–82 (2000).

Deevey, E. S. Life tables for natural populations of animals, *Quart. Rev. Biol.* **22**, 283–314 (1947).

Griffith, B., Scott, J. M., Carpenter, J. W. & Reed, C. Translocation as a species conservation tool: Status and strategy. *Science*, **245**, 477–480 (1989).

Richards P. W., in *The Tropical Rain Forest* (Second edition) (Cambridge University Press, Cambridge, 1998).

Sarukhan, J. & Harper, J., Studies on plant demography-I: Population flux and survivorship. *J. Ecol.* **61,** 676–716 (1974).

Uma Shaanker, R. & Ganeshaiah, K. N. Mapping genetic diversity of *Phyllanthus emblica*: Forest gene banks as a new approach for *in situ* conservation of genetic resources. *Curr. Sci.* **73,** 163–168 (1997).

Whitten, A. J. Recovery: A proposed program for Britain's protected species. *CSO Report No. 1089.* Nature conservancy council, Peter Borough, England (1990).

Tropical Ecosystems: Structure, Diversity and Human Welfare.
Proceedings of the International Conference on Tropical Ecosystems
K. N. Ganeshaiah, R. Uma Shaanker and K. S. Bawa (eds)
Published by Oxford–IBH, New Delhi. 2001. pp. 681–682.

Factors affecting resistance to damage following hurricanes in a subtropical moist forest

R. Ostertag[*,†,‡], *W. L. Silver*[*,†] *and A. E. Lugo*[†]

*University of California, Berkeley, CA 94720, USA
†International Institute of Tropical Forestry, Rio Piedras, PR 00928, USA
‡e-mail: ostertag@nature.berkely.edu

Hurricane winds can cause severe tree damage to tropical forest ecosystems, altering forest structure and resource availability, but the degree of tree damage is not evenly distributed across the landscape. We examined tree damage in long-term permanent plots in a subtropical moist forest in Puerto Rico following Hurricane Georges, which hit the island in 1998. These plots contained a mixture of native and non-native species. Previous surveys in these plots had determined tree size, topography, and damage from an earlier hurricane. Using similar methodology, we resampled this forest to address how tree species, size, and environmental factors affect resistance to hurricane damage. Severe damage included uprooted trees, snapped stems, or crowns with greater than 50% loss. Hurricane-induced mortality after 21 mo was 9.1%, thirteen times higher than background mortality levels during non-hurricane periods. Species differed greatly in their mortality and damage patterns, but there was no relationship between damage and wood density or biogeographic origin. Rather, damage was highly correlated with mean annual increment, with faster growing species experiencing greater damage. The dominant species in this forest, *Tabebuia heterophylla* (Bigno-

Keywords: Large scale disturbance, resistance, resilience.

niaceae), suffered the least damage and very low mortality rates, but was very slow growing, suggesting that there may be tradeoffs between survival, growth, and resistance and resilience to hurricane disturbance. Size (dbh, crown class) was also predictive of damage, with larger trees suffering more damage. Topography had no significant effect on patterns in damage. A strong relationship was noted between previous hurricane damage and present structural damage. We suggest that resistance of trees to hurricane damage is therefore not only correlated with individual and species characteristics but also with past disturbance history, which suggests that in interpreting the effects of hurricanes on forest structure, individual storms cannot be treated as discrete, independent events.

Acknowledgements

We thank I. Ruiz for help in the field and F. Scatena and others at IITF for invaluable logistical support while in Puerto Rico.

Biodiversity of Tropical Urban Areas

Tropical Ecosystems: Structure, Diversity and Human Welfare.
Proceedings of the International Conference on Tropical Ecosystems
K. N. Ganeshaiah, R. Uma Shaanker and K. S. Bawa (eds)
Published by Oxford–IBH, New Delhi. 2001. pp. 685–688.

Diversity of trees and butterflies in forest fragments around Pune city

Vaijayanti Alkutkar, Prajakta Athalye, Shweta Adhikari,
Anagha Ranade, Mandar Patwardhan,
Krushnamegh Kunte and Ankur Patwardhan

Research and Action in Natural Wealth Administration, C-26/1, Ketan Heights,
Kothrud, Pune 411 029, India
e-mail : ranwa@pn3.vsnl.net.in

Drastic changes in land use pattern associated with urbanization have resulted in an immense impact on those fringe areas of human habitations where forests are situated. The present paper discusses patterns in diversity of trees and butterflies in different impact zones across natural and semi-natural vegetation types close to Pune city. We also argue the important role played by the Defence Services and other institutional campuses in protecting the forests and enabling the forests to serve as islands and as migratory corridors.

Pune city, close to which all our study sites were located, is situated at $18^0 31'$N lat. and $73^0 51'$E long., at the junction of the Deccan Plateau and the Western Ghats. For evaluating the role of human influences, the present study, conducted during 1999–2000, compared the diversity of trees and butterflies in the forest patches at Katraj, Parvati-Pachgaon, Vetal Hills, Sinhgarh and National Defence Academy or NDA (abbreviated as K, P, V, S and N, respectively). Katraj hills harbour a mosaic of dry deciduous forests

---

Keywords: Tree diversity, butterfly diversity, Defence Area, conservation.

with 5–10 m tall trees, and grasslands. NDA campus is moist, and houses taller (10–15 m) forest and scrub, prone to fire (Patwardhan and Gandhe, 2000–2001). Sinhgarh hosts the most moist forests in the area, with tall trees (15–20 m) in the valley. The scrub at Parvati-Pachgaon and forest at Vetal Hills have dwarf (3–6 m), scattered wild trees as well as exotic planted trees.

To facilitate comparison, the habitats were arranged in their increasing order of relative human impact as forest, scrub, plantation and habitation. The former two categories host mostly indigenous flora constituting the wilderness zone, i.e. low-impact zone. The latter two represent species that are cultivated, mostly exotics, constituting the influenced zone. The species were classified on the basis of habitat preferred as forest species, scrub species, plantation species or species that are cultivated in house gardens as habitation species. A qualitative abundance scale was used to classify the species encountered as very common (VC), common (C), occasional (O) and rare (R). Species confined to just one habitat was termed exclusive to it.

A total of 131 tree species, of which one was an addition to the flora of Pune, were recorded from the study areas. Maximum number (81%) and density of tree species were recorded from NDA, which is a low-impact area. The least number of species were recorded from Parvati-Pachgaon (58%), a high-impact area. Tree species richness (number of species), number and percentage of species found exclusively at each locality are given in Table 1. Number of species in wilderness and influenced zones are also given. Figure 1 reveals that percentage wilderness of trees increased as the distance between the city and the forest/scrub increased.

A total of 103 butterfly species were recorded from the study sites. The habitat preferences as well as seasonal species richness have been recorded (Kunte, 2000–2001). The number of species in late monsoon and winter was far higher at Parvati-Pachgaon, the high-impact area, than at any other site, including the low-impact areas. However, during other seasons, the low-impact areas had greater species richness. In any case, the low-impact forested areas of Sinhagad and NDA supported more number of habitat-specific butterfly species than any other site.

Though populations of certain forest butterflies may possibly have declined in past few decades due to reduction in forest cover (Kunte, 2000–2001), no species have reportedly gone extinct in the neighbourhood of Pune city. However, Joshi et al. (1992) reported loss of four tree species of butterflies due to impact of biotic factors from Vetal Hills. It is important to note that these species are still common at the sites in the low-impact zone (Sinhagad and NDA). At this backdrop, the importance of the forests which are protected by

Table 1. Tree species diversity and exclusiveness across habitat types within each locality.

Habitat type/zone	Species richness (*a*)					No. of exclusive species (*b*)					% of exclusive species (*b/a*)				
	N	K	S	P	V	N	K	S	P	V	N	K	S	P	V
Forest species	51	46	55	33	34	30	26	33	14	18	59	56	60	42	52
Scrub species	43	36	30	30	29	25	19	11	14	14	58	52	36	46	48
Plantation species	31	30	29	29	31	11	10	10	9	11	35	33	34	31	35
Habitation species	24	25	24	25	23	7	8	8	8	7	29	32	33	32	30
Wilderness zone	77	66	67	48	50	67	56	57	38	41	87	85	85	79	82
Influenced zone	39	39	38	38	38	29	29	28	28	29	75	75	74	74	76
Total no. of tree species	106	95	95	76	79	8	1	11	1	0	8	1	12	1	0

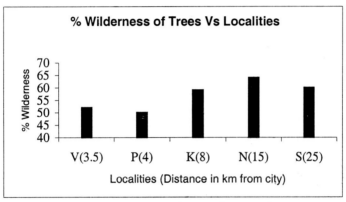

Figure 1. Distribution of trees in forest patches.

Defence Services needs to be carefully assessed. This study reveals that species richness as well as percentage wilderness of trees at NDA are comparable with those of Sinhagad that supports a good forest patch. Though butterfly species richness was higher at the high-impact site of Parvati-Pachgaon, there were more exclusive species at Sinhagad and NDA. This highlights the point that Defence Services-protected forests enjoy a high degree of protection that assist in conservation of various floral and faunal components, and possibly the overall biodiversity. In the light of recent observations that the practice of maintaining sacred groves is degenerating at the face of urbanization, the Defence Services-protected forests may emerge as modern sacred groves. Such forests are distributed throughout the country in various ecological zones, including the extremely biodiverse areas such as the Himalayas and the NE India. Studying further these forests *vis-à-vis* other protected and unprotected forests may elucidate the presently neglected potential of the Defence Services-protected forests as floral and faunal conservation areas.

Acknowledgements

We are grateful to the then Commandant, National Defence Academy, Pune, for permitting us to conduct surveys at Peacock Bay. We acknowledge the cooperation of the following individuals; R. Khunyakari, S. Kavade, S. Punekar, R. Mungikar, S. Bhagwat and M. Pendse. Our sincere thanks are also due to Four Eyes Foundation, Pune, for their infrastructure support.

References

Joshi, V. N., Kumbhojkar, M. S. & Kulkarni, D. K. Changing floristic pattern of Chatushringi-Vetal plateau near Pune – A comparative study. *J. Econ. Tax. Bot.* **16**, 133–139 (1992).

Kunte, K. Butterfly diversity of Pune city along the human impact gradient. *J. Ecol. Soc.* **13/14**, 40–45 (2000–2001).

Patwardhan, A. A & Gandhe, R. V. Tree diversity of Pune urban area: Cosmetic increase? *J. Ecol. Soc.* **13/14**, 21–33 (2000–2001).

Tropical Ecosystems: Structure, Diversity and Human Welfare.
Proceedings of the International Conference on Tropical Ecosystems
K. N. Ganeshaiah, R. Uma Shaanker and K. S. Bawa (eds)
Published by Oxford–IBH, New Delhi. 2001. pp. 689–692.

Urban wilderness: The case of Pune city, Western India

Ghate Utkarsh, Sanjeev Nalavade and Ankur Patwardhan

Research and Action in Natural Wealth Administration, C-26/1, Ketan Heights,
Kothrud, Pune 411 029, India
e-mail: ranwa@pn3.vsnl.net.in

Over two dozen naturalists, largely college students, from Pune city have published millennial biodiversity assessment (Nalavade *et al.*, 2000–2001). The study area is termed as 'Pune Urban Area', measuring about 700 km^2 with a radius of about 25 km from the city post office. The studied land habitat types include: forest (F), scrub (S), grasslands (G), plantations (P), agriculture (A), habitations (H). The first three habitat types constitute the wilderness (W) zone while the latter three constitute the impacted (I) zone. The aquatic ecosystem classification remained at a broad level – low (W) and high (I) impact zones.

Comparison of Pune biota with records of Bangalore and Delhi highlights that Indian cities also host phenomenal levels of biodiversity and few cities like Mumbai or Chennai host even wildlife reserves alongside skyscrapers. Table 1 indicates that various organismic groups differ considerably with respect to their diversity distribution across habitat types and human impact levels. Some groups such as butterflies, reptiles and mammals have almost all their species recorded from wilderness zone while nearly two thirds of them occur in forests. In contrast, diversity of fungi, herbs and trees

---

Keywords: Habitat diversity, landuse changes, human impact.

Table 1. Taxonomic richness of various habitats in the Pune urban habitat.

Group	Unit	Total	F	S	G	A	P	H	W	I
			\multicolumn{8}{c}{% of total diversity}							
Fungi	Genus	65	65				95	40	80	95
Herbs	Species	600	20	10	15	10	20	40	45	55
Trees	Species	350	25	15			65	15	35	65
Aquatic insects	Family	13							75	70
Snails	Species	15							60	70
Ants	Genus	12	45		35	35	65	35	50	70
Butterflies	Species	105	70	75	40	70	55	35	95	70
Fish	Species	70							100	50
Amphibians	Species	14							100	60
Reptiles	Species	50	60	40	40	45	50	15	80	55
Birds	Species	300	35	50	15	30	25	10	95	35
Mammals	Species	65	60	30	15	30	20	20	65	55

seems to be as much or more in impacted habitats than in low-impact zones. However, notwithstanding the maintenance of considerable species richness amidst the urban habitats, the unique species may be wiped out under human impact, only to be replaced by more tolerant species owing to suitable life attributes like dispersal abilities.

The results confirm the widely known trends of increase in species diversity with moderate disturbance, though severe human influence tends to erode the diversity, with the exception of a few stress-tolerant species that may have a cosmopolitan distribution and wide habitat choice. Rather than the total species diversity, human impact critically affects the diversity of unique species like habitat specialists. Another notable trend is lack of strong correlation across organismic groups, in response to similar human influence as reported in the literature (Kunte *et al.*, 1999). For instance, fish species seem to be sensitive to impact levels such as pollution, besides harvest or introduction of exotic species. The higher diversity recorded in some moderately or even considerably human impacted zones must not mislead one to undervalue the less impacted zones like the hill forests surrounding the city. For, these continue to be the biodiversity source while most urban habitats serve merely as sinks, unable to sustain the diversity on their own. For instance, nearly a fifth of the butterfly species emerge from their food plants confined to the hill forests which also exclusively host over a sixth of the bird species, seldom seen elsewhere in the city. Thus, bird or butterfly richness in the city gardens is difficult without the hill forests.

Table 2 presents the past and present landscape composition and ongoing changes, based on perceptions, besides records. The most suffered habitat

Table 2. Habitat dynamics of Pune urban area.

Habitat type	% Area 1950	% Area 2000	Converted into
Forest	7	5	Plantations
Riparian forest	1	–	Agriculture
Scrub/grassland	10	7	Habitation, plantations
Wastelands	2	1	Habitation
Agriculture	60	40	Habitation
Plantations	–	3	–
Wetlands	3	2	Habitation
Habitation	15	40	–

type has perhaps been riverine vegetation, especially babul (*Acacia nilotica*) tree groves along the rivers of Mula and Pawana. Grassland and scrub in the eastern outskirts have also been severely destroyed. Pune city hosts more than one million vehicles, over three-fourths being two-wheelers. The Mumbai–Bangalore bypass constructed recently skirts the city along the western and south-western margin and virtually cuts through the Parvati-Panchgaon forest park which has now turned into an island. This study brings out its impacts in terms of loss of amphibian and mammal populations.

Our efforts were inspired by a pioneering, amateur effort to compile checklists of urban fauna at Bangalore (Karthikeyan, 1999). Later, we chanced upon a more professional endeavour (Anon, 1997). The college/NGO network initiated by IISc along the Western Ghats (Gadgil, 1996; Kunte *et al.*, 1999) has also begun emphasising importance of building local assessments and awareness, besides developing the training material. Such monitoring can help better understand the ongoing process of ecorestoration around the city. The revival of natural trees or herbs amidst plantations, has triggered colonisation of these new habitats by birds, butterflies, etc. Notably, even seasonal puddles formed in these areas now harbour moults of dragonfly nymphs, etc. indicating ongoing colonisation and establishment of even organisms that were highly susceptible to seasonality. Recently, increasing tree cover of suitable species in some parks has probably helped predominantly the population of Western Ghats dweller butterflies. Such ongoing monitoring can easily detect notable declines in certain species like the sparrows and vultures, being noticed and debated currently. If such sudden population fluctuations are any signal of impending calamity, the purpose of monitoring is served much beyond academic interests. Such monitoring can even become quite popular, yet cost effective through internet publicity such as the electronic discussion group of Asian naturalists having thousands of members worldwide <nathistory-india@lists.Princeton. EDU>.

Notably, such publications based on long-term observations might pre-empt the facile environmental impact assessments (EIAs) that are currently mushrooming like a fashion. Unfortunately, environmentalists are not geared with much concrete, scientific data. That scientific data can at times lead to stringent legal action against environmental hazards is proven in case of air pollution at Delhi and Agra. If the recent legal activism takes note of such serious publications, environmental care cannot be easily wished away by the shroudy EIAs.

Acknowledgements

Dr Prakash Gole of Ecological Society greatly encouraged us. Prof. M. Gadgil and the IISc team, including WGBN members have variously motivated and buttressed these efforts. World Wide Fund for Nature-India (WWF-I) had earlier sponsored RANWA ultimately triggering this endeavour. We remain obliged to them all, besides our friends and families that gracefully bore the brunt of our ecological obsession!

References

Anon. *Fauna of Delhi* (Zoological Survey of India, Calcutta, 1997).

Gadgil, M. Documenting diversity: An experiment. *Curr. Sci.* **70**, 36–44 (1996).

Kartikayan, S. *The Vertebrate and Butterfly Fauna of Bangalore: A Checklist* (WWF–India, Karnataka State Office, Bangalore, 1999).

Kunte, K. Seasonal patterns in butterfly abundance and species diversity in four tropical habitats in northern Western Ghats. *J. Biosci.* **22**, 593–603 (1997).

Kunte, K., Joglekar, A., Utkarsh, G. & Pramod, P. Patterns of butterfly, bird and tree diversity in the Western Ghats. *Curr. Sci.* **77**, 577–586 (1999).

Nalavade, S., Dixit, A. & Utkarsh, G. Pune City wilderness: A case for urban biodiversity assessment. *J. Ecol. Soci.* **13–14**, 8–14 (2000–2001).

Tropical Ecosystems: Structure, Diversity and Human Welfare.
Proceedings of the International Conference on Tropical Ecosystems
K. N. Ganeshaiah, R. Uma Shaanker and K. S. Bawa (eds)
Published by Oxford–IBH, New Delhi. 2001. pp. 693–695.

Institutions: Biodiversity hot-spots in urban areas

M. Kulkarni, S. Dighe, A. Sawant, P. Oswal, K. Sahasrabuddhe and A. Patwardhan

Research and Action in Natural Wealth Administration, C- 26/1 Ketan Heights,
Pune 411 029, India
e-mail: ranwa@pn3.vsnl.net.in

Urbanization and habitat fragmentation seem to be increasing world wide. A case study in Pune city, Maharashtra, India reported 25% decline in vegetation cover within 50 years due to encroachment by human habitation (Dixit and Utkarsh, 2000–1). However embedded in most such cities are pockets of campuses, institutes and other expanse of land-use which even today offer a good vegetation cover. These sites act as hot-spots of urban biodiversity and may merit serious attention by conservationists. Here we make an attempt to explore ecological conservation values of institutions in an urban ecosystem. The study constitutes an assessment of species richness, standing biomass, carbon sequestration by tree flora in Pune university campus, with inventories on butterflies. Though the area of university is a mere 0.23% of the city, it harbours 51% (194) of tree species and 40% (40) of butterfly species of total recorded in the city (Patwardhan and Gundhe, 2000–1; Kunte, 2000–1). The total area of the university campus is 166 ha and has 51% of vegetation cover (Moghe, unpublished). The four structural vegetation types found in the campus are:

Keywords: Urban biodiversity, biomass, carbon sequestration.

Table 1.

Vegetation type	Area (ha)	Unit biomass	Total standing biomass (ton)	Percentage biomass	Stored carbon (tons)
Scrub savanna	24.9	14	335	9	167.5
Monoculture	27.07	22	656	18	328
Mixed vegetation	27.79	73	2032	55	1016
Gardens	6.22	107.5	668	18	334
Total	85.98		3691		1845.5

1. Scrub savanna (29% of vegetation area) – Barren land, with dry deciduous trees frequently distributed.
2. Monoculture (32% of vegetation area) – Plantation of dry deciduous species like *Dalbergia melanoxylon* and *Gliricidia* sp.
3. Mixed vegetation (32% of vegetation area) – Quite dense with a greater percentage of evergreen species.
4. Garden (6%) – Highly diverse species, with high percentage of evergreen species. Densest vegetation type.

For each of these vegetation types, species richness and biomass were estimated. Checklist of plants in this area was earlier prepared by Varadpande (1973). There is virtually no difference in species composition even today.

A line transect of 500 m × 20 m was laid in each vegetation type. The corresponding GBH (girth at breast height) and height for each individual tree were noted. Biomass and stored carbon for each vegetation type were calculated (Gadgil *et al.* (unpublished); Ravindranath *et al.*, 1997).

Table 1 provides the estimated value of a few of the parameters. Human influence on areas such as scrub savanna and monoculture show less percentage of standing biomass, while mixed plantation holds more than 50% of total biomass; the latter area also exhibits the highest regeneration. Gardens occupy just 6% of the total area of the campus, but are rich in biomass, holding almost 108 ton/ha. This is quite comparable with semievergreen forest (120–130 ton/ha) (Ravindranath *et al.*, 1997).

Thus our studies suggest that far from being degraded, certain sites within the urban landscape can still contribute significantly towards conservation of biological diversity and provide crucial ecosystem functions. These sites should therefore merit to be conserved in the urban scape.

These aspects definitely add to the conservation value of such institutionally safeguarded areas in the urban biodiversity. In conclusion, we argue that campuses and institutional areas in urban landscape can be significant repositories of biodiversity and provider of ecosystem function.

Acknowledgements

We thank RANWA for giving us this opportunity to conduct this study and for encouragement. We also thank Head, Department of Environmental Science, Pune university for providing computer facility.

References

Dixit, A. S. Nalawade & Utkarsh, G. Pune urban biodiversity: A case of millenium ecosystem assessment. *J. Ecol. Soc.* **13/14,** 8–13 (2000–1).
Patwardhan, A. A. & Gandhe, R. V. Tree Div. of Pune urban area: cosmetic increase? *J. Ecol. Soc.* **13/14,** 21–33 (2000–1).
Kunte Krushnamegh. Butterfly diversity of Pune city along human impact gradient. *J. Ecol. Soc.* **13/14,** 40–45 (2000–1).
Moghe Kaustubh. State of the art of Pune University Campus. Dessertation submitted for M.Sc. degree to Dept. of Environmental Science, University of Pune (unpublished).
Varadpande, D. G. The flora of Ganeshkhind, Poona. *J. Poona Univ. (Sci. & Tech.)* **44,** 97–133 (1973).
Gadgil, Madhav, Ranjit, R. J. & Utkarsh, G. A methodology manual for sc. inventorying, monitoring and conservation of biodiversity vol. I. (unpublished).
Ravindranath, N. H., Madhav Gadgil & Somashekhara, B. S. Carbon flows in Indian forests, in *Climatic Changes* (Kluwer Academic Publishers, Dordrecht, 1997) 297–320.

Tropical Ecosystems: Structure, Diversity and Human Welfare.
Proceedings of the International Conference on Tropical Ecosystems
K. N. Ganeshaiah, R. Uma Shaanker and K. S. Bawa (eds)
Published by Oxford–IBH, New Delhi. 2001. pp. 696–698.

Environmental degradation of an urban lacustrine water body in Pune, India

A. Waran, M. Mhasavade, S. Yewalkar, D. Kulkarni,
P. Kulkarni, T. Vaishampayan, P. Deshpande, S. Manchi,
K. Sahasrabuddhe and A. Patwardhan [*][†]

Department of Environmental Sciences, University of Pune, Pune 411 007, India
*Research and Action in Natural Wealth Administration (RANWA), C-26/1, Ketan
Heights, Kothrud, Pune 411 029, India
[†]e-mail: ankurpatwardhan@hotmail.com

Pashan Lake is situated between 18°32′7″N and 73°46′58″E near Pune in Western India. The present paper assesses the impact of urbanization on the water quality with a focus on ongoing changes in biotic communities as well (Patrick, 1972). Six sampling points were selected of which 2 are point sources and 4 nonpoint sources of pollution to evaluate the spatio-temporal trends in water quality. Surface water samples were collected from these sites once a month and analyzed for different physico-chemical parameters like pH, temperature, dissolved oxygen, free CO_2, chlorides, phosphates, total alkalinity, total hardness (AWWA & WEF, 1992). Composite water samples were subjected to microbial analysis including determination of most probable number, total viable count and plankton analysis (Lackey, 1938). For the quantitative estimation of aquatic fungi the sector analysis method described by Willoughby (1962) was employed. Besides exploring the relationship between birds and surrounding vegetation, the paper also

Keywords: Lacustrine water body, environmental profile, human impact, urban biodiversity.

focuses on the impacts of degrading water quality on fishes and aquatic molluscs.

Pashan Lake is a man-made lake, built by bunding Ram River. The catchment area is 40 km². Since long it has been attracting migratory birds. Gole (1973) recorded the temporal variation in these winter visitors with notes on their ecological niche. Ghate and Vartak (1981) listed angiosperm flora at Pashan lake. Kahnere and Gunale (1999) recorded the trends in water quality with emphasis on phytoplankton studies. The surrounding area has witnessed a tremendous change because of the ever expanding city limits which is reflected in decreased scrub vegetation surrounding the lake.

Table 1 highlights the important factors that have changed drastically during the last two decades owing to urbanization.

Deforestation on nearby hills has caused heavy siltation resulting in decrease in the depth of the lake. This has reflected in reduction in the number of Pochards (deep diving duck) which prefer to occupy the central deep portion of the lake. They are now outcompeted by Pintails and Shovellers (dabbling ducks) which prefer shallow water. The increasing number of these ducks in the central position is an inexpensive indicator of decrease in the depth of the lake. Amphibious/marshy flowering plants are found to dominate the area replacing aquatic species like *Nymphea* and emergents like *Typha*. This has affected the nesting habit of birds like Pheasant Tailed Jacana. The aquatic weeds like *Ipomoea carnea* and *I. aquatica* have covered the banks, nearly invading the fringe vegetation. Birds like Spot Bill Ducks favouring this habitat have increased in number. Little Cormorant and Black Winged Stilts which favour organically polluted water (Gole, 1985) are also increasing in number. This is further supported by the variation in the dissolved oxygen content of water as it is supposed to be the best indicator of the status of surface water quality (Hem, 1970).

Table 2 highlights the alteration in water quality, an indication of ever increasing stress on the water body. The deterioration in the water quality is reflected in an increase in fishes like *Tilapia* which tolerate organic pollution. Introduced fishes such as this outcompete the native fishes like *Cirrhinus*

Table 1. Temporal changes in various environmental attributes

	1980	2001
Depth	30–40 ft	15–20 ft
Vegetation	Purely aquatic species	Amphibious/marshy species increased
Birds	(1) Deep diving ducks abundant	(1) Shallow water ducks abundant
	(2) Waders common	(2) Waders drastically reduced
Fishes	Indigenous abundant	Introduced abundant

Table 2. Physico-chemical and microbial analysis

Parameters	Values
Oxygen dissolved	0–5.0 ppm
Free carbon dioxide	1.99–127.99 ppm
Fungal count	2900 Propagules/l
MPN	1.8×10^4 organisms/100 ml
TVC	4.2×10^4 organisms/ml

fulungee (Syhes) Reba and *Salmostroma boopis* (Geinaces), the former believed to have become locally extinct. *Euglena* and *Oscillatoria* are consistently recovered during plankton analysis. Bivalves which prefer unpolluted water were replaced by pollution tolerant *Bellamya bengalensis* (Raut *et al.*, 2001).

We propose that for the eco-restoration of Pashan Lake, appropriate management strategies including a few of the following needs to be urgently taken up:

(a) Diversion of sewage line from the lake; (b) dredging of silt and removal of weeds; (c) planting native trees like *Acacia nilotica* and *Zizyphus jujube* which are favoured for nesting instead of exotics like Eucalyptus around the lake; (d) allowing controlled grazing by cattle to keep check on *Ipomea*. (e) planting of trees on nearby hills to avoid runoff.

Acknowledgements

We are thankful to the Head of the Department for providing necessary laboratory facilities. Thanks are also due to RANWA for providing technical guidance in conducting the project.

References

APHA, AWWA & WEF. *Standard Methods for Examination of Water and Waste Water.* 18th edn (1992).

Ghate, V. S. & Vartak, V. D. Studies on aquatic flowering plants of Greater Pune area. Part I. Enumeration. *J. Poona Univ. (Sci. and Tech.)* **54**, 121–129 (1981).

Gole, P. *Eka talyat hoti* (Marathi) (Rajas Prakashan, Pune, 1973).

Gole, P. Birds of a polluted river. *J. Bombay Nat. Hist. Soc.* **81**, 613–625 (1985).

Hem, J. D. Study and interpretation of the chemical characteristics of natural water. *US Geol. Surv. Water Supply Paper* **1473**, 1–363 (1970).

Kahnere, Z. D. & Gunale, V. R. Evaluation of the saprobic system for tropical water. *J. Environ. Biol.* **20**, 259–262 (1999).

Lackey, J. B. The manipulation and counting of river plankton and changes in some organisms due to formalin preservation. *US Public Health Rep.* **53**, 2080–2093 (1938).

Patrick, R. Aquatic communities as indices of pollution. in *Indicators of Environmental Quality* (ed. Thomas, W. A.) (Plenum Press, New York, 1972).

Raut, R. N., Desai Shruti, Bapat Rohini. & Kharat, S. S. Aquatic insects and molluscs of Pune city. *J. Ecol. Soc.* **13/14**, 34–36 (2000–2001).

Willoughby. Quantitative estimation of water fungi. *J. Ecol.* **50**, 733–759 (1962).

Animal Ecology

Tropical Ecosystems: Structure, Diversity and Human Welfare.
Proceedings of the International Conference on Tropical Ecosystems
K. N. Ganeshaiah, R. Uma Shaanker and K. S. Bawa (eds)
Published by Oxford–IBH, New Delhi. 2001. pp. 701–706.

Bird community structure along an elevational gradient in a tropical rainforest

T. R. Shankar Raman and N. V. Joshi

Centre for Ecological Sciences, Indian Institute of Science, Bangalore 560 012, India
e-mail: shankar@ces.iisc.ernet.in, nvjoshi@ces.iisc.ernet.in

Understanding the distribution of tropical rainforest birds on elevational gradients is as crucial for ecologists trying to understand principles of community organisation as for conservation biologists striving to conserve these species in montane regions. Following the pioneering work of Terborgh (1971, 1977), most studies have focussed on variation with elevation in species richness, range sizes, and turnover rates of bird species, mainly in the Neotropics (Rahbek, 1997; Blake and Loiselle, 2000). Although species richness is generally considered to decline with elevation, it may peak at mid-elevations (Terborgh, 1977; Rahbek 1997). This 'mid-domain effect' is expected from null models that place species ranges randomly along the elevational gradient between upper and lower boundaries (Colwell and Lees, 2000). Elevational range sizes of species may increase with increasing elevation, as suggested by Stevens (1992) in an extension of Rapoport's rule to elevational gradients. Patterns of species richness and turnover may also vary with elevation according to avian diet-guild and habitat structure (Terborgh, 1977; Blake and Loiselle, 2000).

Here, in a tropical rainforest of the Western Ghats of India, we explore the following questions. How do bird community and guild attributes vary with

Keywords: Species richness and turnover, bird conservation, mid-domain effect, Rapoport's rule, null models.

elevation in this tropical rainforest? Is there evidence for the mid-domain effect and Rapoport's rule? Does elevational distance and degree of change in habitat relate predictably with change in bird community composition or are turnover rates predicted accurately by null models that simulate non-equilibrium dynamics?

The study was carried out in tropical wet evergreen rainforest in the Kalakad–Mundathurai Tiger Reserve (KMTR, 895 km², 8°25' to 8°53'N and 77°10' to 77°35'E). Although around 250 bird species have been observed in and around KMTR, only about 70 resident species (including 12 endemics), and 10 migrants occur regularly in rainforests. Fourteen sites located along an elevational gradient of 500 to 1400 m were selected for the study. The fixed-radius point count method was used to survey bird populations during the breeding season. At each site, 25 point count surveys were carried out, except one where only 18 point counts were sampled because of logistical difficulties. Vegetation parameters (tree, shrub, cane, and bamboo densities, tree species richness, vertical stratification, horizontal heterogeneity, canopy cover and height, and leaf litter depth) were measured using standard methods.

Across 14 sites, we obtained 2900 detections comprising around 6600 individual birds belonging to 67 species (including 9 migrants) during the study. Total bird species richness, rarefaction bird species richness, and Shannon–Weiner diversity were significantly negatively correlated to elevation (Kendall $T < -0.52$, $df = 12$, $P < 0.01$). Resident bird species richness was not significantly correlated to elevation ($P = 0.08$). The abundance of birds (individuals/ha) showed a positive correlation with elevation that was marginally significant ($T = 0.39$, $df = 12$, $P = 0.055$).

The number of species of canopy, understorey, and terrestrial insectivores increased significantly with elevation ($T > 0.60$, $P < 0.005$). In contrast, species richness of frugivores, nectarivore–insectivores, and omnivores decreased with elevation ($P < 0.005$). Carnivores and bark-surface feeders were not related to elevation, although the latter appeared to achieve peak richness in mid-elevations. The abundance of birds in various diet-guilds varied significantly among sites (one-way ANOVA, $F_{13,329} = 1.35$, $P < 0.001$). Patterns of variation in abundance were similar to patterns in species richness, except that nectarivore–insectivores increased and terrestrial insectivores did not vary significantly with elevation. Thus, the proportional representation of birds of different diet-guilds varied with elevation. From the lower to the upper limit, frugivores decreased from c. 30% to 10% of all birds, omnivores decreased from 17% to 1%, while nectarivore–insectivores increased from 20% to 28–61% and canopy insectivores varied from 2% to 6%.

None of the measured vegetation parameters was significantly correlated to elevation, except for bamboo density that showed a positive correlation (as bamboo occurred only in two of the higher altitude sites). Bird species richness and diversity indices were significantly negatively correlated with only leaf litter depth. There were no other significant correlations between birds and vegetation.

The mid-domain effect was tested using the null model of Veech (2000) breaking the elevational range into 18 zones of equal (50 m) width. Around 35 resident bird species were found in each of the eighteen 50-m wide elevational zones. This was contrary to the Veech (2000) null model of the mid-domain effect, which simulated random placement of elevational range widths on the gradient. The null model predicted an inverse U-shaped curve (Figure 1) that was accurately described by a polynomial model ($y = -0.0001x^2 + 0.2027x - 53.065$, $R^2 = 0.99$). The displacement of the observed curve ($d = 6.51$) was significantly different from the displacement expected under the null model (1000 simulations, $P < 0.001$). Range sizes of species showed a quadratic relationship ($R^2 = 0.72$) with their range mid-point. Thus, in contrast to expectation under Rapoport's rule, the mean elevational range of species in a site also showed a quadratic relationship to elevation of the site ($y = -0.0005x^2 + 8.8839x + 216.99$, $R^2 = 0.78$).

Despite the relatively invarying species richness, there was substantial turnover along the gradient, and sites at the two ends of the gradient were less than 5% similar to each other (Figure 2a). Partial Mantel tests indicated that dissimilarity in bird community composition between sites was independently related to both elevational distance and dissimilarity in tree species composition between sites (Figure 2a,b, Partial $T = 0.60$ and 0.22,

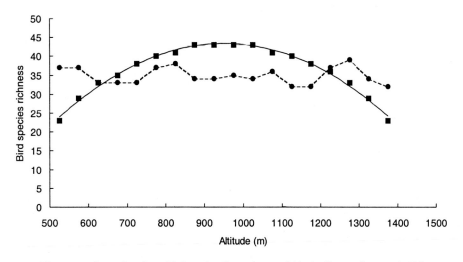

Figure 1. Assessing the mid-domain effect: observed (dashed) vs null curve (solid).

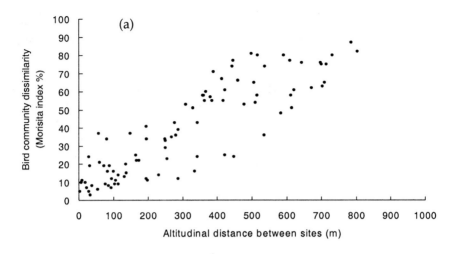

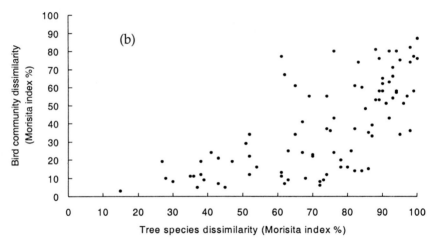

Figure 2. Between-site dissimilarity in bird community composition in relation to (a) elevational distance, and (b) tree species dissimilarity.

$P < 0.013$). Dissimilarity in guild structure between sites showed a similar pattern, however, the magnitude of variation was far less than dissimilarity in species composition, as sites at the two ends of the gradient were only about 10% dissimilar in guild structure.

We also used three null models to assess the likelihood of obtaining the observed average bird community similarity (Jaccard index) between sites by chance: (i) Saturation model: observed species richness of a site was constrained while species distribution across sites varied; (ii) Distribution model: the observed number of sites occupied by a species was constrained; and (iii) Unconstrained model: only the matrix total number of presence was constrained. Significance of differences between observed and expected null model estimates of the average Jaccard index was assessed through 1000 simulations. Null models of species turnover indicated that the saturation and unconstrained models predicted similar values of Jaccard similarity between sites, but the predicted values (0.31–0.32) were significantly lower than the observed values (0.529, $P < 0.001$). The distributional model predicted an average Jaccard similarity (0.516) that was remarkably close to that observed, but was nevertheless significantly lower ($P < 0.001$).

Although there were 58 species in the species pool, only 25–30 species occurred at any given site. Such an almost invarying pattern is contrary to the general pattern of monotonic decline and the mid-domain effect. The remarkably similar species richness and guild structure despite substantial variation in species composition suggest that local diversity is regulated at relatively constant levels. This is possibly due to limited variation in habitat structure, productivity, or resource availability along the gradient (Brown *et al.*, 2001) in contrast to patterns observed in South American rainforests (Terborgh, 1977). This is partially supported by the fact that habitat structure did not show any substantial or consistent elevational trend and only leaf litter depth showed a (possibly spurious, negative) correlation with bird species richness.

Although some of the variation in guild structure could be attributed to elevational turnover and variation in tree species composition, guild structure was maintained by species replacements along the gradient. This suggests a resource- or competitively-limited community (Brown *et al.*, 2001).

The lack of a significant species richness trend coupled with a substantial variation in community composition in relation to elevation and tree species composition has consequences for bird conservation in the region. The lack of support for Rapoport's rule and the nonlinear pattern of range sizes is a result of many species, including endemics such as the White-bellied Shortwing (*Brachypteryx major*) and Black-and-Rufous Flycatcher (*Ficedula nigrorufa*), being specialised to higher altitudes. The null model tests indicate that deterministic influences on bird community structure may be stronger than non-equilibrium factors. The results suggest a need to protect areas at all elevations and representing a spectrum of variation in tree species composition in order to conserve the full complement of rainforest bird communities.

Acknowledgements

We are grateful to the Ministry of Environment and Forests, India, and John D. and Catharine T. MacArthur Foundation for financial support and colleagues at CES, especially R. Sukumar for help. We thank the Tamil Nadu Forest Department, particularly R. P. S. Katwal and V. K. Melkani, for field permits and support. N. M. Ishwar, Divya Mudappa, and K. Vasudevan were inspiring colleagues to work with in the field and we thank them, and Ravi Chellam for unstintingly lending his support and sharing the field station resources. Divya Mudappa also helped enormously in fieldwork, data collection, and manuscript preparation. P. Jeganathan,

M. Jeyapandian, A. Silamban, Sashikumar, and many others are thanked for their assistance in the field.

References

Blake, J. G. & Loiselle, B. Diversity of birds along an elevational gradient in the Cordillera Central, Costa Rica. *Auk* **117,** 663–686 (2000).

Brown, J. H., Ernest, S. K. M., Parody, J. M. & Haskell, J. P. Regulation of diversity: maintenance of species richness in changing environments (2001).

Colwell, R. K. & Lees, D. C. The mid-domain effect: geometric constraints on the geography of species richness. *Trends Ecol. Evol.* **15,** 70–76 (2000).

Rahbek, C. The relationship among area, elevation, and regional species richness in Neotropical birds. *Am. Nat.* **149,** 875–902 (1997).

Stevens, G. C. The elevational gradient in altitudinal range: an extension of Rapoport's latitudinal rule to altitude. *Am. Nat.* **140,** 893–911 (1992).

Terborgh, J. Distribution on environmental gradients: theory and a preliminary interpretation of distributional patterns in the avifauna of the Cordillera Vilcabamba, Peru. *Ecology* **52,** 23–40 (1971).

Terborgh, J. Bird species diversity on an Andean elevational gradient. *Ecology* **58,** 1007–1019 (1977).

Veech, J. A. A null model for detecting non-random patterns of species richness along spatial gradients. *Ecology* **81,** 1143–1149 (2000).

Tropical Ecosystems: Structure, Diversity and Human Welfare.
Proceedings of the International Conference on Tropical Ecosystems
K. N. Ganeshaiah, R. Uma Shaanker and K. S. Bawa (eds)
Published by Oxford–IBH, New Delhi. 2001. pp. 707–710.

Biogeographic patterns of Indian bats: Identifying hot-spots for conservation

Shahroukh Mistry

Department of Biology, Grinnell College, Grinnell, IA 50112, USA
e-mail: saiful@pkrisc.cc.ukm.my

The identification of hotspots for conservation priorities has often been based on the species richness and endemism of plants and vertebrates, primarily because of the extensive data available on these taxa (Dinnerstein and Wikramanayake, 1993; Meyers *et al.*, 2000). Among the vertebrates, much emphasis is placed on birds, especially in areas of the world where there is high endemism (Stattersfield *et al.*, 1998). Yet there are other groups that show levels of endemism and richness equal or greater than that of plants or birds, and the study of these taxa may prove to be valuable for identification of hotspots. India, for example, has 11.6% of the global bat species, which is on par with birds and substantially higher than plants. In terms of endemic species, India has a higher proportion of endemic bat species than either plants or birds (Figure 1). Unfortunately, many of the lesser-known taxa, such as bats, have received little protection, and face considerable threats and higher rates of endangerment. This paper examines the biogeographic distribution of bats in the Indian subcontinent to recognize hotspots, as well as outlines priorities for the conservation of bats in India.

India's bat fauna is rich and diverse. With 110 species, India has more than 11% of the world's bats including 13 megachiroptera and 97 microchiroptera (Bates and Harrison, 1997). These species provide substantial ecological and

Keywords: Bats, conservation, biodiversity, hotspots, India.

economic services via pollination, seed dispersal and agricultural pest control. For example, just the three common species of fruit bats (*Cynopterus sphinx*, *Pteropus giganteus* and *Rousettus leschanaulti*) visit over 114 plant species and act as important pollen and seed vectors. Yet, many bat species in India face numerous threats. Almost 18% of the species are considered endangered or vulnerable, and only 32% are at low risk (Figure 2). Of great concern, however, is that half of the species are considered to be data-deficient, with little or no information available on their status. Many of these species may actually be vulnerable or endangered, but are underestimated because of the lack of studies (Ceballos and Brown, 1995).

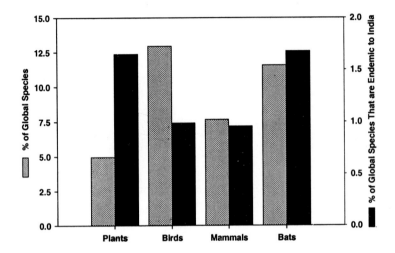

Figure 1.   Bat species diversity.

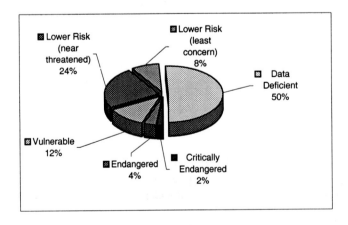

Figure 2.   Endangered and threatened species.

India has a high number of endemic bat species, many of which have limited distribution and are considered threatened. Nine species (8%) are endemic to India and 16 (15%) are endemic to the subcontinent. Five of these endemic species have very limited distribution – they are known from only a single location. These include two of the most unique bats in India, *Latidens salimalii* (an endemic, monospecific genus) and *Otomops wroughtoni*. Most of India's bats are known from only a few locations. Over 60 species (55%) are found in less than 10 localities, and 20 of these are found at only one locality. Only 11 species have a ubiquitous distribution with sightings in more than 50 localities.

The distribution of bat species in India is non-uniform. The Northeast is the most species-rich, with 81 (74%), including 9 of the 16 regionally endemic species. Thirty (27%) of these 81 species are found only in the Northeast region. This area also matches the biodiversity hotspots for plants and birds, and reinforces the need to protect more habitats in this region. Fifty-eight of India's bat species show a biogeographic affinity to the bat fauna of Southeast Asia. The Western Ghats also has significant species richness with 45 (41%) species of which 6 are regionally endemic.

The conservation of bats in India requires effort on many fronts. Most important is the establishment of detailed monitoring programs to assess the status of each species and determine the risks. Particular emphasis needs to be placed on endemic species, endangered species, as well as common species that provide substantial dispersal or pest control functions. Identifying hotspots for conservation should also be a priority, and this is especially true for the Western Ghat region and the Northeast states.

The Wildlife Protection Act does not afford any level of protection to either insect or fruit-eating bats. In fact, the Act specifically identifies fruit bats as vermin and strips them of any protection. Despite repeated attempts over the years to change the protection status of fruit bats in India, the policy remains in effect. Few surveys of bat roosts or population monitoring studies exist, however, one survey of 65 *Pteropus giganteus* roosts indicated that half of these roosts are declining in population size and that only 25% of the roosts are either protected or do not face some form of disturbance. The average colony is small in size (less than 500) and not very old (2–23 years). Tree felling and killing are the major causes of population declines. If these patterns are true for other bat species as well, declines may be widespread.

Additional concerns include the impacts of humans on bats and their roosts. Many bat roosts are destroyed because of their proximity to human dwellings or ancient buildings or ruins. The high rate of deforestation has severe consequences on many bat species, and the increasing use of pesticides may increase mortality. In concert with scientific research, educational efforts are also needed at the grass roots level, academic level and especially at the policy level. The future for bats in India depends on significant endeavours in research, education, and conservation within the next decade, especially the establishment of population monitoring programs and the protection of hotspots.

References

Bates, P. J. J. & Harrison, D. L. *Bats of the Indian Subcontinent* (Harrison Zoological Museum, Kent, UK, 1997).

Ceballos, G. & Brown, J. H. Global patterns of mammalian diversity, endemism and endangerment. *Conserv. Biol.* **9**, 559–568 (1995).

Dinnerstein, E. & Wikramanayake, E. D. Beyond 'hotspots': How to prioritize investments to conserve biodiversity in the Indo-pacific region. *Conserv. Biol.* **7**, 53–65 (1993).

Meyers, N., Mittermeier, R. A., Mittermeier, C. G., Da Fonseca, G. A. B. & Kent, J. Biodiversity hotspots for conservation priorities. *Nature* **403**, 853–858 (2000).

Stattersfield, A. J., Crosby, M. J., Long, A. J. & Wege, D. C. *Endemic Bird Areas of the World, Priorities for Biodiversity Conservation* (Birdlife International, Cambridge, UK, 1998).

Tropical Ecosystems: Structure, Diversity and Human Welfare.
Proceedings of the International Conference on Tropical Ecosystems
K. N. Ganeshaiah, R. Uma Shaanker and K. S. Bawa (eds)
Published by Oxford–IBH, New Delhi. 2001. pp. 711–715.

Species composition and associations of birds in mixed-species flocks at Parambikulam, South India

Robin Vijayan and Priya Davidar*

Salim Ali School of Ecology and Environmental Sciences,
Pondicherry University, Pondicherry 605 014, India
*e-mail: pdavidar@pu.pon.nic.in

Mixed-species flocks of birds have been known to be a common phenomenon in the woodlands in different parts of the world, especially in the tropics. Although a century of work has been done on them (Bates, 1864), the flocks in the Old World, especially Asia, lack attention (Jepson, 1987). Very few studies have been done on flocks in India (MacDonald and Henderson, 1977; Vijayan, 1989; Pramod, 1996) looking at the species composition and other functional aspects of these diverse flocks. This study was conducted to fill a part of this lacuna.

The study area comprises natural forest (moist deciduous forest) and old teak plantations (1916 onwards). Scan sampling (Altman, 1974) was used to record the flocks. Two minutes were spent after sighting every group of birds encountered to confirm it as a flock. One to three observations were made on flocks; each lasting for five minutes followed by a two-minute interval. Different trails and paths were followed everyday in order to obtain sufficient replicates. Occurrences of aggressive behaviour and individuals

Keywords: Mixed-species foraging flock, moist deciduous forest, insectivorous birds, south India, species composition, teak plantation.

closely following another were noted. All analyses were done using SPSS ver 7.5.

Two hundred and three flocks were observed. As equal number of observations on every flock was not available, the first observation of every flock was taken for the analysis of the number of species and individuals per flock. However, all observations were used to analyze the associations. The study revealed 68 species of birds, two species of primates and two species of squirrels in the flocks. The number of species per flock (species richness) was found to be 5.55 (Table 1), with no significant difference in the species richness of the flocks between the two habitats studied ($df = 86$; $p = 0.577$). Although the number of species per flock remained the same between the two habitats, a few species found in the plantations were not found in the moist deciduous forest flocks and vice versa. The overall species richness is not as low as reported from the flocks of Australia (3.93 species per flock, Bell, 1980) and New Guinea (3.2 species per flock; Bell, 1983) but not as high as the Amazonian (10 to 20 species per flock; Powell, 1979), Mexican flocks (6.48, 9.33, and 9.93 species per flock in different habitats; Gram, 1998) or Sarawak flocks (11.3 species per flock; Croxall, 1976). The latter might be because the flocks of this study are of moist deciduous forests and not of the rainforests (where higher species richness can be expected).

Various authors have classified the members of the flocks as 'nuclear' and 'followers' or 'attendants' (Grieg-Smith, 1978). The 'nuclear' species are usually the most common species and are responsible for the formation and cohesion of the flocks (Buskirk et al., 1972; Powell, 1979). The ten most common species (> 11%) were Racket-tailed Drongo (Dicrurus paradiseus), Bronzed Drongo (Dicrurus aeneus), White-bellied Drongo (Dicrurus caerulescens), Scarlet Minivet (Pericrocotus flammeus), Small Minivet (Pericrocotus cinnamomeus), Large Wood Shrike (Tephrodornis virgatus), Great Tit (Parus major), Golden-backed Woodpeckers (Dinopium benghalense and D. javenense) and Jungle Babbler (Turdoides striatus). The three species of Drongo, which are sallying or fly-catching species, had the highest percentage occurrence (> 50%). Winterbottom (1943, 1949) and Croxall (1976) have reported similar trends in the flocks of N. Rhodesia and Sarawak respectively. This phenomenon seems to be in contrast with the Neo-tropical flocks where the gleaners are the most common species and are joined by the fly-catching species.

Table 1. Species richness of mixed-species bird flocks at Parambikulam

	N	Minimum	Maximum	Mean	SE (±)
Species	203	2	12	5.55	0.1475
Individuals	203	1	26	9.55	0.2952

A decrease in species richness and the abundance of birds in the flocks was noted from December to March (Tables 2 and 3). Unlike other studies where winter migrants increase the species richness of the flocks (e.g. Gram, 1998 where there were 37% migratory species), in the present study, it was not so. The species richness was low in March probably due to the onset of the breeding season. Munn and Terborgh (1979) and Vijayan (1989) have documented such reduction in flocking behaviour during the breeding season due to increased territoriality.

The ten common species were subjected to further analyses for associations. The associations between different species of birds were looked into in the light of the (i) predator avoidance hypothesis (Morse, 1977) and (ii) foraging enhancement hypothesis (Greig-Smith, 1978; Krebs, 1973; Morse, 1970; etc.). The associations were tested using Cole's coefficient of association. The Racket-tailed Drongo was associated with Golden-backed Woodpeckers and Jungle Babbler. Bronzed Drongo was associated with Scarlet Minivet and Small Minivet. The data collected from the field corroborates the above results. It was proposed that the vertical stratification of the different species of birds would give an idea of the associations. The Racket-tailed Drongo, Golden-backed Woodpeckers and Jungle Babbler were found in the lower strata, below seven meters and the other seven common species were found above eight meters. Such stratification has been documented in flocks of different parts of the world (e.g. Croxall, 1976; MacDonald and Henderson, 1977; Eguchi et al., 1993). Dicrurus adsimilis adsimilis occurred in 88% of the flocks in N. Rhodesia (Winterbottom, 1943) while Dicrurus aeneus was found in 75% of the flocks in this study and was found to be following two species of squirrels and two species of primates. Most of the associations formed by

Table 2. Seasonal analysis of the number of species per flock.

Month	N	Minimum	Maximum	Mean	SD (±)
December	35	2	12	5.85	2.22
January	58	2	12	5.86	2.02
February	78	2	12	5.62	2.10
March	32	2	11	4.47	1.83

Table 3. Seasonal analysis of the number of individuals per flock.

Month	N	Minimum	Maximum	Mean	SD (±)
December	35	4	22	9.97	4.84
January	58	3	26	10.36	4.34
February	78	1	23	9.52	3.96
March	32	2	15	7.56	3.25

the Drongos were with other gleaning and probing species. There is much ambiguity regarding the classification of Drongo as 'nuclear' species as most 'nuclear' species have been defined on the basis of the Neo-tropical species where the gleaners and probers are usually gregarious and associate in small flocks themselves (Powell, 1979; Eguchi *et al.*, 1993). Drongo have been hence referred to as 'accidental' and 'regular accidental' species (Winterbottom, 1949), as 'catalyst' species (Croxall, 1976), etc. The behaviour of the Drongos might be in accordance with the view that these form the 'sentinel' species, providing anti-predatory services towards the gleaners and probers. Only one instance of a Shikra trying to attack a flock was noticed where the Drongo mobbed it followed by other birds. It is proposed that foraging enhancement occur in all species involved in the flocks, whether sallying, gleaning or probing by some way or the other. The insects flushed by the gleaning or the probing species directly benefit the Drongos while the gleaners and probers are benefited from the protection given by the Drongos, thereby enhancing their foraging efficiency. However, these have not been experimentally proven during this study and are open to further studies.

Acknowledgements

We thank the Kerala Forest Department and the officers and staff at Parambikulam. The study was conducted as a part of the first author's Masters programe and he thanks the Bombay Natural History Society for the support rendered towards the project through Salim Ali-Lok Wan Tho Ornithological Fellowship Co-operation of all the staff and colleagues at Ecology Department, Pondicherry University is gratefully acknowledged. Thanks are due to various people for discussions during different stages of the project, Mr. Shankar Raman, Dr N. V. Joshi, Mr J. C. Daniel, Mr. Pramod Padmanabhan, Dr P. A. Azeez, Dr V. S. Vijayan and Dr Lalitha Vijayan.

References

Altmann, J. Observational study of behaviour. Sampling methods. *Behaviour* **49**, 227–267 (1974).

Bates, H. W. *The Naturalist on the River Amazon* (John Murrays, London, 1864; Reprint of second edition. University of Berkley, 1962).

Bell, H. L. Composition and seasonality of mixed species flocks of insectivorous birds in The Australian Capital Territory. *Emu* **80**, 227–233 (1980).

Bell, H. L. A bird community of lowland rainforest in New Guinea. Five mixed species foraging flocks. *Emu* **82**, 256–275 (1983).

Buskirk, W. H., Powell, G. V. N., Wittenberger, J. F., Buskirk, R. E. & Powell, T. U. Interspecific bird flocks in Tropical Highland Panama. *Auk* **89**, 612–624 (1972).

Croxall, J. P. The composition and behaviour of some mixed species bird flocks in Sarawak. *Ibis.* **118**, 333–346 (1976).

Eguchi, K., Yamagishi, S. & Randriana Solo, V. The composition and foraging behaviour of mixed species flocks of forest living birds in Madagascar. *Ibis.* **135**, 91–96 (1993).

Gram, K. W. Winter participation by Neotropical migrant and resident birds in mixed species flocks in Mexico. *Condor.* **109**, 44–53 (1998).

Greig-Smith, P. W. The formation, structure and function of mixed species insectivorous bird flocks in West African Savanna woodland. *Ibis.* **120**, 284–297 (1978).

Jepson, P. Mixed-species bird flocks. *OBC Bull.* No. 5, 13–17 (1987).

Krebs, J. R. Social learning and the significance of mixed species flocking of chickadees. *Can J. Zool.* **51**, 1275–1288 (1973).

MacDonald, D. W. & Henderson, D. G. Aspects of the behaviour and ecology of mixed species bird flocks in Kashmir. *Ibis* **119**, 481–491 (1977).

Morse, D. H. Ecological aspects of some mixed species foraging flocks of birds. *Ecol. Monogr.* **40**, 118–168 (1970).

Morse, D. H. Feeding behaviour and predator avoidance in heterospecific groups. *Bio Science* **27**, 332–339 (1977).

Munn, C. .A. & Terborgh, J. W. Multispecies territoriality in Neotropical foraging flocks. *Condor* **81**, 338–347 (1979).

Powell, G. V. N. On the possible contribution of mixed species flocks to species richness in neotropical avifaunas. *Behav. Ecol. Sociobiol.* **24**, 387–393 (1979).

Pramod, P. Ecological studies of bird communities of Silent Valley and neighbouring forests. Ph.D Thesis. Calicut University (1996).

Vijayan, L. Feeding behaviour of the Malabar woodshrike at Thekkady, Kerala. *J. Bombay Nat. Hist. Soc.* **86**, 396–399 (1989).

Winterbottom, W. On woodland bird parties in Northern Rhodesia. *Ibis* **85**, 437–442 (1943).

Winterbottom, W. Mixed bird parties in the tropics with special reference to Northern Rhodesia. *Auk* **66**, 258–263 (1949).

Tropical Ecosystems: Structure, Diversity and Human Welfare.
Proceedings of the International Conference on Tropical Ecosystems
K. N. Ganeshaiah, R. Uma Shaanker and K. S. Bawa (eds)
Published by Oxford–IBH, New Delhi. 2001. pp. 716–721.

Feeding and roosting habits of *Petaurista philippensis* in a rain forest fragment, South India

Nandini Rajamani

Salim Ali School of Ecology and Environmental Sciences, Pondicherry University,
Pondicherry 605 014, India
e-mail: rnandini@usa.net

The Western Ghats, recognised as one of the 25 biodiversity hotspots of the world (Myers *et al.*, 2000), have suffered large scale destruction of forest areas due to selective logging and clear felling for plantations and developmental activities. Expanding human settlements and continued pressure for fuel wood and timber have reduced many forest tracts to isolated fragments. Such disturbance gradually changes the forest structure and composition and can have a deleterious effect on arboreal mammals (Sunderraj and Johnsingh, 2001). This study investigated the ecology of a nocturnal arboreal mammal, *Petaurista philippensis*, in a rain forest fragment, in an attempt to understand the response of the species to drastic alteration of natural ecosystems. Flying squirrels are found in the Nearctic, Palaearctic and Oriental regions, with maximum species diversity occurring in South Asia. Of the six genera of flying squirrels found in India, only two species in two genera are found in the Western Ghats, the large brown flying squirrel *Petaurista philippensis* and the Travancore flying squirrel *Petinomys fuscocapillus fuscocapillus*. *P. philippensis* is distributed in the Indian peninsula. Despite its wide occurrence, little is known of the ecology of the species, information only

Keywords: Flying squirrel, nocturnal, Western Ghats, ficus, roost cavities.

being available on the distribution and abundance (Ashraf *et al.*, 1993; Umapathy and Kumar, 2000) and morphometrics (Xavier *et al.*, 1998) of the species. This study aimed at documenting the diet and roosting habits of the species.

The study was carried out in a degraded rain forest fragment, Puduthottam (10°20'N and 76°58'E), situated close to the Indira Gandhi Wildlife Sanctuary in the Anamalai Hills, southern Western Ghats. The site, approximately 50 ha in size, is surrounded by tea and coffee plantations. Trails were walked mostly between 1830 and 2230 h and spotlighting was the primary method of locating squirrels. When a squirrel was observed feeding, the part consumed and the phenophase of the part was noted. In order to ascertain if any specific habitat parameters were important for the species, various characteristics (GBH, canopy cover, canopy contiguity with neighbouring trees, presence/ absence of understorey, distance to four nearest neighbours, girths of the neighbours) of the feeding trees were recorded. The same parameters were quantified for random trees to characterize trees in the general habitat. Characteristics of both roosting hollows (height of cavity on tree, diameter of the tree at nest height, shape of cavity entrance, length and width of cavity entrance, orientation of cavity, possible mode of cavity formation and location of cavity on the tree) and roosting trees (GBH, height of tree, canopy cover, canopy contiguity of the nest tree, presence/absence of understorey, presence of other cavities) were recorded. The same vegetation parameters were recorded for random trees located 50 meters from the nest tree in a random direction.

The density of trees was 270 individuals ha^{-1} in the forest interior, while it was considerably lower at the edge of the fragment (50 trees ha^{-1}). A comparison with two undisturbed rain forests in Anamalais, the Iyerpadi–Akkamalai complex and Karian Shola that have tree densities of 697 trees ha^{-1} and 755 trees ha^{-1} (D. Mudappa, pers. commun.) respectively, indicates the extent of degradation of the fragment. The vegetation at the study site was dominated by pioneer and introduced shade tree species, and a very low percentage of original rain forest vegetation was recorded.

Petaurista philippensis was sighted on 322 occasions and was observed feeding on 103 occasions. During the study period, which was mainly during the dry season, flying squirrels were observed to feed on a total of 10 species of plants from 8 families. Flying squirrels consumed fruit more than any other plant part. All the fruits consumed were of the species *Ficus racemosa*. Leaves were consumed in 39.21% of the observations, of which mature leaves accounted for 8.82% of the observations, and immature leaves in 7.84% of the observations. Leaves of six species were consumed, the leaves of *F. racemosa* being consumed more than any other species (17.65%). Flying squirrels fed on the flowers of *Cullenia exarillata* (8.82%), and on the bark of *Eucalyptus* (7.84%), *Mesua ferrea* (0.98%) and *Dimocarpus longan* (0.98%) (Figure 1).

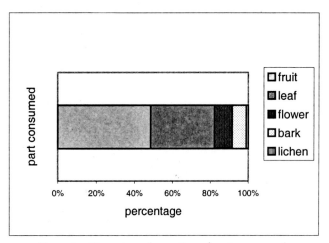

Figure 1. Percentage observation of parts consumed.

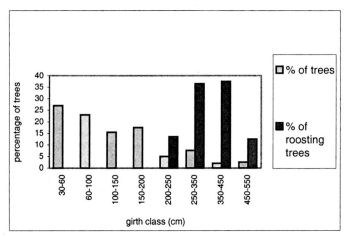

Figure 2. Distribution of GBH of roost trees and random trees.

The average height of the feeding trees was 28.75 m and the average height at which the squirrels were seen feeding was 16.67 m. Comparison of the characteristics of feeding tree plots and random plots revealed that feeding tree plots had greater canopy contiguity (6.78 ± 0.72) than random tree plots (2.82 ± 0.67). Canopy contiguity was significantly different between the actual and random tree plots (Kolmogorov–Smirnov 2 sample test, $z = 1.398$, $p = 0.040$).

Flying squirrels were observed in roost cavities on 10 occasions. All observations of nest exit were 88.2 ± 9.5 minutes after sunset. Five nest cavities were seen on *Mesua ferrea*, one on *Artocarpus heterophyllus*, one on *Dimocarpus longan*, and one on *Toona ciliata*. The average height of roost trees was 41.88 m and the average height of the roost cavities was 18.6 m. The average GBH of roosting trees was 3.35 m and the diameter at cavity height was 1.72 m (Figure 2). Average cavity length and width were 18.58 cm and 14.39 cm respectively. Only two cavities appeared to be caused by excavations by primary hollow nesters, the rest seemed to be natural cavities caused by heart rot where branches had broken away. Three of the roosting trees had branches that were partially dead and it was on these branches that the cavities were seen. Analysis of vegetation characteristics at roost sites revealed that the girth of roosting trees was significantly different from that of the random trees (Kolmogorov–Smirnov 2 sample test, $z = 2.0$, $p = 0.01$).

P. philippensis consumed the fruits of *F. racemosa* more than any other food item. However, the high consumption of this species despite its low abundance and dominance at the study site need not necessarily imply a specific preference for the species. The study was carried out in the dry season when few species apart from *F. racemosa* were in fruit, and hence the consumption of this species might simply be related to availability of food. The role of figs as important resources for frugivore communities during periods of food scarcity has been established in Peru (Terborgh, 1983) and in east Kalimantan (Leighton and Leighton, 1983). Borges (1993), however, maintains that figs are a lean fruiting season resource only for those individuals who have access to figs within their territories, and suggests that figs can be utilized as a major fruit source only by a mobile species with a large home range. *P. philippensis* can cover large distances by gliding, and this would greatly increase their ability to track food resources. At Puduthottam, flying squirrels were seen in pairs on a few occasions, but aggregations of more than two individuals, when seen, were most often on *F. racemosa* than on any other tree species. The congregation of flying squirrels in numbers of up to four individuals per fruiting tree might suggest the exploitation of figs as a major fruit resource.

Flying squirrels were seen to consume more mature leaves than immature leaves, though young leaves are known to be more nutritious and less tough and fibrous (Coley, 1983). Kawamichi (1997) in a study of *P. leucogynys*, used the proportion of mature leaf consumption as a negative index of food availability. It has been hypothesized that flying squirrels obtain the protein requirements of their diet from leaves and obtain high-energy carbohydrates from fruits. *Ficus* fruits are a source of easily assimilated energy with low fat content and are a potential source of animal proteins (provided by the larvae of fig wasps inside the figs) (Vellayan, 1981). Flying squirrels were seen feeding on bark of three species; while bark and woody parts of plants are high in cellulose, the flying squirrels may have eaten the bark of *Eucalyptus* to obtain the resin and exudates.

Flying squirrels showed a preference towards roosting in the cavities of large trees. This may be due to the presence of cavities only in such large trees or may be a selective choice towards large trees. Proportional utilization of trees in the large girth classes was higher than the proportional availability of trees in these girth classes. These large girth classes are composed of native remnant rain forest tree species that have not been felled during selective logging operations. No roosts were located in the plantations or in the shade trees or introduced species, though the density of these trees was high. Flying squirrels seemed to prefer large trees in mature forest stands for roosting purposes. This could be as larger, mature trees might provide greater protection from predators. Most of the trees that the squirrels were nesting in contain sap. The squirrels were also seen to nest in trees that appeared to be partially dead. Gibbons and Lindenmayer (1997) stress the importance of snags or dead trees for hollow dwelling fauna in the Austarlian *Eucalypt* forests and recommend that these trees be retained during selective logging operations. Such trees provide refuge to a wide variety of birds and mammals, and form an important component of the habitat.

It was seen from this study that *P. philippensis* showed a preference for certain habitat features like large mature trees, presence of canopy contiguity, and the presence of keystone species such as *Ficus* that support the population during the lean fruiting season. These factors may be responsible for the successful survival of the species in fragmented habitats.

Acknowledgements

The study was financially supported by the India Program of the Wildlife Conservation Society. We thank the Tamil Nadu Forest Department for the research permission, and the Estate Manager, Puduthottam Estate, for the permission to work in their private forest. I thank Dr N. Parthasarathy, Dr Priya Davidar and Dr Ajith Kumar for their inputs and suggestions during the course of the study. I thank Divya Mudappa for her ideas and encouragement throughout the study. Silamban, Rajamani, Sivakumar and Dinesh provided assistance in the field. T. R. Shankar Raman provided valuable inputs through discussions and his advice helped strengthen the manuscript.

References

Ashraf, N. V. K., Kumar, A. & Johnsingh, A. J. T. On the relative abundance of two sympatric species of flying squirrels of the Western Ghats, India. *J. Bombay Natl. Hist. Soc.* **90**, 158–160 (1993).

Borges, R. M. Figs, Malabar giant squirrels and food shortages within two tropical Indian forests. *Biotropica* **25**, 183–190 (1993).

Coley, P. D. Herbivory and defensive characteristics of tree species in a tropical lowland forest. *Ecol. Monogr.* **53**, 209–233 (1983).

Gibbons, P. & Lindenmayer, D. B. The performance of prescriptions employed for the conservation of hollow-dependent fauna. Implications for the Comprehensive Regional Assessment process. Centre for Resource and Environmental Studies Working Paper 1997/2.CRES, The Australian National University Press, Canberra (1997).

Kawamichi, T. Seasonal changes in the diet of the Japanese flying squirrels in relation to reproduction. *J. Mamm.* **78**, 204–212 (1997).

Leighton, M. & Leighton, D. R. Vertebrate responses to fruiting seasonality within a Bornean rain forest. in *Tropical Rain Forest: Ecology and Management* (eds Sutton, S. L., Whitmore, T. C. & Chadwick, A. C.) 181–196 (Blackwell Scientific Publications Ltd., London, England, 1983).

Myers, N., Mittermeier, R. A., Mittermeier, C. G., da Fonseca, G. A. B. & Kent, J. Biodiversity hotspots for conservation priorities. *Nature* **403**, 853–858 (2000).

Sunderraj, Wesley, S. F. & Johnsingh, A. J. T. Impact of biotic disturbances on Nilgiri langur habitat, demography and group dynamics. *Curr. Sci.* **80**, 428–436 (2001).

Terborgh, J. *Five New World Primates* (Princeton University Press, Princeton, New Jersey, 1983).

Umapathy, G. & Kumar, A. The occurrence of arboreal mammals in the rain forests fragments in the Anamalai hills, South India. *Biol. Conserv.* **92**, 311–319 (2000).

Vellayan, S. The nutritive value of Ficus in the diet of the lar gibbon (*Hylobates lar*). *Malay. Appl. Biol.* **10**, 177–181 (1981).

Xavier, F., Joseph, G. K. & Michael, B. Comparative morphometric indices of the large and small flying squirrels. *Zoos Print. XIII*, 46–47 (1998).

Tropical Ecosystems: Structure, Diversity and Human Welfare.
Proceedings of the International Conference on Tropical Ecosystems
K. N. Ganeshaiah, R. Uma Shaanker and K. S. Bawa (eds)
Published by Oxford–IBH, New Delhi. 2001. pp. 722–723.

Ecological and social correlates of foraging decisions in a social forager, the bonnet macaque *Macaca radiata diluta*

Jayashree Ratnam

Department of Biology, Syracuse University, Syracuse, NY 13244, USA
e-mail: jratnam@mailbox.syr.edu

It is well known that solitary foragers make foraging decisions in response to ecological constraints such as the distribution of food and predators in the environment (Pulliam and Caraco, 1984). However, most primates live and forage in groups. Such social foragers must respond not only to ecological constraints, but also to social constraints. Social constraints may often be imposed in the form of dominance hierarchies that influence the costs and benefits of interactions between individuals of different social ranks. A social forager must not only track ecological information but must also track social information such as where group members are, in which patches dominant individuals or allies are feeding and how much it might benefit or loose from interactions with group members of different social ranks (Giraldeau and Caraco, 2000). How do ecological and social constraints interact to influence decision-making behaviors in social foragers? Do individuals of different social ranks weigh these constraints differently? If so, what are the implications of these differences? This study explores such questions experimentally in free ranging groups of a social primate species, the bonnet macaque *Macaca radiata diluta*.

Keywords: Social foragers, constraints, social structure, habitat change.

Specifically the experiments reported addressed the question of how ecological factors such as habitat structure and potential predation risk in habitat interacted with social factors such as group size and social dominance to influence foraging responses and energetic gains of individual members of social groups. I approached this question experimentally by creating food distributions in different habitats (varying in predation risk and vegetation structure) and analyzing the responses of free ranging groups of the south Indian bonnet monkey. The results suggest that dominant individuals in groups of this species track ecological factors closely, whereas the responses of subordinate individuals to ecological factors are more varied and are contingent upon the strategies of dominant individuals. As a consequence, strategies of subordinates tend to be more opportunistic and variable than that of dominant individuals. These results suggest that individuals of different social ranks within a social group may experience the same habitat differently, with subordinates often experiencing it as being more marginal or unpredictable.

These results may have important implications for the direction of change in social systems in response to habitat changes that occur during events such as fragmentation or degradation. If these findings are indeed more general, then they suggest that there will be a temporal phase lag between dominant and subordinate group members in both perceiving and adjusting to changes in habitat parameters. While the actual outcome of this phase lag will depend on a number of additional factors, its inclusion in our consideration of the responses of group living species to habitat change should greatly improve both our mechanistic understanding and predictive power of such events.

References

Pulliam, H. R. & Caraco, T. Living in groups: Is there an optimal group size? in *Behavioural Ecology: An Evolutionary Approach* (eds Krebs, J. R. & Davies, N. B.) 122–147 (Sinauer Associates, Sunderland, Mass, 1984), 2nd edn.

Giraldeau L. A. & Caraco, T. *Social Foraging Theory* (Princeton University Press, Princeton, NJ, 2000).

Tropical Ecosystems: Structure, Diversity and Human Welfare.
Proceedings of the International Conference on Tropical Ecosystems
K. N. Ganeshaiah, R. Uma Shaanker and K. S. Bawa (eds)
Published by Oxford–IBH, New Delhi. 2001. pp. 724–726.

Changes in freshwater fish fauna in northern Western Ghats, Pune, India

Neelesh Dahanukar*, Rupesh Raut**,
Mukul Mahabaleshwarkar[†,‡] and Sanjay Kharat[§]

*1104/B, Shivajinagar, Shashi Apts, Model Colony, Pune 411 016, India
**136, Budhwar Peth, Pune 411 002, India
[†]24E/15, Pashchimanagari Housing Society, Kothrud, Pune 411 029, India
[§]Department of Zoology, Abasaheb Garware College, Pune 411 004, India
[‡]e-mail: mookool@ip.eth.net

Rivers all over the world have supported the growth of human civilization since the first town appeared some 7000 years ago. As a result of this growth and diversification of activities, most of the rivers have been affected to varying degrees. Mula–Mutha rivers, originating in the Northern Western Ghats of India, are no exception. In this study, we deal with the changing status of the fish fauna of Mula–Mutha river systems passing through Pune city, Maharashtra State, India.

Collection of fish fauna was made with the help of local fishermen and the tribal people at various locations. The specimens were preserved in 5% formalin. Fishes were identified with the help of literature (Day, 1878; Menon, 1987; Jayaram, 1991; Talwar and Jhingran, 1992).

We collected 67 fish species during 1999–2000. Prior to the present study, Fraser (1942a, b), Suter (1944), Tonapi and Mulherkar (1963) and Ghate and

Keywords: Local extinction, decline, new records, fish fauna.

Wagh (pers. commun.) made surveys for the collection of fish species. However out of the 108 fish species recorded so far, 16 have become locally extinct while 8 are declining in their populations.

There seems to be an increasing trend of local migration of fishes from human impacted zones to wild zone (less disturbed) as evident from the site of collection. The probable driving forces for these migration include increase in pollution, heavy harvesting, construction of dams across the rivers and introduction of exotic fish species. Four new species, viz. *Cirrhinus cirrhosus, Danio malabaricus, Rasbora labiosa* and *Xiphophorous hellerii* have been recorded for the first time.

Four consecutive studies were compared to estimate the changes in the species composition of these river systems. Accordingly, the frequency record of each fish species was noted (Table 1). The study area was divided into 'wild' zone and 'impact' zones, based on increasing human interference. The comparison between two previous studies (Table 2) shows that there is local migration of species from impacted zone to the wild zone.

The Mula and Mutha rivers in the Pune city are highly polluted due to the discharge of industrial wastes and sewage. However, the river segment outside the city area, especially in the upstream, seems to be unpolluted. Pollution seems to be one of the probable causes of decline in species number in the impacted zone (Table 2).

Table 1. Frequency record of species and their status.

Recorded frequency from studies	Number of fish species out of 108				
	Total	Doubtful	Locally extinct	Declining	New record
1+	36	12	11	–	4
2+	24	–	5	3	–
3+	25	–	–	2	–
4+	23	–	–	3	–

Table 2. Distribution of fish species in impacted and wild zone as evident from the three studies.

	Number of fish species		
	Fraser (63 sp.)	Tonapi (57 sp.)	Present (67 sp.)
Total no. of species found in impacted zone	48	46	33
Species found only in wild zone	15	11	34
Total no. of species found in wild zone	44	52	66
Species found only in impacted zone	19	5	1

Besides pollution, heavy harvest, construction of dams and introduction of exotic fishes are likely to be responsible for such changes.

Acknowledgements

We are thankful to Dr Hemant Ghate and Dr C. P. Shaji for their valuable suggestions and confirmation and identification of the species. Utkarsh Ghate and Dr Milind Watve encouraged and helped in manuscript preparation. We are thankful to the authorities of Department of Zoology, Abasaheb Garware College, for providing facilities.

References

Day, F. *Fishes of India,Vol. I & II* (William Dawson and Sons Ltd., London, 1878). 778 + CXCV.

Frazer, A. G. L. Fish of Poona – Part 1. *J. Bombay Nat. Hist. Soc.* **43,** 79–91 (1942a).

Frazer, A. G. L. Fish of Poona – Part 3. *J. Bombay Nat. Hist. Soc.* **43,** 452–454 (1942b).

Ghate, H. V. & Wagh, G. K. Proc. First National Symp. on Environ. Hydraulics, CWPRS, Pune (1992).

Jayaram, K. C. Records of Zoological Survey of India, Revision of genus Puntius (Ham.) from Indian Region. (Pisces: Cypriniformes: Cyprinidae: Cyprininae). Occasional paper No. *135,* Zool. Surv. India, Calcutta, 178. (1991).

Menon, A. G. K. *Fauna of Indian and adjacent countries, Pisces, vol. 4 (part I) Homalopteridae,* (Zoological Survey India, Calcutta, 1987) X + 259.

Suter, M. J. New records of fish from Poona. *J. Bombay Nat. Soc.* **44,** 408–414 (1944).

Talwar, P. K. & Jhingran K. C. *Inland Fishes, Vol. I & II* (Oxford IBH, New Delhi, 1992) 1097.

Tonapi, G. T. & Mulherkar, L. Notes on the freshwater fauna of Poona, Part 1, Fishes, *Proc. Indian Acad. Sci.* **58,** 187–197 (1963).

Tropical Ecosystems: Structure, Diversity and Human Welfare.
Proceedings of the International Conference on Tropical Ecosystems
K. N. Ganeshaiah, R. Uma Shaanker and K. S. Bawa (eds)
Published by Oxford–IBH, New Delhi. 2001. pp. 727–731.

Food, sex and society: Social evolution among wild bonnet macaques

Anirban Datta Roy* and Anindya Sinha

National Institute of Advanced Studies, Indian Institute of Science Campus,
Bangalore 560 012, India
*e-mail: ad_roy@rediffmail.com

In primates, ecological pressures have often been invoked to explain not only the formation of societies, but also the size of social groups and the nature of interactions between individuals within and across such groups (Wrangham, 1980; Smuts *et al.*, 1987). The evolution of new stable groups with unique individual social strategies, directly influenced by immediate ecological factors, has, however, only very rarely been observed.

The bonnet macaque (*Macaca radiata*, Geoffroy), an endemic cercopithecine primate found commonly in peninsular India, usually lives in multimale, bisexual troops that range in size from 8 to 60 individuals (Sinha, 2001). We report here the unusual occurrence of small unimale groups, containing only a single adult male each, within a population of the species and explore the possible ecological and behavioural mechanisms that may have led to the evolution of this form of social organisation within a typical multimale primate social system.

This study, involving 21 groups of the macaque, was conducted in the dry deciduous forests of Bandipur National Park, Karnataka, in southern India. The natural diet of the study troops consisted predominantly of fruits,

Keywords: Bonnet macaque, unimale troop, multimale troop, birth sex ratio, demography, social evolution.

flowers, or leaves of several tree species, assorted herbs, several grasses, and a variety of insects. The groups also occasionally fed on high-calorie human food handed out by tourists passing through the sanctuary; such food was, however, of low amounts, and seasonal and unpredictable in distribution.

Demographic data were collected from the study groups at periodic intervals during the period from March to November 2000. The demographic variables analysed in this study include total group size, the number of individuals of the two sexes belonging to different age groups classified as adults, subadults, juveniles and infants, and patterns of male emigration across troops.

An enumeration of 21 bonnet macaque groups in the study area revealed the presence of 11 unimale troops, a surprising finding because bonnet macaques almost invariably live in multimale groups. Earlier studies of the species in this particular area have either not reported the presence of unimale troops (D'Souza and Singh, 1992), or report only a very small proportion of such groups in the population – 12.5% (Simonds, 1965) or 13.8% (Kurup, 1981), as compared to 52.4% in the present study.

The unimale troops (with 10.82 ± 3.25 individuals in each group) were significantly smaller than the multimale troops (with 20.20 ± 9.39 individuals; Mann–Whitney U-test, $p < 0.02$) and had less number of adult females (Figure 1, $p < 0.05$). The adult male/adult female ratio ranged from 0.25 to 1.0 for the unimale and from 0.60 to 2.0 for the multimale groups.

In comparison to the multimale troops, the unimale groups were remarkably depleted in all categories of males, including subadults, juveniles and infants (Figure 1; Mann–Whitney U-test, $p < 0.05$ for all categories). There were, however, no significant differences in the number of female subadults, juveniles and infants (Figure 1), or in the total number of individuals in these categories across the two types of social groups ($p > 0.10$ for all categories).

A substantial re-analysis of the data from an earlier survey carried out by Kurup (1981) indicates that in two populations, from Mysore and Kolar districts of Karnataka, the ratios of adult, subadult and juvenile males to females are also relatively more female-biased in unimale troops than in multimale groups (Figure 2). This is particularly true for subadult individuals where the sex ratio significantly differs across the two social systems in all the three populations (G-test of independence, $p < 0.05$). Unfortunately, Kurup's report does not include enough data on juveniles and infants to permit similar statistical analyses.

Differential infant mortality does not seem to explain the significant bias toward female infants (1:4) displayed by the unimale troops in Bandipur since the birth rate of the unimale groups (0.61 ± 0.36 infant/adult female)

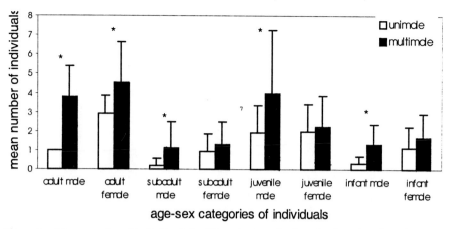

Figure 1. Mean number of individuals in different categories in unimale and multimale groups.

appears to be virtually identical to that of the multimale troops (0.60 ± 0.30). The birth sex ratio within the unimale population must therefore be genuinely skewed in favour of female infants – perhaps the first example of a variation in social organisation leading to variable birth sex ratios in any mammalian species. Although specific proximate mechanisms underlying skews in primate birth sex ratios remain unknown, female primates may themselves facultatively determine the sex of their infants in response to appropriate environmental or social stimuli (Silk *et al.*, 1981).

The observed skew in infant sex ratio in unimale groups could have arisen in response to the possible reproductive monopolisation practised by the resident adult male. In multimale groups, adult males are remarkably tolerant of the reproductive efforts of others (Sinha, 2001). In unimale groups, however, the adult male is rather intolerant of other males, often herding the group females and displaying severe aggression towards subadult males. Moreover, these are invariably involved in territorial defence and often inflict grievous injuries on males from other groups; the alpha males of multimale troops virtually never participate in inter-troop encounters. Finally, adult males in unimale groups are rather successful in preventing immigration of other males into their troops; there were no immigrants into seven unimale troops monitored during the mating season, but a total of 10 males had successfully joined six multimale troops during the same period (Mann–Whitney U-test, $p < 0.02$). Such despotic behaviour of the resident male in unimale troops may lead to increased male emigration from such groups and result in the depletion of subadult males. Since male emigration at a relatively young age could potentially entail heavy costs, females in such troops may have been selected to bear and raise daughters, who would remain in the troop and contribute to the matriline, rather than sons, who would almost invariably be driven away.

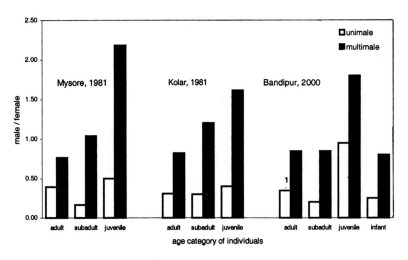

Figure 2. Sex ratio of individuals in unimale and multimale groups in different populations.

The adult sex ratio in most primate groups appears to be related to the length of the breeding season and the degree of oestrus synchrony among the females (Ridley, 1986). Since bonnet macaques generally live in seasonal environments and most females within a troop come into oestrus synchronously, this may have led to a relatively greater proportion of males in this species, and correspondingly, a promiscuous mating system. Is it, however, possible that adult females within unimale bonnet troops exhibit oestrus asynchrony enabling a single adult male to sequentially monopolise each sexually receptive female? The clustered distribution of births within the unimale groups of Bandipur, where all births occurred between February and April, seems to argue against such a hypothesis; females in multimale troops, in contrast, gave birth from January to May.

An ecological factor that may have significantly affected bonnet macaque populations in Bandipur in the recent past is the increasing tourist traffic within this protected area. Since this species appears to naturally gravitate towards human settlements and food sources, many tourists invariably come in contact with bonnet troops and provision them with typically human foods. Such food is often nutritionally rich and provisioning is thus usually marked by intense scramble competition within the troop and among neighbouring troops, especially along the highway. This is accentuated during the dry summer months when natural food resources within the dry deciduous forests are particularly sparse and very patchily distributed, and most of the troops come and space themselves along the highway. Provisioned human food being rather unpredictable in amount and clumped in distribution, however, may possibly support only small groups of macaques. Since bonnet females alone are philopatric and form stable core groups in a particular area, this may have led to the evolution of small troops of closely related females that could then be easily reproductively monopolised by a single adult male.

Acknowledgements

We thank the Karnataka Forest Department for research permission and logistic support in the Bandipur National Park. AS acknowledges a research grant from the Wenner-Gren Foundation for Anthropological Research, New York, USA.

References

D'Souza, L. & Singh, M. Density and demography in roadside bonnet monkeys *Macaca radiata* around Mysore. *J. Ecobiol.* **4**, 87–93 (1992).

Kurup, G.U. *Report on the Census Surveys of Rural and Urban Populations of Non-human Primates of South India* (Man and Biosphere Programme: Project No 124, Zoological Survey of India, Calicut, 1981).

Ridley, M. The number of males in a primate troop. *Anim. Behav.* **34**, 1848–1858 (1986).

Silk, J. B., Clark-Wheatley, C. B., Rodman, P. S. & Samuels, A. Differential reproductive success and facultative adjustment of sex ratios among captive female bonnet macaques (*Macaca radiata*). *Anim. Behav.* **29**, 1106–1120 (1981).

Simonds, P. E. in *Primate Behavior: Field Studies of Monkeys and Apes* (ed. DeVore, I.) 175–196 (Holt, Winehart and Winston, New York, 1965).

Sinha, A. *The Monkey in the Town's Commons. A Natural History of the Indian Bonnet Macaque. NIAS Report 2-01* (National Institute of Advanced Studies, Bangalore, 2001).

Smuts, B. *et al.* (eds) *Primate Societies* (University of Chicago Press, Chicago, 1987).

Wrangham, R. W. An ecological model of female-bonded primate groups. *Behaviour* **75**, 262–300 (1980).

Tropical Ecosystems: Structure, Diversity and Human Welfare.
Proceedings of the International Conference on Tropical Ecosystems
K. N. Ganeshaiah, R. Uma Shaanker and K. S. Bawa (eds)
Published by Oxford–IBH, New Delhi. 2001. pp. 732–735.

Freshwater shrimp population structure and distribution dynamics in a tropical rainforest stream

T. Heartsill-Scalley*†, T. A. Crowl*, M. Townsend* and A. P. Covich‡

*Department of Fisheries and Wildlife and Ecology Center, Utah State University,
Logan, Utah 84322, USA
‡Department of Fisheries and Wildlife Biology, Colorado State University, Ft.
Collins, Colorado 80523, USA
†e-mail: heartsill@cc.usu.edu

Even though Atyid shrimps are the most abundant animals in the headwater streams of the Luquillo Experimental Forest in Puerto Rico, population dynamics of these shrimp are still not well understood. In particular, populations of the shrimp *Atya lanipes* appear relatively constant in abundance over a decade of monitoring, with densities showing little variation after disturbance events (Covich *et al.*, 1991; Covich *et al.*, 1996b). Atyid shrimps are amphidromous, living mostly in the headwater streams (Covich and McDowell, 1996a). Their growth and reproduction occur in the freshwater, and only the larval stage needs saline water to complete their development. As juveniles, they return to the freshwater of the headwater streams where they continue to grow, develop and reproduce. Adult survivorship is high in this species and recruitment of juveniles into the pools of headwater streams seems to occur at very low rates. To better understand

Keywords: Freshwater shrimp, population dynamics, tropical streams, Pureto Rico.

Atya population dynamics, we contrast population structure of pools with different physical conditions along an elevation gradient over a four-year sampling period. We conducted this study in the first order tributary Quebrada Prieta, which is part of the Espíritu Santo River watershed within the Luquillo Experimental Forest, Puerto Rico. The Quebrada Prieta substrata is dominated by boulders and cobbles, and is surrounded by steep slopes which provide run-off during heavy rainfall events (Covich and McDowell, 1996a). Leaf-litter inputs and debris dams are also characteristic of this system. Pools were selected to represent different physical conditions along an elevation gradient within the Q. Prieta, as defined by Covich *et al.* (1991). Six pools were chosen, with replicates of similar physical parameters such as elevation and gradient (Table 1). The pools were sampled every two months, from January 1996 to December 1999.

Shrimp abundance was estimated by using baited wire funnel traps, with the quantity of traps reflecting pool area, at approximately 0.5 traps per m^2. Physical parameters such as pool depth, area and volume were also collected at bi-monthly intervals. For all captured individuals carapace length was measured, and three size class categories were established. Individuals < 10 mm cl were classified as juveniles (size 1), individuals between 10–19.9 mm cl were considered to be reproductive females (size 2), and individuals > 19.9 mm cl were considered males (size 3). A repeated measures ANOVA revealed that the size class structure of pools was not different in terms of juveniles and showed no seasonal patterns ($F = 0.71$, $P = 0.90$). The proportion of females (size 2) was different among pools ($F = 27.25$, $P < 0.0001$) and months ($F = 8.97$, $P < 0.0001$) and there was strong pool x month interaction ($F = 5.71$, $P < 0.0001$). The proportion of adult males (size 3) was similar to that found for females, with differences among pools ($F = 8.97$, $P < 0.0001$) and among months ($F = 3.54$, $P = 0.0015$) and the pool and month interaction was not significant ($F = 1.53$, $P = 0.06$). Within the female size category, the proportion of gravid individuals was not different among pools ($F = 0.50$, $P = 0.77$), but was different among months ($F = 8.72$, $P < 0.0001$) with no pool × month interaction ($F = 0.25$, $P = 1.00$). The lowest density of gravid females occurred from November to February (Figure 1) in all pools.

Table 1. Physical characteristics of individual pools, based on yearly averages.

Characteristics	Low elevation high gradient		Intermediate low gradient		High elevation high gradient	
Pool number	–9	–6	0	8	13	15
Elevation (masl)	305	329	360	381	433	439
Depth (m)	0.54	0.41	0.38	0.30	0.22	0.19
Area (m²)	15.46	5.86	74.20	9.34	4.06	5.42
Volume (m³)	8.65	2.42	29.06	2.85	0.90	1.03

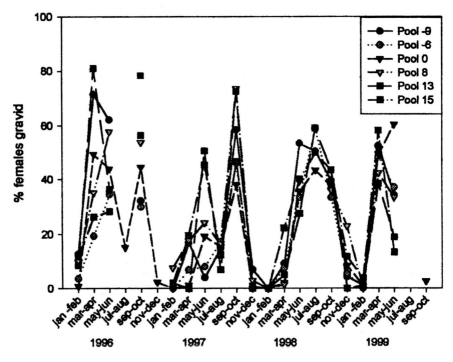

Figure 1. Monthly per cent of gravid *Atya lanipes* females by pool, from 1996 to 1999.

A post hoc comparison (REGWQ test) revealed that the months with the lowest proportion of gravid females were significantly different from the rest of sampled months. Even though there was a strong seasonality of gravid females ($F = 8.72$, $P < 0.0001$), this seasonal pattern was not different among pools ($F = 0.50$, $P = 0.77$). Similar results were found by Johnson *et al.* (1997) although their results did not show such a strong seasonal pattern. It is proposed that because individual *Atya* have low adult mortality and a relatively long life span, the low occurrence of juvenile individuals (size 1) may reflect the lack of large recruitment events and may not be important for the stability of this population. Enough adults are always present in the population to produce and release large numbers of potential new recruits. The high fecundity, long life span and low adult mortality are all factors that are attributed to populations under the 'storage effect' hypothesis (Warner and Chesson, 1985).

Acknowledgements

We thank Sherri Johnson, Ryan Philips, and John Bithorn for help with data collection. Gary Belovsky and Fred Scatena provided valuable discussion and ideas during the development and various phases of the analysis.

References

Covich, A. P., Crowl, T. A., Johnson, S. L., Varza, D. & Certain, D. L. Post-Hurricane Hugo increases in Atyid Shrimp abundances in a Puerto Rican Montane Stream. *Biotropica* **23**, 448–454 (1991).

Covich, A. P. & McDowell, W. H. The stream community. in *The Food Web of a Tropical Rain Forest Stream* (eds Reagan, D. P. & Waide, R. B.) 433–459 (Chicago Univ of Chicago Press, 1996a).

Covich, A. P., Crowl, T. A., Johnson, S. L. & Pyron, M. Distribution and abundance of tropical freshwater shrimp along a stream corridor: Response to disturbance. *Biotropica* **28**, 484–492 (1996b).

Johnson, S. L., Covich, A. P., Crowl, T. A., Estrada, A., Bithorn, J. & Wurtsbaugh, W. A. Do seasonality and disturbance influence reproduction in freshwater Atyid shrimp in headwater streams, Puerto Rico? *Proc. Int. Assoc. Theoret. Appl. Limnol.* **26**, 2076–2081 (1997).

Warner, R. R. & Chesson, P. L. Coexistence mediated by recruitment fluctuations: A field guide to the storage effect. *Am. Nat.* **125**, 769–787 (1985).

Tree Diversity and Phenology

Tropical Ecosystems: Structure, Diversity and Human Welfare.
Proceedings of the International Conference on Tropical Ecosystems
K. N. Ganeshaiah, R. Uma Shaanker and K. S. Bawa (eds)
Published by Oxford–IBH, New Delhi. 2001. pp. 739–743.

Fungal endophytes in neotropical trees: Abundance, diversity, and ecological implications

A. Elizabeth Arnold

Department of Ecology and Evolutionary Biology, 1041 E. Lowell, BSW 310,
University of Arizona, Tucson, AZ 85721, USA
e-mail: arnold@ccit.arizona.edu

Although fungi comprise only ca. 72,000 described species, many authors rank them among hyperdiverse taxa, comparing fungi with nematodes, mites and insects in terms of global species richness (Colwell and Coddington, 1994; Hawksworth *et al.*, 1995). However, the true scale of fungal diversity is a matter of open debate, with estimates ranging from hundreds of thousands of species (Aptroot, 1997) to values as high as 1.5 million species (Hawksworth, 1991) and greater (e.g., Cannon, 1997). Recent assessments of fungal biodiversity in tropical forests, where fungal richness is thought to be greatest, have begun to refine richness estimates, and in the last decade have lent credence to higher values (e.g., Fröhlich and Hyde, 1999). However, several authors argue that such extrapolations either represent dramatic overestimates of fungal diversity (May, 1991), or are suspicious due to a lack of relevant data (Hammond, 1992). Resolution of the fungal diversity debate bears directly upon such human enterprises as medicine, agriculture, and industry, and upon our understanding of the structure and function of terrestrial ecosystems (e.g., Fox, 1993; Lodge, 1997). Mycologists agree that further exploration of tropical fungi is critical for understanding both the scale of fungal diversity (Hawksworth, 1993) and its ecological implications.

Keywords: Abundance, Barro Colorado Island, biodiversity, fungal endophytes, similarity indices.

Tropical fungi are traditionally understudied (Cannon, 1997), and their taxonomic placement has been confounded, at times, by misidentification according to temperate mycota (Lodge, pers. comm.). Extensive training needed for sampling complex tropical habitats (Shivas and Hyde, 1997), inconsistencies in quantitative methods among studies (Arnold et al., 2000), a paucity of newly trained systematists specializing in tropical mycology (Rossman, 1997), and traditional difficulties in delineating species boundaries (Hawksworth, 1991) further inhibit both the accuracy and precision of fungal diversity estimates. Especially challenging to enumerate are tropical microfungi: those cryptic fungi that produce only microscopic fruiting bodies. Rossman (1997) has suggested that as many as 700,000–900,000 microfungal species may exist, the majority of which are likely to occur in tropical regions. Data from several authors (e.g., Lodge et al., 1996; Fröhlich and Hyde, 1999) have suggested that many of these microfungi may occur as fungal endophytes inhabiting tropical woody angiosperms; however, studies of endophytic fungi in tropical forests are yet in their infancy.

Endophytes are microorganisms that colonize and cause asymptomatic infections in healthy plant tissues (Wilson, 1995). Fungal endophytes are considered to be at least as ubiquitous as mycorrhizal associations among temperate-zone plants (Carroll, 1988), having been found in algae (Hawksworth, 1988), mosses (Schulz et al., 1993), ferns (Fisher, 1996), conifers (Legault et al., 1989), and both monocotyledonous (Clay, 1988) and dicotyledonous angiosperms (Petrini et al., 1982). Saikkonen et al. (1998) noted that individual plants in the temperate zone may harbor dozens of endophyte species, and several recent, quantitative surveys of tropical angiosperms have documented remarkable endophyte richness in individual leaves and trees (e.g., Lodge et al., 1996; Fröhlich and Hyde, 1999; Arnold et al., 2000). These surveys suggest that tropical endophytes may contribute substantially to fungal diversity, but both an ecological approach to quantitative sampling, and consistency among methods at a variety of scales, have been lacking. To address these issues, I have explored patterns of endophyte abundance, diversity, host preference, and spatial heterogeneity in the lowland, moist forest of Barro Colorado Island, Panama (BCI).

In surveys of 24 common host species representing 23 families of woody angiosperms, I have shown that endophytes are highly abundant in healthy, mature leaves of understorey saplings, where they typically colonize > 80% of leaf tissue. Unsuccessful attempts to isolate endophytic fungi from both seeds and seedlings grown under sterile conditions, and observations that endophyte-free seedlings quickly accumulate endophytes under field conditions, strongly suggest that endophytic fungi associated with tropical trees are horizontally transmitted. A consistent and significant increase in endophyte abundance with increasing leaf age further corroborates this pattern and appears to result from differences in leaf exposure time, rather than from differences in leaf chemistry between young and mature leaves.

Similarly, high endophyte abundance among forest-grown seedlings relative to clearing-grown seedlings appears to be a common pattern, and can be explained by differences in inoculum level, rather than by differences in leaf chemistry between exposed and shaded hosts. These data, and emerging data from other tropical sites (e.g., Puerto Rico: Lodge *et al.*, 1996; Gamboa and Bayman, in press; Borneo and Australia: Fröhlich and Hyde, 1999) suggest that fungal endophytes may represent a ubiquitous, cryptic, and ecologically interesting component of tropical forests.

Similarly, endophytic fungi appear to be consistently diverse among host species at BCI: mature leaves typically contain > 10 morphospecies of fungal endophytes, and individual host species may harbor hundreds of endophyte morphospecies. However, endophyte assemblages differ with respect to both sampling site and host taxon: using frequency-based similarity indices and randomization tests based on isolation frequencies, I have found strong evidence for both spatial heterogeneity and host preference among nonsingleton morphospecies. In ongoing work, I am exploring the scale of spatial heterogeneity within the forest at BCI, and am empirically assessing endophytes for a mechanism of host preference, concentrating on host-specific growth rates. Concurrently, I am seeking to reconcile endophyte morphospecies with genetically delimited species using sequence data from the nrDNA internal transcribed spacer regions 1 and 2 (ITS1, ITS2), and introns associated with translation elongation factor 1α (EF-1α). Preliminary data suggest general concordance between ITS-delimited species and morphospecies, and demonstrate that endophytes are diverse at high taxonomic levels within the Ascomycota; however, further sampling is needed to assess the number of genetically distinct units recovered from individual leaves, trees, sites, and host species.

Studies of endophyte abundance, diversity, spatial heterogeneity, and host preference are critical for assessing the contribution of tropical endophytes to fungal diversity estimates. Based on evidence from preliminary surveys of endophytes in tropical forests, several authors have suggested that the most commonly accepted estimate for fungal diversity (1.5 million species: Hawksworth, 1991) represents a marked underestimate (e.g., Dreyfuss and Chapela, 1994; Fröhlich and Hyde, 1999; Arnold *et al.*, 2000). Similarly, data regarding endophyte abundance and diversity have important implications for understanding plant–fungus interactions in tropical forests, and accumulating data present several questions of ecological importance: for example, given that endophytes are obligately heterotrophic, and that plants in the shaded tropical understorey are often carbon-limited, why should tropical trees harbor such a high abundance of endophytic fungi within their leaves? Similarly, given that fungal diseases represent a significant cause of plant mortality in lowland tropical forests, that many tropical trees are relatively well defended against a variety of fungal pathogens, and that many fungal endophytes are congeneric with a taxonomically diverse array of pathogens, why do tropical trees harbor such a diversity of fungal endophytes? Using these questions to guide my work, I am assessing the costs and benefits of endophyte infection, and am testing the hypothesis that fungal endophytes may influence host resistance to invasive pathogens.

The ecological importance of fungal endophytes in tropical trees is not yet known, but endophytes appear to be both ubiquitous and highly diverse in tropical forests. Further exploration of tropical endophytes will help to clarify the fungal diversity debate, and will likely lend support to higher estimates of global fungal diversity. At the same time, the study of tropical endophytes seems promising as a means to enrich our understanding of plant–fungus interactions in tropical forests, tropical biodiversity, and tropical ecology.

Acknowledgements

I sincerely thank L. McDade for valuable insights and guidance; E. A. Herre for furthering tropical endophyte research; G. Gilbert for training and advice; Z. Maynard and J. Barnard for technical assistance; and the Smithsonian Tropical Research Institute for logistical support. Funding support from an NSF Graduate Fellowship, NSF-DEB 9902346 (to L. McDade and AEA), the Research Training Group in Biological Diversification at the University of Arizona (NSF-DIR-9113362, BIR-9602246), the American Cocoa Research Institute, the Smithsonian Tropical Research Institute, and the Hoshaw family are gratefully acknowledged. For travel support, I thank the Department of Ecology and Evolutionary Biology at the University of Arizona and the Association for Tropical Biology. Special thanks are given to P. D. Coley and T. A. Kursar for an introduction to tropical ecology.

References

Aptroot, A. Species diversity in tropical rainforest ascomycetes: Lichenized vs non-lichenized; foliicolous vs corticolous. *Abstr. Bot.* **21,** 37–44 (1997).

Arnold, A. E., Maynard, Z., Gilbert, G. S., Coley, P. D. & Kursar, T. A. Are tropical fungal endophytes hyperdiverse? *Ecol. Lett.* **3,** 267–274 (2000).

Cannon, P. Diversity of the Phyllachoraceae with special reference to the tropics. in *Biodiversity of Tropical Microfungi* (ed. Hyde, K. D.) 255–278 (Hong Kong University Press, Hong Kong SAR, China, 1997).

Carroll, G. Fungal endophytes in stems and leaves: from latent pathogen to mutualistic symbiont. *Ecology* **69,** 2–9 (1988).

Clay, K. Fungal endophytes of grasses: A defensive mutualism between plants and fungi. *Ecology* **69,** 10–16 (1988).

Colwell, R. K. & Coddington, J. A. Estimating terrestrial biodiversity through extrapolation. *Philos. Trans. Roy. Soc. London, B* **345,** 101–118 (1994).

Dreyfuss, M. M. & Chapela, I. H. Potential of fungi in the discovery of novel, low-molecular weight pharmaceuticals. in *The Discovery of Natural Products with Therapeutic Potential* (ed. Gullo, V. P.) 49–80 (Butterworth-Heinemann, London, UK, 1994).

Fisher, P. J. Survival and spread of the endophyte *Stagonospora pteridiicola* in *Pteridium aquilinum*, other ferns and some flowering plants. *New Phytol.* **132,** 119–122 (1996).

Fröhlich, J. & Hyde, K. D. Biodiversity of palm fungi in the tropics: Are global fungal diversity estimates realistic? *Biodiver. Conserv.* **8,** 977–1004 (1999).

Fox, F. M. Tropical fungi: their commercial potential. in *Aspects of Tropical Mycology* (eds Isaac, S., Frankland, J. C., Watling, R. & Whalley, A. J .S.) 253–264 (Cambridge University Press, Cambridge, UK, 1993).

Gamboa, M. A. & Bayman, P. Communities of endophytic fungi in leaves of a tropical timber tree *(Guarea guidonia*: Meliaceae) (In press).

Hammond, P. M. Species inventory. in *Global Diversity: Status of the Earth's Living Resources* (ed. Groombridge, B.) 17–39 (Chapman & Hall, London, UK, 1992).

Hawksworth, D. L. The variety of fungal-algal symbioses, their evolutionary significance, and the nature of lichens. *Bot. J. Linn. Soc.* **96**, 3–20 (1988).

Hawksworth, D. L. The fungal dimension of biodiversity: magnitude, significance, and conservation. *Mycol. Res.* **95**, 641–655 (1991).

Hawksworth, D. L. The tropical fungal biota: Census, pertinence, prophylaxis, and prognosis. in *Aspects of Tropical Mycology* (eds Isaac, S., Frankland, J. C., Watling, R. & Whalley, A. J. S.) 265–293 (Cambridge University Press, Cambridge, UK, 1993).

Hawksworth, D. L., Kirk, B. C., Sutton, B. C. & Pegler, D. N. *Ainsworth & Bisby's Dictionary of the Fungi.* 8th edition. 359 pp. (CAB International, Wallingford, UK, 1995).

Legault, D., Dessureault, M. & Laflamme, G. Mycoflore des aiguilles de *Pinus banksiana* et *Pinus resinosa* I. Champignons endophytes. *Can. J. Bot.* **67**, 2052–2060 (1989).

Lodge, D. J., Fisher, P. J. & Sutton, B. C. Endophytic fungi of *Manilkara bidentata* leaves in Puerto Rico. *Mycologia* **88**, 733–738 (1996).

Lodge, D. J. Factors related to diversity of decomposer fungi in tropical forests. *Biodiver. Conserv.* **6**, 681–688 (1997).

May, R. M. A fondness for fungi. *Nature* **352**, 475–476 (1991).

Petrini, O., Stone, J. & Carroll, F. E. Endophytic fungi in evergreen shrubs in western Oregon: A preliminary study. *Can. J. Bot.* **60**, 789–796 (1982).

Rossman, A. Y. Biodiversity of tropical microfungi: An overview. in *Biodiversity of Tropical Microfungi* (ed. Hyde, K. D.) 1–10 (Hong Kong University Press, Hong Kong SAR, China, 1997).

Saikkonen, K., Faeth, S. H., Helander, M. & Sullivan, T. J. Fungal endophytes: A continuum of interactions with host plants. *Annu. Rev. Ecol. Syst.* **29**, 319–343 (1998).

Schulz, B., Wanke, U., Draeger, S. & Aust, H.-J. Endophytes from herbaceous plants and shrubs: Effectiveness of surface sterilization methods. *Mycol. Res.* **97**, 1447–1450 (1993).

Shivas, R. G. & Hyde, K. D. Biodiversity of plant pathogenic fungi in the tropics. in *Biodiversity of Tropical Microfungi* (ed. Hyde, K. D.) 47–56 (Hong Kong University Press, Hong Kong, SAR, China, 1997).

Wilson, D. Endophyte – the evolution of a term, and clarification of its use and definition. *Oikos* **73**, 274–276 (1995).

Tropical Ecosystems: Structure, Diversity and Human Welfare.
Proceedings of the International Conference on Tropical Ecosystems
K. N. Ganeshaiah, R. Uma Shaanker and K. S. Bawa (eds)
Published by Oxford–IBH, New Delhi. 2001. pp. 744–749.

Flowering and fruiting phenology of a tropical forest in Arunachal Pradesh, northeast India

Aparajita Datta and G. S. Rawat

Wildlife Institute of India, Post Bag 18, Dehra Dun 248 001, India
e-mail: loony10@hotmail.com

Phenological studies in Indian tropical forests have been largely restricted to deciduous forests and evergreen forests in south India, while such studies in northeast India are limited. In this paper, we attempt to describe the seasonal and annual variations in patterns of flower and fruit availability of wind and animal-dispersed tree species in a tropical forest in Arunachal Pradesh. While environmental factors may be the primary force influencing phenology (Frankie *et al.*, 1974), strong seasonality in fruiting patterns in forest types with differing rainfall and climatic regimes (Foster, 1982; Leighton and Leighton, 1983; Terborgh and van Schaik, 1993) has suggested that other factors may be also influencing fruiting patterns in forests with a high percentage of animal-dispersed tree species (Snow, 1965; Wheelwright, 1985). It has been predicted that fruiting schedules of tree species that share common dispersers should be staggered to avoid competition for dispersers (competition avoidance hypothesis). We tried to determine if there is any evidence that fruiting patterns of a set of bird-dispersed plant species are driven by competition for dispersers. Alternatively, temporally aggregated fruiting phenologies may occur when abundance of seed dispersal agents

Keywords: Anemochorous, competition avoidance, flowering, fruiting, phenology, tropical forest, rainfall, zoochorous, seed dispersal, north-east India, Arunachal Pradesh, frugivory.

vary seasonally or when synchronous fruiting enhances dispersal (enhancement hypothesis) (Rathcke and Lacey, 1985; Poulin *et al.*, 1999) or to satiate seed predators (Janzen, 1971). Null models have been used to test whether observed phenological overlap in guilds or communities are indeed segregated/aggregated or not different from random simulations, but the evidence is equivocal and results depend on the kind of model used (Ashton *et al.*, 1988; Pleasants, 1990).

Flowering and fruiting patterns of 1899 individuals (GBH ≥ 30 cm) representing 164 tree species were investigated in twenty-one randomly located 0.25 ha study plots in Pakhui Wildlife Sanctuary, a tropical semi-evergreen forest in Arunachal Pradesh. Flower and fruit production was recorded monthly from February 1997 to July 2000. The presence/absence of trees in flower, unripe and ripe fruit was recorded. Flowering and fruiting patterns of all species together, and of anemochorous and zoochorous species were examined graphically. Presence/absence of the fruits of 20 bird-dispersed tree species in the diets of 16 frugivorous bird species was established through field-observations. We calculated the similarity between these 20 tree species in their avian dispersers (Sorenson's Index, Krebs, 1989). This similarity matrix was correlated with a matrix of correlation coefficients of ripe fruit availability (% trees in ripe fruit of each species every month) over 40 months (1997–2000) using Mantel tests (Manly, 1994). Since fruiting patterns of these species were similar between years, a matrix of correlation coefficients based on the mean for all years was also used. We expected a significant negative correlation if the competition avoidance does play a role in influencing fruiting schedules of these species. The analysis was repeated with a smaller subset of 11 tree species and 9 bird species. Cluster analysis was also carried out to determine whether distinct clusters are formed by groups of species with similar fruiting phenologies.

Sixty-three per cent (103 species) of the tree species were animal dispersed, 15% (24 species) were wind-dispersed, while the dispersal mode of 37 species could not be established. Of the animal-dispersed species, 57 were largely bird-dispersed. The dispersal modes of most species were established through field observations/fruit characteristics and for a few from literature.

Overall, flowering was unimodal with a major peak before the monsoon in March–April (relatively dry hot season), though there was a minor peak in November–December (Figure 1). Overall fruiting peaked between April and July (Figure 1). Fruit scarcity occurred during the period between September and January (end of monsoon and winter). Most species had fairly synchronous fruit production and most species produced fruits annually, though a palm *Livistonia jenkinsii* showed supra-annual fruiting. One species, *Polyalthia simiarum* had two fruiting peaks in the year with a major one in June–July and a minor one between December and February. Inter-annual variability in overall fruiting patterns (all species) was high, while flowering patterns between years were more similar. But fruiting patterns of the 20 bird-dispersed species were similar between years, though there was variation in fruiting intensity with a failure of fruiting of several species belonging to the Meliaceae and Myristicaceae in 1999 (Figure 2).

The flowering and fruiting peaks of wind-dispersed and bird-dispersed species were dissimilar. Flowering and fruiting peaks of wind-dispersed species were bimodal, occurring during the relatively dry months, February to April and October to December, while flowering of bird-dispersed species

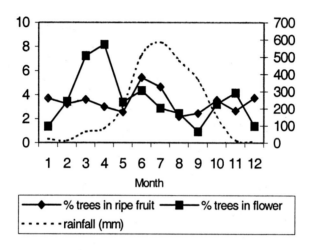

Figure 1. Mean overall flowering and fruiting patterns (all 164 species).

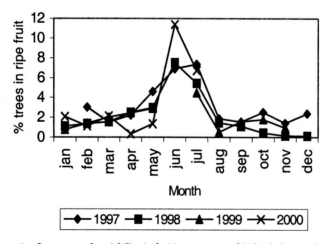

Figure 2. Inter-annual variability in fruiting patterns of 20 bird-dispersed species.

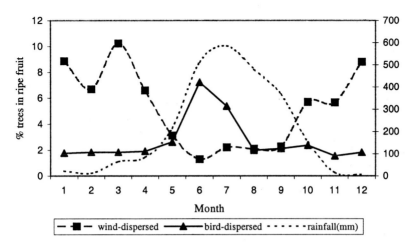

Figure 3. Fruiting patterns of wind-dispersed and bird-dispersed species.

occurred throughout the year. Fruiting of wind-dispersed species was also bimodal peaking in drier months, while the fruiting peak of bird-dispersed species was unimodal, with most middle-storey trees that produced bird-dispersed fruits maturing during the wet season (May–July) (Figure 3). Ripe fruits of several other bird-dispersed species were available throughout the year. All the larger arillate capsular fruit species belonging to the Meliaceae and Myristicaceae ripened between March and May, while fleshy berries of the Lauraceae, Annonaceae and other families ripened between July and December. Peak fruit abundance of bird-dispersed species occurred between May and July, which also coincides with the breeding season of resident frugivorous birds such as Hornbills, Barbets, and Hill myna. This is contrary to findings for temperate birds and understorey frugivores in Neotropical forest (Levey, 1988), where breeding season coincided with peak of insect abundance. But, importantly, fruit availability of bird-dispersed species was more uniform, which suggested that there is some degree of staggering of bird-dispersed species throughout the year that may be driven by extra-climatic conditions. The competition avoidance hypothesis predicts a more uniform fruiting pattern if fruiting schedules are indeed staggered. But a quantitative test of the hypothesis showed that there was no significant negative correlation between similarity in diets of bird species and similarity in fruit ripening schedules ($r = 0.054$, $p = 0.43$, 5000 iterations). Similarly, with a smaller subset of species there was no significant correlation ($r = 0.227$, $p = 0.07$, 5000 iterations), in fact the trend was towards a positive correlation. Cluster analysis also indicated a high degree of overlap in fruiting phenology of the 20 species, with most species forming a single cluster. This does not provide conclusive evidence to reject this hypothesis. We suggest that while climatic factors may be the main force dictating timing of fruiting at a broad-level, fine-scale differences in fruit ripening times of tree species may be difficult to pinpoint with monthly phenological data on number of trees in ripe fruit. It would be easier to detect through weekly data on actual ripe fruit availability as well as diets of bird species that are tracking fruit availability. This seemed to be true from the diet of hornbills, where one or two fruit species were important at any given time, thus showing that in terms of the contribution to the dispersers' diet, species were staggered temporally. Segregated fruiting may be beneficial for plants relying on sedentary dispersers, while aggregated fruiting may benefit plants during periods of high bird abundance. The peak ripe fruit abundance of large-sized fruits coincided with the breeding season of resident frugivores, while smaller-sized fruits were available during winter, when there is an influx of migrants. Therefore, while there may be a broad overlap in fruiting schedules of species (that depends on optimal environmental factors), within that period, species may show a less detectable staggering.

Acknowledgements

This work was part of a project on hornbill ecology funded by the Ministry of Environment and Forests, India and the Wildlife Conservation Society, USA (a grant to the first author). We thank the Arunachal Pradesh Forest Department for permissions and facilitating the fieldwork, especially D. N. Singh, M. K. Palit and C. Loma (DFOs at Pakhui WLS) and the staff of Pakhui WLS. We thank our three field assistants whose dedication and help in the field made this work possible. We thank Dr K. Haridasan, S.F.R.I., Itanagar for sharing his knowledge of the flora and identifying plant specimens. We also thank Pratap Singh for his inputs into the project. We thank T. R. Shankar Raman, Charudutt Mishra and M. D. Madhusudan of the Nature Conservation Foundation, Mysore for discussions and ideas on this paper. We thank Dr Jagdish Krishnaswamy for comments on analysis.

References

Ashton, P. S., Givnish, T. J. & Appanah, S. Staggered flowering in the Dipterocarpaceae: new insights into floral induction and the evolution of mast fruiting in the aseasonal tropics. *Am. Nat.* **132**, 44–66 (1988).

Foster, R. B. The seasonal rhythm of fruitfall on Barro Colorado Island. in *The Ecology of a Neotropical Forest: Seasonal Rhythms and Long-term Changes* (eds Leigh, E. G., Rand, A. S. & Windsor, D. M.) 151–172 (Smithsonian Institution Press, Washington, 1982).

Frankie, G. W., Baker, H. G. & Opler, P. A. Comparative phenological studies of trees in tropical wet and dry forests in the lowlands of Costa Rica. *J. Ecol.* **62**, 1049–1057 (1974).

Janzen, D. H. Seed predation by animals. *Annu. Rev. Ecol. Syst.* **2**, 465–492 (1971).

Krebs, C. J. *Ecological Methodology* (Harper & Row Publishers, New York, 1989) pp. 654.

Leighton, M. & Leighton, D. R. Vertebrate responses to fruiting seasonality within a Bornean rain forest. in *Tropical Rain Forest: Ecology and Management* (eds Sutton, S. L., Whitmore, T. C. & Chadwick, A. C.) 181–196 (Blackwell Sci. Publications, Oxford, 1983).

Levey, D. J. Spatial and temporal variation in Costa Rican fruit and fruit-eating bird abundance. *Ecol. Monogr.* **58**, 251–266 (1988).

Manly, B. F. J. *Multivariate Statistical Methods: A Primer.* 2nd Edition (Chapman and Hall, London, 1994) pp. 215.

Pleasants, J. M. Null-model tests for competitive displacement: The fallacy of not focusing on the whole community. *Ecology* **71**, 1078–1084 (1990).

Poulin, B., Wright, S. J., Lefebvre, G. & Calderon, O. Interspecific synchrony and asynchrony in the fruiting phenologies of congeneric bird-dispersed plants in Panama. *J. Trop. Ecol.* **15**, 213–227 (1999).

Rathcke, B. & Lacey, E. P. Phenological patterns of terrestrial plants. *Annu. Rev. Ecol. Syst.* **16**, 179–214 (1985).

Snow, D. W. A possible selective factor in the evolution of fruiting seasons in tropical forest. *Oikos* **15**, 274–281 (1965).

Terborgh, J. & Van Schaik, C. P. The phenology of tropical forests: Adaptive significance and consequences for primary consumers. *Annu. Rev. Ecol. Syst.* **24**, 353–377 (1993).

Wheelwright, N. T. Competition for dispersers and the timing of flowering and fruiting in a guild of tropical trees. *Oikos* **44**, 465–477 (1985).

Tropical Ecosystems: Structure, Diversity and Human Welfare.
Proceedings of the International Conference on Tropical Ecosystems
K. N. Ganeshaiah, R. Uma Shaanker and K. S. Bawa (eds)
Published by Oxford–IBH, New Delhi. 2001. pp. 750–753.

Effect of tree stratum on flowering synchrony in a seasonal cloud forest in the Bhimashankar Wildlife Sanctuary, India

Hema Somanathan*[†], *Subhash Mali** and Renee M. Borges*[‡]

*F 52A, Anna Nagar East, Chennai 600 102, India
**Foundation for the Revitalisation of Local Health Traditions,
Bangalore 560 032, India
[‡]Centre for Ecological Sciences, Indian Institute of Science,
Bangalore 560 012, India
[†]e-mail: hsomanathan@hotmail.com

Flowering has been found to depend on plant water status (Reich and Bochert, 1984; Bochert, 1994) and light availability (van Schaik *et al.*, 1993). The water and light regimes that a tree experiences can be influenced by the stratum to which it belongs within a forest patch. We studied the effect of strata on flowering phenology in four forest patches within a seasonal cloud forest in the Bhimashankar Wildlife Sanctuary of Maharashtra State, India. The patches vary in type and size and in species composition though some species were common to more than one patch. The forest patches were ranged from the east to the west (Adal, Hindola, Choura and Rai) of the sanctuary along a rainfall gradient with Rai being the wettest and Adal the driest. The presence or absence of flowers was recorded at monthly intervals for 24 months from January 1994 to December 1995 for 579 trees ($n = 18$

Keywords: Flowering, phenology, stratum, synchrony, India.

species) in Adal, 471 trees ($n = 17$ species) in Hindola, 485 trees ($n = 22$ species) in Choura and 4031 trees ($n = 35$ species) in Rai. At each site species were assigned to understorey, middlestorey and canopy strata based on the height that trees achieved in each forest. Since environmental variables (light, water) can vary for different strata within a forest patch, species from the same strata could be more synchronous in the timing of flowering than species from different strata. To test this we asked two related questions: 1. Is average synchrony of species pairs within a stratum greater than average synchrony of species pairs irrespective of stratum? 2. Is average synchrony of species pairs within a particular stratum greater than average synchrony of species pairs within another stratum?

We measured synchrony using Augspurger's index (Augspurger, 1983). Using this index (x_i) synchrony for each tree within a patch with all other species within the patch was computed. This was done by pairing every tree

Table 1. Differences in mean flowering synchrony between pairs of species from the same strata and pairs of species from different strata in four forest patches in the Bhimashankar Wildlife Sanctuary.

Group 1	Group 2	Adal	Hindola	Rai	Choura
(A)					
CC	CU	CC	CC	CC	CC
CC	CM	CC	CC	–	–
MM	MC	MC	MC	MC	MC
MM	MU	–	–	MU	–
UU	UM	UM	UM	UU	UU
UU	UC	–	–	UU	UU
(B)					
CC	CU, CM	CC	CC	–	CC
MM	MU, MC	MU, MC	MU, MC	MU, MC	MU,
MC					
UU	UM, UC	UC, UM	–	UU	UU

The pair with greater mean synchrony is indicated under each site. Only those groups (out of a total of 9 groups that were possible) that were significantly different using t-tests ($P < 0.01$) are indicated in the table.

Table 2. Differences in mean flowering synchrony between pairs of canopy species, middlestorey species and understorey species (t-tests) in four forest patches in the Bhimashankar Wildlife Sanctuary.

Group 1	Group 2	Adal	Hindola	Rai	Choura
UU	MM	–	–	UU	UU
UU	CC	CC	–	UU	UU
MM	CC	CC	–	–	–

The pair with greater mean synchrony ($P < 0.01$) is indicated under each site.

751

in a given patch with individuals of all heterospecific species, taking one species at a time. We calculated the index for all possible combinations of species pairs as $x_i = (1/n-1)(1/f_i)\Sigma e_{j\neq i}$, where n is the number of trees in the heterospecific species, e_i is the number of months both individual i of the first species and individual j of the heterospecific species flowered together and f_i is the number of months individual i of the first species was in flower. The species pairs were then divided into groups based on the stratum to which each species of the pair belonged. All possible pairs of strata yielded nine groups of species pairs, e.g. CC (canopy–canopy), CM (canopy–middle storey), MC (middlestorey–canopy) and so on.

We performed t-tests to test for difference in mean synchrony for same-strata pairs and different-strata pairs. For example, for the canopy strata we did t-tests between canopy–canopy pairs, canopy–understorey pairs and canopy–middlestorey pairs. Similar t-tests were also done for the middlestorey and understorey. Furthermore, in order to test if species in a stratum flower more synchronously than species in another stratum, we performed t-tests between the following pairs: Canopy–canopy, middlestorey–middlestorey and understorey–understorey.

The results are indicated in Tables 1 and 2. In all the patches for the canopy species pairs, the canopy–canopy pairs (CC) were more synchronous than canopy species paired with other strata (CU and CM) (Table 1A). This shows that within the canopy, species are more synchronous in their flowering schedules than species from different strata. However, the middlestorey–canopy pairs were more synchronous than middlestorey–middlestorey pairs in all the patches. This is possibly explained by the fact that some middlestorey species have synchrony values similar to that of the canopy, while other species have their own unique 'middlestorey' synchrony.

Paired species in the understorey flowered synchronously compared to understorey species paired with other strata in the Choura and Rai patches. In the other two sites the understorey–middlestorey pairs were more synchronous than paired species from the understorey. This is again possible if some of the understorey species have synchrony values that are more similar to the middlestorey. The results were largely the same when the t-tests were repeated, this time by comparing canopy–canopy pairs with canopy–understorey and middlestorey pairs grouped together (Table 1B). This was also done for the middlestorey–middlestorey and the understorey–understorey pairs. t-tests to check for differences in flowering synchrony for the three strata indicated that pairs of understorey species were more synchronous in flowering than pairs of species belonging to the middlestorey or canopy in Rai and Choura (Table 2). In Adal the canopy species pairs were more synchronous than the other two strata pairs. The results were not significant in Hindola.

The results suggest that despite basic differences between the four patches with respect to species composition, canopy height, tree density, flowering and history, in all the four patches flowering phenology was largely dependent on the stratum that a species occupied in a patch. This greater synchrony within a stratum is possibly explained by similar environmental conditions which are experienced by species that share the same stratum within a forest patch. Yet for some strata within some patches, a clearly defined stratum effect was absent. This could happen if the stratification of species within a forest patch was continuous rather than discrete with some species falling into transition zones between strata.

References

Augspurger, C. K. Phenology, flowering synchrony, and fruit set of six Neotropical shrubs. *Biotropica* **15**, 257–267 (1983).
Bochert, R. Soil and stem water storage determine phenology and distribution of tropical dry forest trees. *Ecology* **75**, 1437–1449 (1994).
Reich, P. B. & Bochert, R. Water stress and tree phenology in a tropical dry forest in the lowlands of Costa Rica. *J. Ecol.* **72**, 61–74 (1984).
Van Schaik, C. P., Terborgh, J. W. & Wright, S. J. The phenology of tropical forests: adaptive significance and consequences for primary consumers. *Annu. Rev. Ecol. Syst.* **24**, 353–377 (1993).

Tropical Ecosystems: Structure, Diversity and Human Welfare.
Proceedings of the International Conference on Tropical Ecosystems
K. N. Ganeshaiah, R. Uma Shaanker and K. S. Bawa (eds)
Published by Oxford–IBH, New Delhi. 2001. pp. 754–757.

Observations of phenological changes of Mesoamerican tropical dry forests and implications for conservation strategies

Arturo Sanchez-Azofeifa[*,†] *and Margaret Kalacska*[*]

*Earth Observation Systems Laboratory, Department of Earth and Atmospheric Sciences, University of Alberta, Edmonton, Alberta, Canada T6G 2E3
[†]e-mail: arturo.sanchez@ualberta.ca

Tropical Dry Forest Ecosystems (TDF) cover significantly large areas of India, Australia, Central and South America, the Caribbean, Mexico, Africa, and Madagascar. TDF are characterized by bio-temperatures > 17°C and an annual precipitation that ranges from 250 to 2000 mm. It is estimated that 49% of the vegetation of Central America and the Caribbean is covered by TDF, and worldwide about 42% of all intra-tropical vegetation is TDF. Trees of TDF are usually smaller than those in rain forests, and many lose their leaves during the dry season. Although they are still extremely diverse, dry forests often have fewer species than rain forests. Worldwide, TDF ecosystems are cataloged as critical or endangered (Janzen, 1988). They are also heavily utilized and disturbed by human activities. Although TDF ecosystems are considered endangered, their protection is minimal. In Mesoamerica, only 0.9% of the TDF has official conservation status while the rest is under constant pressure of deforestation.

Keywords: Tropical dry forest, land cover change, remote sensing, Mesoamerica.

Methods for monitoring and detecting tropical deforestation have been successfully developed, tested and applied in several countries (Stone and Lefebvre, 1998; Skole and Tucker, 1993; Sanchez-Azofeifa et al., 2001), providing important information on the extent of tropical evergreen forests. However their lack of ability to identify and/or delineate tropical dry forests is notorious (Kramer, 1997). Current remote sensing approaches used to map their area are often erroneous since they often confuse TDF with pasture and shrubs (Pfaff et al., 2000). These errors in general are due to the use of satellite images acquired during the dry season when the trees are leafless.

In this paper, we explore the role of TDF phenology in the estimation of the area of this forest type, as well as the impacts of phenology on estimating TDF deforestation rates. We used a set of radiometric and atmospherically corrected Landsat Thematic Mapper [TM 5, 7 spectral bands and 28.5 m spatial resolution] satellite images acquired between October and April 1986/87 for the Santa Rosa National Park (Path 16–Row 53). The study area is centered on longitude 85°36′54″W and latitude 10°48′53″N. The TDF at the Santa Rosa National Park is characterized by a dry period (January to April, leaf-off season), a transition period (late May to July) and a wet period (August to December, leaf-on season). The precipitation range is from 1100 to 1500 mm/yr, and the annual bio-temperature of the ecosystem ranges from 24°C to 27.8°C. The Tropical Dry Forest at GCA is representative of this ecosystem, as it exists throughout Mesoamerica.

In this paper, we documented observations related to statistically significant changes in spectral reflectance of TDF in the red and near infrared wavelengths as well as changes in the Normalized Difference Vegetation Index (NDVI) during a full leaf growth season (October, December and April). Using a modified version of the Red-Green-Blue-NDVI approach (RGB-NDVI) to map land cover changes (Sader and Winne, 1992), we also documented the impact of phenology on the detection of the extent of TDF in the study area.

In order to detect changes in the spectral signature of the TDF in the dry and wet seasons, we compared the two extreme months: April and October. The October image represents the wet season, where the majority of the species are in full foliage and the April image shows the dry season where most of the species have lost their leaves. With a more traditional first level analysis – using the TM spectral bands 4, 3 and 2 – we classified both images into fifty spectral classes using an unsupervised classification method. Each of the fifty classes was then separately evaluated to determine whether or not it represented an area covered by leaf-on vegetation or not. Our results showed a 35% difference in area between the two images. This 35% difference corresponds with the extent of the TDF spectral signature that was not distinguishable from the pastures and shrubs in April. Using phenology as an independent variable, the area we extracted from the October image is a spectral mixture of evergreen and deciduous species. In contrast, the area extracted from the April image is comprised almost solely of evergreen species because the deciduous (having no or very few leaves) have a different spectral signature during this season. Therefore, using the traditional unsupervised classification on an image of the dry season with TM bands 4, 3 and 2, the area comprised of deciduous trees is not observable and may be erroneously interpreted as having been deforested. Consequently, it is important to use a time series of images in order to detect the TDF and to distinguish it from the tropical evergreen forest.

Our second research method to delineate the TDF consisted of a modified RGB-NDVI approach. For this method, we compared two images: a control image comprised of January, April and October and a test image comprised of January, April and December. For the control image, we created a forest mask from the October RGB image and applied it to the January, April and October RGB images. The NDVI was subsequently calculated for each of the masked images, and the results were joined through a layer stacking method which projects each of the NDVI images in red, green or blue. The resultant image clearly separated the TDF from the tropical evergreen forest. This product was then classified through an unsupervised classification method into fifty spectral classes. Each of the spectral classes was then labeled as either representing the TDF or not. Our results indicated that through this method, 85% of the total forest area identified in October was a TDF with significant changes in phenology during the growth season. The same analysis was run on the test images, with the exception that the whole image was used without a forest area mask. Fifty per cent less TDF was identified without the inclusion of the October image.

Our results from the analyses indicated that phenology could decrease the accuracy of identifying the true area of TDF by as much as 50%. Our observations at the Santa Rosa National Park allow us to conclude that the true worldwide extent of tropical dry forest may be significantly underestimated by regional monitoring systems such as the current UN – Food and Agriculture Organization Forest Resources Assessment 2000 (FRA-2000). More significantly, because of this underestimation and the important role that phenology plays, current deforestation rates in TDF ecosystems may be overestimated.

Acknowledgements

This research note has grown as a continuity of a research agenda on land use, land cover change and carbon sequestration in Costa Rica by the Earth Observation Systems Laboratory (EOSL) at the University of Alberta, Canada. We would like to thank the generous support of the Canada Foundation for Innovation (Grant No. 2041 to Sanchez-Azofeifa), and the U.S. National Science Foundation (Grant No. BCS 9980252).

References

Janzen, D.H. Tropical dry forest: The most endangered major tropical ecosystem. in *Biodiversity* (ed. Wilson, E. O.) 130–137 (Natural Academy Press, Washington DC, 1988).

Kramer, E. Measuring landscape changes in remnant tropical dry forest. in *Tropical Forest Remnants: Ecology, Management, and Conservation of Fragmented Communities* (eds William F. Laurence and Richard O. Bierregaard, Jr.) 386–399 (The University of Chicago Press, USA, 1997).

Pfaff, A. S. P., Kerr, S., Hughes, F., Liu, S., Sanchez-Azofeifa, G. A., Schimel, D., Tosi, J. & Watson, V. The Kyoto protocol and payments for tropical forest: An interdisciplinary method for estimating carbon-offset supply and increasing the feasibility of a carbon market under the CDM. *Ecol. Econ.* **35**, 203–221 (2000).

Sader, S. A. & Winne, J. C. RGB-NDVI color composites for visualizing forest change dynamics. *Intern. J. Rem. Sen.* **13**, 3055–3067 (1992).

Sanchez-Azofeifa, G. A., Harriss, R. C. & Skole, D. L. Insights from remote sensing on the status of conservation of tropical ecosystems in Costa Rica. *Biotropica*, accepted (2001).

Skole, D. & Tucker, C. Tropical deforestation and habitat fragmentation in the Amazon: Satellite data from 1978 to 1988. *Science* **260**, 1905–1910 (1993).

Stone, T. A. & Lefevbre, P. Using multi-temporal satellite data to evaluate selective logging in Para, Brazil. *Intern. J. Rem. Sen.* **19**, 2517–2526 (1998).

Tropical Ecosystems: Structure, Diversity and Human Welfare.
Proceedings of the International Conference on Tropical Ecosystems
K. N. Ganeshaiah, R. Uma Shaanker and K. S. Bawa (eds)
Published by Oxford–IBH, New Delhi. 2001. pp. 758–761.

Ecological adaptations and population structure in teak (*Tectona grandis* L.)

Babu Nagarajan, Mohan Varghese and Abel Nicodemus

Institute of Forest Genetics and Tree Breeding, P.B.1061, Forest Campus,
Coimbatore 641 002, India
e-mail: teakguy@usa.net

Domestication of teak was initiated during the late 18th century, since then it has followed the regular classical breeding methods such as provenance testing (DFSC, 1995), establishment of seed orchards (Pereira, 1961) and progeny testing (Nagarajan *et al.*, 1996). Among the teak-growing countries, breeders and practising foresters have made hundreds of phenotypic selections and assembled sizeable breeding populations. An inherent problem observed in most of these populations is the extremely low seed production that leads to a wide gap between the need and supply of genetically improved planting material (Nagarajan *et al.*, 1996; Palupi and Owens, 1996; Tangmitcharoen and Owens, 1996).

Teak is a deciduous species which sheds off leaves from the months of November to January, and remains leafless from January to March. Usually it has only one flowering season between late May and mid June coinciding with southwest monsoon in peninsular India. Flowers are borne in large terminal inflorescences with 4–5 internodes bearing 500–12,000 flowers. Normally, flowering phase lasts up to 70 days which is followed by a longer fruit maturation phase of 150 to 200 days. Episodic or sequential flowering commonly occurs in teak wherein flowering occurs in discrete episodes of irregular intervals.

Keywords: Breeding population, DNA marker, floral adaptation, flower–fruit ratio, phenology, pollen, seed filling, teak.

Teak pollen is binucleate, very sticky and fertile up to 99%. Though the species has been reported to be self-incompatible (Hedegart, 1976), there is no difference in the pollen–pistil interaction between selfed and cross-pollinated conditions. Style morphology is known to be quite influential in seed setting in several species (Martin, 1972; Baksh et al., 1978; Dulberger et al., 1981). Invariably flowers with longer style are known to show higher fruit set. The early produced flowers have relatively broader corolla cup, longer style and prominently forked stigma favouring pollen receipt. The late flowers on the other hand have smaller styles and reduced stigma forking meant for male function. Flowers that are proximal to the maternal stem axis develop into healthy and bigger fruits (Nagarajan et al., 1996b).

The stigma becomes receptive by 10.00 h and is at its peak between 11.00 and 12.00 h. It is of wet and pappillate type and remains relatively straight throughout the day of receptivity. The peak receptive period is characterised by expansion of the style and stigma, increased turgidity of the unicellular papillae and the presence of stigmatic secretion and nectar. Flowers are bisexual, weakly protandrous (Rawat et al., 1992), anthesise during the early mornings (5.30–9.00) and last for a day. Protandry serves in prolonged pollen presentation, avoids pollen stigma interference and optimally positions pollen for dispatch and reception. Nectar secretion in teak starts from 08.00 h, while anthers start opening by 08.30 h and are fully dehisced by 09.00 h. Nectar is the chief pollinator attractant in teak; it is produced in large volumes and over a long period of time within the day of anthesis. *Apis mellifera* and *Ceratina* bees are the most frequent visitors. Peak visitation of insects and pollen import per flower is between 09.00 and 13.00 h. In teak, fertilisation normally occurs 24 h after flower opening. After pollination , the pollen grain germinates on the stigma and sends its tube through the style to the embryo sac in the ovary. It takes about 8 weeks to develop into a fruit (14–15 mm) and about 16–17 weeks to mature, at which time the fruit water content declines and the fruit diameter decreases (Palupi and Owens, 1996).

Teak has extremely low flower to fruit ratio (Bryndum and Hedegart, 1969; Hedegart, 1973), fruit setting varies from 1 to 2% under natural (Nagarajan et al., 1996a). Such low flower to fruit ratio is typical in many hermaphroditic plants (Stephenson, 1981; Wilson and Burley, 1983; Sutherland and Delph, 1984; Sutherland, 1986; Guitan, 1993). Most fruits produced have only one fully developed seed. A study made on the seed filling from different locations in India showed emptiness as high as 86% (Gupta and Kumar, 1976). A major cause for early seed abortion in teak is because of the late acting gametophytic self-incompatibility (Tangmitcharoen and Owens, 1996).

Though normal development of proembryo is seen in all the four ovaries during the first week, after 14–20 days the development of the embryos is not uniform. In most cases one ovule differentially develops much bigger than its siblings, a feature attributed to intra-ovular competition. Invariably the ovule that develops faster, matures into seed while others abort. The position of the surviving ovule is random. Such a fixed abortion system reduces the sibling competition and possibly provides selection at the zygote level (Casper and Wiens, 1981; Palupi and Owens, 1996).

In nursery conditions teak germinates poorly and the planting stock is of varying sizes due to asynchronous germination. Germination is between 30 and 50% and takes about 30 to 50 days (Suangtho, 1980; Phengduang, 1993). In Thailand, Wellendorf and Kaosa-ard (1988) estimated that the successful plant percentage in a large-scale nursery operation is only 5%. This is because of the extremely low levels of Pre Emergent Reproductive Success (PERS), which is calculated on the basis of flower to fruit, fruit to ovule and ovule to seed ratios (Wiens et al., 1978). Palupi and Owens (1996) observed that in open-pollinated condition breeding populations were found to show lower fruit setting compared to seed production areas and the PERS values were also as low as 0.1–0.5%. Studies on natural selfing of clones show extremely low PERS values ranging from 0.0004 to 0.001, explaining the self-incompatible nature of the species.

The high degree of seed abortion coupled with other features may have considerable effect on the genetic structure of the populations. Genetic variation studies in teak populations using Randomly Amplified Polymorphic DNA assay indicate higher levels of variation within population than between populations. The relationship between the geographical and genetic distance between populations is not greatly pronounced. In general, the RAPD profiles in teak populations are in confirmity to the teak population structure described using isozymes markers (Kertadikara and Prat, 1995).

From the point of view of teak improvement, since there is considerable variation within populations, selections can be made from local populations so that problems concerning excess, scantly or asynchrony of flowering can be easily overcome.

Nursery managers can adopt the strategy of sowing nurseries based on diameter class for obtaining synchronised germination. While developing seed production areas or seed orchards care should be taken to observe the flowering, fruiting behaviour and the age of the trees. In clonal seed orchards site selection, broader espacement (at least 10 m) and intensive management during the first five years should yield better results.

References

Baksh, S., Iqbal, M. & Jamal, A. Breeding system of Solanum integrifolium Poir with an emphasis on sex potential and intercrossability. Euphytica 27, 811–815 (1978).
Bryndum, K. & Hedegart, T. Pollination in teak (Tectona grandis L.f). Silvae Genet. 18, 68–70 (1969).

Casper, B. B. & Wiens, D. Fixed rates of random ovule abortion in *Cryptantha flava* (Boraginaceae) and its possible relation to seed dispersal. *Ecology* 866–869 (1981).

Danida Forest Seed Centre (DFSC). *Second Evaluation of International Series of Teak Provenance Trials* (eds Kjaer, E. D., Lauridsen, B. & Wellendorf, H.) pp. 118 (1995).

Dulberger, R., Levy, A. & Palevitch, D. Androemonoecy in *Solanum marginatum*. *Bot. Gaz.* **142,** 259–266 (1981).

Guitian, J. Why *Prunus mahaleb* (Rosaceae) produces more flowers than fruits. *Am. J. Bot.* **80,** 1305–1309 (1993).

Gupta, B. N. & Kumar, A. Estimation of potential germinability of teak fruits from twenty three Indian sources by cutting tests. *Indian For.* **102,** 808–813 (1976).

Hedegart, T. Pollination of teak (*Tectona grandis* L. f) *Silvae Genet.* **22,** 124–128 (1973).

Hedegart, T. Breeding systems variation and genetic improvement of teak (*Tectona grandis* L.). in *Tropical Trees: Variation, Breeding and Conservation* (eds Burley, J. & Styles, B.) 109–121 (Academic Press, London, 1976).

Kertadikara, A. W. S. & Prat, D. Isozyme variation among teak (*Tectona grandis* L.) provenances. *Theor. Appl. Genet.* **90,** 803–810 (1995).

Martin, F. W. Sterile styles of *Solanum mammosum*. *Phyton* **29,** 127–134 (1972).

Nagarajan, B., Gireesan, K., Venkatasubramanian, N., Rajesh Sharma, A. S. & Mandal, A. K. An early evaluation of gene action in Teak. *My Forest* **32,** 136–139 (1996a).

Nagarajan, B., Varghese, M., Nicodemus, A., Sasidharan, K. R., Bennet, S. S. R. & Kannan, C. S. Reproductive biology of teak and its implication in tree improvement. in *Tropical Tree Improvement for Sustainable Tropical Forestry*, vol I (eds Dieters, M. J., Matheson, A. C., Nikles, D. G., Harwood, C. E. & Walker, S. M.) 265–270 (QFRI - IUFRO Conference, Caloundra, Queensland, Australia (IUFRO Working party S2.08.01) 1996b).

Palupi, E. R. & Owens, J. N. Reproductive biology of teak (*Tectona grandis* Linn. F.) in east Java, Indonesia. in *Tropical Tree Improvement for Sustainable Tropical Forestry*, vol I (eds Dieters, M. J., Matheson, A. C., Nikles, D. G., Harwood, C. E. & Walker, S. M.) 255–260 (QFRI–IUFRO Intern. Conference, Caloundra, Queensland, Australia (IUFRO Working party S2.08.01), 1996).

Pereira, W. R. H. Teak seed orchards stage 1. Plus trees and perfection of a bud grafting technique. *Ceylon For.* **5,** 6–16 (1961).

Phengduang, V. Teak in Laos PDR. in Teak in Asia Technical Document GCP/RAS/134/ASB, FORSPA Publication 4 FAO-RAPA. pp 41–50 (1993).

Rawat, M. S., Uniyal, D. P. & Varkshaya, R. K. Variation studies in the model teak seed orchard, New Forest, Dehra Dun. *Indian For.* **118,** 60–65 (1992).

Stephenson, A. G. Flower and fruit abortion: proximate causes and ultimate functions. *Annu. Rev. Ecol. Syst.* **12,** 253–280 (1981).

Suangtho, V. Factors controlling teak (*Tectona grandis* Linn.F): Seed germination and their importance to Thailand. M.Sc., Australian University, Canberra, Australia (1980).

Sutherland, S. Floral sex ratios, fruit set and resource allocation in plants. *Ecology* **67,** 991–1001 (1986).

Sutherland, S. & Delph, L. F. On the importance of male fitness in plants: Patterns of fruit set. *Ecology* **65,** 1093–1104 (1984).

Tangmitcharoen, S. & Owens, J. N. Floral biology, pollination and pollen tube growth in relation to low fruit production of teak (*Tectona grandis* Linn. F.) in Thailand. in *Tropical Tree Improvement for Sustainable Tropical Forestry*, Vol I (eds Dieters, M. J., Matheson, A. C., Nikles, D. G., Harwood, C. E. & Walker, S. M.) 265–270 (QFRI–IUFRO Intern. Conference, Caloundra, Queensland, Australia (IUFRO Working party S2.08.01), 1996).

Wellendorf, H. & Kaosa-ard, A. Teak Improvement Strategy in Thailand. Forest Tree Improvement No. 21, p. 43 (1988).

Wiens, D., Calvin, C. L., Davern, C. I., Frank, D. & Seavey, S. R. Reproductive success, spontaneous embryo abortion and genetic load in flowering plants. *Oecologia (Berlin)* **71,** 501–509 (1978).

Wilson, M. F. & Burley, N. *Mate Choice in Plants: Tactics, Mechanisms and Consequences* (Princeton University Press, Princeton NJ, USA, 1983).

Tropical Ecosystems: Structure, Diversity and Human Welfare.
Proceedings of the International Conference on Tropical Ecosystems
K. N. Ganeshaiah, R. Uma Shaanker and K. S. Bawa (eds)
Published by Oxford–IBH, New Delhi. 2001. pp. 762–765.

Genetic diversity analysis of mango in India using RAPD markers and morphological characters

K. V. Ravishankar, M. R. Dinesh, Lalitha Anand and G. V. S. Saiprasad*

Indian Institute of Horticultural Research, Hessaraghatta Lake Post,
Bangalore 560 089 India
*e-mail: ravishankar_kv@usa.net

India is the primary center of origin (diversity) for mango. Mango has been cultivated for more than 4000 years. All the cultivated varieties belong to the species *Mangifera indica*. India is also reported to be home of four other species, *M. andamanica*, *M. khasiana*, *M. sylvatica* and *M. camptosperma*. Allopolyploidy ($2n = 40$), out breeding and a wide range of agroclimatic conditions prevailing in the country contribute to the diversity of mango. Besides, there have been widespread hybridization and recombination of characters in mango. Over the hundreds of years of its cultivation and domestication, the genetic diversity of mango has been fixed in many varieties. The present day varieties are mainly seedling selections maintained through vegetative propagation (Mukherjee, 1972; Yadav and Rajan, 1993).

The *Mangifera indica* germplasm is broadly classified into two broad categories: The first category is seedling races, which include wild and cultivated types. In this category the cultivated ones are polyembryonic and generally found in Western Ghats. The second category is horticultural races.

Keywords: *Mangifera indica,* genetic diversity, RAPD markers, morphological characters.

These varieties are monoembryonic types. These varieties are grown throughout India and propagated vegetatively. Although collection and conservation of mango cultivars started in the seventeenth century, it received major emphasis only in the middle of 20th century. In the early part, collection and conservation efforts were primarily based on quality of fruits, while current efforts are for collection of germplasm with distinct desirable traits which can be utilized for crop improvement. Till date more than 1200 named cultivars have been collected and are conserved at different locations in field gene bank.

In India, each mango-growing region has different varieties. These varieties have different fruit characteristics like colour, taste, flavour, size and bearing habit. In this center, an effort has been made to assess the genetic diversity of mango cultivars using RAPD markers and morphological characters. Eighteen mango cultivars commercially grown in different parts of India were selected. These are cvs Alphonso, Goamankurd, Raspuri, Kesar, Borsha (from Western India), Totapuri, Neelum, Padiri, Kalapadi, Janardhana Pasand, Rumani, Banganapalli, Suvarnarekha (from Southern India); Fazli, Bhutto Bombay, Bombay Green, Rataul and Himasagar (from the Northern and Eastern parts of India).

DNA was extracted from the leaves of these cultivars and RAPD amplification was performed using 30 arbitrary decamer primers. Of the 30 primers, only 27 primers amplified mango DNA. From these, data from only 19 primers which amplified maximum number of bands and showed high polymorphism, were used for statistical analysis. These primers amplified 178 bands out of which 130 were polymorphic (Ravishankar et al., 2000). Based on presence/absence of data, a squared Euclidean distance matrix was calculated to estimate all pair-wise differences in the amplification product of all varieties. Based on the distance matrix, cluster analysis was done using a minimum variance algorithm (Ward, 1963). Data on morphological characters like fruit, floral and leaf characteristics were documented and subjected to cluster analysis.

Cluster analysis indicated two major groups of mango cultivars (Figure 1). The first group comprised mainly cultivars from western, northern and eastern regions of India. The second group contained cultivars indigenous to south India. It can be seen that cvs Alphonso, Raspuri and Goamankurd which originate from same node are indigenous to western parts of India. In the second group, south Indian cultivars clustered together. Based on panicle emergence, canopy density and the emergence of a second flush of panicles, Indian cultivars were classified into two groups, southern and northern cultivars representing two different ecotypes of Mangifera indica (Yadav and Singh, 1985). Principal component analysis (PCA) using morphological data, however, was not able to separate the cultivars into different groups (Figure 2).

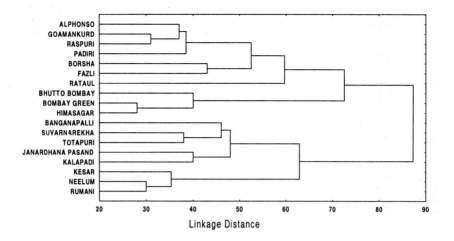

Figure 1. Dendrogram of Indian mango cultivars constructed by cluster analysis of RAPD markers.

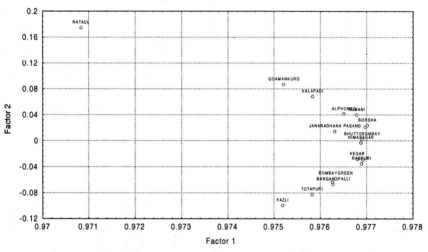

Figure 2. Principal component analysis using morphological characters.

Since the cultivars used here are seedling selections, the results from this study show that cultivars might have evolved from the existing mango gene pool in that geographical location. Therefore they show a high degree of relatedness. These cultivars were selected by local people and later domesticated by cultivation in large areas. Therefore it appears that the majority of present day commercial cultivars in different parts of the Indian subcontinent have originated from mango germplasm existing in that particular geographical area.

References

Mukherjee, S. K. Origin of mango. *Econ. Bot.* **26,** 260–264 (1972).

Ravishankar, K. V., Lalitha Anand & Dinesh, M. R. Assessment of genetic relatedness among mango cultivars of India using RAPD markers. *J. Horticul. Sci. Biotech.* **75,** 198–201 (2000).

Ward, J. H. Hierachic grouping to optimize an objective function. *J. Am. Stat. Assoc.* **58,** 236–239 (1963)

Yadav I. S. & Rajan S. Genetic resources of *Mangifera.* in *Advances in Horticulture,* vol. 1 (eds Chadha, K. L. & Pareek, O. P.) 77–93 (Malhotra Publishing House, New Delhi, 1993).

Yadav I. S. & Singh, H. P. Evaluation of different ecological groups of mango cultivars for flowering and fruiting under subtropics. *Progr. Hortic.* **17,** 165–175 (1985).

Tropical Ecosystems: Structure, Diversity and Human Welfare.
Proceedings of the International Conference on Tropical Ecosystems
K. N. Ganeshaiah, R. Uma Shaanker and K. S. Bawa (eds)
Published by Oxford–IBH, New Delhi. 2001. p. 766.

Conservation of *Citrus* species in Arunachal Pradesh

B. Singh, P. Rethy, S. Kalita, A. Kar and P. Gajural*

Department of Forestry, North Eastern Regional Institute of Science and Technology,
Nirjuli, Arunachal Pradesh, India
*e-mail: bs@nerist.ernet.in

Economically, *Citrus* species constitute one of the very important horticultural crops of Arunachal Pradesh. A large number of both wild and cultivated species of *Citrus* are reported from the state. Till recently only 8 species and a few genetic resources are being conserved through their cultivation in farmer's fields and in some *Citrus* research stations. Scientific approaches for the conservation of the wild species as a whole have not been as of yet developed. Deforestation and other anthropogenic pressures in the state constitute a major threat to the *Citrus* species. In the wake of these threats, we discuss strategies aimed at conserving the genetic resources of *Citrus*. We present a few methodologies that could be followed for the conservation of *Citrus* genetic resources.

Keywords: *Citrus* species, Arunachal Pradesh, conservation strategy.

Tropical Ecosystems: Structure, Diversity and Human Welfare.
Proceedings of the International Conference on Tropical Ecosystems
K. N. Ganeshaiah, R. Uma Shaanker and K. S. Bawa (eds)
Published by Oxford–IBH, New Delhi. 2001. pp. 767–770.

Phenology of neem tree, *Azadirachta indica* A. Juss in South India

J. Jayappa and A. R. V. Kumar

Department of Entomology, University of Agricultural Sciences, GKVK,
Bangalore 560 065, India

Neem, *Azadirachta indica* A. Juss. (Meliaceae) originated in Indo-Myanmar region and is now distributed in all parts of the world. Adapted to tropical semi-arid conditions, it is largely an evergreen tree that can grow to a height of up to 35–40 m under favourable conditions. But in extreme hot locations, it might shed leaves for a short period during February–March (Bahuguna, 1997). Neem is an important source of many active principles that are effective as insecticides, fungicides, bactericides, viricides and nematicides. Traditionally different parts of neem are being used in the treatment of innumerable ailments (Varma, 1976) and for the management of insect pests of crops, store insects and mosquitoes. The active principles such as azadirachtin, salannin, nimbin, etc. are now commercially exploited for insect pest management. The seeds have the highest concentration of these chemicals (Schmutterer, 1995). A homemade insecticide, neem seed kernel extract, is widely recommended in lieu of commercial synthetic insecticides in India (Srivastava *et al.*, 1984). Seeds are also collected for propagation and extraction of these chemicals for commercial purposes. Short seed viability and exponential decay of insecticidal property of seeds make it necessary to collect the seeds early for use either as propagules or as a source of insecticide. Therefore the present study was taken up to understand the

Keywords: Neem, *Azadirachta indica*, trees, phenology, flowers and fruits.

phenology of the plant to identify the time of availability of seeds for harvesting.

Fifteen trees were marked at the GKVK campus of the University of Agricultural Sciences, Bangalore. The marked trees were followed year round to record the different phenological features such as, time of occurrence of new flush of leaves, flowering and fruiting. The data were collected at fortnightly interval on each of the fifteen marked trees during 1998. All phenological events were quantified. From each tree 4 branches were collected from each of the four compasss directions. Observations were recorded on the numbers of old (dark green) leaves, new leaves (light green or light brown), inflorescences for each terminal branch. Similarly, number of flowers and fruits were recorded for each inflorescence. Raw data were then used to work out the means (± SD) for each of the characters studied for both individual trees and by pooling the data for all the fifteen trees sampled. These means (± SD) were plotted across time. The mean time of each of the phenological events was calculated by considering the standard days starting from January 1 as the first day and December 31 as the 365th day.

$$\text{Mean time} = \frac{\Sigma(x)\,(fx)}{\Sigma(fx)}$$

x = standard day of observation; fx = number of phenological features.

The number of old leaves per branch was lowest (4.10 ± 1.94) during late February. The peak density of old leaves was observed during June and late September–October. These leaves remained on the tree even when new flush began to sprout. There was minimal variation among the trees for leaf density as was observed previously (Mahadevan, 1991; Bahuguna, 1997). The standard mean date of old leaves ranged from 182.73 to 198.47 days between the trees with an overall mean (± SD) of 191.16 ± 3.95 days.

The new leaves started appearing in the tree during late February and peaked during March. New leaves were found on the trees until after September. The mean (± SD) standard date of new flush varied from 140.00 to 160.36 days for the fifteen trees sampled with an over all mean (± SD) of 146.27 ± 30.24 standard days. Clearly the data indicated much greater variation for occurrence of new leaves compared to the old neem leaves.

The inflorescences started appearing on the tree during March and the density was at its peak in April. These appeared on the tree with the initiation of buds and disappeared with elimination of ripened fruits by the end of October. The mean standard date of occurrence ranged from 140.37 to 154.19 days between the trees with an overall mean (SD ±) of 145.75 ± 4.26 standard days. However, there was considerable variation in the initiation of inflorescence between the trees sampled and this variation narrowed with time (Figure 1*b*).

Flowers, initiated during early and mid March, were the shortest lived parts of the neem tree. The peak density was during April and May. The mean (± SD) days of occurrence of flowers ranged from 104.18 to 113.67 with an overall mean of 108.79 ± 2.76 standard days (Figure 1*c*).

Young fruits were first observed on the trees during April. The density of fruits reached maximum during June–July, when fully developed and some ripened fruits were available on the tree (Figure 1*d*). The mean (± SD) standard day of fruiting time ranged from 103.64 to 186.52 between trees and the mean for the fifteen trees was 174.79 ± 6.10 standard days. Fruit density decreased owing to fruit dropping and predation by birds, bats, ants, etc. The fruit fall continued up to September–October and by the end of October the trees were completely devoid of fruits. These results closely match those of Bahuguna (1997) and Mahadeva (1991). The study revealed that July-August is the ideal time for harvesting neem seeds for propagation and for the purpose of use in pest management practices.

All the phenological events observed were found to follow largely unimodal patterns of occurrence across time. Other casual observations made in and around Bangalore during this period did not reveal any other alternative patterns. Guahabakshi (1984), however, observed bimodal fruiting pattern in

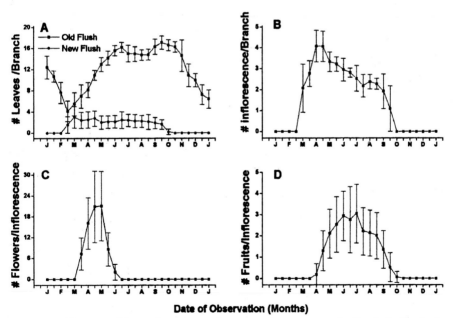

Figure 1. Pattern of four phenologocal events of neem trees in South India during 1998. Data from fortnightly observations on 15 trees at Bangalore.

North India in a small proportion of the neem population that he studied. The occurrence of new branches of neem along with new flush of leaves during the early part of the year can be used as a guide for predicting the seed availability on the trees.

Further, from the point of view of pest management, it can be seen that the fruits are available on the trees from April to October with a peak during July–August. Consequently, the ripened fruits are available on the trees from June to October with a peak in July. This calls for regular collection of fruits from the trees and fallen fruits and seeds from July through October. Regular harvesting is a dire need as proper processing is required to maintain higher biological activity of the neem seeds. Yet it is likely that fruits harvested in July would have lost considerable amount of their biological activity by October. Therefore, by mixing the seeds from freshly harvested fruits during September–October with seeds of July it is possible to maintain higher biological activity of seeds used for the purpose of pest management.

References

Bahuguna, V. K. Selviculture and management practices for cultivation of *Azadirachta indica* (neem). *Indian For.* 379–386 (1997).

Guahabakshi, D. M. *Flora of Mushidabad district, West Bengal, India* (Scientific publications, Jodhpur, 1984).

Mahadevan, N. P. Phenological observation of some forest trees as an aid to seed collection. *J. Trop. For.* **7**, 243–247 (1991).

Schmutterer, H. The tree and its characteristics. in *The Neem Tree: Source of Unique Natural Products for Integrated Pest Management, Medicine, Industry and Other Purposes* (ed. Schmutterer, H.) 1–34 (VCH, Weinheim, 1995).

Srivastava, K. P., Agnihotri, V. Z. & Jain, H. K. Relative efficacy of fenvalerate, quinolphos and neem seed kernel extracts for the control of pod fly, *Melanagromyza obtuse* Molloch and pod borer, *Heliothis armigera* Hubner, infesting red gram, *Cajanus cajana* (L.) Millsp. together with their residues. *J. Entomol. Res.* **8**, 1–4 (1984).

Varma, G. S. *Miracles of the Tree* (Rasayana Pharmacy, 3 Darya Ganj, New Delhi, India, 1976).

Tropical Ecosystems: Structure, Diversity and Human Welfare.
Proceedings of the International Conference on Tropical Ecosystems
K. N. Ganeshaiah, R. Uma Shaanker and K. S. Bawa (eds)
Published by Oxford–IBH, New Delhi. 2001. pp. 771–773.

Patterns of life history traits of endangered Indian orchids

R. Lokesha and R. Vasudeva***

*Department of Genetics and Plant Breeding, College of Agriculture, P. B # 24,
Raichur 584 101, India
**Department of Forest Biology, College of Forestry, Banavasi Road,
Sirsi 581 401, India

Orchids form the largest group among flowering plants comprising over 18,000 species. In India about 1,300 species exist, of which nearly 250 species are either rare, endangered and/or threatened (RET). It is feared that more than fifty native species are either extremely rare or possibly extinct. The orchid wealth of India is found in four geographical regions, viz. the Himalayas (the richest), the Western Ghats (next richest), the Eastern Ghats and the Andaman and Nicobar islands. The Indian contribution to global orchid industry is remarkable and about 200 species have been identified to have commercial importance (Table 1). In an attempt to examine the life history traits that may predispose the orchid species to becoming rare, endangered or threatened and possibly extinct, we have developed a database of 748 orchid species. We present here the general pattern of life history traits associated with the RET orchids and discuss the potential of such analysis and results in managing the RET species in general.

A significantly higher proportion (40.7%) of commercially exploited orchid species were RET; in contrast, only 19.48% of the non-commercial species were RET (Table 1).

Keywords: Orchids, rare, endangered, threatened, life history trait, database, India.

On an average, RET orchids possessed large floral contrivances. Mean flower length and size was least for the non-RET species and high for rare, threatened and endangered/extinct species, in that order (Table 2). Epiphytic orchids tend to be less prone to be RET than the terrestrial orchids (Table 3).

These results indicate that there are well defined life-history features that might be associated with and often drive a species to becoming RET (Lokesha and Vasudeva, 1992, 1993, 1997). Based on such studies, it is possible to predict the fate of a species and accordingly institute measures to prevent the species from becoming RET (Lokesha and Vasudeva, 1999).

Table 1. Distribution of orchid species into categories of RET status and commercial exploitation

Status	Commercial	Non-commercial	Total
RET	53 (40.77)	120 (19.48)	173
Non-RET	77 (59.23)	496 (80.52)	573
Total	130	616	746

$\chi^2 = 27.74$; $P < 0.01$; $df = 1$.
Values in parentheses are percentages computed for each column total.

Table 2. Comparison of length and size of flowers in different RET status of Indian orchids.

Category	Flower length (mm)		Flower size (mm)	
	N	Mean ± SD	N	Mean ± SD
Non-RET	385	17.65–14.41	414	17.83–17.48
Rare	193	27.71–27.45	211	22.87–22.53
Threatened	17	35.32–27.68	23	34.26–30.10
Endangered/extinct	12	46.08–26.08	12	35.71–21.23
F test, df		3,602		3,656
F ratio		19.44		9.54
P level		< 0.001		< 0.001

Table 3. Distribution of RET and non-RET into habit classes.

Habit class	RET	Non-RET	Total
Epiphyte	102 (124)	280 (258)	382
Terrestrial*	128 (128)	200 (222)	328
Total	230	480	710

$\chi^2 = 8.46$; $P < 0.01$; $df = 1$; values in parentheses are those expected from a random distribution.
*Includes saprophytes and lithophytes.

References

Lokesha, R., Vasudeva, R. & Yellappa Reddy, A. N. Do rare/endangered/threatened plant species of south India have specific reproductive syndromes promoting their extinction? *The proceedings of the symposium on Rare, Endangered and Endemic Plants of the Western Ghats*, 30 and 31 August 1991, Thiruvananthapuram. pp. 128—134 (1991).

Lokesha, R. & Vasudeva, R. Commercial exploitation – a threat to Indian orchids? *Curr. Sci.* **63**, 740–744 (1992).

Lokesha, R. & Vasudeva, R. Do the endangered orchids differ from the common ones in their habit, distribution, and phenological patterns? *J. Orchid Soc. India* **7**, 53–60 (1993).

Lokesha, R. & Vasudeva, R. Patterns of life history traits among rare/endangered flora of south India. *Curr. Sci.* **73**, 171–172 (1997).

Lokesha, R. & Vasudeva, R. Association of floral features with endangered status of Indian orchids. *J. Orchid Soc. India* **13**, 25–27 (1999).

Tropical Ecosystems: Structure, Diversity and Human Welfare.
Proceedings of the International Conference on Tropical Ecosystems
K. N. Ganeshaiah, R. Uma Shaanker and K. S. Bawa (eds)
Published by Oxford–IBH, New Delhi. 2001. pp. 774–777.

Leaf litter nutrients and soil N availability do not affect short-term mass loss of mixed-species litter in two Puerto Rican forests

Heather E. Erickson*, Gaddiel Ayala, Patricia Soto
and Waleska Rivera

Department of Science and Technology, Universidad Metropolitana, San Juan,
Puerto Rico, USA
*e-mail: um_herickson@mail.suagm.edu

Initial stages of litter decomposition are often correlated to litter N (Sullivan *et al.*, 1999), litter P (Vitousek *et al.*, 1994), carbon:nutrient (Taylor *et al.*, 1989) or lignin:nutrient ratios (Melillo *et al.*, 1982) reflecting the putative role of nutrients, especially N, in facilitating early decay processes. Soil N availability may also regulate mass loss, though the effect remains controversial (Hobbie, 2000). Using litterbags with mixed-species litter, we conducted a reciprocal litter transplant to examine interactions between litter quality and soil N availability on leaf litter mass loss in two humid successional forests in Puerto Rico.

Both forests are located within the sub-tropical moist forest zone of Puerto Rico and have similar climate. Previous work indicated very low soil N availability and a high litter C:N ratio in a successional forest that lacked legumes (Erickson *et al.*, 2001); the presence of legumes in a second forest

Keywords: Decomposition, successional forests, lignin, nitrogen availability, C:N ratio.

suggested higher N availability and lower litter C:N ratios. Thus, in our case, comparing mass loss of same-source litters between sites evaluates soil-mediated effects (likely N availability) on decomposition, while comparing mass loss of different-source litters within a single site evaluates the effect of litter quality on mass loss. We hypothesized (1) if soil N availability limits mass loss, decomposition would be faster at the site with greater soil N availability, and (2) litter with greater N (or P) concentrations and/or lower C:nutrient ratios would decompose faster.

We made two sets of litterbags using mixed-species native litter from each site and systematically located them on the ground at both the sites. Additionally, we hung common litterbags 1.4 m above ground at each site to test our assumption of lack of climatic differences between the two forests. For each sampling, five litterbags were collected from each of the three treatments (native, non-native, hung) at each site after approximately 10, 20, 40, 80, and 160 days of decomposition. The negative exponential decay model $(\ln(M_t/M_o) = -kt + b)$ was fitted to oven dry (70°C) mass data to determine the decay constant (k). Initial litter was analyzed for total C, N, P, K, Ca, and proximate C fractions, including lignin. Soil N availability was assessed once, using KCl extractions of inorganic N and 7-day laboratory incubations to determine net N mineralization.

Soil N availability and litter quality were low (Table 1) for the forest lacking legumes (hereafter 'low N site'). Net nitrogen mineralization was negative, indicating microbial immobilization of N during incubations; nitrate was not detectable (Table 1a). Litter C:N, C:P and lignin:N ratios were relatively high (Table 1b). In contrast, at the forest containing legumes (hereafter 'high N site'), soil N availability was high (Table 1). Net N mineralization was positive and nitrate was 13.6 µg N/g. Litter C:N, C:P and lignin:N ratios were relatively low at 31.5, 921 and 15.3, respectively (Table 1b). Litter from the high N site had greater concentrations of N, P and Ca than litter from the low N site (Table 1b), while K was greater in litter from the low N site. Lignin did not differ between different source litters.

There was no significant difference in decay rates for suspended litters between the sites suggesting that any differences in the ground litters between sites would be due to factors other than climate. Despite the wide variations in soil N availability and several measures of litter quality between the two sites, we found little effect of location or litter source on short-term mass loss (Figure 1). Decay rates did not depend on the gradient of N levels between the originating and non-native sites, although litter from the high N site decomposed slightly faster when moved to the low N site. Comparing mass loss of the different litters at a single site (Figure 1) – at the high N site, there was no difference in mass loss between the different-source litters. At the low N site, the k-value was slightly higher for the high N litter, but again, the results from a single date are responsible for the difference.

Table 1. Soil (0–10 cm) N cycling properties (a) and initial carbon and element chemistry for mixed-species litter (b) at 2 successional forests in Puerto Rico. Different superscripts for a given variable indicate significant differences (T-test, $P < 0.05$).

a

Site	Net N mineralization	Net nitrification	NH$_4$-N	NO$_3$-N	Total N	Total C	pH
	μg N/g soil/7 d		μg N/g		%		
High N	11.7[a]	15.9[a]	5.5[a]	13.6[a]	0.42[a]	4.38[a]	4.38
Low N	−2.84[b]	1.14[b]	6.3[a]	0.0[b]	0.28[b]	3.69[b]	4.86

b

Site	N	Lignin	Lignin:N ratio	C:N ratio	C:P ratio	P	K	Ca
	%					%		
High N	1.79[a]	27.3[a]	15.3[a]	31.5[a]	921[a]	0.062[a]	0.054[a]	1.01[a]
Low N	1.05[b]	26.0[a]	24.8[b]	49.5[b]	1244[b]	0.042[b]	0.061[b]	0.72[b]

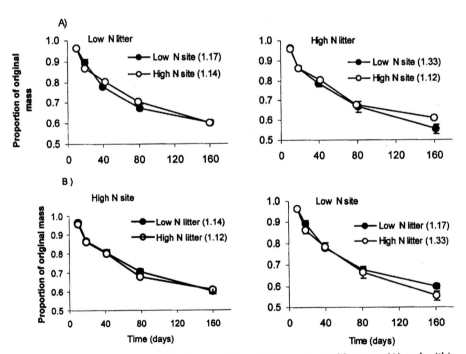

Figure 1. Mass loss (means ± SE) of low and high N litter compared between (A) and within sites (B). Shown in parentheses are k-values for each treatment.

Mixed-species litter from the low and high N sites represent very different substrates with large and significant differences in N, P and Ca concentrations and in C:N, C:P and lignin:N ratios. However, both sources had similar and high (26.6%) mean lignin contents. We suggest that the high lignin concentration explains the lack of differences in mass loss for the two otherwise very different litters. Lignin control of decomposition typically characterizes later phases of decay when recalcitrant compounds remain (Berg and Staff, 1980). Here, with high initial lignin, decomposition seems to be controlled by lignin rather than nutrient content (see also Gallardo and Merino, 1993; Taylor et al., 1989; Singh, 1969). LaCaro and Rudd (1985) working in this forest zone, noted that secondary forest species with relatively high lignin, decomposed slower than primary forest species with lower lignin and with no effect of litter nutrients. Furthermore, we found no site effect on decay rate despite major differences in soil N availability, and suggest that this finding may also be a result of the high lignin. Hobbie (2000) found that decay rates of high lignin litter only increased slightly in response to experimental N additions in contrast to the large response of low N litter. Thus decomposers appear limited by low C-quality rather than by nutrients. These studies suggest that lignin may play a significant role in short-term litter decomposition in tropical forests.

References

Berg, B. & Staff, H. Decomposition rate and chemical composition of Scots pine litter. II. Influence of chemical composition. *Ecol. Bull. (Stockholm)* **32**, 363–372 (1980).

Erickson, H. E., Keller, M. & Davidson. E. Nitrogen oxide fluxes and nitrogen cycling during secondary succession and forest fertilization in the humid tropics. *Ecosystems* **4**, 67–84 (2001).

Hobbie, S. Interactions between litter lignin and soil nitrogen availability during leaf litter decomposition in a Hawaiin montane forest. *Ecosystems* **3**, 484–494 (2000).

Gallardo, A. & Merino, J. Leaf decomposition in two Mediterranean ecosystems of southwest Spain: influence of substrate quality. *Ecology* **74**, 152–161 (1993).

LaCaro, F. & Rudd, R. L. Leaf litter disappearance in Puerto Rican montane rain forest. *Biotropica* **17**, 269–276 (1985).

Melillo, J. M., Aber, J. D. & Linkins, A. E. Nitrogen and lignin control of hardwood leaf litter decomposition dynamics. *Ecology* **63**, 621–626 (1982).

Singh, K. P. Studies in decomposition of leaf litter of important trees of tropical deciduous forests at Varanasi. *Trop. Ecol.* **10**, 292–311 (1969).

Sullivan, N. H., Bowden, W. B. & McDowell, W. H. Short-term disappearance of foliar litter in three species before and after a hurricane. *Biotropica* **31**, 383–393 (1999).

Taylor, B. R., Parkinson, D. & Parsons. W. F. Nitrogen and lignin content as predictors of litter decay rates: a microcosm test. *Ecology* **70**, 97–104 (1989).

AUTHOR INDEX

SUBJECT INDEX